Light Symposium 2023: Architecture Lighting Environments - Space With(Out) Light

IOP Conference Series: Earth and Environmental Science
Volume 1320

Stockholm, Sweden
4-6 December 2023

ISBN: 978-1-7138-9576-3
ISSN: 1755-1307

Printed from e-media with permission by:

Curran Associates, Inc.
57 Morehouse Lane
Red Hook, NY 12571

Some format issues inherent in the e-media version may also appear in this print version.

Institute of Physics
Dirac House, Temple Back
Bristol BS1 6BE UK

Phone: 44 1 17 929 7481
Fax: 44 1 17 920 0979

techtracking@iop.org

Additional copies of this publication are available from:

Curran Associates, Inc.
57 Morehouse Lane
Red Hook, NY 12571 USA
Phone: 845-758-0400
Fax: 845-758-2633
Email: curran@proceedings.com
Web: www.proceedings.com

TABLE OF CONTENTS

IOP Conf. Series: Earth and Environmental Science 1320 (2024) 011001 doi:10.1088/1755-1315/1320/1/011001

The 9th Light Symposium, December 4-6, 2023 · Stockholm, Sweden

Light Symposium 2023: Architecture Lighting Environments - Space With(Out) Light

Daylight and electric light are an integral part of our built environment, and the sustainable design of lighting in this context, its presence, absence, and variability over time, requires research and practice that is both trans- and interdisciplinary.

Although the presence of light is intimately linked to the life of humans, flora, and fauna, research and practice have not yet fully explored how light and dark can shape the built environment in a sustainable and holistic way.

Research and practice into the affects and effects of light combine lighting design with other fields, including, for example, architecture, engineering, computer science, visualization, arts, humanities, social sciences, environmental psychology, health sciences, cognitive sciences, and more. The Light Symposium is an annual, international event to facilitate discussions and the exchanging of ideas; currently, it is coordinated by four academic partners; the Lighting Design Division at KTH in Sweden, Hochschule Wismar in Germany, AAU Copenhagen in Denmark, and SNU Kongsberg in Norway.

The 9th Light Symposium, LS2023, was held in Stockholm from the 4th-6th of December 2023; it was hosted by the Architectural Lighting Design Division of KTH Royal Institute of Technology, in collaboration with the Perception Studio of Konstfack, the University of Arts, Crafts, and Design. During the 2,5 days of the symposium, more than 200 participants came together to share and learn about current research and design across different fields within architectural and urban lighting.

In addition to invited keynote lectures, the conference program consisted of sessions with talks, panel discussions and poster presentations from researchers and practitioners worldwide, selected through a two-step peer-review process. The conference topics included daylight and electric light in architecture, art, and urban design, organized into four broad categories: Light and Health for all Species, Designing for Darkness, Sensory and Spatial Perceptions, Variability and Rhythmicity.

These proceedings include 32 conference papers that were presented either as talks or posters during LS2023. All papers were reviewed by an international panel of reviewers. We thank all reviewers and our partner organizations for their time and help.

We also thank the following organizations for their support of LS2023, and for enabling documentation through recordings and proceedings: The Swedish Energy Agency, Bertil & Britt Svenssons Stiftelse för Belysningsteknik, as well as Helvar, Selux, Fergin, Nordic LED, Stockholm Lighting, Annell, Fox Belysning, Optoga.

The LS2023 Local Organizing Committee and Proceedings Editorial Team:
Ute Besenecker, Foteini Kyriakidou, Gerhard Rehm, Hamidreza Eizadi, Federico Favero, Rodrigo Muro, Johanna Enger, Stavroula Angelaki, Seren Dincel

IOP Conf. Series: Earth and Environmental Science 1320 (2024) 011001 doi:10.1088/1755-1315/1320/1/011001

List of Reviewers:

#	First name	Last name	Country	Organization
1	Myriam	Airies	Sweden	Jönköping University
2	Stavroula	Angelaki	Sweden	KTH, Royal Institute of Technology
3	Ken	Appleman	USA	Precognetics
4	Thanos	Balafoutis	Greece	Hellenic Open University
5	Rigmor C.	Baraas	Norway	University of South-Eastern Norway
6	Ute	Besenecker	Sweden	Royal Institute of Technology (KTH)
7	Henrik	Clausen	Sweden	Fagerhult
8	Lambros T.	Doulos	Greece	Hellenic Open University
9	Hamidreza	Eizadi	Sweden	Royal Institute of Technology (KTH)
10	Johanna	Enger	Sweden	Konstfack
11	Federico	Favero	Sweden	Royal Institute of Technology (KTH)
12	Natalia	Giraldo Vasquez	Denmark	Technical University of Denmark
13	Hillevi	Hemphälä	Sweden	Lund University
14	Mette	Hvass	Denmark	Aalborg University
15	Maria	Johansson	Sweden	Lund University
16	Ellen	Kathrine Hansen	Denmark	Aalborg University
17	Panagiotis	Kontaxis	Greece	University of West Attica
18	Foteini	Kyriakidou	Sweden	Royal Institute of Technology (KTH)
19	Thorbjörn	Laike	Sweden	Lund University
20	Stine	Louring Nielsen	Denmark	Universal Design Hub - Bevica Foundation
21	Michael	Mullins	Denmark	Aalborg University
22	Rodrigo	Muro	Sweden	Royal Institute of Technology (KTH)
23	Mihkel	Pajuste	Denmark	Aalborg University
24	Henrika	Pihlajaniemi	Finland	University of Oulu
25	Derek	Porter	USA	Parsons School of Design
26	Bipin	Rao	Germany	University of Wismar
27	Siobhan	Rockcastle	USA	University of Oregon
28	Michael	Rohde	Germany	University of Wismar
29	Are	Roysamb	Norway	University of south-eastern Norway
30	Georgios	Triantafyllidis	Denmark	Aalborg University
31	Helga	Waseth	Norway	University of south-eastern Norway
32	Emmanouil	Xylakis	Malta	University of Malta
33	Veronika	Zaikina	Norway	University of South-Eastern Norway
34	Karolina M.	Zielińska-Dabkowska	Poland	Gdansk University of Technology (GUT)

Peer Review Statement

Proceedings: The 9th Light Symposium
December 4-6, 2023 · Stockholm, Sweden

Peer Review Statement:
All papers published in this volume have been reviewed through processes administered by the Editors. Reviews were conducted by expert referees to the professional and scientific standards expected of a proceedings journal published by IOP Publishing.

Type of Review:
Double-blind, anonymous

Conference submission management system:
Conference system by Axaco / Axacoair with upload of final information to Morressier

Number of submissions received:
81 initial abstracts, 53 final papers

Number of submissions sent for review:
81 initial abstracts, 53 final papers

Number of submissions accepted:
32

Acceptance Rate (Submissions Accepted / Submissions Received × 100):
40 (from initial abstract submissions), 60 (from final paper submission)

Average number of reviews per paper:
3-4

Total number of reviewers involved:
34

Contact person for queries:
Ute Besenecker, PhD
KTH Architecture, Lighting Design
uteb@kth.se

Theoretical Foundations of Designing for Darkness

Taylor Stone[1]

[1] Institute for Science and Ethics, University of Bonn, Bonn, Germany

Corresponding author's email: taylor.stone@uni-bonn.de

Abstract. This paper sketches the theoretical foundations of designing for darkness. Drawing from (lighting) design theory, philosophy of technology, and environmental ethics, it argues that we have an opportunity to re-examine the meaning and experiences of darkness, exploring why and how we should protect or re-introduce darkness into urban nightscapes – not as a constraint, but a constructive goal for urban lighting. It is further positioned as a formative, not prescriptive, framework: bounding and guiding a flexible design process, and able to inform different strategies and approaches. Framed in this way, three core principles are outlined that serve as moral and aesthetic goals for urban lighting. The first is aspirational, situating 'darkening cities' as a type of urban (ecological) restoration akin to greening cities. This requires seeing urban darkness as a critical natural infrastructure that can bring both material and social benefits. The second is evaluative, specifying the value-level design requirements to be fostered in urban nightscapes. Lighting strategies should, as a prima facie obligation, promote and preserve the values of darkness. The third principle is experiential, exploring how strategies can create the conditions for positive experiences of – and engagement with – urban darkness. In combination, they offer a means to specify darkness as a design requirement, as well as a reflexive tool to assess and refine the overarching theory of designing for darkness.

1. Light pollution and the future of cities after dark

The growing recognition of light pollution has spurred research into how policy and design can incorporate new, transdisciplinary approaches to mitigate adverse effects [1-4]. However, despite evidence and awareness of the ecological, economic, and cultural impacts of light pollution, our urban nights continue to get brighter. A multi-year citizen science study has shown that previous estimates of a 2-3% annual increase are incorrect; due largely to the rapid introduction of LED outdoor lighting, the current rate of increase is likely closer to 10% [5]. This seems to confirm worries that the rapid proliferation of LEDs could lead to an increase in brightness and energy use, as it offers a more efficient means of producing artificial light and is therefore susceptible to rebound effects [6-8].

This points to the continued need for ongoing inter-and-transdisciplinary research into the effects of artificial light at night, alongside new policies at municipal, regional, and national levels to curb light pollution. It also speaks to the need for innovative approaches to urban lighting, and more generally novel visions for the future of cities at night. Otherwise, we risk becoming locked-in to a new generation of outdoor lighting that exacerbates negative impacts and fails to realise sustainable urban (nighttime) futures.

One idea that has been gaining momentum is the re-conceptualisation of *darkness* as a positive and desirable feature of nightscapes – coalescing around the theories of *dark design* and *designing for darkness* [9-15]. This represents a profound shift in thinking, for while architecture and lighting design have a long history of designing *with* darkness and shadow, darkness is here imbued with positive moral and aesthetic significance. Dark design has been proposed as a strategy that strives to both reduce the impacts of light pollution, as well as raise awareness about this environmental issue [10]. Designing for darkness has the added layer of positioning darkness as a constructive target within urban lighting policy and design. It can be understood as a broad and inclusive design theory that works to align and inform a variety of methods and practices.

The goal of this paper is to offer a first sketch of the concepts that unite and drive designing for darkness as a design theory. It will not analyse specific design methods or case studies or delineate rigid boundaries. Nor will it delve into practical considerations, such as the role of stakeholder engagement or its translation into actionable policies. Rather, it will focus on what exactly it means to conceptualise and specify darkness as something to design *for*. Designing for darkness is in many ways an exercise in problem framing [16] – of re-framing the problem of light pollution away from what is bad about artificial light and towards what is good about darkness. It is an opportunity to re-examine darkness, exploring why and how we should protect or re-introduce it into our urban nightscapes. In this way is it focused on developing a foundation for designing for darkness, which can inform downstream policymaking and design practice.

Towards this goal, insights will be drawn from the philosophy of technology, which offers tools for translating (environmental) values into design requirements. Such an approach moves away from a constraint-oriented framing, instead asking how darkness can become a constructive and generative target for technical, policy, and design innovations. Critically, this positions darkness as not just as a quantifiable item, but a guiding value with moral and aesthetic weight. To operationalize as a design theory, designing for darkness is defined via three interrelated principles that are aspirational, evaluative, and experiential. In combination, they offer a theoretical foundation for different strategies aiming to re-imagine the role of darkness in urban lighting (and more generally cities at night), and provide a means to critically and creatively advance sustainable lighting strategies.

2. Ethics by design: values and technology

Understanding the formative role lighting technologies play in shaping cities at night, and in turn shaping urban darkness, allows for a better understanding of how lighting strategies can attend to darkness in rich and nuanced ways. For this, recent developments in philosophy of technology – in particular, *technological mediation* and *design for values* – provide useful starting points for interrogating human-technology-environment relations.

Both approaches are concerned with the ethical dimensions of technologies, and the relationship between artifacts and moral values. While technology design and development has historically been viewed as a value-neutral endeavour, this has been questioned since at least the seminal article 'Do Artifacts Have Politics?' by Langdon Winner [17] in which he argued that "... technical devices and systems important in everyday life contain possibilities for many different ways of ordering human activity. Consciously or unconsciously, deliberately or inadvertently, societies choose structures for technologies that influence how people are going to work, communicate, travel, consume, and so forth over a very long time..." Winner showed that seemingly benign artifacts, such as a highway overpass, can in fact be used as a political tool – causing social impacts well beyond any narrow instrumental function. While the specific examples presented by Winner have been subject to debate, he nevertheless opened a new way of investigating technological artifacts: not via technical functions, but rather their relation to social values.

More recently, *technological mediation* has further developed these insights to highlight the profound influence that technologies have in our lives. It interprets our actions and perceptions as "always closely interwoven with the material environment in which they play out" [18]. Technology is not something *out there* to be dealt with. Instead, our interactions and uses of technological artifacts play an active role in constituting our reality. Hence, we experience the world – and morality – *through* or *by way of* technological artifacts. In co-constituting our understanding of the world, technologies can make explicit issues that were not apparent (or even possible) before, moving them from the taken-for-granted backdrop of moral decision-making and into ethical debates.

While mediation provides a description of the moralizing role of technology, a normative and action-oriented approach has been developed through the interrelated theories known as *value sensitive design* and *design for values* [19-20]. This includes a variety of different methodologies but is united by a few key tenets. First, that technologies are not value-neutral, but can embody and/or impact social and environmental values. Second, that value considerations should be explicitly and proactively included in technology development, design, and associated policymaking. As such, value-sensitive approaches actively strive to situate moral values as "supra-functional" design requirements, effectively striving to operationalise ethics *through* design processes [21]. Third, they take an interactional stance with regards to technology and values – rather than being deterministic, value-sensitive approaches posit that "individuals, organisations, or societies shape the tools and technologies they design and implement; in turn, those tools and technologies shape human experience and society" [19].

Value sensitive design was originally developed in relation to information and communication technologies and individual artifacts. More recently, it has been expanded and refined as a means to surface values in urban technologies, showing that it is necessary to understand the historical developments and inherited symbolism that urban technologies carry, as new innovations are layered over past choices and material conditions [22]. For urban lighting, this means appreciating the legacy of values – such as safety, social order, and progress – that were foundational to the formalisation of public lighting and continue to influence lighting choices well beyond any quantifiably 'needed' lighting. It is also critical to appreciate that urban technologies often do not have discrete boundaries but are influenced by other artifacts, systems, and infrastructures. For example, the introduction of autonomous vehicles into cities will impact the uses and need of streetlighting. Through such analyses, we can appreciate what is valuable at a certain time and place, and how to respond in conscientious and creative ways. The proliferation of artificial light and the rising awareness of light pollution necessitates a reconsideration of the meaning of urban darkness and its presence and qualities in cities at night.

To summarize, the philosophy of technology reveals two key ideas. First, that our perceptions and experiences of (urban) darkness are co-constituted by our artificial lighting. This means that designing for darkness requires attention to lighting, as it will be formative to our experiences of, and relationship with, darkness. The second is that we can proactively strive to design *for* specific environmental values, which requires a conceptualisation of said values as part of an iterative design process.

3. Principles of designing for darkness

Designing for darkness as conceptualised here is not a prescriptive framework. It does not impose specific steps or actions, nor does it provide a checklist or minimum thresholds to satisfy. It is not meant to be instrumentalized as an add-on to existing approaches. Rather, it is a formative framework, bounding and guiding a flexible design process [19]. It should be a reflexive and evolving approach that can encapsulate different strategies, technologies, and outputs. Designing for darkness can therefore be understood as heuristic tool for re-thinking the relationship between lighting and darkness in urban nightscapes. Framed in this way, three core principles are proposed to serve as moral and aesthetic guides for urban lighting strategies: *darkening cities as urban restoration*, *darkness as value-level design requirements*, and *rewilding nocturnal atmospheres* (table 1). Each will be elaborated in turn, and finally their interrelation within design processes will be briefly discussed.

Table 1. Three principles outlining the theoretical foundations of designing for darkness.

Principle	Purpose	Description
Darkening cities as urban restoration	Aspirational	Overarching goal for urban futures; darkness positioned as a natural infrastructure, and darkened cities as a new ideal
Darkness as value-level design requirements	Evaluative	Normative framing of darkness; identifying the values that can be achieved by urban darkness, formulated as *prima facie* obligations
Rewilding nocturnal atmospheres	Experiential	Phenomenological and design considerations; bring positive experiences of darkness into cities, use lighting to foster an ecological and cosmological sense of place

3.1. *Darkening cities as urban restoration*

As an overarching aspirational goal, darkness can be situated as a type of urban (ecological) restoration. In this sense, 'darkening cities' can be seen as akin to 'greening cities.' More generally, this aligns with contemporary approaches that seek to bring nature back into cities and collapse built-natural dichotomies, such as biophilic design [23]. Darkness and/or dark skies can be situated as a 'natural infrastructure' to be integrated into urban fabrics – a perspective arguably underlying the concepts of dark infrastructures and dark ecological networks [24-26]. Critically, darkening cities is about a more foundational ideal underlying such approaches; it is about finding a new story about urban darkness.

Surfacing new narratives around urban darkness requires shifting thinking away from negative associations with danger or fear, and towards seeing darkness as a source of positive environmental value [12,15,27]. To do so, we can consider darkening cities as an act of repair, with both a material and social function [28]. Materially, darkening cities can restore or preserve urban ecologies in the service of urban sustainability. Like other natural infrastructures, such as green spaces, darkened cities can reduce energy consumption while improving biodiversity and the well-being of citizens. Yet, a darkened city also carries an interrelated social dimension, offering the possibility of restoring an ecological and cosmological sense of place – something elaborated further in the third principle. Importantly, this makes the goal of darkening cities more than simply repairing an infrastructure, or even of restoring nocturnal ecosystems; it is also repairing our *relation* to urban darkness and the night sky [29].

This requires a conscientious and judicious use of lighting, but can also be a source of inspiration and creativity. Just as attempts to green cities explore how urban innovations can create the conditions for urban ecosystems to flourish at a variety of scales – or perhaps just step back to 'let nature back into cities' – aspiring towards a darkened city can likewise stimulate new innovations and interventions into urban nights. It can be situated as an ideal to guide our thinking about what sorts of urban futures we aim to strive towards, and why. The following two principles work to further articulate how such a vision can be further specified.

3.2. *Darkness as value-level design requirements*

Darkening cities provides an aspirational goal for the future of urban nightscapes. A next step is the specification of value-sensitive design requirements in order to consider what values can be fostered or preserved in different contexts. This can provide an evaluative tool that further delineates what we should be achieving with our lighting plans and strategies in order to move towards a darkened city. For this, the impacts of ecological and atmospheric light pollution can be reframed and categorised into ways in which value is derived from darkness [15]. This process surfaces nine values that encapsulate the environmental goods we strive to realise in urban nightscapes. Each value can be considered as a *prima facie* obligation to promote, protect, or pursue (table 2). This means that there is a morally relevant reason to uphold or strive to achieve these values, and in the absence of other considerations it would be a responsibility to do so. However, *prima facie* obligations are not categorical obligations – it can be that they are not achievable in every circumstance. For example, while efficiency should always be a consideration, a lighting plan for a public square in a densely populated city likely cannot itself achieve stellar visibility. Yet, creative uses or design interventions in such a space *could* at least offer modest ways to connect with nature via darkness – a motivation for the third principle. We therefore must consider which values are achievable and relevant, and to what degree, on a case-by-case basis. The values of darkness can then be used to proactively create site-specific design requirements [30], as well as inform novel frameworks for evaluating realized projects [31].

Positioning the values of darkness as moral obligations creates an important re-framing in effectively flipping the burden of proof. Rather than arguing for light pollution reduction as an add-on to existing policies and practices (i.e., minimising negative effects), it makes darkness a foundational consideration. The proof must be to show why it is good or better *not* to uphold or promote these values, or alternatively why it is not possible to achieve them due to contextual limitations.

Table 2. Nine values of darkness as *prima facie* obligations, adapted from [15].

Value of darkness	Design obligation for lighting policy and design
Efficiency	The efficient and optimal use of illumination; avoiding waste or excess
Sustainability	The responsible use of lighting for future-generations (preserving non-renewable resources)
Ecological conservation	Protect and preserve of nocturnal habitats; create conditions to promote biodiversity
Healthiness	Promote and foster human health and well-being via benefits of dark spaces and light-dark cycles
Happiness	Promote and foster emotional well-being through access to dark spaces and the natural night sky
Connection to nature	Create opportunities to connect with the more-than-human world
Stellar visibility	Preserve conditions for accessing and experiencing the starry night sky
Heritage	Preserve the cultural heritage and traditions associated with the night sky for future generations
Nocturnal sublime	Preserve and foster the wonder and beauty of both urban nightscapes and the natural night sky

3.3. Rewilding nocturnal atmospheres

The values of darkness offer a means to specify the ideal of a darkened city into design requirements. The final principle is concerned with the lived experiences of urban nightscapes, and how lighting technologies could be utilised creatively. This moves closer to design practice, asking what qualities should motivate lighting strategies and design interventions. Building on the first two principles, *rewilding nocturnal atmospheres* focuses on creating conditions for positive experiences of darkness, and, if possible, dark skies. For this, a re-imagining of urban nocturnal ambiances is required that strives to look beyond binary or dualistic thinking, instead exploring different qualities and relations between light and dark [32].

Drawing from the insights of technological mediation, we can appreciate that how we choose to design the qualities of urban nights and the relationship between darkness and lighting therein is a formative act. It can either reify the negative connotations of darkness or challenge them. It can reinforce the view that dark skies are only possible 'out there,' or subtly question built-natural dualistic thinking [32-35]. Such interventions can not only create more convivial and nature-inclusive cities; they can create conditions for city inhabitants to see a taken-for-granted aspect of the natural world otherwise within the spaces inhabited daily, and in the process foster new relations with the more-than-human world.

Critically, this points towards a re-thinking of the meaning and qualities of urban darkness. It is not only about the reduction or absence of artificial illumination, but rather how lighting reacts to the qualities and conditions of a specific time and place. A particularly interesting avenue is how novel lighting technologies could be utilised to respond to changing ambient conditions. This can include the rhythms and temporalities of natural cycles [36] or attuning lighting to natural light sources [37]. What these approaches offer is a means of bringing some wildness back into our cities in a way that extends current thinking about urban rewilding [38]. First, it does not position technologies as antithetical to nature, but rather strives for an alignment of urban technologies with ecological and cosmological rhythms and processes. Second, by focusing on urban spaces it can aid in the process of *human* rewilding [39]. Many rewilding efforts effectively treat the symptom and not the cause. Humans need to be included in these processes – not just as stakeholders, but requiring some rewilding themselves. In doing so, lighting can create conditions for *synergistic rewilding* of urban nights, as they are human interventions that contribute to flourishing ecosystems [40]. Further, they will open novel aesthetic experiences that can serve to re-imagine how and why we illuminate our world after dark. Such interventions, if successful, reinforce the relational aspect of repairing urban nights central to the notion of a darkened city.

3.4. Designing for darkness as iterative process

As a formative framework and heuristic guide to design processes, each principle is purposely left open-ended. They can be further refined based on the scale and scope of projects, as well as the stakeholder engagement and site analysis methods utilised in individual design processes. Internally, the three principles can be understood as embedded and layered. *Darkening cities* outlines an aspirational goal to strive towards; the *values of darkness* offers a means to proactively assess and specify site-specific requirements; and *nocturnal atmospheres* focuses on the aesthetic experience and associated design choices that can realise identified values. Used in combination, these three principles offer an iterative approach that allows for a back-and-forth between specification and reflection regarding the successes and/or impacts of designing for darkness (figure 1).

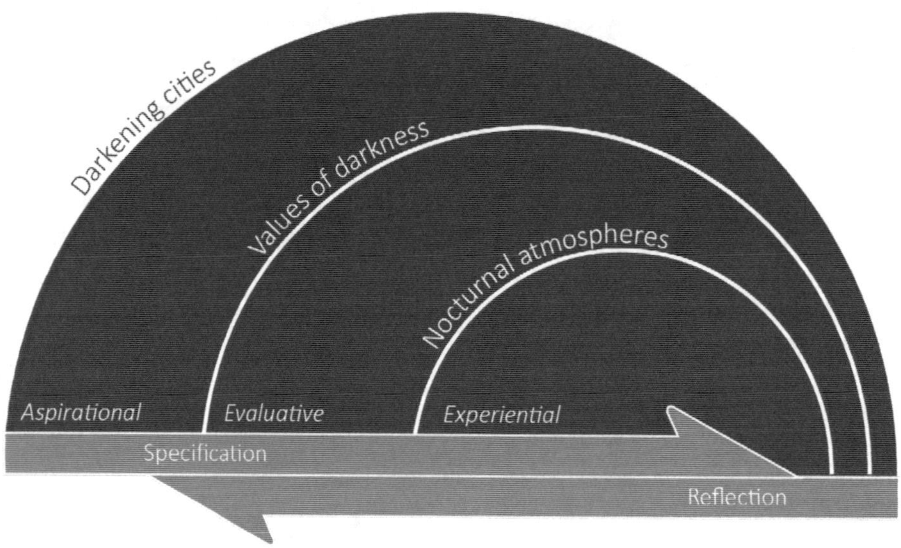

Figure 1. The principles of designing for darkness as a layered and iterative design process.

4. Conclusion: searching for darkness

This paper sketches the foundations of designing for darkness as a lighting design theory. Importantly, this is not meant to be a final, definitive statement; rather, a first step that can be refined and iterated upon as new scientific knowledge emerges and relevant methods and practices evolve. It draws from the philosophy of technology to situate designing for darkness as a formative framework and heuristic guide that can inform a variety of methods and approaches. Three principles are presented which, in combination, can allow for an iterative process of specifying design requirements and reflexively examining their success in achieving identified goals. As the theoretical foundations and downstream practical interventions of designing for darkness continue to evolve, it will ideally offer a tool for re-imagining of the future of cities after dark.

Aligned with the notion of a formative framework, but perhaps somewhat counter-intuitively, no formal definition of darkness has been provided. Darkness is typically defined as the partial or total absence of light, which says very little about the qualities of darkness itself. Here an alternative means of understanding darkness is offered – not as essential or static, but rather mediated by technologies, inherited symbolism, and social context. This highlights a continued need for reflexivity in the theory and practice of designing for darkness. Lighting technologies, as with many other large-scale infrastructures that are arguably responsible for the Anthropocene, now actively shape their own operating environment [28]. This creates a feedback loop, as any interventions into urban nightscapes will necessarily change the relative meaning and importance of darkness. In adopting an interactional stance to the co-evolution of technology and values, this is to be expected. Artificial lighting, and more generally cities at night, will continue to shape – and be shaped by – how we evaluate and experience darkness.

References

[1] Meier J, Hasenöhrl U, Krause K and Pottharst M 2014 *Urban Lighting, Light Pollution and Society* (New York: Taylor & Francis)

[2] Pérez Vega C, Zielinska-Dabkowska KM, Schroer S, Jechow A and Hölker F 2022 A systematic review for establishing relevant environmental parameters for urban lighting: translating research into rractice *Sustainability* **14** 1107

[3] Pérez Vega C, Zielinska-Dabkowska KM and Hölker F 2021 Urban lighting research transdisciplinary framework—a collaborative process with lighting professionals *Int J Environ Res Public Health* **18** 624

[4] Zielinska-Dabkowska KM, Schernhammer ES, Hanifin JP and Brainard GC 2023 Reducing nighttime light exposure in the urban environment to benefit human health and society *Science* **380** 1130–35

[5] Kyba CCM, Altıntaş YÖ, Walker CE and Newhouse M 2023 Citizen scientists report global rapid reductions in the visibility of stars from 2011 to 2022 *Science* **379** 265–68

[6] Gandy M 2017 Negative Luminescence *Ann Am Assoc Geogr* **107** 1090–107

[7] Kyba CCM, Kuester T, Sánchez de Miguel A, Baugh K, Jechow A, Hölker F, et al. 2017 Artificially lit surface of Earth at night increasing in radiance and extent *Sci Adv* **3** e1701528

[8] Kyba CCM, Hänel A and Hölker F 2014 Redefining efficiency for outdoor lighting *Energy Environ Sci* **7** 1806–09

[9] Dunn N 2016 *Dark Matters: A Manifesto for the Nocturnal City* (Winchester: Zero Books)

[10] Dunn N 2020 Dark design: a new framework for advocacy and creativity for the nocturnal commons *Int J Des Soc* **14** 19–30

[11] Dunn N and Edensor T 2022 *Rethinking Darkness: Cultures, Histories, Practices* (New York: Routledge)

[12] Edensor T 2017 *From Light to Dark: Daylight, Illumination, and Gloom* (Minneapolis: University of Minnesota Press)

[13] Edensor T 2022 Dark designs: creating shadow, gloomy spaces and enchanting light *Lighting Design in Shared Public Spaces* ed S Sumartojo (New York: Routledge) pp 195–215

[14] Eklöf J 2022 *The Darkness Manifesto: How light pollution threatens the ancient rhythms of life* (London: The Bodley Head)

[15] Stone T 2018 The value of darkness: a moral framework for urban nighttime lighting *Sci Eng Ethics* **24** 607–28

[16] Dorst K 2015 *Frame Innovation: Create New Thinking By Design* (Cambridge: The MIT Press)
[17] Winner L 1980 Do artifacts have politics? *Daedalus* **109** 121–36
[18] Verbeek P-P 2011 *Moralizing Technology: Understanding and Designing the Morality of Things* (Chicago: University of Chicago Press)
[19] Friedman B and Hendry D 2019 *Value Sensitive Design: Shaping Technology with Moral Imagination* (Cambridge: The MIT Press)
[20] van den Hoven J, Vermaas PE and van de Poel I 2015 *Handbook of Ethics, Values, and Technological Design: Sources, Theory, Values and Application Domains* (Dordrecht: Springer)
[21] van den Hoven J 2017 The design turn in applied ethics *Designing in Ethics* ed J van den Hoven, S Miller and T. Pogge (Cambridge: Cambridge University Press) pp 11–31
[22] Stone T 2021 Design for values and the city *J Responsible Innov* **8** 364–81
[23] McDonald R and Beatley T 2021 *Biophilic Cities for an Urban Century* (Cham: Palgrave MacMillan)
[24] Challéat S, Barré K, Laforge A, Lapostolle D, Franchomme M, SIRAMI C, et al. 2021 Grasping darkness: the dark ecological network as a social-ecological framework to limit the impacts of light pollution on biodiversity *Ecol Soc* **26** 15
[25] Narboni R 2017 Imagining the future of the city at night *Architect Magazine*
[26] Sordello R, Busson S, Cornuau JH, Deverchère P, Faure B, Guetté A, et al. 2022 A plea for a worldwide development of dark infrastructure for biodiversity – practical examples and ways to go forward *Landsc Urban Plan* **219** 104332
[27] Edensor T 2015 The gloomy city: rethinking the relationship between light and dark *Urban Stud* **52** 422–38
[28] Henke CR and Sims B 2020 *Repairing Infrastructures: The Maintenance of Materiality and Power* (Cambridge: The MIT Press)
[29] Stone T 2023 Darkening cities as urban restoration *Dark Skies:Places, Practices, Communities* ed N Dunn and T Edensor (New York: Routledge) pp 149–60
[30] Stone T, Dijkstra I and Danielse T 2021 Dark acupuncture: a design strategy for sustainable lighting *Int J Sustain Light* **23** 70–87
[31] Dugar A and Slade J 2022 Dovetailing darkness and light for designing sustainable outdoor spaces *Int J Sustain Light* **23** 88–99
[32] Stone T 2021 Re-envisioning the nocturnal sublime: on the ethics and aesthetics of nighttime lighting *Topoi* **40** 481–91
[33] Dunnett O 2015 Contested landscapes: the moral geographies of light pollution in Britain *Cult Geogr* **22** 619–36
[34] King R 2000 Environmental ethics and the built environment *Environ Ethics* **22** 115–31
[35] Light A 2001 The urban blind spot in environmental ethics *Env Polit* **10** 7–35
[36] Griffiths R and Dunn N 2020 More-than-human nights: intersecting lived experience and diurnal rhythms in the nocturnal city *Int Conf on Night Studies* ed M Garcia-Ruiz and J Nofre (Lisbon: ISCTE, Instituto Universitário de Lisboa) pp 203–20
[37] Rathsach CM and Hvass M 2022 The moon as a light source: new sustainable ways of lighting up cities at night *IOP Conf Ser Earth Environ Sci* **1099** 012046
[38] Owens M and Wolch J 2019 Rewilding cities *Rewilding* ed N Pettorelli, SM Durant and JT du Toit (Cambridge: Cambridge University Press) pp 280–302
[39] Maffey G and Arts K 2023 Human rewilding: practical pointers to address a root cause of global environmental crises *Routledge Handbook of Rewilding* ed S Hawkins, I Convery, S Carver and R Beyers (New York: Routledge) pp 374–82
[40] Dill K 2021 In defense of wild night *Ethics Policy Environ* **25** 153-77

Light formed through urban morphology and different organism groups: First findings from a systematic review.

S Dincel[1], U Besenecker[1], D Koch[2], KM Zielinska-Dabkowska[3]

[1] Lighting Design Division, School of Architecture, KTH Royal Institute of Technology, Osquars backe 5, 114 28 Stockholm, Sweden;
[2] School of Architecture, KTH Royal Institute of Technology, Osquars backe 5, 114 28 Stockholm, Sweden;
[3] GUT LightLab, Faculty of Architecture, Gdansk University of Technology, Narutowicza 11-12, 80-233, Gdansk, Poland
Corresponding author's e-mail: dincel@kth.se

Abstract. The prevailing implementation and usage of contemporary lighting technologies and design practices in cities have created over-illuminated built environments. Recent studies indicate that exposure to electric lighting effects formed through spatial characteristics has visual, physiological, and behavioural effects on both humans and non-humans, such as wildlife. In order to gain a better understanding of the impact that electric lighting has on space and different organism groups, a comprehensive literature review was conducted applying PRISMA 2020 systematic review guidelines. Results of the searches from various databases, such as Web of Science, PubMed and Scopus, identified 5260 related studies. A total of 55 papers connected to four themes: (1) urban morphology; (2) human visual impressions; (3) ecological impacts; and (4) design approaches and methods were analysed with a focus on urban morphology. The review provided the following general findings: lighting properties alone are inadequate to depict visual impressions of pedestrians, patterns formed through light interacting with spatial characteristics can contribute to understanding how spaces are visually perceived and help characterising the exposure of wildlife organisms to potential disturbances.

1. Introduction

1.1. Background

Built environments, whether indoor or outdoor, affect people's experiences through visual impressions. Electric lighting supports people's movements after darkness falls and engages them with their surroundings [1]. Outdoor spaces are utilised in different ways and for different purposes after dark, including recreation, socialisation, commuting, training, and a range of other activities. For most everyday purposes, good visibility of relevant objects and the overall environment is essential for the readability of public space. Some activities, for example, walking in an urban park, may be restricted by the lighting conditions this demands, and lighting conditions are also part of constructing habits and behavioural patterns [2, 3]. The process of vision and perception is mediated by the anatomical and psychophysical responses to light. Human vision is influenced by optical and spectral characteristics of light sources and visual perception and interpretation processes [4]. Human vision adapts and operates differently in environments of varying light levels and spectral content. The three states of vision are photopic (at light levels higher than 5 candelas per square metre, cd/m^2), scotopic (at light levels less than 0.005 cd/m^2), and mesopic (between light levels of 0.005 and 5 cd/m^2); they operate respectively during daytime, night-time, and transitional periods [5]. Visual perception processes occur in connection with the interplay between illuminated objects and the illumination of the areas surrounding them [4]. The visibility of objects located in immediate and distant surroundings depends on the object's size and the contrast ratios created by the varied light levels [6]. The material form of the city affects lighting and vision in many ways, be it through, e.g., size, shape, density, articulation, or surface materials. This suggests that the integration of urban morphology, that is, studies of e.g., the growth and form of settlements by considering land use, building structures, and plot and street patterns [7, 8, 9], with lighting design research should be of interest. Aside from how the shapes of buildings and infrastructure organise, open and limit space and vision, the choice of surface materials is also a determinant factor on regulating where and how light can pass or be reflected [10]. Visual impressions created through the interaction of light with spatial characteristics unfold an architectural dimension that is not limited to easily measurable physical lighting characteristics in urban contexts [11]. This review aims to inform a better understanding of the interconnectedness between lighting, spatial characteristics, and the states of vision that play a role in the actual visual experience of illuminated after-dark environments.

Environmental issues related to lit urban areas call for urban lighting research to engage with sustainability. With an increasing interest in the well-being of humans and the preservation of ecology, particularly two of the Sustainable Developments Goals (SDGs) put forth by the United Nations become relevant in this context of lighting research and practice. These are SDG11: Sustainable Cities and Communities and SDG15: Life on Land, which address the needs for sustainable, inclusive human settlements and the preservation and restoration of terrestrial ecosystems. The health-related impacts of exposure to electric lighting in human and wildlife organisms can range from physiological changes to survival capabilities [12]. Implementation of contemporary lighting technologies, e.g., conversion to energy-efficient LEDs, came with increased adverse effects on ecological habitats [13]. Besides the direct effects from light sources, atmospheric conditions, such as cloud cover and weather, alter the environmental impact when combined with the reflected illumination [14]; this can generate skyglow, leading to disruptions on some wildlife organisms [11, 15]. The human-centric perspective of planning practices resulted in the human-ecology dichotomy as these entities have been only addressed respectively, lacking an overall holistic consideration in research and practice [16]. The intersection between ecological responsibility and the visual experiences of pedestrians would entail a comprehensive design approach for simultaneously embracing human and wildlife organism groups' needs in urban green environments. The authors hypothesise that connecting lighting and spatial characteristics could support reassessing contemporary lighting technologies and their potential to minimise the ecological impact of electric lighting.

To explore possibilities and collect knowledge on bridging the human-ecology dichotomy created by urban lighting applications, this paper reviews research on the impact of LED illumination in urban parks. This literature review was conceptualised to help identify the relevant lighting parameters and spatial characteristics within green areas that influence the pedestrian visual impressions and wildlife organisms to be later tested in a full-scale field experiment. The objective was to understand the state of the art (and gaps) in lighting technology and to interpret the interplay between lighting properties and urban morphology impacting different organism groups from an architectural perspective. This review is based on papers that include relationships between lighting and spatial characteristics, visual experiences of pedestrians, and impacts on wildlife; papers with other focus areas that did not inform the review topic were excluded based on a set of criteria, see section 2. for more details.

1.2. Research questions

To explore and inform a comprehensive urban lighting design approach by involving considerations on both pedestrian's visual needs and ecological concerns in designing and planning processes, the following questions have been formulated to guide the review and research. To explore the link between urban morphology and lighting design:

1. How do spatial characteristics in urban form influence the effects created by electric lighting that impact different organism groups?

To more specifically define these effects and properties, the additional questions:

2. What lighting properties can support pedestrians' visual impressions after dark?
3. What lighting properties can be tuned to minimise the impact on local wildlife organism groups?

Overall, the research that this review paper is part of aims to explore the broad inquiry:

4. How can responses in human and wildlife organism groups to electric light exposure inform urban lighting design approaches and methods?

These four questions, or themes of inquiry, were used to categorise the review (Fig.3). Note that this paper focuses mainly on discussing the research questions 1–2 and broadly addresses the questions 3–4. The latter questions (3–4) in the reviewed studies will be respectively analysed and discussed in-depth in separate, dedicated papers.

2. Materials and methods

This systematic review paper includes research studies published from 2000 onwards in English. The review process follows PRISMA 2020 guidelines to identify and screen studies (Fig.1). Initially, keyword combinations in English were established by forming categories of areas and organism groups for the literature search. Keyword combinations for searching literature were arranged into four categories, and three separate searches were performed by combining categories. The defined keywords were searched, combining the areas of lighting and urban morphology with the respective organism groups of humans and non-humans (Fig.2). "AND" is used to bind keywords (e.g., lighting and urban morphology with humans), and "*" to broaden the search finding by including word stems or plural forms of the keywords. The three-stage review process included identification, screening, and inclusion phases.

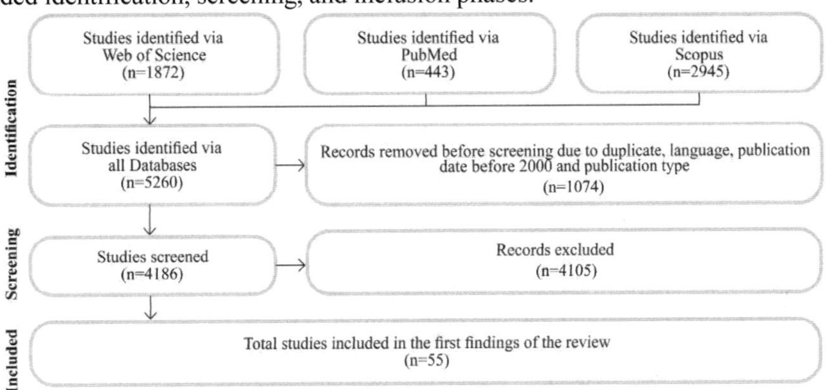

Figure 1. Flow chart for the literature search process and selected studies for review using PRISMA 2020 guidelines.

Areas	Lighting	"urban light*" OR "artificial light*" OR "electric light*" OR "street light*" OR "outdoor light*" OR "exterior light*" OR "city light*" OR "lighting design" OR "urban illumination"
	Urban morphology	"urban env*" OR "built env*" OR "urban plan*" OR "urban design" OR "urban space" OR "architectur*" OR "park*" OR "green area*" OR "green space*" OR "greenery" OR "vegetation*" OR "tree*" OR "pathway*" OR "pedestrian space" OR "spatial*" OR "material*" OR "reflect*" OR "weather" OR "snow" OR "rain" OR "asphalt"
Groups	Humans	"human*" OR "people" OR "pedestrian*" OR "anthropogenic" OR "visibility" OR "visual" OR "percept*" OR "vision" OR "accessibil*"
	Non-humans	"wildlife" OR "animal*" OR "species" OR "bird*" OR "mammal*" OR "rodent*" OR "shrew*" OR "hedgehog*" OR "mole*"

Figure 2. The keywords for the category of areas and organism groups.

The literature searches were performed using Web of Science, PubMed, and Scopus search engines during the period of February to March 2023, and the papers were reviewed from March to July 2023. The literature search was aimed to detect and include only peer-reviewed articles, proceedings papers, and book chapters. In the identification stage, a large number of articles that were not relevant to our inquiry were eliminated. 1074 papers were excluded due to duplication, language, publication date, and type. Because our review concerns the specific case of pedestrian lighting for urban green spaces, screening was performed by applying exclusion criteria to the studies, including the following aspects:

- Outdoor lighting that concerns only motorised vehicles or/and cyclists.
- Outdoor lighting focusing exclusively on people's perception of safety, health-related or social aspects, atmosphere, artistic installations, and energy efficiency.

- Lighting for visually impaired people.
- Street lighting for roadways or areas adjacent to buildings and facades.
- Non-human organism groups that are not relevant to wildlife research in the Nordic countries.
- No inclusion of the use or comparison of LED light sources.
- Unidentified non-human organism groups and undefined light sources, e.g., only described broadly as ALAN ("artificial light at night" or "anthropogenic light at night").

	Amoruso et al.	Bandopadhyay et al.	Cafuta, M.R.	Capron et al.	Choi et al.	Davies et al.	Dominoni et al.	Dunn N.	Dwimirnani et al.	Esposito et al.	Gaston et al.	Feng et al.	Fotios et al.	Hvass et al.	Karamouzi et al.	Koen et al.	Johansson et al.	Li et al.	Lin et al.	Liu et al.	Maccheroni et al.	Markvica et al.	Masullo et al.	Moeck M.	Navvab et al.	Ogando-Martínez et al.	Pan et al.	Patching et al.	Pedersen et al.	Pérez Vega et al.	Portnov et al.	Pracki et al.	Rahm et al.	Rea et al.	Ross et al.	Sakici et al.	Schroer et al.	Toet et al.	Tripathy et al.	Valetti et al.	Viikari et al.	Vitasek et al.	Wänström Lindh et al.	Zák et al.	Zhang et al.	Zielinska-Dabkowska, KM.
Theme 1 Light and Urban Morphology			x		x			x	x			x		x	x			x		x	x	x				x	x	x			x				x	x			x	x		x	x			x
Theme 2 Light and Visual Impression	x	x	x	x									x	x			x		x				x	x	x				x				x	x	x			x		x	x		x	x	x	
Theme 3 Light and Non-humans						x	x				x					x																			x		x									
Theme 4 Design Approaches and Methods			x					x		x				x			x						x							x		x			x			x		x					x	x

Figure 3. Papers corresponding to the four themes (research questions) based on the first findings of the literature review.

The exclusion criteria were constructed to focus on reviewing papers on light parameters that influence pedestrians' visual experiences concurrently causing ecological impact in urban green areas. The aim is to establish an ecological understanding for embracing a more responsible use of contemporary lighting technologies designed to enable human activities after dark. After scanning titles and abstracts in the screening phase, full texts were reviewed for identifying and assembling results. The results were categorised based on four themes related to the research questions: (a) light effects formed through urban morphology, (b) pedestrian visual impressions, (c) responses in wildlife organisms, (d) design approaches and methods (Fig.3). A total of 55 studies were included in the first stage of the systematic review.

3. Results

This review on urban lighting design is aimed at the integration of how urban morphology conditions are affected by light, human perception and ecological considerations. The detailed review results are described and discussed below. The results indicate that urban morphology, together with the physical properties of LED light sources, is a contributing factor in shaping patterns of light and dark, thus influencing pedestrian visual impressions and impacting wildlife. Moreover, physiological and behavioural effects of exposure to direct and indirect electric lighting properties on wildlife organisms were identified.

3.1. Interplay between urban morphology and electric lighting

3.1.1. City planning and morphology. An indirect result of the exclusion criteria applied for this paper demonstrates what is generally true also for the remaining included papers: planning and morphological research tends to focus on analysing lighting as a means to an end and rarely engages directly with light perception. In the included papers, arguments regarding the sense of safety [17-19], effects on use intensity [17, 19-22], attraction and comfort [17, 19, 22, 23], and energy efficiency [19, 20, 24] are dominant. In line with morphological research traditions, historical reflections are included in several papers [3, 20-22] but remain brief and broad rather than specific, while one can find it central for the discussion [3] or specifically related to the analysed area [22]. The papers primarily concern methods for improving lighting and fall into either empirical study of lighting effects on behaviour or perceived quality [17-19, 21, 22] or methods to optimise lighting based on sourced or assumed priorities [20, 23, 24]. Only a few directly address relations between morphology, lighting, and the evaluation criteria of the respective paper—e.g., attractiveness, behaviour, use intensity, etc.—and most empirical papers used either perceived or reported light quality, intensity of use, or observed behaviour of pedestrians. Almost all papers treat lighting as complementary: as an addition to urban form rather than integrated into urban design thinking. Importantly, while most papers report little direct links between lighting and behaviour [19, 21, 22], perception and evaluation are consistently reported to be impacted [17, 19-21]. The paper on Dark Design [3] stands out in this regard. While lighting is generally considered intrinsically good, two papers [3, 20] question lighting not only from efficiency or ecological perspectives but also from human and social perspectives and highlight how electric lighting has and will change how individuals and societies relate to light and darkness.

3.1.2. Scale and edges. Brightness, contrast, uniformity, distribution, and surfaces are significant factors for altering impressions of scale, boundaries, and form of the urban landscape from a human perspective [25-27]. Jan Gehl's human-scale concept is referred to in lighting design literature, supporting social activities by looking into architectural proportions and urban contexts [1,11]. Design aspects such as lighting layout, luminaire mounting height, the amount and direction of luminaires, and their distribution play a role in changing the scale perception of spaces. Light patterns that are horizontally or vertically arranged can enlarge the width and height impression when overlapped effects increase the brightness regardless of low-level ground lighting [25-27]. Overlapping the light patterns can also positively influence the impression of lighting uniformity, hence strengthening spatial continuity. Furthermore, eliminating the dark spots that break off the continuity makes the space look broader [26, 27]. Detaching the light patterns and illuminating background surfaces distinctively escalates depth perception [27]. A lower luminaire mounting height of four and a half metres is preferred for pedestrian lighting over a higher mounting position of six metres and above, typical in roadway lighting, providing a smaller-scale and personal impression of space despite the non-uniform distribution of lighting at a lower height [26].

3.1.3. Visual permeability. Several studies include Kevin Lynch's five elements: paths, edges, districts, nodes, and landmarks, as a valuable approach to illuminating key urban features [1, 26], constructing a visual representation of the city after dark [11]. Decreasing contrast ratios enhances the readability of architectural characteristics [26, 28]; however, visual coherence between day- and night-time appearances of structures is diminished [29], and spatial order among architectural layers is blurred [11]. Variations in brightness-contrast levels can create either a visual enclosure or openness in the urban context. Thus, perceiving illuminated objects, such as tree trunks, as detached single elements generates visual boundaries, whereas observing them as being part of the background gives a broad and open space impression [25]. Visual boundaries reinforce the space's legibility by applying contrast and non-uniform, peripheral, and vertical illumination [26], thereby supporting the perception of safety [11].

3.1.4. Soft and hard landscape elements. The interaction between electric lighting and soft landscape elements, such as trees and bushes, results in varying effects due to light being reflected or diffused [26, 30]. Illuminated buildings and spatial forms located within a height range of 0-20 metres can produce intrusive and spilled light [30]. The texture of trees, dense or light, determines the visual quality of the surroundings and should be taken into account when planning illumination. Dense-textured trees can effectively be illuminated by fixtures from a distance due to their low light transmittance [31]. Dense and tall vegetation cover at pedestrian height can contribute to an improved perception of the light distribution [26, 30], and broad-leaved trees can act as reflectors and diffusers, leading to the impression of a "large-scale outdoor living room" [30]. In natural surroundings, non-uniformly distributed lighting and sharp contrasts are softened due to tall trees; therefore, large pole spacings of 40 metres become unnoticeable [26]. Low light levels in highly vegetated urban parks have a better uniformity than in areas with high building coverage ratios [30]. Discomfort glare can be reduced by changing the light source visibility and direction towards greenery [32].

The composition of surface materials, such as being specular or uneven, affects the light behaviour by projected light onto these surfaces being reflected or absorbed [10, 11, 33]. Standards that are defined for road pavement reflectance are outdated for obtaining accurate results in computer-aided simulations.

Especially reflectance values of contemporary materials which are developed to combat the heat island effects, such as cool asphalts, as well as porous and stone mastic pavements, are insufficient [34]. Material surface properties can contribute to light uniformity and low luminance values, e.g., a porous pavement in dry conditions performing well at low light levels, such as 2 cd/m^2. High luminance levels that emerge through specular properties of surfaces lead to decreased uniformity and discomfort glare [35]. Lowering luminance levels also supports improved visibility of the surroundings in connection with diminished reflections from glossy surfaces, such as glass panels [28]. Besides visibility issues, reflections from illuminated surface materials can pose an ecological risk due to altered skyglow [11].

3.1.5. Weather conditions. Light pollution is produced by electric lighting and can be increased by weather conditions, cloud thickness, air quality, and lunar phases. Liu [36] measured different light intensities during rainy weather in clear and cloudy sky conditions and demonstrated the highest brightness in the evening hours between 22:00 and 23:30 under clear sky conditions in which the sky colour was described as orange. Cloudy sky conditions amplify artificial illumination effects the most compared to a clear sky and lunar phase. Due to the reflectance properties of snow, sky brightness escalates more than it does under rainy weather. Severely polluted air exacerbates light pollution effects 10 times higher, while lunar phases contribute least to environmental impacts [36]. Despite the minimal impact, illuminance levels of 0.0001 lux on a starry night and 0.3 lux on a full-moon night can interfere with the melatonin production of some of the light-sensitive wildlife organism groups [37]. On the other hand, Vitasek [38] showed that varying fog intensities reduce light distribution. Three separate layers are proposed for assessing electric lighting outdoors: ground surface, canopy, and sky [36].

3.2. Impact of electric lighting on human visual processes and impressions

3.2.1. Human vision under photopic, mesopic and scotopic light levels. Visual perception is influenced by light source characteristics, the properties of object surfaces that the light reflects from, and the receivers and their visual system [39]. Cone photoreceptors in the eye facilitate detail and colour vision at high light levels (photopic vision) and guidelines and tools for outdoor lighting systems [40]. Light levels of 3-5 cd/m^2 are the threshold for mesopic vision, meaning that cone vision decreases with lowering light levels, and the rod photoreceptors become active [39, 41]. Rods are the photoreceptors that enable low light, scotopic vision at levels lower than 0.001 cd/m^2. Rods are distributed throughout the visual periphery and tuned to shorter wavelengths. Therefore, with decreasing light levels, the sensitivity of human vision moves towards shorter wavelengths, colour perception diminishes, and peripheral vision becomes heightened over central vision. While these mechanisms have been known, and the scotopic to the photopic ratio (S/P ratio) is an established measure to evaluate the ability of light to support low-light vision, there is no agreement on designing lighting for the urban and suburban outdoors [42-44]. Early models proposed look-up tables for factors to adjust photopic luminance to mesopic [44], but one of the complexities is that the spectral sensitivity of low light vision changes with light levels [39]. Researchers have explored mesopic vision models to digitally simulate visibility in responses to different light spectra and light levels [43] and related affective responses. Mesopic systems of photometry integrate scotopic and photopic vision models to estimate visual performance; the development of brightness models, however, turned out to be more complex [42]. The role of the intrinsically photosensitive retinal ganglion cells, a recently discovered photoreceptor, is also being researched, and evidence suggests involvement in scene brightness perception [45]. Visual performance is important to detect obstacles and recognise information [42], and brightness perception has been associated with perceived safety [46, 47], although, as highlighted by other research, such effects are also personally, culturally, and historically bound [3]. With decreasing light levels, human visual systems tune increasingly to shorter wavelength, and to take advantage of that notion, earlier research suggested strategies to install short wavelength-rich (cool-white appearing) lighting which would allow the lowering of light output and energy consumption [42]. However, the short-wavelength light contributes to skyglow [48], which raises ecological concerns, and, especially under certain weather conditions, decreases visual comfort and performance by hindering dark-light adaptation and producing more glare than visually warmer lighting [49]. Higher brightness does not necessarily support vision, as it can create high contrast ratios with darker surroundings, challenging visual adaptation [50]. When transitioning from bright to dark environments, it can take several minutes to return to normal visual sensitivity. Lowering light levels and providing gradual transitions could be a strategy to support dark adaptation, increase visibility, and minimise energy consumption and skyglow. Adaptation luminance changes with gaze direction, especially when walking in non-uniformly illuminated outdoor spaces [43]. Research using eye tracking has investigated adaptation luminances in cities and their simulations after dark (0.1 to 10 cd/m^2) [40, 43]. Generally, more light is needed for people over 60, but at the same time, higher uniformity and less contrast. The ability to light/dark adapt decreases with age, while the sensitivity to glare increases [40].

3.2.2. Visibility of surroundings. Navigating the environment includes needs to distinguish risks and recognise objects, that is why having good visibility of the outdoor surroundings is important [11]. Besides the lighting properties, position, angle, and direction of light sources, surface materials influence object visibility by altering luminance levels and distribution patterns [26, 33]. Visibility of the surroundings shifts in relation to the proximity to lighting installations and ambient lighting conditions [51]. Variations of light levels that occur in immediate and distant areas result in contrast ratio differences. Moreover, uniformly distributed high light levels in a particular area increase the contrast ratios to the surroundings [52]. Balancing the ratios through the means of using vertical objects, such as trees, can support peripheral vision by acting as reflectors [27, 32]. However, lower light levels are perceived as insufficient when individuals enter a darker area from brighter environments [28]. Adaptation can be supported; decreasing illuminance levels in immediate adjacent areas don't need to worsen the object visibility except at very low levels, e.g., 20% of the surroundings at ground level [51]. Using an LED with equivalent chromaticity (colour) characterisations but lowering output in the short and mid wavelengths region (metamer) showed no indications of worsened visibility [32].

3.2.3. Visual accessibility and comfort. Walking behaviour [51, 53, 54], readability of signs [51, 53, 54], facial expression identification [53, 54] and obstacle detection [42, 54], as well as visual comfort [55], are dependent on the quality of urban lighting. Parameters connected to lighting that influence both visual accessibility and visual comfort are typically characterised in illuminance [53-55, 56, 57] and correlated colour temperature (CCT) [54, 55, 56, 57]; meanwhile, luminance [53, 54], colour rendering index (CRI) [54, 56], distribution [53, 54] and SPD [54] are distinctively relevant for accessibility. The perceived outdoor lighting quality (POLQ) questionnaire was developed to assess the relationship between quantifiable and perceived qualities of outdoor lighting [56]. Perceptual tasks that were performed on a constructed walking path inside a lab successfully demonstrated a high rating on the perceived strength quality (PSQ) dimension when high overall illuminance was combined with a CCT around 3810K, and they showed a deficiency in pleasantness [54]. In a virtual reality study, it was suggested that warmer CCT of around 2700K and high overall illuminance lead to better visual accessibility of areas covered with vegetation [57]. Brightness is linked to ground-level illuminance and CCT [55], while gaze behaviours have no direct association with these parameters [57].

In controlled lab studies, walking speed varies, dimming light levels decrease the speed, and it remains constant even after shifting back to higher light levels [51]. In another study, it was shown that the constant speed persists under light installations that have different photometric characteristics [53]. Readability of signs [51] and facial recognition [53] become the most demanding in low light levels therefore, increasing light levels and uniformity improves readability at longer distances, especially by young users (20-35 yrs) compared to elderly users (60-75 yrs) [53, 54]. Visibility of details and facial recognition is determined by the distance between an observed subject and an observer; hence, a distance range between 12-25 metres is given for identifying facial expressions sufficiently [2]. Despite the adequate ground light levels, facial recognition at the proposed distance of 15 metres is limited, implying the importance of sufficient eye-height light levels. Detecting obstacles that are located outside of a constructed path becomes easier under a high S/P ratio due to the considerable amount of short wavelength content in the spectrum that eases rod-driven peripheral vision [54]. A study suggests that by applying illuminance levels between the range of 4.08-6.99 lux and a CCT of 3126K-4498K, physiological fatigue can be reduced, and visual comfort is enhanced when compared to a higher ground-level illuminance of 10 lux and a CCT of 5600K [55]. Moreover, decreasing excessive light levels [11], such as, high contrast ratios between an immediate area of 200 lux and a distant surrounding of 5 lux [52], directing light sources away from the path towards vertical surfaces [32], and shielding luminaires [11] has been shown to improve visual comfort by reducing discomfort glare.

3.2.4. Overall impression of the environment. Perceived safety [11, 46, 47, 53, 56, 58] and atmosphere [26, 29, 51, 52, 54] are influenced by lighting conditions. In studies, light levels, uniformity, CCT, and CRI are important factors that support perceived safety [47], while contrast ratios and perceived brightness contribute to visual impressions of individuals, thus interrelating lighting quality [56] with the usage of spaces [11, 58]. Spatial brightness and perceived safety have a positive correlation in some studies [46, 47, 56], although evaluating perceived safety impartially is difficult in field studies where other people are present [53] and is affected by a range of interconnected aspects [3]. Unmaintained vegetation cover and mesh fences can enclose pathways and restrict the view, thereby increasing the feeling of entrapment. Such effects regularly lead to avoidance of areas and route changes after dark, especially by women. On the other hand, well-maintained vegetation and illuminated areas can improve perceived safety [58]. It has been shown that lowered contrast ratios between immediate and distant zones lead to a sensation of pleasantness and harmony [26, 29, 52], whereas high contrast ratios that contain uniformly distributed lighting [29] and high overall light levels combined with cooler CCT [54] negatively impacts the overall impression of the environment [29].

3.3. Impact of electric lighting on wildlife organism groups

Technological advances in public lighting are changing the spectral characteristics of illuminated habitats. Reviewed studies indicate physiological and behavioural impacts in birds and non-human mammals, including rodents, shrews, hedgehogs, and moles. The first findings from the systematic review present an overview of the ecological impacts produced by electric lighting without providing impact details for the specific species. For wildlife organisms, light is an influential factor in regulating reproduction, migration, sense of direction, and signalling within and between species [15]. Light levels, spectral content, timing, and duration have varied effects on diverse groups of wildlife organisms [15]. Moreover, the broad spectrum of street and pedestrian lighting allows animals to perceive objects that reflect light over more of the spectrum they are perceptive to, thereby changing the stability of species' interactions [59]. In environments such as urban forest parks and green spaces where wild animals exist, light pollution is correlated to alterations in their circadian behaviour, reproduction, and predator–prey relations [12], as wildlife organisms react differently to the spectral content of light sources and light level thresholds [37]. Skyglow can have an impact over long distances with low light levels, modifying the movement and distribution of other organism groups [60]. Analysed studies provide meaningful data indicating that light levels are increasing where biodiversity is high, demonstrating a global threat to non-human organisms, demanding urgent actions [61]. Increased intensity of ALAN from pedestrian lighting in darker areas can be problematic as thresholds for fauna at night are often associated with the intensities of moonlight and far below new levels [60].

3.4. Urban lighting design approaches and methods

3.4.1. Assessment methods and parameters. The reviewed studies employed both quantitative and qualitative methods for evaluating lighting and data were collected in forms of hand-held light measurement instruments [17, 18, 21, 22, 26, 28-30, 34, 36, 51-54], computer-aided simulations [17-20, 23, 24, 26, 34, 35, 38, 62], virtual reality (VR) [55, 46, 57], eye-tracking devices [40, 57], wearable electroencephalograms (EEGs) [55], participatory simulation interfaces [32], scale models [27], questionnaires [11, 18, 19, 25, 29, 31, 51-56], interviews [17, 25, 58], observations [19, 21, 22, 25-27, 30], photographs [25, 26, 29, 31, 36, 43, 53, 54, 65], and sketching [25, 26]. Some of the theories used in the studies were inspired by Gordon Cullen [26], Jan Gehl [1, 11], Kevin Lynch [1, 25, 26], space syntax [21, 22], and landscape aesthetics [29]. Furthermore, empirical studies were conducted in the forms of field works [11, 17-19, 21, 22, 25, 26, 28-30,

IOP Conf. Series: Earth and Environmental Science 1320 (2024) 012002 doi:10.1088/1755-1315/1320/1/012002

36, 40, 43, 53, 58], lab experiments [38, 46, 51, 52, 55, 57, 58, 63], and computer simulations [26, 27, 34, 35].

Some of the methods presented in the studies to support design and planning processes are: semantic scales describing impressions of illuminated environments, which were established by Flynn et al. and served as a basis for the creation of a considerable amount of questionnaires for rating subjective qualities of spaces [52, 56, 57]; Three-Dimensional Evaluation SEC (suitable for everyone, environmentally accepted and cost-effective) based on site-specific spatial qualities [11]; POLQ questionnaire for assessing the perceived outdoor lighting qualities of pedestrians [56]; immersive virtual reality simulations of urban parks [57]; evaluation of historical architecture features by combining subjective and objective lighting qualities [29]; and multi-criteria assessment procedure, an integrated tool supporting the assessment of aspects related to safety, visual comfort, energy efficiency and environmental sustainability [63]. The papers indicate a need to better understand alignments and differences between effects on perception and behaviour [3, 19, 21, 22] and their relation to light and darkness [3, 20]. Present light measurement parameters, such as illuminance, CCT, and uniformity ratios and instruments targeted to appraise pedestrian lighting fall short of assessing LED applications using high-intensity, non-uniform lighting [37]. Therefore, different metrics are required to analyse the impactful properties of lighting on various species [37].

3.4.2. Ecological design approach in urban lighting applications. To integrate an ecological perspective into projects, the application of environmental lighting zones (E1-E4), including distinctive ambient brightness levels that protect environmentally sensitive settlements, the use of an overall urban lighting plan focused on darkness preservation, and environmental and health impact assessment procedures are proposed [37]. Today, research has established that short wavelength content should be minimised in outdoor lighting solutions to not contribute unnecessarily to skyglow or light-related biological triggers [37, 48]. "Amber" LED lighting and its characteristics are currently being explored in more detail for recommendations, meaning a warm-appearing light with reduced short wavelength content, typically between 1800K and 2200K CCT [37, 48]. Depending on the spectrum, objects and surface colours can still be rendered. However, no concrete short-wavelength radiation thresholds and recommendations are yet in place. Also, amber lighting might negatively impact some species, e.g., the mating of firefly beetles, so it should be carefully applied based on site-specific conditions [64]. Decreasing light levels, directing luminaires attentively, limiting operational times, and implementing adaptable lighting control systems are necessary besides spectral considerations [37]. Mitigating ecological impact can be possible using metamers (light of the equivalent chromaticity but mixed with different spectral compositions) [32]. The ecological impact measurement method of Longcore et al., in which response curves of wildlife organisms can be transformed into "action spectra", can enable a first estimation of an SPD's potential effects on species. The total effective irradiance can be estimated based on the plotted action spectrum of an animal group to the respective SPD of the light; lower effective irradiance represents a lower ecological impact [32]. Metamer mixing using multi-channel LEDs is intended to dampen undesired effects produced by phosphor-converted (PC) LEDs through decreasing short wavelength content in the spectrum [32, 37]. However, this method can currently compromise energy efficiency and CRI (depending on configuration) as opposed to white PC LEDs [32]. Furthermore, present standards and guidelines only address impacts that occur over the horizon, such as light pollution [65]. At the regulatory level, environmental aspects are managed in the form of soft laws, e.g., guidelines, that exclude technical parameters about spectrum, flicker, peak luminance, and peak radiance that are relevant to wildlife organisms or encompassing generic information that makes ecological strategies harder to be applied in specific local conditions [37].

4. Discussion - Towards a comprehensive urban lighting design approach

Existing knowledge related to urban lighting design considers various factors and impacts, e.g., perceived light qualities from a human perspective [56] and physiological and behavioural alterations on species [15]. Few studies concentrate on the interaction between spatial characteristics and lighting [17-27, 28-38]; others focus exclusively on either human vision [39-50, 62], perception [51-54, 58] or environmental impacts [12-15, 59, 61, 64]. Thus, the overall knowledge on the contribution of spatial characteristics in transforming the light effects in urban contexts, impacting pedestrians and ecology, appears fragmented and not holistically integrated. Together with the changing abilities of the human eye and vision under different light conditions, the interplay between material forms of built environments and electric lighting transforms spaces in different ways. For instance, illuminated spaces can be perceived as small, large, personal, visually accessible, or safe from a human perspective. Such experiences are embodied beyond quantifiable metrics, highlighting the importance of social, spatial and temporal situatedness for the visual qualities of spaces. This review shows that subjective impressions can be altered in relation to spatial characteristics, physical properties of lighting, luminaire configurations (e.g., layout and mounting height) and contextual conditions. In natural settings, vegetation, specifically dense and tall trees, can ease sharp contrasts and glare issues from non-uniform distributions and large spacings between light sources. Luminance levels of surfaces in immediate and distant areas can have more effect upon the visibility of objects than the illuminance levels at those areas or surfaces. On the other hand, surface characteristics of materials altered by light and weather conditions result in either reflected or diffused illumination and can wield unexpected light patterns, potentially causing increased discomfort glare. In addition to disturbing humans, produced light effects can expose wildlife organisms to uncontrolled ecological disruptions due to direct exposure, generated light trespass in their habitats and skyglow, including physiological and behavioural changes. Although the impact of light properties varies among species, to mitigate ecological disturbances, some general considerations are: lowering light levels, avoiding light trespass into unwanted areas, and minimising short-wavelength (blue-green) content in the light spectrum. In a design process, overlooking the qualities of space, lighting, human vision, or wildlife habitat is problematic, as impacts and perceived lighting effects are interdependent and interrelated. Therefore, a comprehensive design approach is needed to respond to the needs of pedestrians and wildlife organisms, at a minimum, allowing the coexistence of both, especially in urban green spaces. To enable this, integrative design thinking is necessary where properties of electric lighting are planned and specified in regard to project-specific considerations of the spatial structure, visual objectives of pedestrians, and local wildlife. By determining thresholds necessary for people's visual purposes in urban parks, design strategies can be incorporated into lighting applications to minimise the ecological impact.

With this in mind, we note that in the reviewed studies, theoretical inspirations from urban design are mostly limited to the approaches of Kevin Lynch and Jan Gehl, leaving the potential to transfer different urban theories and concepts into lighting research for supporting meaningful analysis and design of illuminated spaces. Furthermore, it needs to be asked to what extent perception or use of outdoor spaces after dark should be evaluated by how much they deviate from daytime or peak activity patterns; this may be both unrealistic and limiting for lighting design work. Conversely, this study also, albeit indirectly, suggests that urban morphology and urban design studies can better integrate advanced lighting design knowledge; we are aware there is a wider inclusion of "lighting" discussions in urban design discourse, which is either not captured, or excluded, by the criteria of this review. The authors noted that several reviewed papers, especially shorter conference papers, could have used more accurate terminology, a conclusion based on findings and clarity of illustrations. While we did not exclude such papers if they introduced interesting research in the area, terms were paraphrased and corrected where appropriate.

5. Considerations

The results of this review indicate that knowledge is fragmented in the domain of lighting design-related research that conjoins the perspectives of humans and wildlife organisms in shared outdoor spaces. Urban lighting research calls for more extensive, integrated knowledge and research-based design guidance for developing new approaches, assessment methods and design methodologies for tackling today's sustainability challenges, hence providing inclusive and ecologically sustainable illuminated outdoor environments. It also points to a need to further develop an integrated view of design fields, allowing lighting design knowledge to inform work with the structure and shape of the built environment, and vice-versa, rather than working with light where the environment is largely seen as fixed. This systematic review aims to support identifying, selecting, summarising, and critically assessing the relevant research which is connected to lighting and urban morphology for responding to the needs of both humans and wildlife of urban green areas after dark to be directly applicable to the lighting practice.

Acknowledgements: This work is financially supported by NordForsk and the Swedish Energy Agency.

References

[1] Maccheroni D et al. 2021 *6th Junior Conference on Lighting*
[2] Fotios S and Johansson M 2019 Appraising the intention of other people: Ecological validity and procedures for investigating effects of lighting for pedestrians *Lighting Research & Technology* **51** 111–30
[3] Dunn N 2020 Dark Design: A New Framework for Advocacy and Creativity for the Nocturnal Commons *The International Journal of Design in Society* **14** 19-30
[4] Cuttle C 2015 Lighting design: A perception based approach (New York: Routledge)
[5] Boyce PR 2014 *Human Factors in Lighting*, Third Edition (New York: Taylor & Francis)
[6] Davoudian N 2011 Visual saliency of urban objects at night: Impact of the density of background light patterns *LEUKOS* **8** 137-52
[7] Hillier B and Hanson J 1984 *The social logic of space* (Cambridge: Cambridge University Press)
[8] Moudon A V 2022 Urban Morphology as an emerging interdisciplinary field *Urban Morphology* **1** 3-10
[9] Oliveira V 2021 *Urban Morphology: An Introduction to the Study of the Physical Form of Cities* (Cham: Springer)
[10] Gibson JJ 2014 *The Ecological Approach to Visual Perception* (New York: Taylor & Francis)
[11] Rozman Cafuta M 2021 Sustainable City Lighting Impact and Evaluation Methodology of Lighting Quality from a User Perspective *Sustainability* **13**
[12] Dominoni D M, Borniger J C and Nelson R J 2016 Light at night, clocks and health: from humans to wild organisms *Biol Lett* **12** 20160015
[13] Pawson S M and Bader M K F 2014 LED lighting increases the ecological impact of light pollution irrespective of color temperature *Ecol Appl* **24** 1561-8
[14] Gaston K J, Visser M E and Hölker F 2015 The biological impacts of artificial light at night: the research challenge *Philos Trans R Soc Lond B Biol Sci.* **5**
[15] Schroer S and Hölker F 2017 *Handbook of Advanced Lighting Technology,* ed R Karlicek, *et al.* (Cham: Springer) pp 957-89
[16] Erixon H, Borgström S and Andersson E 2013 Challenging dichotomies – exploring resilience as an integrative and operative conceptual framework for large-scale urban green structures *Planning Theory & Practice* **14** 349-72
[17] Patching G R, Rahm J, Jansson M and Johansson M 2017 A New Method of Random Environmental Walking for Assessing Behavioral Preferences for Different Lighting Applications *Front Psychol* **8** 345
[18] Portnov B A, Saad R and Trop T 2021 Interactive Scenario-Based Assessment Approach of Urban Street Lighting and Its Application to Estimating Energy Saving Benefits *Energies* **14**
[19] Markvica K, Richter G and Lenz G 2019 Impact of urban street lighting on road users' perception of public space and mobility behavior *Building and Environment* **154** 32-43
[20] Feng X and Murray A T 2017 Spatial Analytics for Enhancing Street Light Coverage of Public Spaces *LEUKOS* **14** 13-23
[21] Dwimirnani P, Karimi K and Palaiologou G 2017 Space after dark: Measuring the impact of public lighting at night on visibility, movement, and spatial configuration in urban parks. In: *11th Space Syntax Symposium,* (Lisbon, Portugal)
[22] Choi AS et al. 2006 Application of the space syntax theory to quantitative street lighting design *Building and Environment* **41**, 355–366
[23] Karamouzi A, Papalexopoulos D, Stavridou A, Tzimopoulou S and Varoudis T 2013 D.L.D., Dynamic Lighting Design: Parametric Interactive Lighting Software in Urban Public Space. In: *2013 9th International Conference on Intelligent Environments,* pp 71-6
[24] Tripathy A K, Mishra A K and Das T K 2017 Smart lighting: Intelligent and weather adaptive lighting in street lights using IOT. In: *International Conference on Intelligent Computing, Instrumentation and Control Technologies (ICICICT)*
[25] Wänström Lindh U and Billger M 2021 Light topography and spaciousness in the urban environment *Nordic Journal of Architectural Research* **33** 103-35
[26] Wänström Lindh U and Jägerbrand A K 2021 Perceived lighting uniformity on pedestrian roads: From an architectural perspective *Energies* **14**
[27] Wänström Lindh U and Billger M 2021 Light Distribution and Perceived Spaciousness: Light Patterns in Scale Models *Sustainability* **13** 1-21
[28] Hvass M and Hansen E K 2022 Balanced Brightness Levels. In: *Light Symposium,* (Copenhagen, Denmark: IOP Conf. Series: Earth and Environmental Science)
[29] Valetti L, Pellerey F and Pellegrino A 2022 A Novel Approach for the Assessment of the Nocturnal Image of the Cultural Landscape *Leukos* **19** 71-93
[30] Pan W and Du J 2021 Impacts of urban morphological characteristics on nocturnal outdoor lighting environment in cities: An empirical investigation in Shenzhen *Building and Environment* **192**
[31] Sakıcı Ç and Var M 2014 A visual perception effect assessment of some large and broad-leaved trees under different lighting arrangements *Architectural Science Review* **57** 139-46
[32] Ross P R and Rutten N Light Sketching for Ecology: A cooperative design tool for balancing human experience and ecological impact. In: *Light Symposium,* (Copenhagen, Denmark: IOP Conf. Series: Earth and Environmental Science)
[33] Moeck M 2000 Lighting design based on luminance contrast *Lighting Research & Technology* **32** 55-63
[34] Ogando-Martínez A, Troncoso-Pastoriza F, Eguía-Oller P, Granada-Álvarez E and Erkoreka A 2019 Model Calibration Methodology to Assess the Actual Lighting Conditions of a Road Infrastructure *Infrastructures* **5**
[35] Li H, Xie N, Harvey J, Liu J, Zhang H and Zhang Y 2022 Exploring the effect of pavement reflection and photometric properties on road lighting performance *Lighting Research & Technology* **55** 170-88

[36] Liu M, Li W, Zhang B, Hao Q, Guo X and Liu Y 2020 Research on the influence of weather conditions on urban night light environment *Sustainable Cities and Society* **54** 1-15
[37] Zielinska-Dabkowska K M 2022 Healthier and Environmentally Responsible Sustainable Cities and Communities. A New Design Framework and Planning Approach for Urban Illumination *Sustainability* **14** 1-22
[38] Vitasek J, Latal J, Stratil T, Jargus J and Wilcek Z 2018 Fog Influence Simulation on Light Distribution from Street Lighting. In: *20th International Conf. on Transparent Optical Networks (ICTON),* (Ostrava, Czech Republic)
[39] Žák P and Zálešák J 2016 The influence of spectral properties of light in street lighting on visual perception. In: *IEEE Lighting Conference of the Visegrad Countries (Lumen V4),* (Karpacz, Poland)
[40] Navvab M, Bisegna F and Cellucci L 2016 Defining pedestrian's visual adaptation field under night lighting in Venice. In: *16th International Conf. on Environment and Electrical Engineering (EEEIC),* (Florence, Italy)
[41] Lin Y, Chen D and Chen W 2006 The significance of mesopic visual performance and its use in developing a mesopic photometry system *Building and Environment* **41** 117-25
[42] Rea M S, Radetsky L C and Bullough J D 2010 Toward a model of outdoor lighting scene brightness *Lighting Research & Technology* **43** 7-30
[43] Decuypere J, Capron J-L and Renglet M 2009 Influence of Mesopic Lighting Conditions on Pedestrian Visual Field in Urban Environment. In: *Lux Europa,* (Istanbul, Turkey)
[44] Viikari M, Ekrias A, Eloholma M and Halonen L 2008 Modeling spectral sensitivity at low light levels based on mesopic visual performance *Clin Ophthalmol.* **2** 173-85
[45] Rea M S, Bullough J D and Brons J 2014 Spectral considerations for outdoor lighting: Designing for perceived scene brightness *Lighting Research & Technology* **47** 909–19
[46] Toet A, Houtkamp J M and Vreugdenhil P E 2016 Effects of personal relevance and simulated darkness on the affective appraisal of a virtual environment *PeerJ* **4** e1743
[47] Liu M, Zhang B, Luo T, Liu Y, Portnov B A, Trop T, Jiao W, Liu H, Li Y and Liu Q 2022 Evaluating Street Lighting Quality in Residential Areas by Combining Remote Sensing Tools and a Survey on Pedestrians' Perceptions of Safety and Visual Comfort *Remote Sensing* **14**
[48] Esposito T and Radetsky L C 2023 Specifying Non-White Light Sources in Outdoor Applications to Reduce Light Pollution *Leukos* **19** 269-93
[49] Bandopadhyay S, Kole A and Das P 2016 Review and Studies on the Effect of Spectral Composition of LED Based Lighting System over Its Scotopic-Photopic Ratio. In: *2016 International Conference on Intelligent Control Power and Instrumentation (ICICPI),* (Kolkata, India)
[50] Amoruso C E, Larsen M H, Hvass M, Triantafyllidis G and Hansen E K 2022 Dark Adaptation in Urban Environments: An Innovative Design Framework for Pedestrian Lighting. In: *Light Symposium,* (Copenhagen, Denmark: IOP Conf. Series: Earth and Environmental Science)
[51] Pedersen E and Johansson M 2016 Dynamic pedestrian lighting: Effects on walking speed, legibility and environmental perception *Lighting Research & Technology* **50** 522–36
[52] Hvass M, Van Den Wymelenberg K, Boring S and Hansen E K 2021 Intensity and ratios of light affecting perception of space, co-presence and surrounding context, a lab experiment *Building and Environment* **194** 1-11
[53] Rahm J and Johansson M 2021 Assessment of Outdoor Lighting: Methods for Capturing the Pedestrian Experience in the Field *Energies* **14** 1-15
[54] Rahm J and Johansson M 2018 Assessing the pedestrian response to urban outdoor lighting: A full-scale laboratory study *PLoS One* **13** 1-20
[55] Zhang J and Dai W 2021 Research on Night Light Comfort of Pedestrian Space in Urban Park *Computational and Mathematical Methods in Medicine* **2021** 3130747
[56] Johansson M et al. 2014 Perceived outdoor lighting quality POLQ. A lighting assessment tool *Journal of Environmental Psychology* **39**:14–21
[57] Masullo M et al. 2023 Urban Park Lighting Quality Perception: An Immersive Virtual Reality Experiment *Sustainability* **15**(3):2069, 1–15
[58] Rahm J, Sternudd C, Johansson M 2021 "In the evening, I don't walk in the park": The interplay between street lighting and greenery in perceived safety *Urban Des Int* **26**, 42–52
[59] Davies T W, Bennie J, Inger R, de Ibarra N H and Gaston K J 2013 Artificial light pollution: are shifting spectral signatures changing the balance of species interactions? *Glob Chang Biol* **19** 1417-23
[60] Gaston K J, Ackermann S, Bennie J, Cox D T C, Phillips B B, Sanchez de Miguel A and Sanders D 2021 Pervasiveness of Biological Impacts of Artificial Light at Night *Integr Comp Biol* **61** 1098-110
[61] Koen E L, Minnaar C, Roever C L and Boyles J G 2018 Emerging threat of the 21st century lightscape to global biodiversity *Glob Chang Biol* **24** 2315-24
[62] Decuypere J, Capron J-L and Dutoit T 2013 Simulation of the retina response to mesopic visual scenes. In: *CIE Centenary Conference "Towards a New Century of Light",* (Paris, France)
[63] Pracki P and Skarżyński K 2020 A Multi-Criteria Assessment Procedure for Outdoor Lighting at the Design Stage *Sustainability* **12** 1-19
[64] Owens A C S, Van den Broeck M, De Cock R and Lewis S M 2022 Behavioral responses of bioluminescent fireflies to artificial light at night *Frontiers in Ecology and Evolution* **10** 1-16
[65] Pérez Vega C, Zielinska-Dabkowska K M and Holker F 2021 Urban Lighting Research Transdisciplinary Framework-A Collaborative Process with Lighting Professionals *Int J Environ Res Public Health* **18** 1-18

Coping with darkness: residents' experience of reduced outdoor lighting in neighbourhoods

P Mattsson[1], B Smidt-Hart[2]

[1]Department of Architecture and Built Environment, Lund University, Box 118, 221-00 Lund, Sweden
[2]LEE Consultants, Box 7737, Petit, 1512, South Africa

Corresponding author's e-mail: pimkamol.mattsson@abm.lth.se

Abstract. Despite ongoing replacement of old lighting installations with new energy-efficient light emitting diodes, the energy crisis has led to the implementation of strategies for reducing energy use for public outdoor lighting in neighbourhood environments. This would save money and reduce light pollution; however, the situations should be evaluated from the residents' perspective. This paper presents a pilot study, which aimed to explore how residents experience and respond to reduced outdoor lighting in their neighbourhoods during dark hours, by applying the adapted Human-Environment Interaction model to understand the relation between outdoor lighting quality, neighbourhood quality and local independent mobility from the residents' perspective. The pilot study was based on a short questionnaire (N =76, 29 males, 47 females, Mage = 66.66 years) and telephone or online interviews (N = 13, 5 males, 8 females, Mage = 68.08 years) with the residents living in the neighbourhood areas, in Southern Sweden, where strategies were employed for reducing the use of public outdoor lighting. Data were collected between February and March 2023. Overall, the results affirmed the important role of outdoor lighting quality in residents' experience of neighbourhood environments and local independent mobility (i.e. the ability of residents to walk or move around on their own in their neighbourhood area), whereas the effect of reduced outdoor lighting was rather insignificant. The interviews further showed different experiences of the reduced outdoor lighting and how the residents coped with the lighting conditions by using solutions (i.e. headlamps, flashlights, bike lamps and reflectors), and that high-level perceived safety from crime in the neighbourhoods was found to play an important role in local independent mobility during dark hours. The adapted conceptual model allowed for the evaluation of outdoor lighting conditions considering both internal and external factors from the residents' perspective, and could provide a basis for discussion about potential changes in public outdoor lighting.

1. Introduction

Of the UN's 17 sustainable development goals (SDGs), six of them (3: Good health and well-being, 7: Affordable and clean energy, 11: Sustainable cities and communities, 13: Climate action, 14: Life below water and 15: Life on land) have been identified as having relevant synergies with outdoor lighting [1]. This includes various aspects such as human health and well-being, safety, restoration, socialization, ecosystems, biodiversity, and energy efficiency. Besides the replacement of old outdoor lighting with energy-efficient light emitting diodes (LEDs), adaptation strategies, i.e. switching lights off, reducing the number of 'light-on' hours (part-night lighting), and dimming have been employed to reduce the amount of light [2]. There is however, concern that reduced outdoor lighting may influence perceived safety, fear of crime, and mobility [2]. This paper presents a pilot study investigating how different adaptation strategies affect residents' experience of their neighbourhood environments and local independent mobility after dark.

1.1. Background

Reducing energy use for public outdoor lighting is considered as a solution to the current energy crisis [3]. This could reduce light pollution, thereby having positive effects on both human health and the environment. At the same time, this may trigger perceived lack of safety and security, thereby influencing individuals' independent mobility which has broad implications through its impacts on physical, psychological, cognitive, social, and economic parameters [4]. Well-designed outdoor lighting is associated with perceived accessibility and perceived safety [5] [6], and is expected to support local independent mobility as well as daily commuting [7] [8]. Regarding this, measures applied to reducing public outdoor lighting should be evaluated by considering the user perspective.

Previous research found that white light enhanced perceived safety among pedestrians [9] whereas reduced street lighting may not always decrease perceived safety [2]. Today, economic and resource issues represent a major challenge, as they have limited the availability of public outdoor lighting. For example, in South Africa, where load shedding limits the availability of public outdoor lighting in residential areas. In particular, the lighting is implemented, in accordance with the national standards [10] [11], where the safety of the movement of the public is of prime importance, thereby allowing safe and comfortable movement of residents after dark while reducing incidences and criminal activities in the area. Due to the need to reduce energy consumption of municipal assets [12] [13], energy saving strategies consider changing high-intensity discharge (HID) lighting with LED lighting. However, public outdoor lighting of residential areas is exposed to the effects of load shedding (i.e. the lighting is programmed to operate at night within set minimum hours for specific days during the week); as a result of the strained network, municipal areas are forced to disconnect from the national electrical grid to assist in stabilising it [14]. Load shedding can be scheduled for periods of 2 hours up to 6 hours at a time per day, depending on the status of the national electrical grid. Together, lower lighting levels for more energy savings and periods of darkness caused from load shedding do impact residents' experiences. Regarding this, the effect of reduced outdoor lighting is seen to be of importance in residential areas, particularly where residents have limited service infrastructure, limited access to electricity to homes, have to commute long distances on foot to public transport depots (leaving homes before sunrise and returning home after sunset), and where residents are exposed to criminal activities. Furthermore, the absence of light would limit residents' social activities and/or community activities after dark while increasing criminal activities that may result in damaging the electrical infrastructure to sustain darkness in such areas.

Though it may not always occur, the reduced outdoor lighting due to load shedding could, based on Steinbach et al. [2], be considered similar to the part-night strategy which has generally been applied to outdoor lighting reductions. By increasing knowledge about the role of reduced public outdoor lighting in local independent mobility, it would serve as a basis for better understanding and discussing the potential changes of outdoor lighting considering both human experience and energy use.

1.2. Conceptual framework

In this paper, residents' local independent mobility is defined as the ability of residents to walk or move around on their own in their neighbourhood area. Based on Mattsson et al. [15] (after Küller [16]), local independent mobility results from individuals' interaction with their environment. Such individual-environment interaction involves characteristics of the individuals' internal factors (e.g. abilities and psychological determinants) and physical and social characteristics of the environment (external factors), and is mediated by the individuals' emotional responses to the environment. The environment's physical characteristics are associated with outdoor lighting and its effects on the environment, and the social characteristics are associated with the perception of safety and security. It is suggested that no direct link between the environmental factor like a certain light level and local independent mobility can be expected. Instead, this relationship rather involves the emotional process from *activation* (i.e. the situation where a certain stimulus activates the individual's emotional system) to *attention/orientation* and *evaluation* (i.e. paying attention to the stimulus and its quality), and to *control* over the situation (i.e. adopting coping strategies) [17] [16]. The interplay of internal and external factors influencing residents' experience of their neighbourhood area and local independent mobility during dark hours could be described through the Human-Environment Interaction model adapted from Marcheschi et al. [17] (Figure 1).

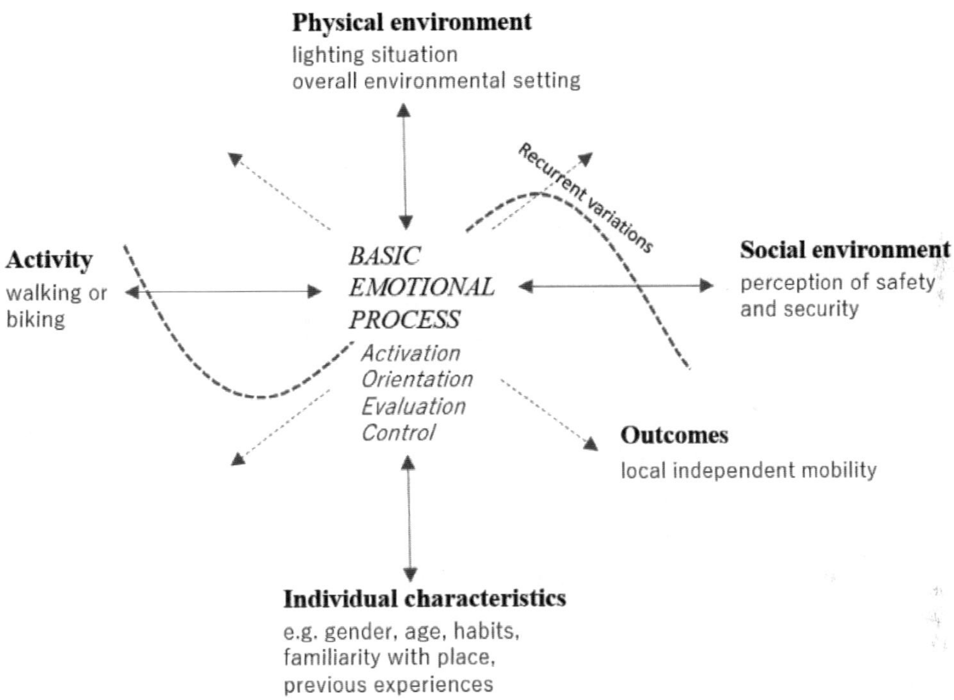

Figure 1. The conceptual model for understanding residents' local independent mobility during dark hours (adapted from Marcheschi et al. [17])

1.3. Aim

This paper presents a pilot study investigating residents' experience of reduced public outdoor lighting in neighbourhoods considering different adaptation strategies. The aim was to explore how the lighting situations affect residents' experience of their neighbourhoods and local independent mobility during dark hours. It was assumed that through the emotional process, humans would have vast capabilities to adapt to and cope with changes in the physical environment, and manage their everyday life, and there would be potential changes in the outdoor lighting due to limited resources.

The underlying purpose of this pilot study was to evaluate the conceptual framework for further studies that may be conducted in different physical locations and social contexts.

2. Method

This pilot study was based on a short questionnaire and telephone or online interviews with residents living in four neighbourhood areas in two municipalities in Southern Sweden where two adaptation strategies have been employed to reduce the use of public outdoor lighting during autumn-winter 2022/2023. The strategies were i) part-night lighting or reducing 'light-on' hours (one hour less than usual, i.e. switching on 30 minutes later and switching off 30 minutes earlier) implemented in one neighbourhood area, and ii) switching off one-third of street lamps in another three neighbourhood areas.

The two municipalities informed the residents about the outdoor lighting reductions through their respective official municipal websites and/or local newspapers. The sources of public outdoor lighting in the areas were HPS light sources (with a CCT of about 1,700 Kelvin and a CRI of about Ra11, and a CCT of about 1,700 Kelvin and a CRI of Ra 30) and LED light sources (with a CCT of 2,500 K, and a CRI of Ra70, and a CCT of 4,000 K and a CRI of Ra 75) which replaced the existing HPSs or were recently installed (Figure 2). The average horizontal illuminance (Eh) values on paths at ground level were 4.63 lx for the HPS lighting and about 3.43 lx for the LED lighting in the neighbourhood area with the part-night lighting strategy, and 3.33 lux in the neighbourhood areas where one-third of street lamps were switched off. The minimum Eh value measured on the paths was 0.00 lx. (The measurements were conducted by the researcher (the first author) using Konica Minolta CRI Illuminance Meter CL-70F).

The invitations to participate in the study were sent by post together with the questionnaires and return envelopes (residents' names and addresses were retrieved through a public online database). Data were collected between February and March 2023, and the participation was voluntary.

(a)

(b)

(c)

Figure 2. Examples of public outdoor lighting in the studied neighbourhoods: HPS lighting (a) and LED lighting with a CCT of 2,500 K, and a CRI of Ra70 (b), and with a CCT of 4,000 K and a CRI of Ra 75 (c) (the first author)

2.1. Questionnaire

The questionnaire (Table 1) was in Swedish and contained five questions with nine rating scale items. The questions were related to residents' (i) perceptions of lighting quality regarding brightness, pleasantness and visibility, (ii) experience of neighbourhood quality after dark regarding safety, security, pleasantness and friendliness, and (iii) local independent mobility, i.e. avoidance of being outside alone and preference for company when going outside in one's own neighbourhood during dark hours in winter 2022/2023. The responses were given on five-point rating scales.

Table 1. Questionnaire questions, items and response scales (translated from Swedish)

	Question	Rating scale
(i) Perceptions of lighting quality	1. How do you perceive the outdoor lighting in your neighbourhood during this winter (winter 2022/2023)?	(1) too dark – too bright (5) (1) unpleasant – pleasant (5)
	2. How well can you see with the outdoor lighting in your neighbourhood after dark, during this winter (winter 2022/2023)?	(1) very bad – very good (5)
(ii) Experience of neighbourhood quality	3. How do you experience your neighbourhood when it is dark, during this winter (winter 2022/2023)?	(1) unsafe – safe (5) (1) unpleasant –pleasant (5) (1) insecure – secure (5) (1) unfriendly – friendly (5)
(iii) Local independent mobility	4. Would you avoid being outside alone in your neighbourhood when it is dark, during this winter (winter 2022/2023)?	(1) definitely wouldn't, (2) probably wouldn't, (3) neither yes or no, (4) probably would, (5) definitely would
	5. Would you prefer to be accompanied for being outside in your neighbourhood when it is dark, during this winter (winter 2022/2023)?	(1) definitely wouldn't, (2) probably wouldn't, (3) neither yes or no, (4) probably would, (5) definitely would

2.1.1. Procedure. The invitation and questionnaire were sent to 407 residents (excluding undelivered mail). A total of 76 questionnaires were filled in and returned, giving 18.67% response rate. The response rate for each neighbourhood area ranged from 8.33% to 29.07% (Table 2). The respondents were 29 males, 47 females with ages between 30 and 86 years, and mean age of 66.66 years (SD = 12.91).

2.1.2. Data analysis. By using IBM SPSS statistics 28, Spearman's correlations (with exclude cases pairwise) were performed to examine associations between perceptions of lighting quality, experience of neighbourhood quality, and local independent mobility during dark hours among the residents in the four neighbourhood areas. Further, a Mann-Whitney U test was used to test whether they were any differences between the two neighbourhood areas having different adaptation strategies to reduce the use of outdoor lighting. The non-parametric tests were applied since the p value obtained in normality tests, Shapiro-Wilk tests, was < 0.001 for all variables indicating non-normal distributions of the data.

Table 2. Questionnaire and response rate

Neighbourhood area	Adaptation strategies	No. of questionnaires sent	No. of undelivered questionnaires	No. of fill-in and returned questionnaire	Response rate (%)
1	Reducing the 'light-on' hours	86	0	25	29.07%
2	1/3 of street lamps were off	104	1	28	27.18%
3	1/3 of street lamps were off	112	2	14	12.73%
4	1/3 of street lamps were off	109	1	9	8.33%
All		**411**	**4**	**76**	**18.67%**

2.2. Interview

Nine interviews were conducted with a total of 13 participants (five males, eight females, ages between 40 and 82 years, mean age was 68.08 years) who agreed to participate (four of the interviews were conducted with two participants at the same time). Of 13 participants, four lived in the neighbourhood area where the part-night lighting strategy (i.e. reducing light-on hours) was employed (Neighbourhood 1). The rest lived in neighbourhood areas where the switching off strategy (i.e. switching off one-third of street lamps) was employed (eight lived in Neighbourhood 2, and one lived in Neighbourhood 4).

2.2.1. Procedure. All the participants gave oral informed consent to the participation and the audio recording before being interviewed. The participants were first asked about their impression of their respective neighbourhoods during dark hours in winter 2022/2023, whether they notice good or bad things, whether there are pleasant and/or unpleasant passages or places, and whether and to what extent the light conditions affect their daily life, particularly in their decision to walk, bike, or be outside during dark hours, and how do they manage their daily activities (Table 3). The interviews lasted for between 15 and 35 minutes.

Table 3. Interview guide (translated from Swedish)

	Question
1.	Could you tell about your impressions of your neighbourhood when it is dark, during this winter (winter 2022/2023)?
2.	Have you noticed anything particularly good or bad in your neighbourhood when it is dark, during this winter (winter 2022/2023)? - Are there any particular unpleasant places or passageways in your neighbourhood in which you avoid after dark, during this winter? - On the other hand, are there any pleasant places or passageways in your neighbourhood after dark, during this winter?
3.	Do the situations of outdoor lighting in the neighbourhood during this winter (winter 2022/2023) have any effects on your daily routines or activities? - How? - Do the situations of outdoor lighting affect your decision to go, bike, or be outside in your neighbourhood after dark?
4.	How do you manage your daily activities when it is dark, during this winter (winter 2022/2023)? - Have you changed any of your daily activities due to the situations of outdoor lighting in your neighbourhood? - Could you further explain about the change(s) you have made?

2.2.2. Data analysis. The audio recordings of the interviews were transcribed (by the first author). The textual material was analysed by means of a deductive thematic approach [18] to map the participants' responses and discussions about their neighbourhood areas during dark hours onto the adapted Human-Environment Interaction model. The data were coded manually based on their alignment with the internal and external factors influencing the experiences of their own neighbourhoods and local independent mobility (Figure 1). When statements aligned with several factors, they were coded under each of the factors. Then, the codes were analysed by examining patterns on how the participants expressed their views and/or discussed in relation to each factor and the relationships between them.

3. Results

The results obtained from questionnaires and interviews are presented separately. The residents' perceptions of lighting quality, experience of neighbourhood quality, and local independent mobility during dark hours obtained from the questionnaires are first presented, followed by the internal and external factors influencing local independent mobility obtained from the interviews.

3.1. Lighting quality, neighbourhood quality and local independent mobility
Spearmans' correlations yielded strong associations among lighting quality, neighbourhood quality and local independent mobility (the correlations were at the 0.01 level). In particular, there was a very strong association between perceived safety of neighbourhoods and local independent mobility i.e. the residents would not avoid going or being outside in their neighbourhoods alone during dark hours (r = -0.71, N = 76, $p < 0.001$).

Between the two neighbourhoods with the different adaptation strategies to reduce the use of public outdoor lighting, there were statistically significant differences in the scores for the residents' perceptions of lighting quality and experience of neighbourhood quality, whereas the differences in local independent mobility were not statistically significant (Table 4). In general, the residents in both neighbourhoods reported that they would not avoid going or being outside during dark hours and their preferences for company were rather neutral. The lighting quality and the neighbourhood quality of Neighbourhood 1 (where the 'light-on' hours were reduced) were rated as significantly higher than those of Neighbourhood 2 (where one-third of street lamps were turned off).

Table 4. Mean *(M)*, standard deviation *(SD)*, mean rank, and Mann-Whitney test results for the two neighbourhoods

		Neighbourhood 1				**Neighbourhood 2**				
		N	*M*	*SD*	*Mean rank*	*N*	*M*	*SD*	*Mean rank*	*Z*
Lighting quality	bright	25	2.88	0.93	36.06	28	1.75	0.70	18.91	-4.22***
	pleasant	25	3.08	1.04	20.02	27	2.32	1.00	32.50	-3.08**
	good to see in	25	3.28	1.10	35.86	28	2.00	0.72	19.09	-4.15***
Neighbourhood quality	safe	25	3.76	1.13	33.42	28	2.61	1.42	21.27	-2.92**
	pleasant	25	3.40	1.29	33.44	28	2.25	1.32	21.25	-2.94**
	secure	25	3.44	1.23	33.50	28	2.32	1.31	21.20	-2.96**
	friendly	25	3.52	1.16	33.46	28	2.39	1.37	21.23	-2.96**
Local independent mobility	avoidance behaviour	25	2.28	1.31	23.60	28	2.86	1.35	30.04	-1.56
	preference for company	25	3.00	1.35	24.94	28	3.32	1.44	28.84	-0.94

*** $p < 0.001$, ** $p < 0.005$ (2-tailed)

3.2 Factors influencing local independent mobility during dark hours
In line with the adapted Human-Environment Interaction model, the results of the deductive thematic analysis describing internal and external factors influencing local independent mobility during dark hours are presented below.

3.2.1. Activity. Taking a daily walk, biking to work, and *running* are activities in relation to local independent mobility mentioned by the participants. A couple of the participants usually take their dogs for daily walks when it is dark outside, two participants often bike, and two participants mentioned about going out for a run after dark during the winter season.

3.2.2. Individual characteristics. Most of individual characteristics that the participants addressed could be coded as *age, habits, familiarity with the place, gender*, and *previous experience*. One participant (aged over 80 years) emphasized *ageing* as an important internal factor limiting local independent mobility, i.e. s/he always remains home after dark, which could be considered as a habit, and only goes outside in the neighbourhood when necessary. Two participants (aged over 70 years) mentioned worse night vision and balance due to old age. In relation to local independent mobility after dark, participants mentioned different *habits*, such as going for a walk, taking their dogs for an evening walk, and, on the other hand, rarely going outside after dark because of having no dog or old age, biking and rarely biking, driving to get groceries, and running every evening. Most of them have habits of having reflectors and/or own flashlights and bike lights when going outside.

Familiarity with own neighbourhood was expressed by participants, mostly through perceived safety from crime, which is somewhat associated with the length of residence. The role of *gender* was reflected by one female participant (aged 40 years) in relation to previous experience of living in a big city with nightlife, socio-economic disadvantaged areas, and unmaintained parks and paths. A couple of participants mentioned their *previous experience* of burglary in their own place or a neighbour's that could be associated with outdoor lighting situations and concealed the view of the house.

3.2.3. Physical characteristics of the environment. In this study, physical characteristics are associated with outdoor lighting and its effects on the neighbourhood area after dark during the winter season. With the focus on outdoor lighting, physical characteristics mentioned and discussed by the participants could be coded as *uneven lighting distribution, strength quality and comfort quality of the lit environment,* and *dim light and darkness*.

Uneven lighting distribution was mentioned and discussed both positively and negatively. Dark areas between light poles were perceived as hindering visibility, particularly in case of obstacles or holes on the way. Conversely, such areas were also perceived as providing opportunities to see contrasts in the environment in the dark and to rest the eyes. *Strength quality and comfort quality of the lit environment* were often mentioned with regard to two main light sources giving cool lighting (LEDs) and warm lighting (HPS) in the neighbourhoods. Most participants stated that they prefer the latter yellow lighting, which was described as more pleasant, softer, cosier, evenly distributed, no glaring, and at the same time, providing acceptable lighting levels. LED lighting installed in the areas was, on the other hand, generally described as giving more direct, brighter, and cooler lighting, glaring as well as limiting the opportunity to experience greenery and natural surroundings. Similarly, participants' responses to *dim light and darkness* are both positive and negative. A couple of participants (aged over 70 years) mentioned that the dim light in their neighbourhood has resulted in reduced visibility (particularly when they need to observe along the way and see ahead). Increased risk of falling was often stated by most participants. However, positive effects of dim light and darkness were also brought up (i.e. opportunities to experience nature at dawn, dusk and night). Further, one participant addressed the need to allow darkness to promote nature experiences and relaxation, which are considered as important to well-being, and to protect biodiversity.

Further, the physical characteristics could be coded as *pleasantness* of the neighbourhoods and *special features* including both built and natural features. *Pleasantness* was reflected in relation to both positive perceptions (i.e. cosiness and harmony of the neighbourhood) and negative perceptions (i.e. perceived risks of falling and encountering obstacles and/or other road users). *Special features,* both built and natural features of the neighbourhoods, were discussed in relation to the current situations and future design of public outdoor lighting considering physical contexts and everyday users. For example, a participant living in Neighbourhood 2, with narrow roads and alleys, pointed out that public outdoor lighting should be planned by taking appearances of the built environment as well as residential density into account, and expressed his/her satisfaction with the reduced outdoor lighting in the neighbourhood. A couple of participants living in Neighbourhood 1, close to nature, have recognised the interplay between outdoor lighting and greenery and/or natural features, and mentioned that their perceptions of the lit environment and visual accessibility were negatively affected by unkempt greenery in their neighbourhood.

3.2.4. Social characteristics of the environment. The social characteristics are associated with the perceptions of *safety* as well as *security* in the neighbourhoods. Most participants stated that they feel very safe (from crime). Some participants emphasised that *calmness* and *friendliness* of the neighbourhoods play important roles in their perceived safety from crime. However, some participants mentioned their previous experiences of burglary, either in their own place or a neighbour's, that could be associated with the reduced outdoor lighting, thereby being worried that the number of burglaries could increase.

3.2.5. Basic emotional process. Based on the Human-Environmental Interaction model, the relation between internal factors, external factors, and local independent mobility is mediated by basic emotional process from *activation* to *attention/orientation, evaluation* and *control* over the situation, which, in this study, is the reduced outdoor lighting.

Participants experienced the changes in outdoor lighting in their neighbourhoods and were *activated* differently by the lighting conditions. For Neighbourhood 2, where one-third of street lamps in the areas were off, participants stated that they have experienced their neighbourhood area as 'darker'. In other word, their *attention* has been paid to the dimly lit environment which, for a couple of participants, has influenced mobility choices (e.g. driving a car, biking or walking).

"For me, to walk is easier than to bike under this lighting condition /..../ because even with bike lamps, ones can still miss if some (other road users) are coming, bike lamps also have their limits" (Interview 2, female, aged over 70)

Their *evaluations* of the reduced outdoor lighting were found to vary from uncomfortable, unpleasant, to pleasant. Participants (aged over 70) often expressed their concern about limited vision due to the reduced outdoor lighting, making it unpleasant to go outside. However, a couple of them experienced the reduction of public outdoor lighting as sufficient and better in combination with outdoor lighting from house facades, yards, as well as with indoor lighting from windows, since the houses in the area are placed along narrow roads and alleys. One participant also observed that the neighbours had Christmas lights for a longer period than usual, and mentioned that this might be due to the reduced outdoor lighting.

"I do not have any problem about this 'dark' (because of the reduced public outdoor lighting), it feels good. /..../. I always get disturbed by so many lights, façade lights, it is too bright so I think that it is better when it gets darker in the area" (Interview 6, male, aged over 70)

In general, participants stated that the reduced public outdoor lighting has no significant role in their daily activities and decision to go outside after dark. Participants who usually go outside for evening walks have headlamps or flashlights with them and therefore are able to exert *control* over the situation. One participant has also observed that others who take their dogs to walk when it is dark usually use this solution. Moreover, a couple of participants (aged over 70) stressed that they have paid more attention to their surroundings, and the importance of being able to see and be seen (by having reflectors and light colour jackets). Participants who have dogs usually put on reflectors or lamps on their dogs. This could also be considered as a regular practice or habit but the participants have become more careful.

"I have a flashlight with me because I am afraid of falling when I cannot see well /..../. I become more careful to have reflectors and 'well-seen' clothes when going outside, the same solutions have been applied to the dog (laughing)*, so the dog is shining, we usually do this and now we think more about it"* (Interview 1, male, aged over 70)

For Neighbourhood 1, where 'light-on' hours have been reduced, two participants stated that they have experienced this change in their neighbourhood area. Also, *attention* has been paid to the walkway in which LED lamps have newly been placed in the area. Regarding *evaluations*, the participants did not express any negative emotional responses to the reduced light-on time, whereas there were both positive and negative responses to the path with new LED lamps. With the cooler and brighter lighting, this path has been experienced by the participants (aged over 65 years) as well-lit. The participants (aged between 50 and 60 years) perceive the lighting as too bright, cool, and glaring, making the environment unpleasant. Moreover, one participant has perceived the bright lighting as focused and, thereby, making the surroundings disappear from her/his view.

"When lighting is bright like along the path with new LED lamps, you see only the lighting on the path well but instead the surroundings are cold dark so you do not have 'distance vision' and personally, I think that it is scarier. I want to see my surroundings but cannot do it with bright lights" (Interview 5, female, aged between 50 and 60 years)

Further, the reduced light-on time has no significant role in the residents' daily activities and decision to go outside after dark. One participant mentioned that there are not many weeks to go through this situation. Another participant specifically expressed that s/he appreciates the darkness when biking to work in the morning, and running outside using bike lamps and headlamps respectively, to have *control* over the situation. The path with new LED lamps was mentioned as having bright-cool lighting, and the participant usually avoid this path when biking. The darkness has been appreciated and is not negatively perceived with respect to neighbourhood safety (from crime).

"I usually get up early and bike to work, and I have come to appreciate the darkness that is why I avoid that path (with new LED lamps). It feels strange to have bright lights in this darkness" (Interview 8, male, aged between 50 and 60 years)

"I understand the argument about the role of (outdoor) lighting in (neighbourhood) safety but I feel really safe in this area /…./. This is different from traffic safety that requires you to see well" (Interview 8, male, aged between 50 and 60 years)

4. Discussion

This pilot study aimed to explore how reduced public outdoor lighting affects residents' experiences of neighbourhoods and local independent mobility during dark hours. Using the Human-Environment Interaction model (adapted from Marcheschi et al. [17]) as a framework, the factors influencing local independent mobility were investigated and discussed from the residents' perspective. It is expected that the framework can be applied for investigating the effects of outdoor lighting situations in different physical locations and social contexts. The study was carried out through a short questionnaire, followed by qualitative interviews. Together, the findings could serve as a basis for discussion to identify potential adaptation strategies for reducing public outdoor lighting and improving outdoor lighting design in residential areas to face the current energy crisis and environmental challenges.

The questionnaire results affirmed associations between outdoor lighting quality, neighbourhood quality and local independent mobility, particularly between perceived safety of the neighbourhood and local independent mobility. Considering the adaptation strategies employed to reduce the use of public outdoor lighting, it was obvious that switching off one-third of street lamps resulted in greater effects on lighting quality and neighbourhood quality compared to reducing light-on hours. When switching off one-third of the lamps, the neighbourhood area was generally perceived as darker (than usual) and the lighting was perceived as unevenly distributed. However, there were no significant effects on local independent mobility between the two adaptation strategies. The participants generally reported that they would not avoid going or being outside during dark hours and their preferences for company were rather neutral. Reducing light-on time (switching on 30 minutes later and switching off 30 minutes earlier than usual) did not seem to have significant impacts, however longer 'light-off' hours (which could be considered similar to the lighting situation affected by load shedding) may give different results.

The questionnaire results on the residents' perceptions of neighbourhood quality are, to some degree, in line with the interviews with respect to the perceived safety. The darker lit environment and unevenly distributed lighting due to the switching off strategy (i.e. switching off one-third of street lamps) have affected the residents' perceived safety (i.e. risk of falling). Though these lighting conditions have limited visual accessibility, a couple of participants in Neighbourhood 2 were also satisfied with the outdoor lighting in combination with the lighting from private residences in their neighbourhood. This is because the area has detached or semi-detached houses along narrow roads and alleys. There were negative effects of unkempt greenery on perceived lighting quality and visual accessibility mentioned by a few participants. This finding supports a previous study [19] which suggested a relationship between outdoor lighting conditions and greenery, and the effects of the interaction between them on neighbourhood experience with regard to perceived safety and local independent mobility. In general, the results indicate that the two adaptation strategies employed for reducing public outdoor lighting have not significantly affected the residents' local independent mobility during dark hours. Despite no significant effects on local independent mobility, the lighting conditions seem to influence mobility choices and also walking route choices in this pilot study. To improve visual accessibility and have control over the reduced outdoor lighting, it was found that the residents use headlamps, flashlights, bike lamps, and reflectors as solutions. It can thus be assumed that the individuals have adapted to and coped with the changes in the physical environment, and managed their daily activities. It should also be noted that the neighbourhoods in this study have been perceived as safe (from crime) which possibly promotes local independent mobility. However, there have been concerns around burglary that may increase due to the reduced outdoor lighting, and whether solutions such as headlamps or flashlights are sustainable considering environmental and economic aspects.

In the interviews, a number of participants expressed concern about LED lighting in the neighbourhoods with both adaptation strategies since the lighting was perceived as unpleasant (e.g. too bright, cold, glaring, as well as limiting an opportunity to experience nature and surroundings). This finding is in line with previous field studies in Northern European contexts (e.g. [15] [20] [21]) that reported a mismatch between perceived comfort or pleasantness and perceived strength of LED lighting. As mentioned by the participants, special features and characteristics of the neighbourhoods, residents' daily activities, and the importance of nature experience and biodiversity should be considered when designing public outdoor lighting. However, it has been suggested that switching to LED lighting may facilitate crime prevention in an area [13]; this is of high relevance to high crime areas.

Through the lens of the conceptual framework (adapted from Marcheschi et al., [17]), this pilot study showed that the residents have coped with the changes (i.e. reduced outdoor lighting) in the physical environment (the respective neighbourhood areas) to support their local independent mobility as well as daily activities as a result of the interplay between individuals, emotional processes, and physical and social characteristics of the neighbourhoods. High-level perceived safety from crime associated with the social characteristics was found to play an important role in local independent mobility during dark hours, while the perceived risks of falling and encountering obstacles and/or other road users could be coped with through the different solutions. By applying the conceptual framework, the outdoor lighting conditions could be evaluated from the residents' perspective, taking internal and external factors influencing local independent mobility during dark hours into account. This pilot study was able to shed light on the effects of different adaptation strategies and to provide a basis for discussion about potential changes in public outdoor lighting considering limited resources. To improve public outdoor lighting, more studies are needed on the effects of different adaptation strategies in different physical locations and social contexts. This would pave the way towards new outdoor lighting design approaches to satisfy the needs as well as preferences of users while simultaneously balancing the different aspects of sustainability.

Acknowledgments
The study was supported by a grant from the Swedish Energy Agency (Project number P2022-00400).

References

[1] Tavares P, Ingi D, Araújo L, Pinho P and Bhusal P 2021 Reviewing the role of outdoor lighting in achieving sustainable development goals *Sustainability.* **13** 12657

[2] Steinbach R, Perkins C, Tompson L, Johnson S, Armstrong B, Green J, Grundy C, Wilkinson P and Edwards P 2015 The effect of reduced street lighting on road casualties and crime in England and Wales: controlled interrupted time series analysis *J. Epidemiol. Community Health.* **69** 1118–1124

[3] Zimmermann A 2022 Europe's street goes dark to save energy: Local measures will cut down on light pollution but some worry that could make cities less safe. *POLITICO.* [cited 2023 August

3]. Available from: https://www.politico.eu/article/europe-cities-streets-dark-lights-off-save-energy/

[4] Waygood EOD, Friman M, Lars E, Olsson LE and Taniguchi A 2017 Transport and child well-being: An integrative review *Travel Behav. Soc.* **9** 32-49

[5] Johansson M, Rosén M and Küller R 2011 Individual factors influencing the assessment of outdoor lighting of an urban footpath *Light. Res. Technol.* **43** 31–43

[6] Johansson M, Pedersen E, Maleetipwan-Mattsson P, Kuhn L and Laike T 2014 Perceived outdoor lighting quality (POLQ): A lighting assessment tool. *J. Environ. Psychol.* **29** 14-21

[7] Rosenberg DE, Huang DL, Simonovich SD and Belza B 2013 Outdoor Built Environment Barriers and Facilitators to Activity among Midlife and Older Adults with Mobility Disabilities *Gerontologist.* **53** 268–279

[8] Fotios S, Liachenko Monteiro A and Uttley J 2019 Evaluation of pedestrian reassurance gained by higher illuminances in residential streets using the day–dark approach *Light. Res. Technol.* **51** 557–575

[9] Peña-García A, Hurtado A and Aguilar-Luzón MC 2015 Impact of public lighting on pedestrians' perception of safety and well-being *Saf. Sci.* **78** 142-148

[10] The council of the South Africa Bureau of Standards *SANS 10098 - 1: Public lighting Part 1: The lighting of public thoroughfares, Part 2: Public Lighting*

[11] The council of the South Africa Bureau of Standards *SANS 10389 -1: Exterior lighting Part 1: Artificial lighting of exterior areas for work and safety*

[12] South African Government 2008 *South Africa's National Energy Act 34 of 2008* Available from: National Energy Act 34 of 2008 (www.gov.za)

[13] Department of Energy, Republic of South Africa 2009 *National Energy Efficiency Strategy* Available from: National Energy Efficiency Strategy of the Republic of South Africa (www.gov.za)

[14] The city of Johannesburg 2018 *What is load shedding?* [cited 2023 November 3]. Available from What is load shedding? (joburg.org.za)

[15] Mattsson P, Johansson M, Almén M, Laike T, Marcheschi E, and Ståhl A 2020 Improved usability of pedestrian environments after dark for people with vision impairment: An intervention study *Sustainability.* **12** 1096

[16] Küller R 1991 Environmental assessment from a neuropsychological perspective *Environment, Cognition, and Action* eds T Gärling and GW Evans (New York, NY: Oxford University Press) p 78–95

[17] Marcheschi E, Ståhl A, Almén M, Johansson M 2020 A theoretical model for urban walking among people with disabilities *Front. Psycho.* **11** 156

[18] Braun V and Clarke V 2006 Using thematic analysis in psychology. *Qual. Res. Psychol.* **3** 77-101

[19] Rahm J, Sternudd C and Johansson M 2021 In the evening, I don't walk in the park: The interplay between street lighting and greenery in perceived safety *Urban. Des. Int.* **26** 42–52

[20] Rahm J and Johansson M 2018 Assessing the pedestrian response to urban outdoor lighting: A full-scale laboratory study. *PLOS ONE.* **13** e0204638

[21] Juntunen E, Tetri E, Tapaninen O, Yrjänä S, Kondratyev V, Sitomaniemi A, . . . and Heikkinen V 2015 A smart LED luminaire for energy savings in pedestrian road lighting. *Light. Res. Technol.* **47** 103-115

IOP Conf. Series: Earth and Environmental Science 1320 (2024) 012004 doi:10.1088/1755-1315/1320/1/012004

Co-designing safe and sustainable darkness for northern urban places

R Vuorenmaa[1], H Pihlajaniemi[1] and O Parhankangas[1]

[1] School of Architecture, Faculty of Technology, University of Oulu, Oulu, Finland
Corresponding author's email: riikka.vuorenmaa@oulu.fi

Abstract. The technological development, quantitative increase and qualitative design of urban lighting have succeeded in improving the safety and comfort of urban spaces. In this context, dark and dimly lit places appear as undesirable, even dangerous. However, the disadvantages of overlighting are increasingly being identified. On the other hand, darkness has well-being effects and aesthetic values in urban environments. The Safe and Sustainable Darkness project studies the frameworks for dimmer urban lighting: how to find room for lower lighting levels and darker places, respecting the diverse use of spaces and the values and experiences of city residents. In the article, we present participatory process for design and implementation of safe and sustainable urban darkness. The expertise and understanding of two different groups is of interest. Interviews of various professionals in the city organization reveal how illuminated and un-illuminated places are perceived from an administrative perspective. Inhabitants and users of urban space share their insight of the meanings, potential and context of the existing and future dark places. By combining the research material from interviews, crowdsourcing, workshops and evaluated pilot installation, we propose a co-design methodology to create enjoyable dimly lit and un-illuminated urban spaces.

1. Introduction

The technological development, quantitative increase and qualitative design of urban lighting have succeeded in improving the safety and comfort of urban spaces [1–2]. In this context, dark and dimly lit places often appear as undesirable, even dangerous. However, the disadvantages of overlighting are increasingly being identified. Excessive lighting has a negative impact on the well-being of people and other organisms, as well as on ecological systems [3–4]. This is leading to a paradigm shift of lighting design to better balance the amount of light brought into environments with the actual needs for lighting. Several recent papers identify the values of darkness as a necessary starting point for urban lighting design [5–7]. Current research and practical applications of lighting design and intelligent lighting technologies such as presence sensitive lighting [8] seek methods to adapt lighting to the environmental conditions and needs of the users, also considering the requirements for different traffic forms. These include, for example, creating balance between brightness levels and reducing contrasts [9]. The dark adaptation process of the human eye can also be taken as one of the design criteria [10–11]. Aesthetic and cultural values of darkness are articulated in research and art [12–13].

This paper presents a case study developing a methodology of co-designing dark or dimly lit places in urban contexts, which are experienced as pleasant and safe. The case study is part of the Safe and Sustainable Darkness research project of the School of Architecture, University of Oulu, Finland. The general aim of the project is to develop and pilot a new and sustainable approach to the planning and implementation of lighting in northern urban environments. The goal is to optimize the use of low illuminance levels with the help of researched design solutions and intelligent lighting control technology. Understanding users' experiences in dark environments and inclusion of participants in the process are in focus. The solutions for well-designed darkness and complementary lighting aim to reduce energy consumption in cities and provide tools to change design strategies towards carbon neutrality. The designed urban darkness gives space to the expressions of artistic light that support the experience of a safe and comfortable city. In the following sections, rationale for designing urban lighting and designing urban darkness is discussed briefly, as well as the concept of co-design. In the last section of the introduction, the aims of the study are presented.

1.1. Rationale for designing urban lighting and designing urban darkness

Artificial lighting enables urban activities to continue during dark period of the day, which is especially important in northern cities where the active part of the day extends to dark hours half a year. The task of urban lighting is to enable working, outdoor recreation, playing and other everyday activities even in the darkest months of winter. In a nocturnal city, the urban space is structured through lighting: landmarks, spatial hierarchies and conception of spaces can differ a lot from daylit situation. The lighting of the passageways improves movement safety and enables orientation for those moving in different ways and at different speeds [14]. A large proportion of urban space users are afraid of using dark places, and lighting, on the other hand, is strongly justified by arguments related to street safety and sense of security [15]. These apply besides streets, also places to stay, play, and spend time, like parks and recreation areas.

In urban lighting, the concept of sustainability concerns three aspects: ecological, economic, and social [16]. From the ecological sustainability point of view, decreasing the amount of light in urban environments reduces both energy consumption and the recognized, harmful effects of excess light to nature and people [3]. Economically sustainable lighting can be created only by evaluating the economic effects during the life cycle of different solutions. The role of lighting in supporting nocturnal economy must be recognized. To design socially sustainable dark and dimly lit urban environments, the principles of social sustainability of lighting solutions must be identified. Information based on up-to-date research on the connection between lighting quality and safety and the feeling of security is necessary. The essential questions are: What kind of darkness can be perceived as safe? What kind of places can be dark, when can it be dark, and what kind of interplay of light and darkness makes a place feel pleasant and safe?

1.2. Co-design

The social acceptability and sustainability of urban darkness can be approached through the theory of open city [17]. The central idea of the theory is that a flexible and adaptable city – open to opportunities and flexible in challenges – can only be created by combining expert and non-expert understanding, as a collaboration of the entire community. Co-design or participatory design can be defined as design practice that involves different non-designers in various co-design activities throughout the design process. By non-designers it is referred to potential users, other external stakeholders, and people on the development team who are from disciplines other than design. Participatory design spans across a broad spectrum of domains and makes use of a broad repertoire of tools and techniques in commercial, community-oriented, and research contexts. Co-design activities can, optimally, be seen as multi-directional communication and interactive learning processes. Additionally, participation has been an important strategy in art. In contemporary participatory and community engaged art, some of the aims are: 1) to educate, activate and empower audiences; 2) to cede authorship in pursuit of a more egalitarian art; and 3) restoration of the social bond in a community [18–19]. These are relevant viewpoints also in co-designing socially sustainable darkness.

1.3. Aims of the study

In this article, we present and reflect a case study of a participatory process for design and implementation of safe and sustainable urban darkness. The research process can be seen as an iterative cycle including methods for context and user experience study, ideation, scenario work, design, implementation, and evaluation. In this paper, we present a piloted model of all the phases in one chosen design site, but more research and iteration are needed to tune it according to context specific needs. The multidisciplinary background of the research group (arts and design, lighting design, architecture, humanities) gives us several viewpoints on expert and lay person co-operation and development of methods.

The key research question for the article is: What kind of methods are useful in combining lighting professionals' expert knowledge and citizens and other stakeholders' lay understanding of urban spaces, lighting conditions and the use of the city in the process of designing urban darkness? More specific research questions include the following: How to gather and use location specific understanding and local meanings, values, emotions, and imagination, and combine them with the more universal design principles and standards of high-quality urban lighting? How to fit together the unique needs and preferences of the individual city inhabitants with the demand for impartiality and equality? What kind of research and artistic methods are applicable? The process and the methods in each stage are discussed. The aim is that the proposed design process will result in valued urban places where artificial lighting is used only minimally and where is room for safe experiences of darkness.

2. Methods and materials

Methodologically our study situates within the field of design research, containing practical experimentation and design processes in a real-world setting including participation of local people and other stakeholders. Our research approach is exploratory [20] in-the-wild research [21]. Exploratory studies aim to widen the scope of research by producing knowledge and insights about novel phenomena and by uncovering new research questions and hypotheses. In-the-wild user studies operate in a naturalistic, real-world settings and aim to capture everyday life user experiences. In our study, a mixed-methods approach [22] was followed, where qualitative, ethnographic inspired methods were combined with piloting, artistic methods, and design process, as well as qualitative and quantitative studies with luminance camera measurements.

The aim in our research is to develop a method for designing safe and comfortable dark urban environments by piloting a whole process of co-designing, implementing, and evaluating an intervention at one location. The time limit for project financing was set to year 2023. The lighting conditions in the north brought another imperative time frame for the implementation: half of the year would be without darkness. The changing lighting and snow conditions caused pressure on the process, but they were also essential from the point of view of research and design: these are the circumstances the proposed solutions must relate to. At the time of writing this article, the pilot project is nearing its end, but is not yet finished. Most of the actions have been conducted both in the snowy winter and in the snowless autumn. The most visible outcome of the process – the intervention during and after the Lumo Light Festival – was realized in November–December 2023. Thus, in this paper we are discussing still a work-in-progress and its preliminary findings. Analysis of the data collected during this project is still ongoing.

The design research project was divided into the following stages: 1) recognizing the dark places, 2) analysing the places, the user experiences, and needs, 3) co-designing the change, 4) realization of an intervention, and 5) evaluating the user experiences of the intervention, as well as the whole process and tested methods. The stages overlap in time. It was also evident from the beginning that a research design project is, by its nature, iterative. The understanding obtained during the process feeds not only the future steps but also loops backwards by reflections.

Designing urban darkness is tied to both urban planning (strategic planning, city structures and regulation) and urban space design [23]. Finding socially accepted places for urban darkness can be seen as task of urban planning, whereas designing a socially accepted dark place in urban environment can be seen as urban design. The following chapters present the methods tried and planned in each stage of the case study research design project with the addition of the research of designing darkness in an administrative context.

2.1. *Understanding the process in the city planning*

The city of Oulu, which functioned as the research area, does not have – besides the lighting master plan – an actual city lighting strategy, nor a planning unit that focuses solely on city lighting. Planning and decision-making regarding light and lighting in urban space takes place as part of other construction processes, in several different units processing city planning, streets and traffic, building supervision, public buildings and large-scale city events. Forming a comprehensive understanding of the processes of urban lighting design in Oulu requires data collection from all these units. For this purpose, we selected 14 experts who are involved in planning and implementing the dark time urban environment for an interview. Since the main goal of this research has been to identify and understand previously unknown ways of thinking and acting related to planning processes and dark time urban environment, as well as the values behind them, we used an approach inspired by ethnography. The interviews were carried out as theme interviews, where the central themes were the design and decision-making processes included in the interviewee's work and the interviewee's conceptions of urban lighting as well as dim and dark urban spaces. In addition to this, we included in the interview such broader themes that we estimate are connected to the phenomenon under study and which can be used to create an interpretive context for the material.

2.2. *Recognising the Dark Places*

The first stage of the design research process was to recognize present and potential dark places in the urban areas of Oulu. We started to look for the dark places with the help of our own local knowledge and using different kinds of maps. Google Maps satellite images and National Geoportal [24] with its map layers were used to get an overall image of the city structure. Light pollution heat maps based on VIIRS/DMSP satellite data [25], and most importantly regional map service [26] with information on zoning and routes, current and near-future design and construction projects were studied to find areas where the amount of lighting has changed or might be changing significantly.

The research group carried out a mapping walk in each area selected for closer look in January–February 2023. A list of attention-directing questions was prepared for the walks. We discussed, took photos, described our observations, and wrote notes from the walks. Along all the walks there were both conventionally and particularly heavily illuminated sites and various dimly lit and dark places.

Using GIS, or geographic information system, provides an intuitive way to ask questions about specific locations and to respond on a map. Map-based questionnaires have been used in urban darkness development in the northern Europe at least by the cities of Helsinki [27–28] and Tallinn [29]. In our research project, the first of two GIS surveys is aimed primarily at recognizing dark places. City dwellers are asked to mark on the map places that are important to them in daylight and in dark hours; pleasantly or unpleasantly dim and dark places; places that are pleasantly or unpleasantly luminous during the dark hours; and places where they have noticed disturbing or unnecessary light. Verbal justifications are requested for the entries. The aim of the survey is to identify dark places that are relevant to local people and to gather understanding of the experiences of darkness in the city. The survey also includes thematic questions on the meanings of darkness. Some of the questions are designed to give context for the interpretation. Both GIS surveys are timed so that there will be responses given in the autumn before snow falls, and also during the winter with the snow. In this way, comparison data on the effect of snow cover can be obtained. Surveys are carried out on the Maptionnaire platform. (www.maptionnaire.com)

2.3. *Analysing the dark places*

After locating some interesting dark places using maps and walks, we started the context study. The aim was to understand the current situation and characteristics of these locations, analyse their current and potential usage, define the needed change, and find possible limitations for design. This included acquisition of both expert and lay understanding of the dark places under study.

2.3.1. Walking interviews. To gain a better understanding of the city dwellers' perceptions of the dark and dim urban environment, we created three different routes for walking interviews based on our own mapping walks. Two of the walking interviews were carried out in March, while the snow was still on the ground, one had to be cancelled. Interview on the most interesting route was repeated in October before the arrival of new snow. Walking with the subject has a long tradition in ethnography. It is a method that can be used to obtain versatile, multisensory information about places [30–31]. In this study, the walking interviews were carried out in groups of three to eight participants. Each of the researchers had a pre-agreed role during the walk and the walks were captured as audio recordings. The walking interview participants were asked context-creating questions about their relationship to the area, conceptions of the area in different seasons and different light conditions, as well as personally experienced meanings of darkness and lighting. In addition to this, there were short stops at pre-planned locations, where the participants were asked to point out and tell about things related to lighting and darkness that they found pleasant, unpleasant or difficult to comprehend in the environment. The indications were photographed. There were also two longer stops along each walk, where the participants were encouraged to discuss what kind of atmosphere the lighting creates in the place, how they experience it in relation to the functionality of the place, and how the observed lighting could be developed or changed. Around the middle of the route, we stopped at an unlit place, where one of the researchers guided a multisensory perception exercise. The purpose of this was to give time for the vision to adapt to a dim lighting level and to evoke perception and interpretation of the environment beyond everyday habits. The walking interviews lasted about one and a half hours. At the end of the walk, participants were offered refreshments and the opportunity to ask questions, discuss the experience and give feedback.

2.3.2. Luminance camera measurements and visual-spatial analysis. Besides the qualitative mapping walks by researchers and walking interviews with participants, we developed and tested a luminance-based method of measuring and analysing the visual-spatial light appearance in dimly lit environments in urban contexts. Luminance images were taken and produced with a calibrated digital camera and relating software in selected locations along the walking routes. The images were shot in March when a new layer of snow had just fallen over the already snowy ground and then again in October with dark snowless ground. We carried out visual-spatial analysis of the luminance images applying concepts inspired by morphological studies of essential elements of architecture [32] and the theory of space and light [33]. Our method was based on the recognition of certain parts and elements of space in the views made visible by distribution of luminances in the images, as well as the relation of the user of the space to the dark and light areas within the view. In addition, several tools of the analysis program were applied for the quantitative analysis of the images, framed by the visual-spatial analysis. For detailed information of the procedure, technology, and analysis method, see [34].

2.3.3. GIS crowdsourcing. The second GIS survey is tied even more directly to experiencing a unique place. This survey not only refines and concretizes the information produced by the more general survey, but also collects ideas for change. Respondents are given six predefined routes with certain places to evaluate. There is also a possibility to draw an own route. The survey is designed to be quick and smooth to answer as the respondents use smartphones to fill it during the walk, while on location. The questions at the places to evaluate concern the current and potential use of the place; the atmosphere; perception of the lighting condition; feeling of safety; and ideas for improving the place. Responses can be enriched with photos.

2.4. Co-designing and realising the intervention

Realizing a temporary intervention is a way to test and demonstrate public acceptance of urban darkness [6]. Co-designing an intervention ensures that ideas, needs, hopes and fears of people using the place are met in the design. In our design research the intervention serves also to test the participatory methods. A newly renewed park in the vicinity of the city centre was chosen as the location for the intervention. The area is bounded by the sea on one side, and by residential area on the other sides. A brightly illuminated fast-cycling lane cuts the park. There are several functional places in the park, namely children's playground and traffic park, outdoor gym, parkour and skate parks, small marina, art pieces, and barbecue place. Not counting the playground, none of these is in active use during the wintertime – a great deal of the potential of the park remains untapped. Co-operation with the city of Oulu made it possible to start the intervention during the annual Lumo Light Festival, a context that gives it publicity and raises wide interest, and continue it for a month after the festival.

Figure 1. Intervention and workshops focus areas in the Hollihaka park: 1) un-lit Sonnisaari island, 2) children's playground, 3) traffic park and 4) parkour and skate parks and 5) fast-cycling lane and pedestrian walkway. Background map from [24], photos and image processing by the authors.

To better understand the use of the park and to identify different user groups, we did some background research using open online sources, interviews, and ethnographical observation of the park. Different workshop concepts for the co-designing process were planned next: Playing with Light, Dark Adaptation and Co-Design Workshop Series. Each of these has its own goal, and they aim to involve different groups of people. The workshops were held in October-November, mostly outdoors at evening twilight and dark hours. One of the researchers directed the workshops, and the others documented and observed the work.

Playing with Light Workshop was organized in the playground, first for a group of children from a nearby kindergarten and then twice during the Lumo Light Festival. The children, accompanied by their parents, explored the familiar environment in the dark, helping the researchers find darkness and play with it. They were then encouraged to illuminate the playground with portable lights. Dark Adaptation Workshops were planned to be held on an unlit island in front of the park. The idea was to let participants explore moving between brighter and darker environments, to experience dark

IOP Conf. Series: Earth and Environmental Science 1320 (2024) 012004 doi:10.1088/1755-1315/1320/1/012004

adaptation and sensations it evokes and to be together and co-operate in darkness. Unfortunately, unexpected construction work started at the island and the planned workshops had to be cancelled.

The goal of the Co-Design Workshop series was to develop the intervention together with the participants. The series begun with the creation of scenarios related to the use of the site and proceeded through on-site experimenting to an implementation influenced by the process. After studying multiple possibilities and aspects, the researchers decided to place the intervention in the parkour area of the park. The permanent lighting of the area was switched off and 27 individually controlled RGB architectural lighting fixtures were installed instead. Temporal dynamics of light and darkness were worked on in the series' last workshop. The final intervention consists of 15 smoothly looping lighting scenes that present alternative ways to light the place. There are three different sets of scenes: shades of white, inspiration from wintertime natural light, and dashing colours.

2.5. *Evaluation of the intervention*

The intervention was to be evaluated mostly by using the same methods that have been in use earlier in the design process. There was a digital survey that visitors can launch using QR-codes while on location. The context of light festival gave the intervention and the survey a possibility to reach wide publicity and audience. Deeper thoughts about the change in lighting of the place were gathered by organizing a walking interview for urban planning and lighting experts. Luminance measurement photographs were taken and will be analysed in relation to the measurements taken before the intervention. All evaluation is tied to the actual place and experience as closely as possible. Themes of interest are functionality and experientiality, social dimensions, atmosphere, and aesthetic qualities of the dimmer place. All these include the element of safety and security.

3. Results of the tested design process

The different methods of design research have been described above. Most were tested during the first quarter of the year and the research continued in the autumn when nights got darker again. Results of the methods tested in spring and preliminary results of those tried in autumn are given below.

3.1. *Incremental understanding of the processes in the city planning*

One essential goal of the expert interviews was to identify key points and roles in the decision-making process regarding the dark time urban environment. The material collected provides good tools for this. The interviews also opened us to understand that it is not only essential *who* makes decisions and *when*, but also *how* decision-making takes place. The interviewed experts often described their work as a negotiation between parties representing different interests. Because of this, the resulting urban space is always some kind of compromise. Consistency in decision-making appears as a central value guiding the work of experts. Since there are no written instructions for everything, experts look for support, coherence, and equality for their own solutions by examining previous implementations, reasonings, and decisions. It seems that these practices significantly guide the formation of urban space. Practices evolve over a relatively long period of time and slowly change by repeating and producing examples of different decisions and their consequences. The observation strengthens the view that experiments and piloting of research projects can really serve as necessary examples that experts can lean on in their decision-making. Conducting the interviews also made us realize that asking questions was probably the most effective way to get the experts' attention to the project's goals, even though we did not primarily aim for that with the interviews. Asking the city representatives to tell us their own perspectives made them think intensively and tangibly about the theme of light and dark.

3.2. *Recognising the dark places*

Based on satellite images and other map data, it appeared that in some parts of the coastal area the lighting had become more intense as a result of the construction of a fast-cycling lane. On the other hand, the sea element brought the dark areas right next to the urban centre, and the inhabited islands in front of the city were clearly darker than the rest of the urban area. The fast-cycling lane will continue through these islands, and a completely new district is being built near them. We selected three different areas for more detailed mapping: the waterfront area of the city centre, where the intensity of lighting varies greatly, some of the islands with many currently dark places, and thirdly a typical Nordic suburb inland, that stood out slightly less bright than other similar districts. The suburb's lighting will be renewed in the next few years. The researcher mapping walks highlighted how moods stagger and change, how places relate to each other, and how the same places appear different when examined from within, compared with how they seem from further away. Based on our experience it is essential to consciously reflect on these perspectives formed during the mapping walks, as they are used in many ways at different stages of the process. In a research situation, such as walking interviews, researchers' own interpretations and preferences should not be transmitted to the subjects, while in the co-design process, these perceptions are needed as researcher's contribution to the shared design. The choice of the coastal park for the place of the intervention was based on the mapping walks. The walks affirmed that continuing research in the other pre-selected areas is justifiable and gave us understanding to create the framework for qualitative analysis of luminance measurements.

3.3. *Walking interviews*

With walking interviews organized for lay persons, we tested several different research methods. This made the experiments multi-level and partly complex. Determining the route and locations of stops in advance, as well as the implementation of walks in groups, have been central to the experiments. The walking interview itself is a method that brings uncontrollable events and disturbances from the environment into the interview. A group interview further increases this uncontrollability, as the group dynamics that arise between the subjects affects the general atmosphere and the subjects' desire and courage to express themselves. In the conducted walking interviews, a good method for reducing the influence of the group was found by occasionally dividing the walking interview into a conversation between one or two subjects and a researcher. By combining an individual interview with the group situation, even the more withdrawn group members got the opportunity to speak their views freely.

Conducting the walking interviews confirmed that placing different research methods in suitable environments and encouraging subjects to examine the environment with an open mind requires a precise script. A pre-selected route with pre-selected stops does not produce the same information as a walking interview where the subject chooses the route and the stops. In this case, however, the choice proved fruitful, as it took the subjects to places they would not normally go during the dark. Interestingly, these places along the route were not obscure back alleys, but unlit small forests by the sea – actively used areas that are perceived as pleasant and appreciated in the light of day. It is also interesting that going into the dark small forest or "thicket", as one participant called it, was not necessarily talked about as a scary thing. Instead, the expressions used, and the tone of the speech convey the perceived irrationality of this act. Going into an unlit small forest seems to require a special reason or explanation, such as walking a dog or a route that is deliberately designated as a dark area. It seems that in addition to possible fear, moving in unusual places at unusual times can make the person themselves appear odd and therefore potentially threatening to others. Taking the research subjects to an unlit small forest also brought out that city dwellers do not necessarily have any experience of moving in an unlit environment. Even those who move around the areas, for example jogging or cycling, often carry a headlamp or other strong light source with them, and therefore have never tried moving around the areas without lighting. Viewed from a strongly lit area, the surroundings can easily look pitch dark. However, a joint perception exercise in the small forest showed that the unlit place was not dark, but quite pleasantly dim. The observation surprised the participants and gave us an idea of demonstrating dark adaptation to city dwellers as part of the later intervention.

We will continue to develop walking interviews by categorizing the different aims and by creating different concepts for different needs: 1) general, ethnography-influenced research walks to find out local cultural, historical, climatic and environment related meanings and values of darkness, 2) more specific design walks for analysing specific places and the needs, hopes, fears, ideas, and attitudes people using these places carry, 3) evaluation walks in places where lighting conditions have been changed and comparative information about the conditions before and after the change is available, and 4) less profound and more entertaining evaluating event walks making it possible to host much larger groups of participants. To gain more versatile understanding of the experiences of dark urban environments dedicated walks should be arranged also for groups with special needs or backgrounds.

3.4. *Qualitative analysis of luminance measurements*

According to the preliminary results of the luminance analysis, the chosen visual-spatial concepts – enclosure, open view, interspaces, framing, scale, detailing and user relation – that were tested in creating the analysis images, seem to be useful conceptual tools for framing quantitative analysis of complex luminance images of urban darkness. Comparing the measured luminance values to the visual-spatial analysis and on-site experiences helps a designer to understand the relation of the measured luminances to the design intentions and possible positive and negative experiences within complex urban environments. The method seems promising as a research tool, as well, in analysis and evaluation processes to understand users' experiences of dark environments. Additionally, it could be applied in co-design processes to illustrate participants' concepts of urban darkness [34].

Figure 2. On the left: children trying to catch escaping shadows in Playing with Light workshop. In the centre and on the right: two of the 15 scenes of the intervention. Photos by the authors.

3.5. *Preliminary results of the workshops, intervention, and evaluation*

When writing this, the experimenting and co-designing workshops have just been conducted and the intervention launched. The analysing process has just begun. Background research played an important role in identifying the target groups for workshops and creating contacts with them. We see the Playing with Light Workshops for children as possibilities to gather physically smaller and culturally not-so-informed persons' experiences of dark urban space and see if some products of their playful imagination could be implemented in the design. For the Dark Adaptation Workshops, which were unfortunately cancelled, our expectation was that they could have given us insight on solutions for transitional zones between bright illumination and darkness [10] and ideas on how to encourage people to explore dark adaptation. Testing these two themes, dark adaptation and playful imagination, in separate workshop concepts also gave us more practical understanding on how to include them in the progressive Co-Design Workshop Series in which the temporary intervention was developed. The workshop series proved to be an interactive learning process we shared with the participants. Concrete working on darkness enabled previously unconscious attitudes to be put into words more easily. All parties generated new local understanding and new ways to experience, think and utilize darkness. Co-creation in a playful setting also deepened our understanding of how to work with darkness and light on multiple layers of human comprehension, such as analytical, sensory, bodily, and social.

4. Discussion: Forming a proposal for an urban darkness design method

Darkness is relative. Brightness, dimness, and darkness form a nuanced scale that could be utilized in urban lighting much more efficiently than today. The lighting conditions in northern outdoor spaces are in constant change. They are affected by the time of year and day, weather, and snow condition.

People's lighting needs, on the other hand, are influenced by personal characteristics and preferences, the local cultural and historical aspects, and the rhythms of society.

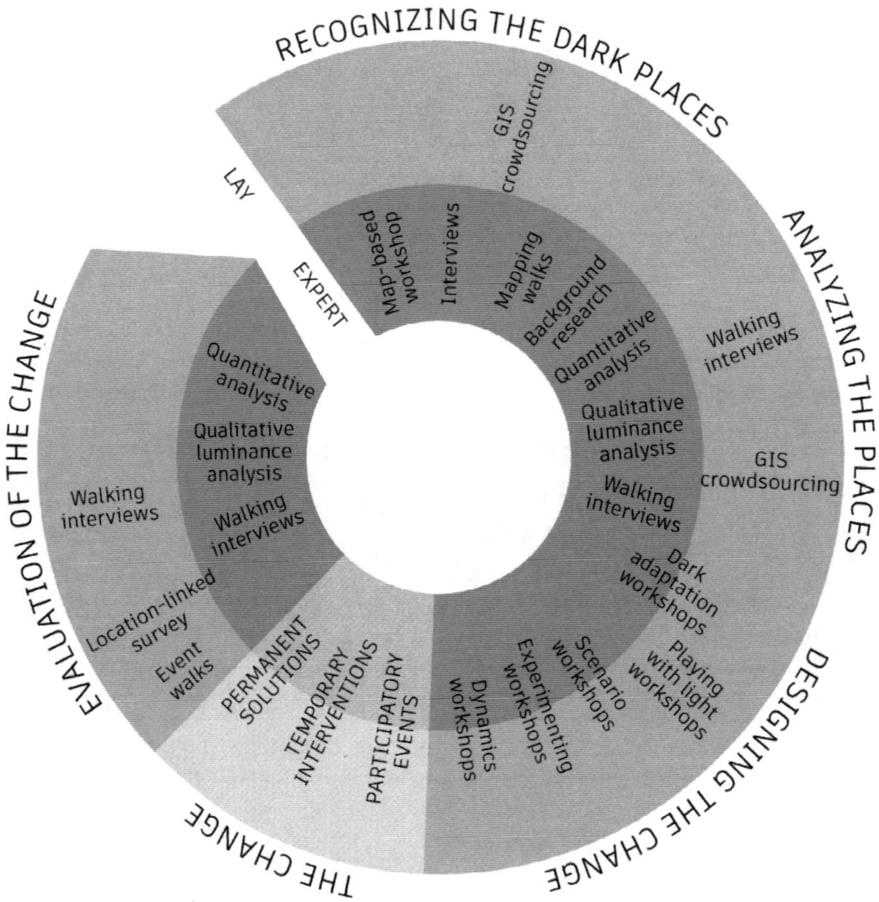

Figure 3. The proposed methods situated on different stages of an iterative design research project and utilizing either or both expert and lay understanding for design purposes. Diagram by the authors.

Based on our case study, we propose a process where, in different stages, assumptions about darkness are challenged and peoples' experiences of and relation to darkness are investigated. This means gathering local and in-depth understanding of what kind of darkness – or lighting – creates feelings of unsafety and insecurity, and in what kind of situations and places people enjoy darkness. We see it as important to look at the question from different points of view, thus, on-location methods, GIS crowdsourcing, quantitative studies, qualitative luminance analysis and ethnographical understanding complement each other. To gain sustainable design solutions that authentically serve multifaceted user needs, it is important to use several forms of information and knowledge. Later in the research we will compare the results of these participatory and qualitative methods with the current quantitative design principles and standards. The aim will be to evaluate the appropriateness of quantitative regulation regarding sustainable lighting and sustainable darkness.

The values of darkness have been proposed as central guiding principles for designing sustainable lighting. We see values not only generic and abstract, but strongly bound with the local environment and society. The co-design and ethnographic approach we suggest here keep the values of darkness under discussion throughout the whole process, reflected and negotiated in every stage of the design process. The strategy is to apply multiple methods in multiple stages to raise both analytical and experience-based awareness of the importance of safe and sustainable darkness. Efforts to combine expert knowledge and layperson knowledge appear multi-layered, especially when discussing the dark and dim urban spaces. In cities where there is no specially designated urban darkness coordinator, urban lighting is basically something planned, but the absence of lighting is usually unplanned or a by-product of lighting design. Therefore, darkness as an asset is an issue that is rarely or not at all addressed in the work of experts responsible for planning urban environments. Because of this, it is difficult to name experts in the dark and dim urban spaces. Even though an expert has knowledge of planning processes or a systematic way of looking at the urban environment that differs from laypeople, in their experiences of the dark, experts are also laypersons. Therefore, the expert knowledge and lay knowledge collected in this study can be viewed as a continuum and not as opposites. Personal experience can be seen as a valuable part of expertise when used as a tool for empathic understanding of users. More conscious deployment of this perspective could be to bring user orientation earlier into the design processes. However, we believe that the continuity between lay knowledge and expert knowledge should also be developed through defining and strengthening expertise, as a sustainable dark urban environment can only be the result of conscious planning.

Designing urban darkness is an iterative process that is implemented on different levels. Our case study was conducted mostly on a very concrete level, leading to an intervention in a park. However, there is yet another, more strategic level which needs to be more deeply scrutinized in the future. In our study, this level problematics was revealed especially in the expert interviews. The importance of long-term objectives and principles that have been confirmed at various policy levels was evident. We argue that this should be advanced by bringing darkness to conscious attention alongside lighting in the design and planning processes. We assume that undulating the theme between the very concrete level of urban place design and the more strategic levels of urban planning, regulation, and governance, helps to consolidate sustainable darkness and sustainable lighting in the societies. Furthermore, we see piloting as a powerful tool to change practises. This underlines the need for co-operation between cities and research institutions, as well as engaging new agents like artists and consultants with ethnographic understanding into design processes.

Acknowledgements
We gratefully acknowledge the funders of the Safe and Sustainable Darkness project (the project code A78790): European Regional Development Fund, Council of Oulu Region, City of Oulu, and University of Oulu. We also thank the participants of our study and Lumo Light Festival.

References

[1] Davoudian N (ed) 2019 *Urban Lighting for People: Evidence-based Lighting Design for the Built Environment* (London: RIBA Publishing)
[2] Edensor T 2017 *From light to dark: Daylight, illumination, and gloom* (Minneapolis: University of Minnesota Press)
[3] Hölker F et al. 2010 The dark side of light: A transdisciplinary research agenda for light pollution policy *Ecology and Society* **15**(4): 13
[4] Vega CP, Zielinska-Dabkowska KM, Schroer S, Jechow A and Hölker F 2022 A Systematic Review for Establishing Relevant Environmental Parameters for Urban Lighting: Translating Research into Practice *Sustainability* **14**(3): 1107
[5] Dugar A and Slade J 2021 Dovetailing darkness and light for designing sustainable outdoor spaces *International Journal of Sustainable Lighting* **23**(2): 88–99
[6] Stone T, Dijkstra I and Danielse T 2021 Dark Acupuncture: A Design Strategy for Sustainable Lighting *International Journal of Sustainable Lighting*, **23**(2): 70–87
[7] Challéat S and Poméon T 2019 'And what do you do with five-hundred million stars?': Assessment of darkness and the starry sky, values and integration in regional planning in Saville S and Hoskins G (eds) *Locating Value: Theory, Application and Critique* (London: Routledge)

[8] Juntunen E, Sarjanoja E, Eskeli J, Pihlajaniemi H, Österlund T 2018 Smart and dynamic route lighting control based on movement tracking *Building and Environment* **142**: 472–483
[9] Hvass M and Hansen E K 2022 Balanced Brightness Levels: Exploring how lighting affects humans' experiences of architectural and social urban contexts *IOP Conf. Ser.: Earth Environ. Sci.* **1099** 012017
[10] Amoruso C E, Larsen M H, Hvass M, Triantafyllidis G and Hansen E K 2022 Dark Adaptation in Urban Environments: An Innovative Design Framework for Pedestrian Lighting *IOP Conf. Ser.: Earth Environ. Sci.* **1099** 012044
[11] Rathsach C M and Hvass M 2022 The Moon as a Light Source: New sustainable ways of lighting up cities at night *IOP Conf. Ser.: Earth Environ. Sci.* 1099 012046
[12] Edensor T 2013 Reconnecting with darkness: gloomy landscapes, lightless places *Soc Cult Geogr* **14**(4)
[13] Meier J, Hasenöhrl U, Krause K and Pottharst M 2014 *Urban Lighting, Light Pollution and Society* (New York: Routledge)
[14] Brandi U and Geissmar-Brandi C 2007 *Light for Cities: Lighting Design for Urban Spaces. A Handbook* (Basel: Birkhäuser)
[15] Jedon R, Haans A and De Kort Y 2022 The effect of non-uniform urban illumination upon pedestrians' alertness and sense of safety *IOP Conf. Ser.: Earth Environ. Sci.* **1099** 012010
[16] Tavares P, Ingi D, Araújo L, Pinho P and Bhusal P 2021 Reviewing the Role of Outdoor Lighting in Achieving Sustainable Development Goals *Sustainability* **13**(22) 12657
[17] Sennett R 2018 *Building and dwelling: ethics for the city* (New York: Farrar, Straus and Giroux)
[18] Muller L and Loke L 2010 Take Part: Participatory methods in art and design *Proc. 11th Biennial Participatory Design Conf.* 283–284
[19] Bishop C ed 2006 *Participation* (Cambridge, Mass.: MIT Press)
[20] Swedberg R 2020 Exploratory Research in Elman C, Gerring J and Mahoney A eds *The Production of Knowledge* (Cambridge: Cambridge University Press)
[21] Rogers Y and Marshall P 2017 *Research in the Wild* (Morgan & Claypool Publishers)
[22] Creswell JW 2014 *Research Design: Qualitative, Quantitative, and Mixed Methods Approaches* (Thousand Oaks, CA: SAGE Publications)
[23] Abd Elrahman A S and Asaad M 2021 Urban design & urban planning: A critical analysis to the theoretical relationship gap *Ain Shams Engineering Journal* **12**(1): 1163–1173
[24] *Paikkatietoikkuna* (Helsinki: Maanmittauslaitos) https://kartta.paikkatietoikkuna.fi (accessed Aug. 06, 2023)
[25] Stare J *Light pollution map* https://www.lightpollutionmap.info (accessed Aug. 06, 2023)
[26] *Karttatie – Oulun seudun karttapalvelu* (Oulu: City of Oulu) https://kartta.ouka.fi (accessed Aug. 06, 2023)
[27] Larsen A and Kaanaa L 2021 *Helsingin häiriövaloselvitys* (Helsinki: City of Helsinki)
[28] *Helsingin ranta-alueiden valaistuksen periaatteet* 2020 (Helsinki: City of Helsinki and Ramboll Finland)
[29] *Lucia - Lighting the Baltic Sea Region: Tallinna pilootala dokumentatsioon* 2021 (Tallinn: City of Tallinn)
[30] Pink S 2015 *Doing Sensory Ethnography* (London: SAGE Publications)
[31] Lee J and Ingold T 2020 Fieldwork on foot: Perceiving, routing, socializing in Collins P and Coleman S eds. *Locating the Field: Space, Place and Context in Anthropology* (New York: Routledge)
[32] Ching F D K 1979 *Architecture: Form, Space & Order* (New York: John Wiley & Sons)
[33] Michel L 1996 *Light: The shape of space: Designing with space and light* (New York: Van Nostrand Reinhold)
[34] Pihlajaniemi H, Vuorenmaa R and Parhankangas O 2023 Measuring darkness for safe and sustainable experiences in northern cities *Proc. CIE 2023 Conf.*

IOP Conf. Series: Earth and Environmental Science 1320 (2024) 012005 doi:10.1088/1755-1315/1320/1/012005

Daylight Diagram – A Method to Map and Analyse the Temporal Conditions of Daylight Intensity

I W Foged

The Royal Danish Academy – Architecture, Design, Conservation
Philip de Langes Alle 10, 1435 Copenhagen, Denmark

iwf@kglakademi.dk

Abstract. This study proposes and examines a new method and diagram for revealing and analysing temporal and transitional daylight conditions based on computational modelling and simulations. For a human to experience and decipher the world, daylight is a predominant resource to engage with the surroundings. With humans spending the prevalent time inside buildings, it is important to understand and design the daylight environments in the early phases of design processes. The aim is to retain and reveal the information of high-resolution simulations and to explore, test and verify the capacities of the method in comparison to Illuminance, Daylight Factor and Daylight Autonomy values. The research design uses digital modelling and representation techniques combined with computational simulation methods for daylight analysis, evaluation and communication. Eight digital test cases show that the proposed Daylight Diagram enables the observation of temporal daylight dynamics. The proposed method and diagram employ existing simulation systems, allowing a fast integration and use in early-phase design processes as a key instrument to design advanced daylight phenomena and conditions focusing on transitional and temporal daylight intensities.

1. Introduction

Daylight is central to human existence. For a human to experience and decipher the world, daylight is the predominant resource to engage with the surroundings. As humans spend the prevalent time inside buildings, with Western populations reaching 90% occupancy indoors [1,2], it is important to understand and design the built environment with daylight in the early phases of design. With the advancements in computing power and methods, simulations have become a potent investigation and design approach to analyse and predict daylight conditions inside buildings. The rise of paired digital modelling and simulation processes over the last two decades has provided architects, designers and engineers with new instruments to explore and create design propositions [3–6]. While computational methods have increased in diversity, speed and accuracy, national and international standards and industry certification systems appear to maintain an 'on-principle-level' or 'low-resolution' approach to daylighting analysis and design. Building Research Establishment Environmental Assessment Methodology (BREEAM) states that at least 80% of the floor area in occupied spaces has an average daylight factor of 2% [7]. Deutsche Gesellshaft für nachhaltiges Bauen (DGNB) states that 50% of the usable area throughout a building has a Daylight Factor (> 3% very good, >2% medium, >1% slight, < 1% none) [8]. Leadership in Energy and Environmental Design (LEED) states that Daylight Illuminance levels should be minimum 270 Lux and maximum 5400 Lux in clear sky conditions on September 21 at 9 a.m and 3 a.m. [9]. The Danish and European building code (DS/EN 17037) criteria for daylight provision is that a space is considered to provide adequate daylight illuminance if a target illuminance level is achieved across a fraction of the reference plane within a space for at least half of the daylit hours [10,11]. This is a bit convoluted, but the typical target illuminance level is 300 Lux, the reference plane's fraction is 50% of the space, and half of the daylit hours is 50% of the occupancy time of the space. The recommendation and relations are captured in the Daylight Autonomy (DLA) calculation method, which again is promoted by the Illuminating Engineering Society of North America [12] to evaluate the daylight quality of spaces. Due to the relative complexity of the DLA calculation method, computational simulation methods are used for this process. A simpler method, adopted from manual measurements, is the Daylight Factor (DF) method [13] which calculates the indoor light level in relation to the outdoor light level, providing a percentage of the outdoor available light. Due to the simplicity of the factors and model, the Daylight Factor is often found in industry guides [14], where an average DF > 2% is considered daylit. In contrast, an average DF > 5% is strongly daylit.

The challenge with all these evaluation and benchmark numbers is that they do not identify, represent, or communicate the temporal and dynamic conditions of daylight in a space over time. If a space has an average (annual) DF = 3%, it could lack daylight in winter and be prone to glare conditions in summer. The same problem could be caused during a singular day, as the solar vector and sky conditions (clear, haze, cloudy) change significantly across 24 hours. In a similar concern, a DLA > 50% could present suitable conditions in the morning and the afternoon, not during midday. While the state-of-the-art simulations calculate across the year, hourly, the resulting output, scores and representation of the daylight environment are reduced to singular numbers that do not reveal the dynamics of the specific daylight conditions and temporal conditions related to design and context.

The previous solutions for simulating, representing and decoding temporal daylight conditions are based on false colour rendering, applying a gradient colour map, where the distribution of daylight is shown on the reference plane, figure 1. This method works well to show the spatial distribution of daylight, where numbers and colours typically are based on an annual average value. Most simulation methods allow the specification of singular hours (or even sub-hour intervals) or specified periods of the year, such as Summer, Winter, October, week 42 etc. Other recent mapping methods show the percentage of time, or hours, on each simulation node (colour mesh field) to indicate the prevalence of the daylight value [15]. While this adds an understanding of time into the map, it does not show or analyse the temporal and dynamic conditions of daylight hourly, daily and annually.

The potential for new methodological solutions is to include multiple time domains in one diagram, thereby creating a representation of the daylight simulation procedure, which retains and reveals detailed daylight intensity for every hour of the year. In acoustics, sound energy is always mapped in relation to octave bands (frequencies), as the spreading of sound waves is dependent on the wavelength [16]. An expanded mapping of this relation includes time, with the three axes including sound energy, frequency and time. Together this forms the FFT Waterfall Diagram, that allows advanced analysis of temporal acoustic conditions. In daylight studies, wave frequency is typically not in focus, which allows two time domains to be plotted together with daylight intensity, creating the potential of a three-dimensional diagram, which shows the temporal conditions of daylight across every hour of a whole year.

This study attempts to construct a new Daylight Diagram, which retains and reveals the information of high-resolution simulations, and to explore, test and verify the capacities of the method in comparison to Illuminance, Daylight Factor and Daylight Autonomy values. The Daylight Diagram is intended as a complementary method to daylight analysis, and design focused on temporal daylight intensity conditions.

The paper presents the idea and inspiration for the Daylight Diagram, the computational methods used to construct and evaluate the proposed method, experimental studies based on parametric modelling and simulation processes, and results description based on a comparative analysis between what the Daylight Diagram shows in difference to Daylight Factor and Daylight Autonomy values and associated colour mappings. A discussion of the methods and results is then included providing the basis for a brief conclusion to the study and its contributions.

2. Methods

The research design uses digital modelling and representation techniques combined with computational simulation methods for daylight analysis, evaluation and communication. Results from the simulation studies are compared to daylight standards and as a comparative analysis between the experiments within this study.

Design Computation and Simulation Studies

Digital design models for testing the proposed Daylight Diagram are created in two case groups of A) generic rectangular space geometries with generic apertures (window geometries) and no ground and context conditions, and B) bespoke space geometries with varied apertures (window geometries) and no ground and context conditions. For each of the two groups, four space-aperture variations are created, figure 2. In group A) the space geometry has dimensions 8x5x3 meters. Apertures are positioned in the space to examine an East/West facing composition, an East facing composition, a North facing composition, a South facing composition. The size of the apertures in each of the four test compositions is defined by gradually modifying the opening area to reach 4% Daylight Factor of each test space, figure 1. The generic rectangular geometries study aims to understand and explore the temporal daylight variance between the test space when the Daylight Factor is constant.

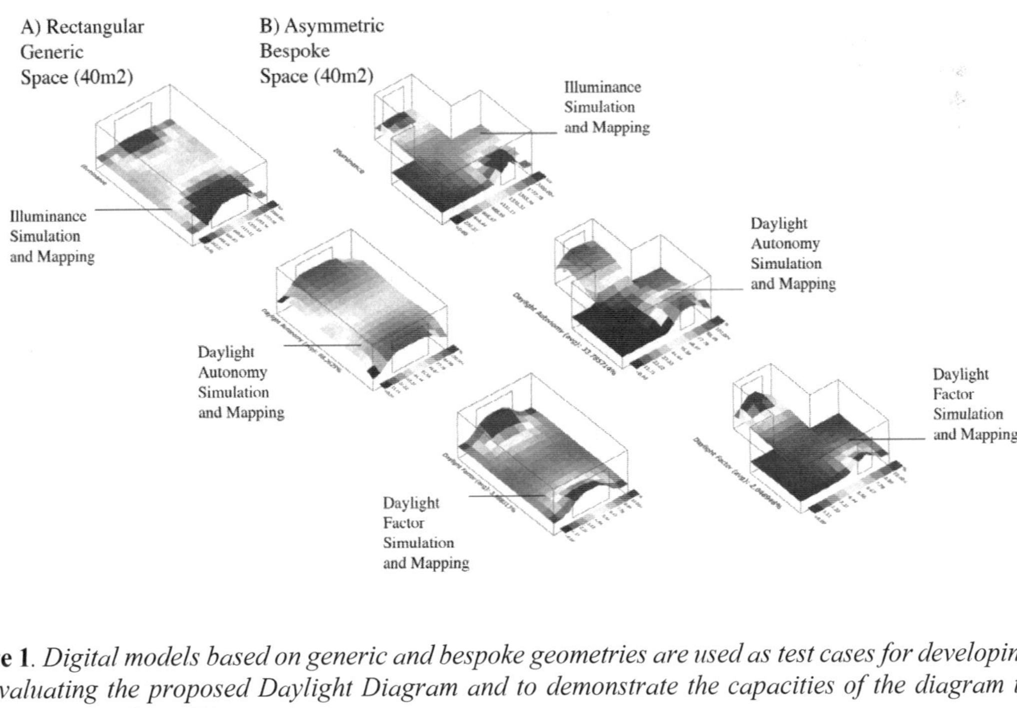

Figure 1. *Digital models based on generic and bespoke geometries are used as test cases for developing and evaluating the proposed Daylight Diagram and to demonstrate the capacities of the diagram to illustrate temporal conditions.*

IOP Conf. Series: Earth and Environmental Science 1320 (2024) 012005 doi:10.1088/1755-1315/1320/1/012005

In study group B), four bespoke test spaces and apertures, figure 1 and 2, are defined by varied plan geometries while maintaining the 40m2 footage of the rectangular test spaces in study A). Apertures are positioned with similar orientations to study A). Alike to study A) apertures are modified gradually, albeit to reach 2% Daylight Factor, then observe and analyse the temporal daylight conditions exposed in the Daylight Diagram. The decreased DF factor, compared to study A, is chosen experimentally to reach a better comparative analysis between B cases.

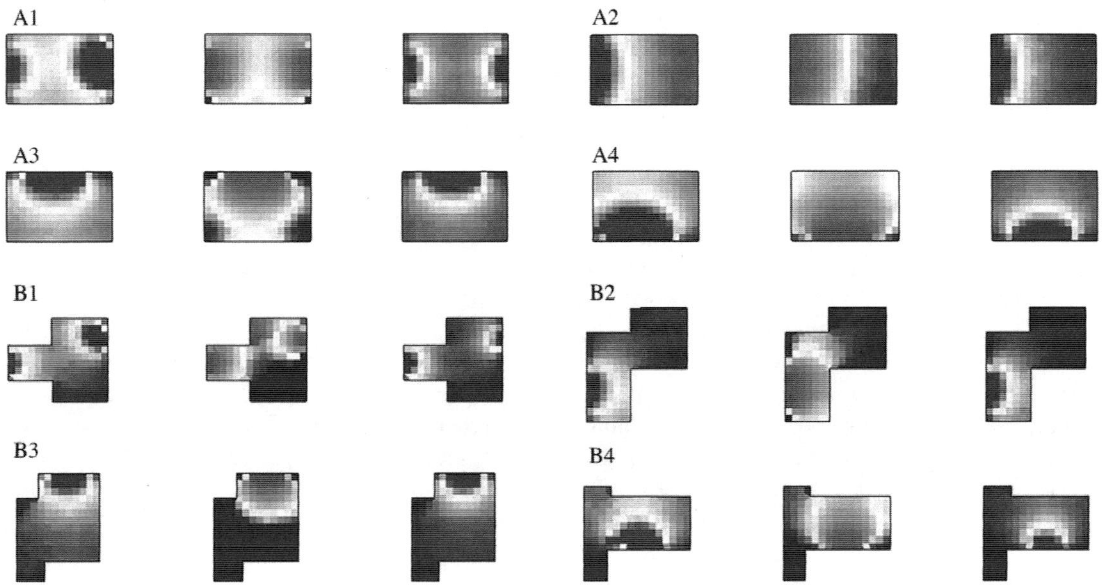

Figure 2. *Eight digital models in two groups, each with three simulation processes for Illuminance (left), Daylight Autonomy (center) and Daylight Factor (right) in each test case.*

The Daylight Diagram is based on the idea of revealing the temporal daylight conditions by mapping hourly light intensity in relation to the time domains of a day (24 hours) and a year (365 days). Time domains are mapped on the x-axis and y-axis, hours and days, respectively, and light intensity on the z-axis. The correlation between time (every hour of the year) and light intensity creates a cloud of points in the diagram. To make light intensities and their temporal and transitional properties eligible, a line is drawn from point to point in the x-axis, creating a wave-form that can be understood and analysed by the investigator/designer, figure 3. As mentioned, this method is inspired by the FFT Waterfall diagrammatic method from acoustics, where sound energy is mapped in relation to time and frequency domains. The resulting curves provide the possibility to analyse the diagram three-dimensionally. Two threshold values can be added to the diagram, providing lower and upper light intensity (Lux) limits. This can be dynamically set and has been fixed to 300 Lx (lower limit) and 2000 Lx (upper limit) in these studies. Values above the 300 Lx threshold are plotted in yellow on the diagram's xy-plane and values above 2000 Lx are plotted in white on the diagram's xy-plane. This technique attempts to create a direct reading of which hours a year the lux level exceeds the threshold of sufficient light and what hours a year may be problematic for glare conditions.

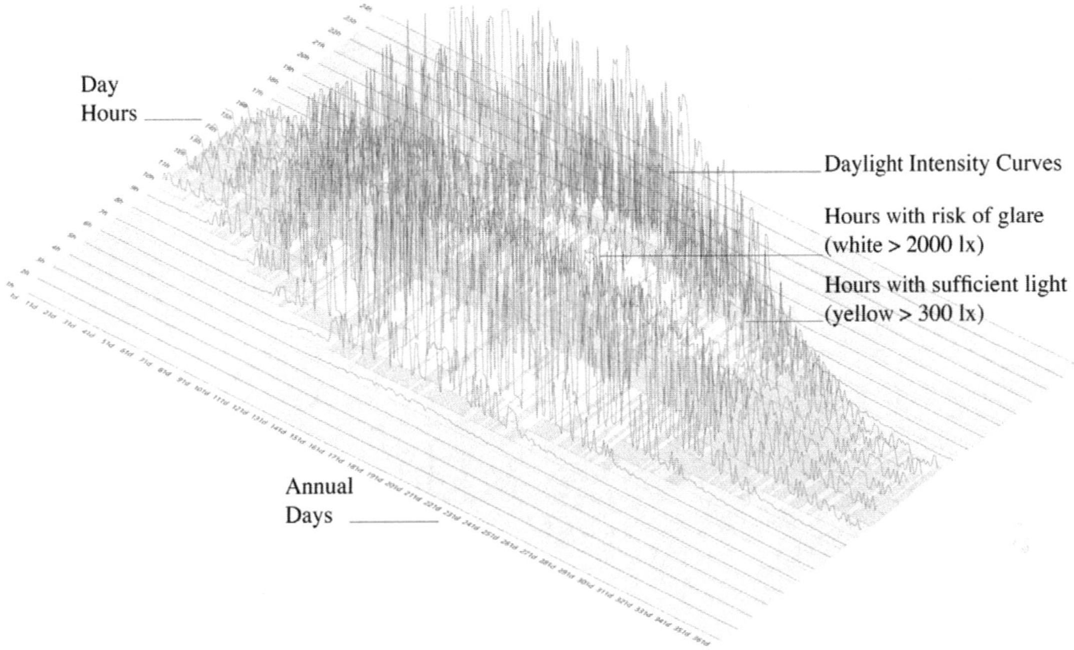

Figure 3. *The anatomy of the proposed Daylight Diagram. The diagram is represented in a parallel projection but can be viewed and rotated as necessary within the 3D model for detailed qualitative analysis.*

The Daylight Diagram is based on simulated hourly data for an entire year. The dataset is in this study based on a climate-based point-grid simulation procedure, using the Rhinoceros-Grasshopper-Ladybug software framework, which utilizes the Radiance simulation engine [17] and Energy-Plus weather data [18]. The weather data used for the studies are for Copenhagen, Denmark. The computational model schematic structure includes, 1) geometric/material models of test spaces/materials, 2) weather data, 3) simulation model properties (test point-grid, sky conditions etc.) 4) daylight analysis procedure (Illuminance, Daylight Factor, Daylight Autonomy), 5) Colour mapping of analysis procedures inside test space and 6) Daylight Diagram, figure 4.

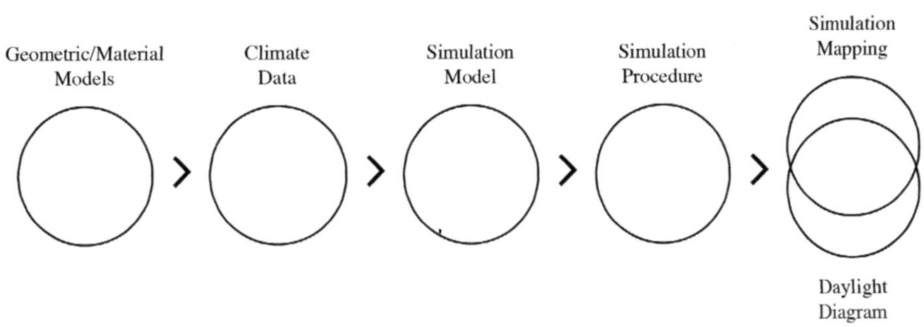

Figure 4. *Schematic structure of the simulation model used for the study. For each experimental study, simulating the test spaces and apertures, Illuminance analysis, Daylight Autonomy analysis, Daylight Factor analysis and Daylight Diagram analysis are conducted.*

IOP Conf. Series: Earth and Environmental Science 1320 (2024) 012005 doi:10.1088/1755-1315/1320/1/012005

3. Results

The test spaces of the A case experiments are all the same, simple in form and apertures. The apertures of each space are modified for the average DF to be 4% (+/- 0.1%). While the DF remains the same, the Daylight Diagram reveals large variances across the year and day between the cases. The DF alone would in such cases be both misleading and non-informative. The difference between case A3 and A4 are striking, just as the difference between A1 and A2 have distinctively different daylight conditions across the day. The characteristics of each test space and apertures are clearly revealed in the diagram, showing, e.g. high afternoon/eve intensity of A2 and morning/afternoon intensity of A1.

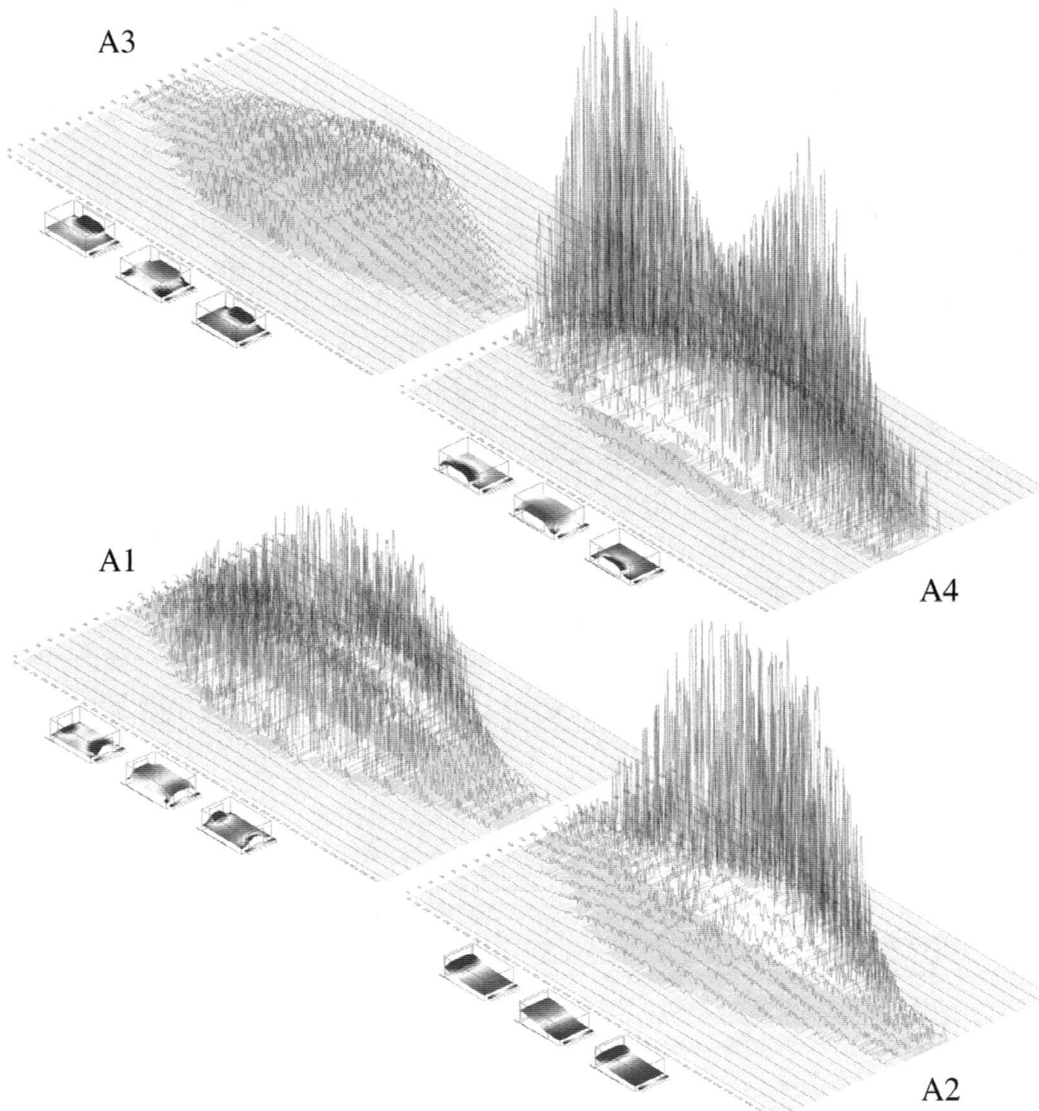

Figure 5. *Four A test case based on a simple, rectangular 40m2 test space with varying apertures, all with DF = 4%, yet with highly varied temporal daylight intensities between the test cases.*

Similar results are found for the B test cases, where the apertures of the different test space geometries are modified to DF = 2% (+/- 0.1%). Not surprisingly, the temporal daylight intensities revealed in the Daylight Diagram reduced compared to the A test cases, but the variance is similar, despite the increased spatial complexity. While some standards propose that DF => 2% is sufficient in daylight level, the B3 case shows that only a limited period exceeds the 300 Lux threshold, as shown by the yellow mapping. In contrast, while limited to DF = 2% (+/- 0.1%) the B4 test case experience significant glare around midday for large parts of the year, as shown by the white mapping. The contrast between the two daylight conditions and the temporal conditions of the light intensities could not have been identified through singular DF or DLA values.

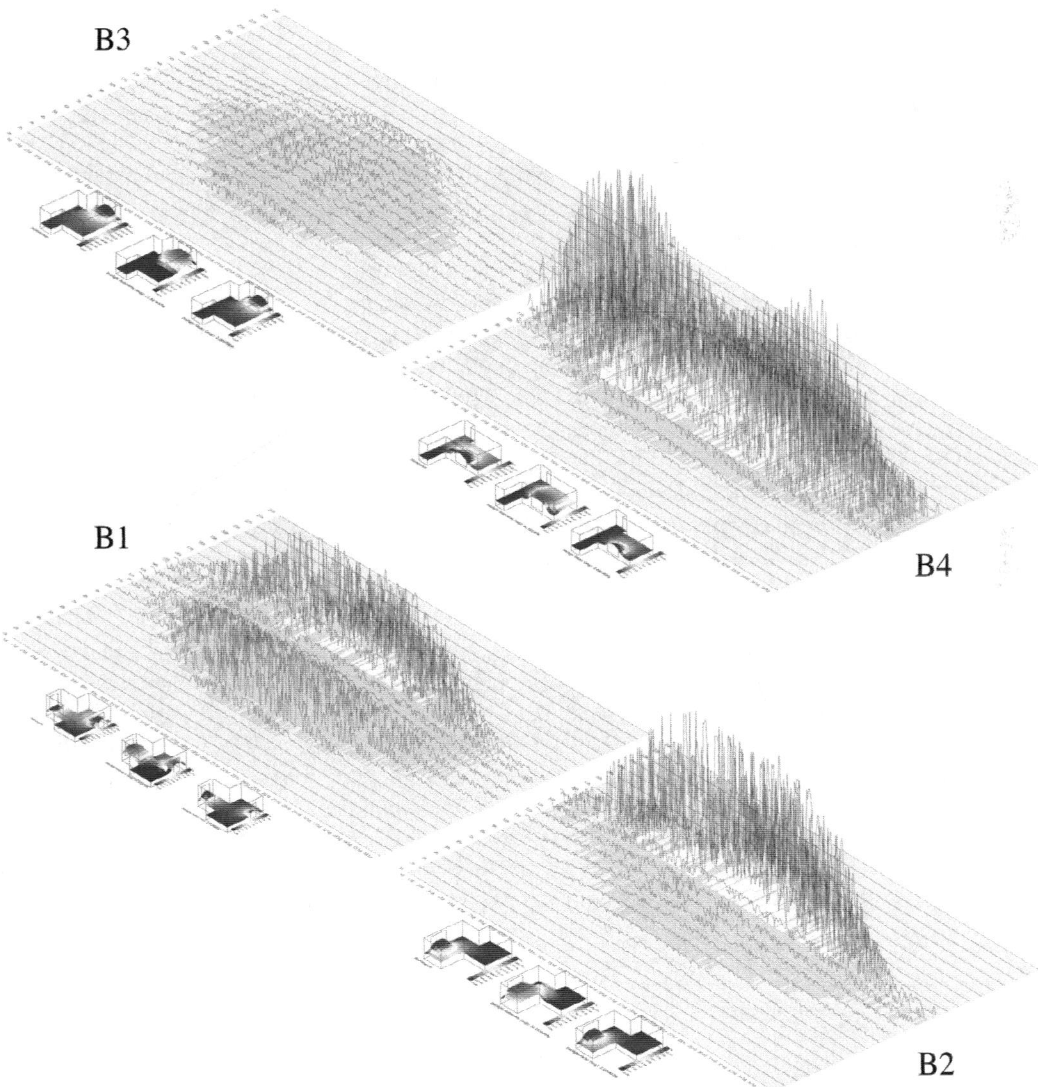

Figure 6. *Four B test cases with varied spatial compositions and apertures, but all with an average DF = 2%, show a high variance in temporal daylight conditions, particularly for the B1, B2 and B4 cases.*

4. Discussion

Daylight predictions in design processes using simulations are based on numerical models. The evaluation of daylight sufficiency and quality are largely based on these simulations, and they are today focused on delivering numbers that encapsulate the daylight condition of a space, by the use of a number, such as DF and DLA. While these are widely acknowledged and used, they do not reveal the temporal conditions of daylight intensities. However, we know this is how the real world is experienced. It does not allow daylight designers to investigate and propose advanced daylight phenomena and experiences based on temporal dynamics, where space, material, human, and environment can be explored and designed for bespoke and context-specific conditions. The proposed Daylight Diagram is a step towards retaining a higher fidelity of the computational simulations, providing new and expanded understandings of the light environment designed or to be designed. The Daylight Diagram should not be seen as a replacement for other methods, but as complementary, particularly with the spatial colour gradient mappings used in this study.

With the quality and quantity of daylight design directly impacting artificial lighting sources, a better understanding of daylight dynamics through the Daylight Diagram could be suggested to improve the conditions, analysis and design of alternative light sources. Such influence could lead to better management of electrical light sources for better light experiences and a reduction of the energy used for electrical light systems, thereby aiding design decision processes for sustainable architectures based on advanced daylight architectures.

Similarly, when developing the material composition of a project, it could be imagined that the Daylight Diagram would provide insights to support how surfaces increase/decrease reflected light energy in general and the solar vector in specific, providing a direct relationship between the dynamic path of the sun at a given location with the resulting daylight mapping. The knowledge of such analysis would support material choice which both increases light quality and the selection of sustainable materials supporting the daylight design intent.

Also, further advancements of the Daylight Diagram could include spatial information. Such inclusion may, however, blur the clarity of the proposed diagram and counteract the ability to observe and design from the current Daylight Diagram information. More research is required, potentially coupled with how such methods are used in practice and what that means for design decision-making.

5. Conclusion

The idea, basis and structure for a Daylight Diagram are proposed in this research with the aim of revealing temporal daylight intensities based on computational simulations. The test cases show that the proposed Daylight Diagram reveals temporal daylight dynamics that are difficult or potentially impossible with the current use of singular benchmark values and methods. The proposed method and diagram employ existing simulation routines, allowing a fast integration and use in early-phase design processes as a key instrument to design advanced daylight phenomena and conditions with a focus on transitional and temporal daylight intensities. It is argued that the Daylight Diagram may also inform better design decision processes for artificial lighting design and material development, as part of increasing the sustainable capacity of daylight design in the built environment.

References

[1] Klepeis N E, Nelson W C, Ott W R, Robinson J P, Tsang A M, Switzer P, Behar J V., Hern S C and Engelmann W H 2001 The National Human Activity Pattern Survey (NHAPS): A resource for assessing exposure to environmental pollutants *J Expo Anal Environ Epidemiol* **11** 231–52

[2] Ortiz-Ospina E 2020 *How do people across the world spend their time and what does this tell us about living conditions?*

[3] Havgaard M V. 2015 *Using Building Simulations to Evaluate Thermal Comfort* PhD (Aalborg University)
[4] Winsberg E 2001 Simulations, Models, and Theories: Complex Physical Systems and Their Representations *Philos Sci* **68** S442
[5] Shaviv E, Yezioro A, Capeluto I G, Peleg U J and Kalay Y E 1996 Simulations and knowledge-based computer-aided architectural design (CAAD) systems for passive and low energy architecture *Energy Build* **23** 257–69
[6] Malkawi A 2005 Performance Simulation: Research and Tools *Performative Architecture - Beyond Instrumentality* ed B Kolarevic and A M Malkawi (Spon Press)
[7] BREEAM 2023 BREEAM *BREEAM Certification*
[8] DGNB 2023 DGNB *DGNB Certification - https://www.dgnb.de/en*
[9] US Green Building Council 2023 LEED Certification *LEED Certification*
[10] Boubekri M 2004 A Overview of The Current State of Daylight Legislation *J Hum Environ Syst* **7** 57–63
[11] DS/EN 17037 2022 *DS/EN 17037:2018A1:2021 - Daylight in Buildings*
[12] Illuminating Engineering Society of North America. Daylight Metrics Committee. 2012 *Approved method : IES spatial daylight autonomy (sDA) and annual sunlight exposure (ASE)*
[13] Hopkinson R 1963 *Architectural Physics Lighting* (Her Majesty's Stationary Office)
[14] CIBSE 2006 *Environmental Design - CIBSE Guide A* (London)
[15] Solemma 2023 Climate Studio Simulation and Mapping *Solemma - Climate Studio*
[16] Long M 2014 *Architectural Acoustics* (Routledge)
[17] Radiance 2023 Radiance *https://www.radiance-online.org*
[18] Energy Plus Weather 2023 Energy Plus Weather *Energy Plus Weather Database*

IOP Conf. Series: Earth and Environmental Science 1320 (2024) 012006 doi:10.1088/1755-1315/1320/1/012006

Poetic Daylight – a pavilion for the perception of daylight

L Grønlund[1], N Mathiasen[2], P Sørensen[1] and A K Frandsen[2]

[1] Royal Danish Academy – Architecture, Design, Conservation, Philip de Langes Allé 10, 1434 Copenhagen, Denmark

[2] BUILD, Aalborg University, A. C. Meyers Vænge 15, 2450 Copenhagen, Denmark

anch@kglakademi.dk

Abstract. Daylight provides us with an understanding of time and space, specifying where we are in the world, every day. We delight in experiencing the changeability of daylight – from morning to evening perceiving how light and darkness affect the spaces surrounding us. This paper highlights daylight as one of the central elements in architecture, by emphasising its qualitative potentials: creating healthy, aesthetic, and poetic spaces. The setting for this study is a 1:1 pavilion, *Poetic Daylight*, built for the UIA Conference in Copenhagen. Presenting studies of scale models 1:10 and studies of the built 1:1 pavilion, this study lays out the possibilities and potentials in the use of scale models when designing spaces, with a focus on the qualitative appearances of daylight. Bearing the distinction laid out by Vitruvius in mind, i.e. *firmitas*, *utilitas* and *venustas*, the pavilion focuses on *venustas* (beauty or delight) in order to form spaces where the atmosphere and the beauty of daylight can be perceived and experienced. The Vitruvian concept of beauty, together with J. J. Gibson's definition of the perception of an environment, constitutes the theoretical framework. The analysis and the description of the three different spaces in the pavilion relates to the theory on defining light as presented by Anders Liljefors in his compendium: "Seende och Ljusstrålning" where he describes seven variables of light and Sophus Frandsen's definition of the four different kinds of shadows. The methodology involves the development of models, observations, and representations. This study shows that using 1:10 scale models in the design process to include qualitative aspects of daylight is practicable and effective, producing tangible and transferable knowledge that will influence the experience and perception of daylight in the built 1:1 pavilion, *Poetic Daylight*.

1. Introduction: Poetic daylight

This paper presents a study carried out to determine how an aesthetically pleasant environment in the pavilion *Poetic Daylight* situated in Copenhagen might be established. The pavilion was built for the UIA Conference in Copenhagen, July 2023. The pavilion aims to illustrate how daylight affects the perceived atmosphere within spaces. The pavilion has no 'function' in a traditional sense – its sole intention is that of creating spaces where one can perceive and experience various daylight scenarios. It intends to illustrate how aesthetically stimulating environments emerge through the interplay of daylight and architectural space. The scope of this paper is to describe the methodology of designing and testing such spaces. The study presented in this paper is a continuation of the study of the paper "*Architectural scale models: Methodologies for studying daylight quality*" [1].

The natural concern of architectural design is to design pleasant spaces for people, but this is sometimes outmatched by demands of documentation on technical aspects of the architectural design schemes regarding durability, energy efficiency, etc. The same holds true for the daylight design that must meet standards described through quantitative data. This may reduce the attention given to the qualitative aspects of daylight design. The current project highlights the immense potential of daylight on the perceived architectural spaces and suggests how to keep focusing on the qualitative aspects of light and space. It presents a study which explores ways of including the qualitative aspects of daylight throughout the design process. It discusses the pros and cons of investigating the perceived atmosphere of a daylit space through scale models and questions whether it is possible to make design decisions regarding the qualities of light based on scale models. As the pavilion *Poetic Daylight* was erected during spring 2023 it gave the unique possibility to combine the scale models with the actual built spaces and to examine whether the intention of the concept was met in the experience of the real space. These investigations are in a direct continuation of a previous study, presented in the paper "*Architectural scale models: Methodologies for studying daylight quality*" [1] where only scale models and simulations were studied without the possibility of compare it to the actual on-site experience. The observations in scale models, the on-site registrations and the photographic representation were conducted by the research team, the authors of this paper.

The concept of the pavilion *Poetic Daylight* is to dedicate a series of spaces to each their daylight character. The three light characteristics, as they are defined in this project, relate to how science of meteorology describe light [2] and the terminology used when simulating light [3, 4]. The three spaces are linked by a dark corridor guiding the visitors from one room to another.

2. Theory: Beauty – in daylight

The scope of the pavilion is to point towards the impact of daylight qualities in an architectural context and to describe how various daylight openings affect the architectural space and the perceived atmosphere of the space.

Architecture or the body of architecture, *corpus architecturae* [5], was described by the Roman architect and engineer Vitruvius as a balanced relationship between *durability, convenience, and beauty* (Book 1, 2, 3) [6]. This Vitruvian triad is probably the most quoted architectural theory in history; it stands out as an important text being the oldest and only surviving doctrine from classical antiquity. The German art historian and librarian Bernd Evers writes: "*Vitruvius, the first of all architectural theoreticians, distinguishes between firmitas (firmness, robustness, structural strength), utilitas (usefulness or commodity), and venustas (beauty or delight). Firmitas relates to all technical matters such as the selection of a site, the materials, and the foundation, utilitas to the function of the building, and venustas to its aesthetics*" [7]. According to Vitruvius these three interrelated approaches produce the best architecture. And when architecture fulfils all three attributes, the architectural design achieves perfection. Vitruvius himself underlines that both laymen and architects can "recognize a good piece of work" (Book 6, 8, 10) [6]. But the difference between the layman and the architect is that the layman will see this when the building is finished, and the architect will know during the design process: "Whereas the architect, as soon as he has formed the conception, and before he begins the work, has a definite idea on the beauty, the convenience, and the propriety that will distinguish it" (Book 6, 8, 10) [6].

Distinguishing between the three attributes of architectural design has inspired architects and architectural theorists throughout the ages, and new versions of the Vitruvian treatise have emerged, including various interpretations of the architectural design process. The organisation of architectural approaches into three relatively straightforward groups is an inspiring tool for architects to adopt and to relate the building design to. With reference to the distinction made by Vitruvius, the focus in the development of the pavilion *Poetic Daylight* was on *venustas* (beauty or delight), to form spaces where the atmosphere and the beauty of daylight can be perceived and experienced.

Navigating daylit spaces is also about sensing light. You register and understand it instantly and intuitively. You do not need an explanation of why and how the light is in the space and how it is distributed – the light is in the space, and you sense it. This immediate feeling and the intuitive registration of the surroundings, based on learned experiences in which the phenomenon is registered, can be linked to a phenomenological way of approaching the world [8]. In *The Ecological Approach to*

Visual Perception, the American psychologist James J. Gibson has described the perceptual approach to our visual environment as follows: "*With what does one see the world (...) One sees the environment not with the eyes but with the eyes-in-the-head-on-the-body-resting-on-the-ground*" [9]. The connection between the visual perception and the 'whole' bodily perception of our world forms the basis for the methodology of the daylight studies presented in this study, both in regard to the scale models and the built pavilion.

The theoretical point of departure for the study of daylight in scale models and the experience of the daylit spaces in the pavilion are based on Anders Liljefors' approach to defining the perceived light [10] and Sophus Frandsen's categorisation of spatial shadow pattern [11].

Liljefors distinguishes between light described through the physical definitions and the visual properties of light, respectively. Highlighting the different approaches to describing and defining light as both physical luminosity and perceived luminous impressions. This study has its main focus on the visually perceived luminous impressions of a space and refers to Liljefors compendium "*Seende och Ljusstråling*" [10] where he pinpoints seven variables to describe the luminous impression: "*level of lightness, spatial distribution of brightness, shadows, reflections, glare, colour of light, and colours*".

Sophus Frandsen has developed a categorisation of shadows based on the shadows size and location within a space. In his paper "*The Scale of Light - a New Concept and Its Application*" [11] Frandsen firstly groups the shadows into four categories: "*big spatial shadows, big attached shadows, small attached shadows, and small shadows*" [9]. These four categories describe the types of shadow experienced within a space. And Frandsen unfolds how big and small has to be interpreted: "*The terms big and small are to be seen in relation to the diameter of the human trunk*" [11]. In that way Frandsen positions people within the space perceiving light and space on site. Secondly, Frandsen defines how objects interact with light and creates various shadow types. He divides them into 11 types of shadows, ranging from 0 to 10 where all shadow types are defined depending on their relation between object, space and daylight opening, resulting in each their shading character. [11]

3. Methodology: Observations and representation of daylight

The purpose of the pavilion is to display and provide the framework for three different appearances of daylight. Daylight studies realised through drawing and modelling were the means to achieve this objective [12] together with the use of several types of observation and representation, with a focus on observation (human and scale models/spaces) and representation (human/ photographs).

The studies of the scale models were conducted in the Daylight Laboratory at the Royal Danish Academy – Architecture, Design, Conservation. The Daylight Laboratory facilitate studies of daylight at various latitudes and in two different daylight condition: an overcast sky and sunlight [13]. The examination of the various designs in the scale models was made under the exact same daylight condition to ensure comparability of the various models. The surrounding conditions were included in the model to represent the light obstructions of the buildings in the context.

The 1:1 studies of the daylight were carried out in the pavilion at Gråbrødretorv in Copenhagen, where the pavilion is located.

The observation of the pavilion in both scale model and 1:1 was made by members of the research group and the design team. The photos in of both scale models and 1:1 was made taken by the research assistant. Based on these observations' conclusions and decisions were made on the design of the pavilion in relation to the overall concept: perceiving three different appearances of daylight.

3.1. Study parameters

The daylight studies were conducted using the following types of observation and representation:

- Daylight observations in the scale models, 1:10
- Photographic daylight registrations in the scale models, 1:10
- Daylight observations in the built pavilion, 1:1
- Photographic daylight registrations in the built pavilion, 1:1

In developing the design for the three main spaces in the pavilion, many scale models in 1:10 have been produced to test and illustrate the purpose of the daylight. After analysing the model in the Daylight Laboratory, new models were made until the desired result was reached.

3.1.1. Daylight observations in a scale model. All studies were conducted within the Daylight Laboratory, ensuring consistent lighting conditions and camera configurations, while observations and registrations were made from predetermined fixed perspectives. The scale model, crafted from wood-pulp board, ensured a uniform surface across all studies. Considering its spatial characteristics, the model allowed examination from multiple angles, with a specific emphasis on interior daylight conditions. Its size enabled the observation of these conditions through a smaller opening, effectively preventing the intrusion of extraneous light [14].

3.1.2. Photographic daylight registration in a scale model. Similar scale models were used to register daylight across several series of model photographs. The registrations were captured using a wide-angle 26 mm f 1.6 lens, with the camera positioned at the eye level of a standing person, equivalent to 160 cm. Ensuring uniformity in comparison, all model photographs were captured from identical positions. The series of photos showcases the interplay between the aperture and the appearance of daylight within the space [15]. These photographic representations can be used as a tool for understanding the effects of the light and their perceptual qualities within the space.

3.1.3. Daylight observations in the built pavilion, 1:1. All observations in the pavilion were made over several days, at different times of day and under different weather conditions.

3.1.4. Photographic registration of daylight in the built pavilion, 1:1. The photographic registration of the pavilion made use of a wide-angle 26 mm f 1.6 lens. The level of the camera was placed at the same height as the eyes of a standing person, i.e. 160 cm. At the outset a series of photographs from different positions in the spaces were taken to start the process of capturing the daylight in the space. The same position for all the photographs are used in the documentation of the daylight conditions in the space in order to provide photos that are comparable to the photos of the scale model.

3.2. The pavilion, Poetic Daylight

The pavilion is 80 m^2 and consists of three main spaces connected by a dark corridor to create a pause and a contrast to the three main daylit spaces, which are characterised by three different appearances of daylight: light from the sky, light from the sun, and reflected light. When first entering the corridor of the pavilion, the eye adapts to the darkness so that the experience of the daylight in the three main spaces is even more striking and powerful. The surfaces in the pavilion vary and have been chosen to emphasise different types of atmospheres, created by the interplay between surface materials and daylight.

Given its location in Copenhagen, Denmark, the pavilion resides beneath a Nordic sky. The characteristics of Nordic light are defined by ample diffuse light originating from the sky and the relatively low position of the sun [2]. These light conditions form the foundational framework for the perception and experience of the spaces within the pavilion.

In the design phase, as mentioned earlier, the focus was on *venustas*, the delight/beauty, of the space and how daylight interacts with and affects the experience of the space. *Firmitas*, the construction of the pavilion was realised with a view to the fact that the pavilion will be only temporary and moved to a new site later on. Furthermore, *utilitas*, an actual function of the spaces, was not part of the programme for the pavilion. Instead, the ‘purpose’ is to explore the potential of daylight in an architectural space focusing on the perceived atmosphere of the spaces. The focus on the beauty of the space is seen in relation to an intention of (re)introducing the various ways daylight effects spaces and the power of daylight to create inspiring, aesthetic, long lasting, and visually stimulating spaces. The design process has been characterised by an interdisciplinary collaboration between architects and engineers in praxis and researchers at the Royal Danish Academy – Architecture, Design, Conservation and Aalborg University, Build.

IOP Conf. Series: Earth and Environmental Science 1320 (2024) 012006 doi:10.1088/1755-1315/1320/1/012006

	Space for light from the sky	Space for sunlight	Space for reflected light
Room size	3.8 x 3.8 m	3.8 x 3.8 m	2-3 x 5 m
Room height	5 m	4 m	3.5 m
Opening orientation	Up	South-east	Up
Opening size	1 x 1 m	1.7 x 3.8 m	2-3 x 5 m
Opening placement	Skylight	Low	Skylight

Figure 1. Overview of the three daylit spaces in the pavilion.

4. Studies and observations in the pavilion Poetic Daylight: Three different appearances of daylight in three different spaces

In the following, the design of the three spaces is described in relation to the daylight characteristics, the observations conducted in the scale models and the actual 1:1 experience of light and space. The study is structured so that it first describes each space in general, in relation to the geometry of the specific space, and lastly in relation to how the daylight is experienced in the pavilion.

4.1. Light from the sky

The light from the sky above us envelops us. Resembling a vast luminous surface that gently bathes the surroundings and objects in a soft, diffused glow from all directions, thereby generating diffuse daylight. The light emanating from the sky is uniformly distributed, creating gentle shading which draws little attention and is therefore often overlooked compared to the sunlight.

Depending on the design and orientation of a skylight, it provides a view of a particular part of the sky. On sunny days the space is illuminated by the blue sky, whereas on overcast days, it is illuminated by the white-grey sky. While the characteristic of the daylight and its even distribution within the space remain consistent, only the colours vary. When it is cloudy the colours are colder and bluer as opposed to when it is sunny, and the daylight takes on a warmer colour.

4.1.1. Light from the sky in the pavilion. In the skylit space of the pavilion, we encounter a precisely positioned skylight placed to channel the diffuse illumination from the sky into the space. This aperture is carefully aligned with the sun's path, allowing sunlight to illuminate the upper part of the aperture while preventing it from fully entering the space. Therefore, when entering the space only the diffuse light from the sky is visible. The surfaces of the space are characterised by a rough, matte texture and are finished with a cool white lime paint, to which a bit of blue has been added, to accentuate the effect of the diffused skylight.

4.1.2. Light from the sky in the scale model. The photograph capturing the daylight within the scale model conveys the experience of the gentle, diffuse light from the sky emanating from the funnel. The space appears bright, with subtle variations in shade without stark contrasts. The variations in levels of luminosity result from the intersecting surfaces of the ceiling, each emitting its distinct shade of light. The shadows within the scale model are perceived as gentle, diffuse shadows resulting in a wide variety of shadow effects in the space. The inclined surfaces of the ceiling create the darkest shadows in the space. The shadows accentuate the geometric forms, the intersecting surfaces that shape the funnel, and the diverse illuminated areas in the space. The scale model's wood-pulp surface forms a bright, matt surface which creates a light uniformly reflected from all surfaces. The material in the model is not representative, and it is therefore not relevant to describe the light's reflections, the glare of the surfaces, the colour, and the colour of light based on the scale model.

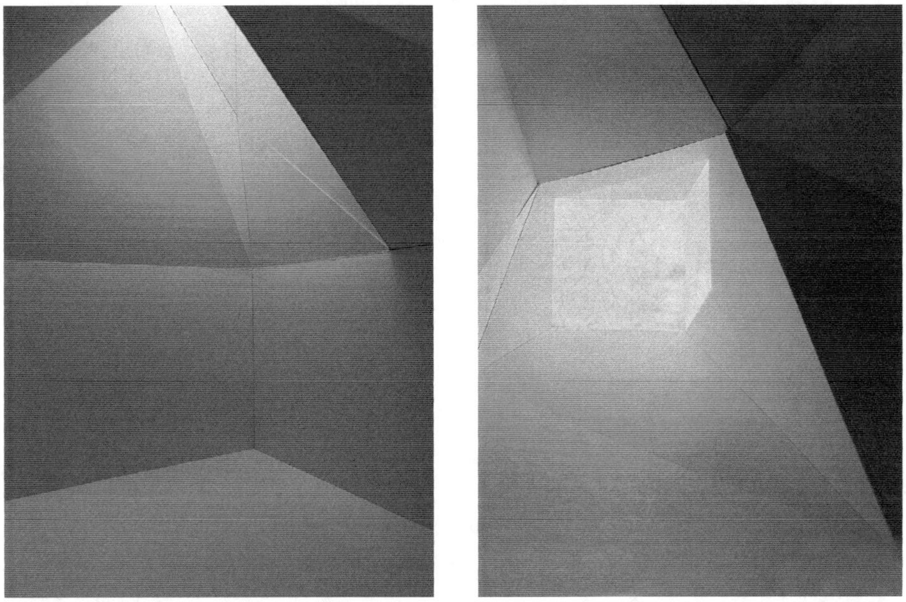

Figure 2. Photos of a scalemodel in an artificial daylight laboratory, 30th of July at 15.00

Figure 3. Photos of the 1:1 pavilion in Copenhagen, 2nd of July at 14.00.

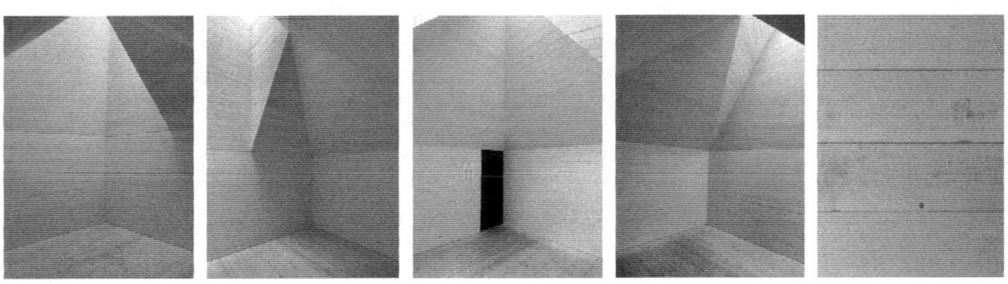

Figure 4. Photos of the space in the 1:1 pavilion in Copenhagen, 7th of July at 13.00.

4.1.3. Light from the sky and how it is perceived. When entering the space, the amount of daylight is overwhelming. The daylight is cold, diffuse and evenly distributed. If the sky goes from blue to overcast when experienced in the space, it transforms the space because the colour becomes colder. When observing the surfaces closely, all the small flecks appear and create their own shadows. This is where one becomes aware of the fact that daylight does not fit into any scale [1]. The shadows are created below every small irregularity in the space. When observing the space for a whole day at the end of June, the funnel will be filled with sunlight, and because of the geometry of the skylight the sun will move round, dancing on the edge of it without crossing the lines, as the lines of the funnel are defined by the path of the sun.

4.2. Sunlight

All daylight originates from the sun. Sunlight is the most intense source of daylight and may be compared to a point-shaped light source that provides directed and warm light, which is why sunlight casts precise and distinct shadows with large contrasts. As the sun moves across the sky, it serves as an indicator of the time of day and of the season, thus contributing to our perception of a space in motion and allowing us to experience significant variations in the daylight in a space throughout the day.

If a daylight opening faces the sun, sunlight can move through the space in the course of the day. Sunlight is dynamic as its position on the sky varies throughout the day and because it creates contrasts between lighted and shadowed areas. The orientation, size, and design of an opening determine how sunlight enters a space.

4.2.1. Sunlight in the pavilion. The sunlit space in the pavilion showcases the interplay between a southeast-facing light aperture and the warm, golden wooden surfaces of the space. In the space for sunlight, the surface is untreated pinewood, so the warmth of the wood enhances the warmth of the sunlight. The profiled boards add texture to the surface. The grooved texture of the pine boards emphasise the shadows in the space and how they change as the sun moves throughout the day. The profiles facing the space are sharp, but rounded at the bottom, thus creating a graduating shadow. The boards are placed vertically so that they create a pattern between the profiles and the joints of the boards which are narrower and create a darker shadow. The cladding in the space is uniform in order to create the feeling that the space is a complete and coherent space. The opening from the space to the corridor occupies the full width of the space to allow the daylight to flow to the darker corridor. The warmth of the sunlight and the untreated pinewood boards warms up the dark space. This changes the perception of the corridor and creates a sense of anticipation of the space to follow.

4.2.2. Sunlight in the scale model. The space is experienced as a bright space characterised by big contrasts between the sunlight in the area by the light opening and the rest of the room. The surface above the opening shows the graduated lighting effect caused by the tilting of the surface. The shadows occur in the corners of the space where surfaces meet. The material of the model is not representative, and it is therefore not relevant to describe the light's reflections, the glare of the surfaces, the colour, and the colour of light based on the scale model.

4.2.3. Sunlight and how it is perceived. The space has a relatively low opening; it is as wide as the entire space, but only 1.7 m high. This, initially, is an invitation for the visitors to sit down on stools placed in the space to let them experience the warm pleasure of occupying the space created by the sunlight in front of the window. The sunlight entering the space will change according to the time of day and year, and therefore the stools can be moved to where the sun is. The profiled cladding catches the daylight and reflects the sunlight further into the space, creating a brighter space. Also, by reflecting the daylight further into the space the daylight can reach all the way to the dark corridor.

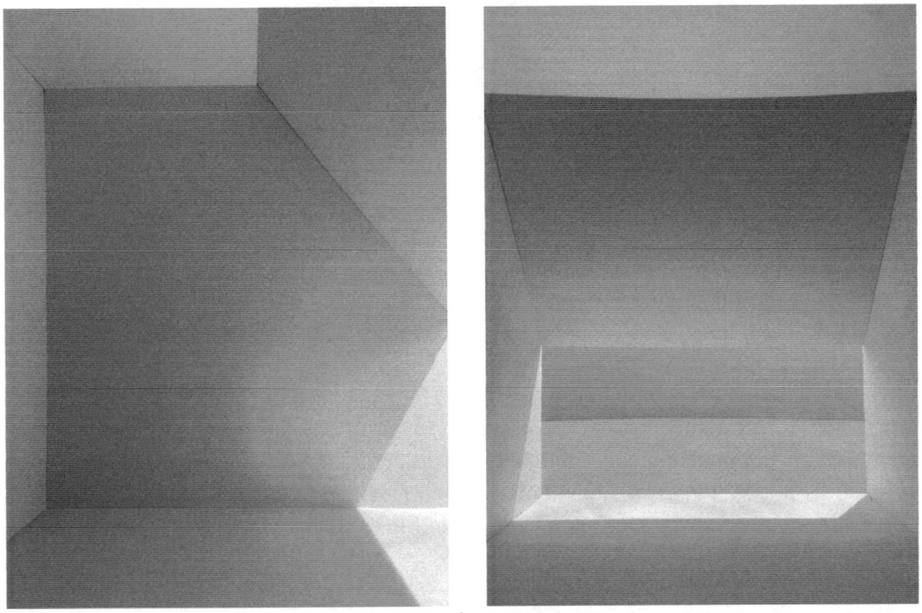

Figure 5. Photos of a scalemodel in an artificial daylight laboratory, 30th of July at 12.00.

Figure 6. Photos of the 1:1 pavilion in Copenhagen, 7 th of July at 13.00.

Figure 7. Photos of the space in the 1:1 pavilion in Copenhagen, 7th of July at 13.00.

4.3. Reflected light
Materials can absorb, transmit, and reflect light. This means that it is not just the daylight itself but also the interaction between daylight and materials that creates the daylight that surrounds us. Daylight interacts with materials which causes reflection, and the brighter the material, the larger the amount of daylight reflected.

4.3.1. Reflected daylight in the pavilion. The third space of the pavilion is dedicated to reflected daylight and is fully exposed to the sky. The concept of the space draws inspiration from the small narrow courtyards found in the old parts of Copenhagen, where the glossy windows can reflect the daylight in the courtyard. Positioned beneath a grand, old sycamore tree, the space is characterised by the tree's branches and leaves, which act as a filter for the daylight while shielding the space from excessive heat during the summer months. When the leaves fall off the tree, the sun will enter and warm up the space in the cold winter. The surfaces in the space have a white, glossy finish. Glossy surfaces possess the ability to reflect and channel the rays of daylight in a particular direction, as they maintain a consistent angle of incidence and reflection. The glossy surfaces are perceived as both shiny and shimmering, changing in intensity based on our viewpoint. This showcases that the surfaces of our surroundings significantly influence our perception of daylight and contribute to the shaping of our luminous environment.

4.3.2. Reflected daylight in the scale model. As the space is fully open towards the sky, it is bright and well illuminated. Shadows are big and full of contrast, but there are not many shadows, only from the edge of the space. As in the other spaces the material used for the model is not representative, and it is therefore not relevant to describe the light's reflections, the glare of the surfaces, the colour, and the colour of light based on the scale model.

4.3.3. Reflected daylight and how it is perceived. When walking from the space for sunlight to the darker corridor the eye adjusts to the darkness, and when turning around the corner the daylight from the third space lights up the corridor, and already here the light is perceived as overwhelming. When entering the space for reflected light it is virtually impossible to keep one's eyes open. The light is reflected from all surfaces and is perceived as coming from all sides, not just from above. When the sky changes in the space, from overcast to clear blue, all the different shadows appear on the surfaces. The leaves cast shadows that move in the wind and the light is perceived as being even more powerful when the sun shines. This is evident in figures 8 and 9 in the comparison of the space in the scale model, without the context and the tree, and the picture of the actual space in the pavilion. The geometry of the space is simple; the surfaces are smooth and even, showing just the holes of the natural knots in the wood, so that the changes in the light are what changes and creates the space. The space holds the light, reflecting it all around.

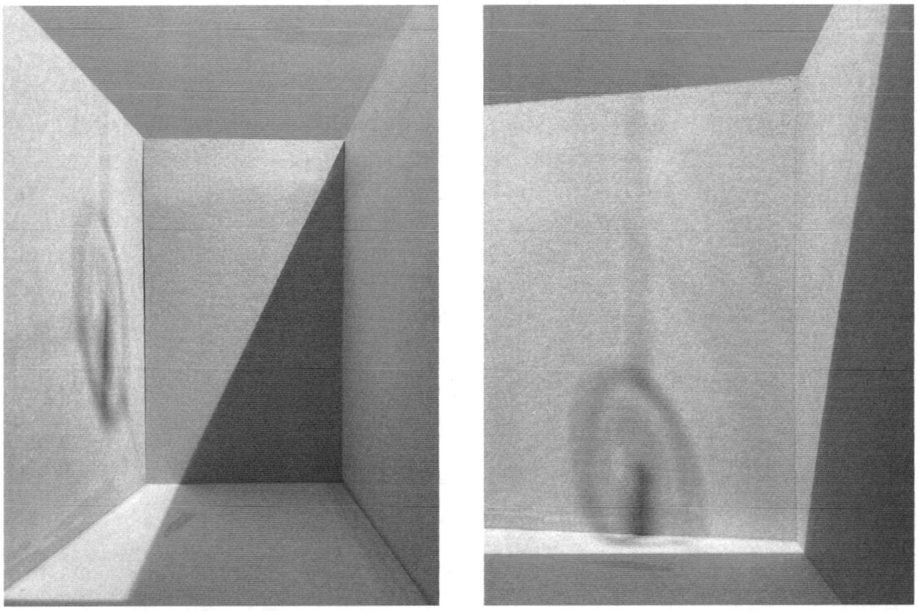

Figure 8. Photos of a scalemodel in an artificial daylight laboratory, 30th of July at 12.00.

Figure 9. Photos of the 1:1 pavilion in Copenhagen, 7th of July at 13.00.

Figure 10. Photos of the room in the 1:1 pavilion in Copenhagen, 7th of July at 13.00.

5. Comparison: Three spaces – three different appearances of daylight

5.1. Space for light from the sky

The photographs of the space for light from the sky in the scale model show that the space generally has a high level of brightness with contrasts between the opening and the ceiling, but otherwise it is a soft and diffuse daylight. This is also evident in the photograph of the space in the pavilion. A bright room with variations in the intensity of daylight. This is due to the geometry which is the same in the two photographs. Therefore, the large interior shadows are the same in the two photographs. Though the only shadows in the scale model stem from the geometry, the texture of the material casts no significant shadows, and the scale model has no representative surface either in terms of colour or texture. When observing the photograph from the scale model the colour of the light is warm due to the warmth of the material from which the scale model is made. The material has impact on the colour of the light and therefore what we perceive in the photographs of the pavilion has a significantly different effect. In the pavilion both the colour and the interior surfaces are different from the scale model. In the pavilion the wooden boards create horizontal shadows where they are joined, and they have a rough, granulated surface whose texture casts shadows; moreover, the surface is painted in a slightly blue-toned colour which enhances the coldness of the light from the sky. Therefore, we cannot, in the model, perceive the colours of the light or the shadow cast by texture.

The scale model can therefore be used as an effective and precise analogue tool to indicate the large interior shadows, the level of lightness, and the variation of daylight in the pavilion, but the details of the materials used in the pavilion add to the perception of the daylight.

5.2. Space for sunlight

As in the case of the space for light from the sky, the space for sunlight in the scale model is experienced as a bright space. The wide opening in the facade creates large contrasts. The light is perceived as having a warm tint due to the material of the scale model, but as mentioned earlier the material casts no significant shadows. In the space in the pavilion the wooden surface is also warm but has a different colour because of the untreated wooden boards. Furthermore, the surface is profiled and therefore the light is distributed further into the space than in the scale model. Also, both the profiles and the joints of the wooden boards create shadows while also creating brighter surfaces in the profiles which are not present in the scale model.

5.3. Space for reflected light

In the space for reflected light the difference also lies in the surface. The white, glossy paint highlights the effect of the reflected light, which is not the case in the scale model. Also, the context has an impact on the light in the space, as the light filtered through the foliage of the tree in the square influences the way the daylight is perceived. The scale model illustrates knowledge of the large interior shadows, but otherwise the space is simple in terms of materials, and the shadows are only created by the geometry. The photographs of the space in the pavilion make you aware that, in this context, the sycamore tree creates another world of light and shadows. This is what makes the difference from observing the scale model under controlled conditions in the daylight lab, where the circumstances of daylight are fully controlled. In the pavilion we cannot control the weather, nor can we prevent the sun from disappearing behind a cloud leaving the room in a totally different light.

6. Conclusion: Poetic daylight

With the Vitruvian triad, i.e. *firmitas*, *utilitas* and *venustas*, in mind, the design for the pavilion has focused on *venustas* (beauty or delight) in order to form spaces where the atmosphere and the beauty of daylight can be perceived and experienced. The study of the daylight characteristics of the three different spaces in the pavilion are based on Liljefors' seven variables for describing light quality. And the characteristics of the shadows are based on Frandsen's definition of four different types of shadow. The Vitruvian quality of *beauty* in conjunction with the definition from J. J. Gibson of the perception of an environment as well as the theory of perceived light by Liljefors and the definitions of shadow characters by Frandsen make up the theoretical framework for this paper.

Traditionally, architects' tools include scale models and photographs of scale models, and it is therefore relevant to compare them with the 1:1 space of the pavilion in regard to the focus of this study: the beauty of the daylit space. By observing the qualities of daylight in, respectively, the photographs of the 1:10 scale models and the built pavilion these comparative studies demonstrate that it is the large spatial shadows and attached shadows, in particular, that are most visible in the scale models, whereas the small attached shadows and those cast by details and textures are invisible. These results align with the experience of daylight as not fitting into a scale, whereas the perceived quality of daylight fits a scale [1]. This is one of the challenges when attempting to predict the beauty and the qualities of perceived daylight in architecture. Acknowledging this, it is still very useful to include scale models in the design phase to inform qualitative aspects of daylight, since this study illustrates that the photos of the scale models can give tangible and transferable knowledge that will enrich the experience and perceived atmosphere of the daylight in the 1:1 built pavilion. The study explicates and illustrates that it is possible to make design decisions regarding the qualities of daylight based on scale models.

Acknowledgements
This project was funded by Dreyers Fond. The authors would like to thank research assistant, architect Yasmin Kokseby and the design team: Claus Pryds Architects, Holst engineering, Micki Aaen Engineering, Camilla Hornemann colour specialist, and FRYD carpenter, in the process of constructing the Poetic Daylight pavilion.

References
[1] Grønlund L, Mathiasen N, Frandsen A K and Sørensen P 2023 *Envisioning Architectural Scales in the Analogue and Virtual Representation of Architecture: Proceedings of the 16th EAEA conference (Copenhagen)* Architectural scale models: Methodologies for studying daylight qualities ed A Kreutzberg (Copenhagen: Royal Danish Academy - Architecture, Design, Conservation)
[2] Mathiasen N 2015 Nordisk lys og dets relation til dagslysåbninger i nordisk arkitektur *PhD Thesis* Det Kongelige Danske Kunstakademis Skoler for Arkitektur, Design og Konservering, Copenhagen
[3] Johnsen K *et al* 2008 *Dagslys i rum og bygninger* (Hørsholm: Statens Byggeforskningsinstitut)
[4] Christoffersen J *et al* 2002 *Beregning af dagslys i bygninger* (Hørsholm: Statens Byggeforskningsinstitut)
[5] Isager J 2016 *Vitruv Om Arkitektur* (Odense: Syddansk Universitetsforlag)
[6] Vitruvius M 1960 *The Ten Books of Architecture* (New York: Dover Publications)
[7] Evers B *et al* 2011 *Architectural Theory From The Renaissance to the Present* (Cologne: Taschen)
[8] Tøgersen U 2010 *Krop og fæmonologi. En introduktion til Maurice Merleau-Pontys filosofi* (Copenhagen: Academica)
[9] Gibson J J 2015 *The Ecological Approach to Visual Perception* classic edn (New York, Hove, England: Psychology Press)
[10] Liljefors A 1997 *Seende och Ljusstrålning* (Stockholm, Sweden: Royal Institute of Technology)
[11] Frandsen S 1989 *2nd European Conference on Architecture (Paris)* The Scale of Light – A New Concept and Its Application (Dordrecht: Kluwer Academic Publishers)
[12] Popovic O 2020 *Physical Modelling for Architecture and Building Design: A design practice tool* (London: ICE Publishing)
[13] Christensen E 1976 *Sol på akademiet* (Copenhagen: Arkitekten)
[14] Egan M and Olgyay V 2001 *Architectural Lighting* 2nd edn (USA: McGraw-Hill Science/Engineering/Math. Print)
[15] Grønlund L 2017 Lysvirkninger – fremkaldelser af rummets arkitektur *PhD Thesis* Aarhus Aarhus School of Architecture

IOP Conf. Series: Earth and Environmental Science 1320 (2024) 012007 doi:10.1088/1755-1315/1320/1/012007

Uniform vs. uneven lighting: Accommodation of multiple users' preferences in study rooms

Evdoxia Iro Gkolompia

Division of Lighting Design, Department of Architecture, KTH Royal Institute of Technology, 100 44 Stockholm, Sweden

evdoxiagolobia@gmail.com

Abstract. Uniform lighting in public spaces like study rooms in libraries, is still a common practice. However, uniformly lit spaces are unable to simultaneously meet the needs of multiple users and can create dull environments. The objective of this research is to analyze users' behavior under different lighting conditions in study rooms and investigate whether alternative lighting designs, which step away from uniformity, could be more beneficial than uniform lighting, from a human performance and well-being perspective. A uniformly lit study room in the library of KTH was chosen as the main case study of this investigation and was thoroughly analyzed and evaluated through qualitative and quantitative research. Under the initial hypothesis that the lighting in this study room does not meet its users' needs and fails to create an inspiring and comfortable environment, three additional study rooms in different Swedish libraries were comparatively analyzed through observations, interviews, and questionnaires, so as to better understand individual users' needs and lighting preferences in study rooms. The research revealed the existence of various types of users with different personal preferences regarding lighting in order to feel comfortable and alert. A new conceptual lighting proposal is presented for the study room in KTH library, following an uneven lighting approach that utilizes different lighting zones to accommodate individual needs and affect the sense of comfort, alertness, and ultimately, the well-being of multiple types of users.

1. Introduction and background

The quality of lighting depends on the interaction between lighting, place, and person [1]. Architecture can serve as a canvas to showcase the complexity of natural light [2]. Light has the power to highlight architectural features, promote historical and aesthetic qualities, and elicit emotional responses. The satisfaction and preference of a space also involve an emotional component, influencing how it makes viewers feel [3]. Some spaces are perceived as more inspiring than others, and light plays a significant role in creating such an atmosphere. Different lighting typologies and distributions can lead to varying perceptions and emotional responses within a space. For instance, uniform lighting can create a sense of detachment in space as well as less tension [4] compared to non-uniform lighting. Studies have shown that spaciousness can be enhanced with uniform lighting, while non-uniform and peripheral lighting can reinforce the experience of privacy [4, 5].

Nowadays, it remains common to use artificial lighting to create even, uniform and sometimes characterless light effect in space. Standards promote this as well. This is probably a "relic" of the older types of lighting, like fluorescent lights, that had a typology and an illuminance output that supported bright and uniform lighting atmospheres. Although artificial lighting was invented to support visibility, during the last few years, more and more examples of "recommended illuminance levels being far in excess of levels required to satisfy visual performance needs" [6] are appearing. The norms and recommendations for illuminance levels, as well as common practice, may need to be reevaluated since in many cases they "exceed what is necessary for effective visibility" [7]. Moreover, research has shown that rooms with non-uniform luminance distribution require less light to achieve equivalent brightness than those with uniform distribution, suggesting potential energy savings through lighting design [8].

Uniform lighting, although providing consistent illumination and accurate color representation, can feel static and unnatural, lacking variation and scale of importance [9]. Pallasmaa points out that homogenous bright light can diminish the imagination and weaken the sense of place, affecting mental well-being and feeling of privacy [10]. In contrast, non-uniform and varied lighting is common in natural environments, where sunlight produces dynamic patterns of illumination, color, and intensity. Humans are naturally connected to nature and have a collective memory of the effects of natural light. The preference for seeing without being seen and spatial arrangements that support movement and exploration can be attributed to this connection [4, 11, 12].

Throughout history, electric lighting has been associated with happiness, knowledge, cleanliness, and health. Light symbolizes truth, wisdom, and progress, while darkness is often linked to ignorance, oppression, and lack of control. Moreover, the transition between light and dark is associated with imagination, intimacy, and the supernatural [13]. All these symbolic meanings, combined with the physiological effects of lighting, make the interplay between light and dark a fascinating subject.

Contrast and dark zones play an important role in highlighting the unique characteristics of architectural spaces, supporting visibility, and aiding orientation. Contrast is fundamental to vision and visual quality [14], while shadows and darkness create ambiguity, inviting peripheral vision and tactile fantasy [10]. Focusing on light at vertical surfaces, brightness contrasts, and departure from uniformity can enhance spatial experiences and orientation [15].

However, contemporary office spaces and educational environments often employ mostly uniform lighting, which might be counterproductive to individual well-being, productivity, and cost-effectiveness. Exploring alternative lighting designs that move away from uniformity could be helpful for fulfilling the vision of creating sustainable and comfortable lighting environments that cater to various needs. By understanding the interplay between light and dark and embracing the natural variations found in nature, architectural spaces can be enhanced, promoting beauty, comfort, and inspiration.

This investigation aims to shed light on the benefits of varied lighting options that can support individual well-being, productivity, and energy savings in contemporary educational spaces like libraries. The author's close connection with educational and library environments, as a recent student, influenced the choice to focus on library environments and study rooms. The hypothesis is that uniform lighting is inadequate to create sustainable and comfortable lighting environments that accommodate diverse needs. To explore this hypothesis, a uniformly lit study room in the KTH library in Stockholm will be thoroughly analyzed and evaluated. A new conceptual lighting proposal will be presented to correspond with the lighting preferences, visual comfort, and circadian rhythms of the space's diverse user types.

1.1. Research question

Main question: Is uniform or uneven lighting better for study rooms in libraries, in terms of well-being, concentration, alertness and comfort?

Secondary questions: What is adequate light? Why do we overlight interior spaces? Why do we avoid shadows, dark zones? What do users need to feel comfortable and alert?

1.2. Sustainable statement

In order to promote sustainable practices, it is essential to prioritize the well-being of users and be aware that there is not one specific, correct way to study. In order to design quality study environments, the consideration of the multiple needs and preferences of users is crucial. Using as much natural light as possible and lighting control systems that follow the circadian rhythm and provide artificial lighting when and where is needed, can lead to sustainable designs that better support physical and mental health (correlation with UN Sustainable Development Goals 3 and 4 [16]). Implementing varied, energy-efficient illumination with motion sensors in study rooms, using eco-friendly, recyclable materials for fixtures and setting illuminance thresholds responsibly after correct evaluation of standards (correlating with UN SDG 7) can lead to a more sustainable future.

2. Method and procedure

The study focused on a study space in the ground floor of the northern gallery in KTH library (Figure 1) to explore the possible negative aspects of uniform lighting. The space appeared uninteresting and had fewer visitors than other parts of the library. In order to better evaluate the chosen case study area, three additional study areas in Swedish libraries were visited and evaluated: the annex in Kungliga biblioteket, the southern wing of Stockholms Stadsbibliotek and the study hall at Ekonomikum, in Uppsala University. Personal observations and questionnaires to users were used to gather information concerning user preferences and the relationship between light distribution and comfort, alertness, and concentration. The research aimed to provide data for a new lighting proposal for the main study area in KTH library, which was assumed to be less advantageous than other study spaces (Figure 2). Both qualitative and quantitative research in the four study areas were expected to validate this assumption.

2.1. Areas of analysis

Main Case Study Area - KTH Library:
The KTH library was renovated in 2002 by Ahrbom & Partners and was originally a machine laboratory designed by Erik Lallerstedt in 1917. The main study area of analysis on the ground floor of the northern gallery in KTH library, formerly a water gutter, can now accommodate around 30 seated people and is adjacent to the library's emblematic main hall, which receives ample daylight through skylights.

Figure 1. The KTH library study room. Left: Ground floor plan of KTH Library (the analyzed area is marked), middle and right: existing situation, plan and lighting fixtures of the study room.

Secondary Case Study Areas (see Table 1):
Kungliga biblioteket's annex, designed by Murman Arkitekter and completed in 1997, is a glass-covered space with study desks around a staircase that receives abundant daylight. The southern wing of Stockholms Stadsbibliotek's historical building is a double-height room with desks for up to 10 users each. High windows provide direct and diffused daylight. Finally, the Ekonomikum study hall in Uppsala University, created in 2019 by Sweco Architects, aims to provide a peaceful and dimmed space with carefully placed light sources to accommodate diverse user preferences and promote concentration.

2.2. *Qualitative study and participants*
Four different study rooms in four Swedish libraries were visited and evaluated, first by the author and then by available users of the spaces. The evaluation involved observations, Visual and Physical lighting theory [14], questionnaires, and small interviews. The author's subjective evaluation included notes on available fixtures, light distributions, adjustability of the fixtures, connection with daylight, and general behavior of the users. The main topics of the questions asked to the users were visual comfort, concentration, and general lighting preferences between bright/uniform and dimmed/uneven light scenarios. The participants in the questionnaires were users of space present at the moment, mostly students aged between 18 and 35 that voluntarily agreed to be engaged in a conversation and anonymously be part of the research. In Kungliga biblioteket, Stockholms stadsbibliotek and Uppsala Ekonomikum, 8 users of each study space answered a questionnaire and evaluated the lighting conditions of the space through V/P Lighting theory [14]. 5 males and 3 females in Kungliga biblioteket, 3 males and 5 females in Stockholms stadsbibliotek and 3 males and 5 females in Uppsala Ekonomikum. Additionally, 31 people, 16 males and 15 females answered a questionnaire and evaluated the lighting conditions of the study room on the ground floor of northern gallery in KTH library, which is the study space that will be further analyzed in order to develop an improved lighting scenario. The evaluation took place on four different sunny dates in May at around 2:00 PM each day. Finally, 4 users of KTH library study room, 2 males and 2 females, were extensively interviewed about their subjective impressions concerning the study area.

2.3. *Quantitative study*
The quantitative study was conducted solely in the main case study area of the KTH library. Illuminance was measured on a formed grid in the space, with an illuminance meter placed at desk height levels (0.72m) during daytime (1:00 PM) to assess the overall uniformity (Figure 8). Additionally, luminance was checked on the desks, floor, and ceiling using the FusionOptix app and a luminance meter during daytime (1:45 PM). Finally, CCT and flicker from existing fixtures were measured using a spectrometer.

2.4. *Limitations*
The three secondary case study areas in Kungliga biblioteket, Stockholms Stadsbibliotek, and Uppsala Ekonomikum were not analyzed in the same level of detail as the main case study area in the KTH library, resulting in an uneven comparison. Fewer users participated in the questionnaire and interviews and less formal analysis of the lighting conditions was conducted.

3. Results
Through observations, the four analyzed study rooms were categorized into four different types based on their connection with daylight (skylight type, embracing type, uniform and neutral type, introverted and concentric zoning type, see Table 1). The qualitative research conducted in these four types of spaces revealed that different lighting conditions, with varied connections to daylight, attract different types of users with different needs (Figures 3-4). Users appreciate the lighting conditions and light distribution individually in the study areas they choose to work in, and they can be classified into different types based on their preferences for a comfortable and inspiring environment that helps them concentrate and stay alert while studying (Table 2). Evaluation of the case study rooms by their users indicates that they are mostly content with the lighting conditions. Natural daylight is highly appreciated by all.

Among the users, those using the study room in the KTH library were the most dissatisfied with the lighting. They commented that the space feels oldish, uninspiring, dimmed, small and oppressive due to the low ceiling, and it lacks diversity or smaller private zones (Figure 7). Consequently, it is not their preferred choice compared to other rooms in the KTH library. However, it is frequently chosen as a seating area simply because it is rarely crowded.

The quantitative study in the KTH library's study room showed that the desks have lower horizontal illuminance than what is recommended by standards.

Finally, the conceptual design proposal presented for the study room in KTH library, which is the least pleasant and appreciated by its users among the four analyzed library areas, aims to incorporate multiple and adjustable lighting conditions in order to affect positively the users of space and create a varied and uneven light atmosphere that simultaneously accommodates the individual needs of multiple users in terms of concentration, alertness and comfort.

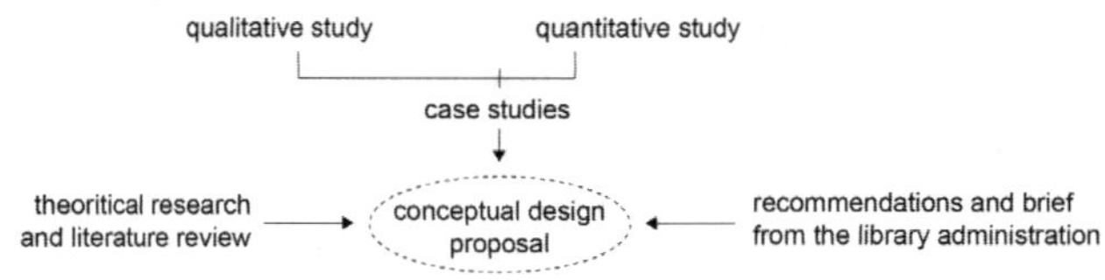

Figure 2. Factors that affected the formation of the design proposal.

Table 1. Observation of space and lighting conditions in the four case study areas.

Kungliga biblioteket (northern annex with skylight)	**Stockholms Stadsbiblioteket** (southern wing)	**KTH Library** (northern gallery ground floor)	**Uppsala Ekonomikum** (lower floor - level 1 study room)
the skylight type	*the peripheral lighting and "embracing" type*	*the uniform and neutral type*	*the introverted and concentric zoning type*
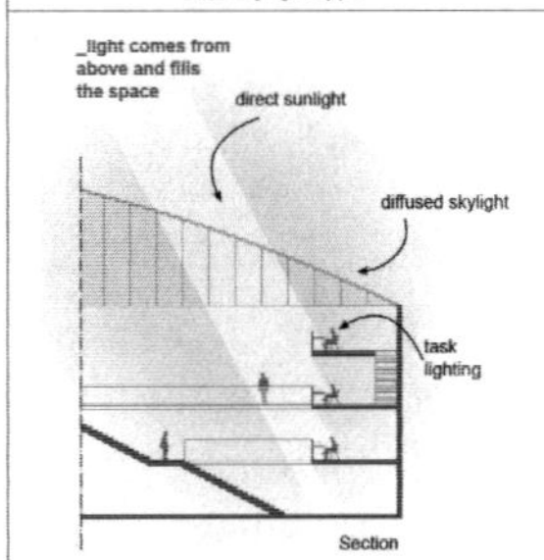	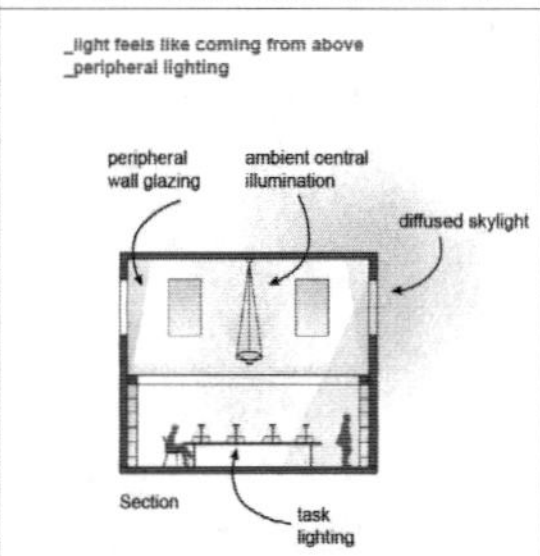	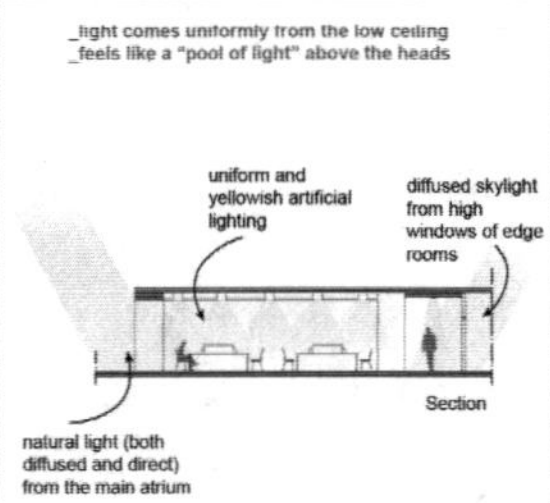	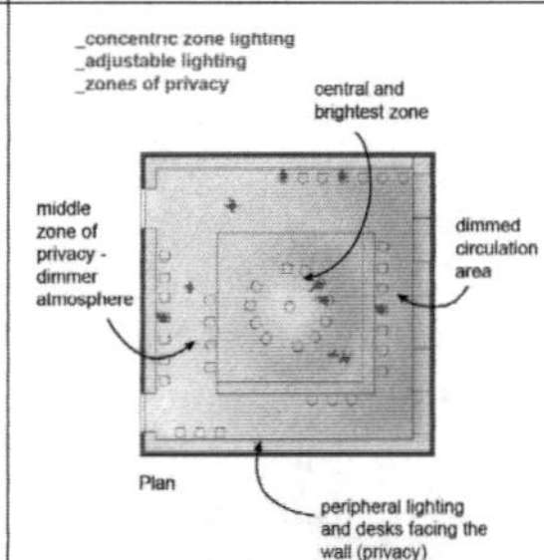
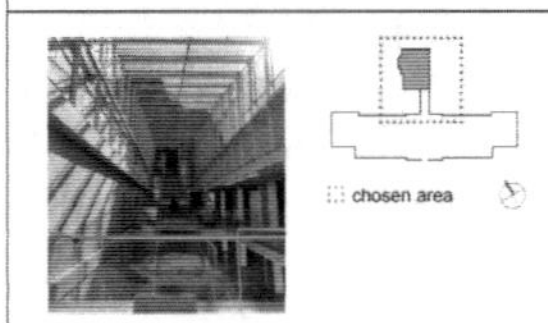	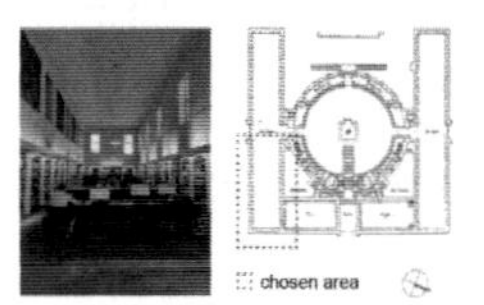	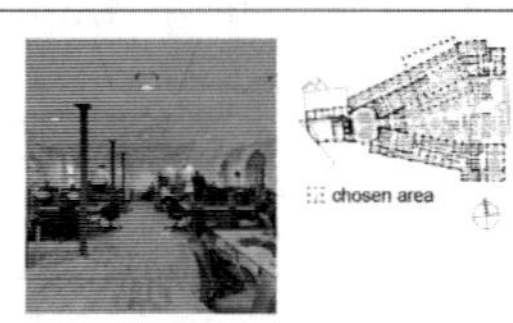	

3.1. *Types of users and evaluation of case study areas*

Table 2. Analysis of six identified users' types with different preferences in lighting conditions.

Case study area	Type of user	needs	comments	
Kungliga biblioteket (*the skylight type*)	type of user a'	needs bright and cold light to concentrate; likes dynamic lighting; likes to socialize in the study area	*"when there is natural light the place feels welcoming, energizing and spacious" "sometimes not so much privacy" "I am awake and focused with bright light"*	
Stockholms Stadsbiblioteket (*the peripheral lighting and "embracing" type*)	type of user b'	needs not too bright, not too dimmed light; likes light to "feel natural" and "homey"; appreciates task lighting	*"I want to see clearly" "It's better to study and concentrate when I feel accompanied" "I feel relaxed to study when the light reminds me of home"*	
KTH Library (*the uniform and neutral type*)	type of user c'	needs cold light to concentrate and be productive; appreciates natural daylight; likes bright environments	*"A bright space feels more spacious" "Bright and uniform light helps me structure my thoughts" "Dim lighting makes me feel claustrophobic"*	
	type of user d'	needs warm light to feel calm and concentrated; appreciates natural daylight; appreciates task lighting	*"I would like reflections of trees or sky to feel calmer" "I dislike the office and hospitalish lighting type" "Dim light and work on laptop is easy" "Dim light doesn't create stress"*	
Uppsala Ekonomikum (*the introverted and concentric zoning type*)	type of user e'	needs warm and focused light to study and concentrate; appreciates adjustable lighting; appreciates privacy when studying	*"I like I can choose between an area with more light or less light" "Darker lighting conditions help me focus" "Warm light is easier for the eyes" "I prefer cozy areas when studying"*	
	type of user f'	sometimes prefers no light at all (working in laptop); appreciates adjustable and warm lighting; appreciates privacy when studying	*"I like I can choose whether I want light or no in a public study room when working in laptop" "Different areas with different light conditions or intensities in the same room is great"*	

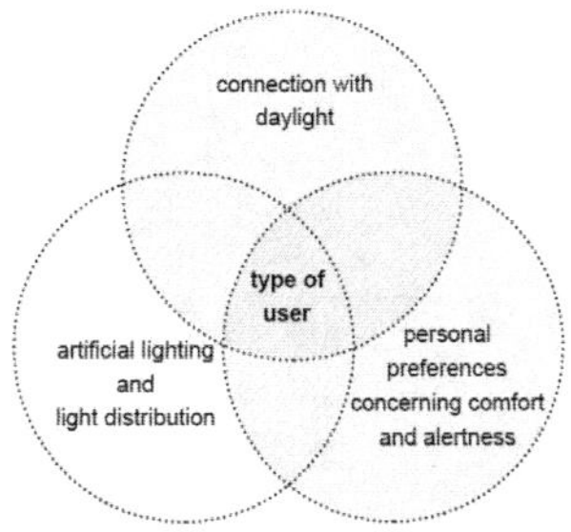

Figure 3. Factors that affect the formation of a type of user in study rooms.

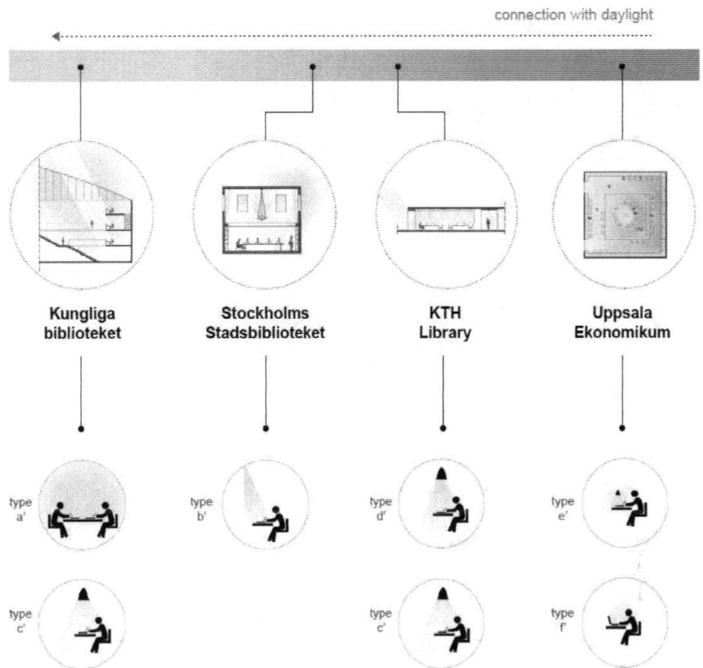

Figure 4. Connection of the four different study rooms with daylight and observation of how this affects the formation of different types of users with different needs concerning lighting.

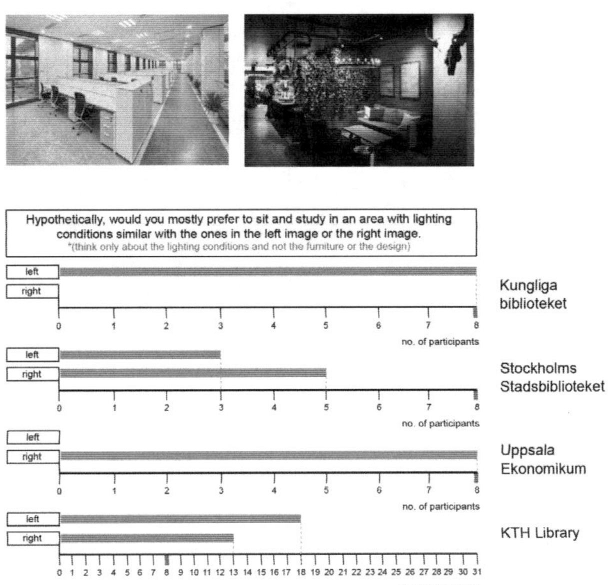

Figure 5. Results derived from questionnaires answered by users of the study spaces in the four different case study libraries. Preference between uniform - bright - cold and varied - dimmed - warm lighting atmospheres in study environments.

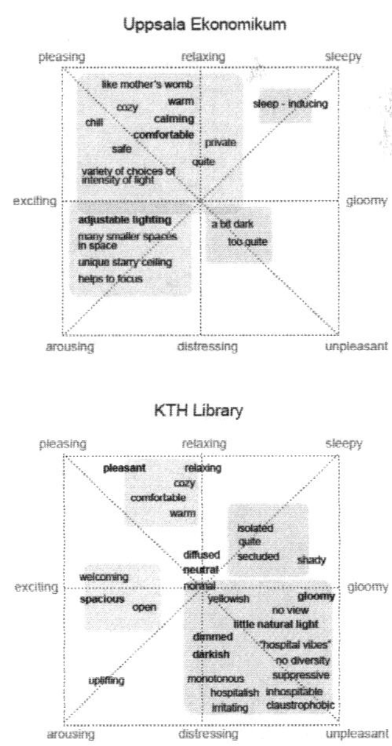

Figure 6. Diagrams with users' opinion about the luminous atmospheric quality in Uppsala Ekonomikum's and in KTH library's study space. Green: positive comments, grey: negative comments.

3.2. Quantitative study in KTH library's study room

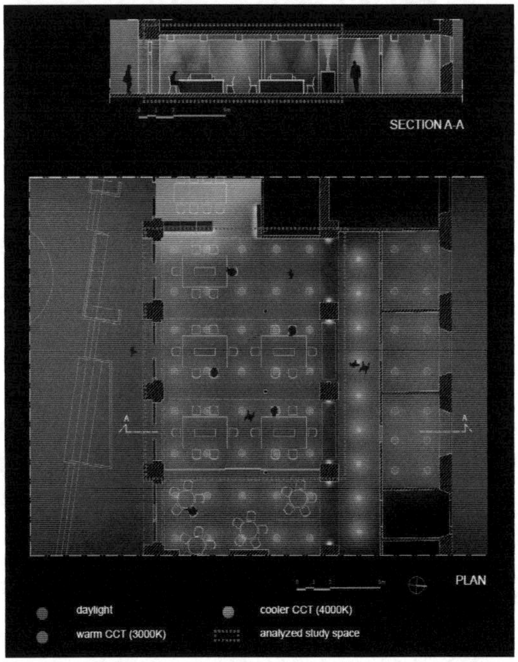

Figure 7. Section and plan showing the perceived uniform light distribution.

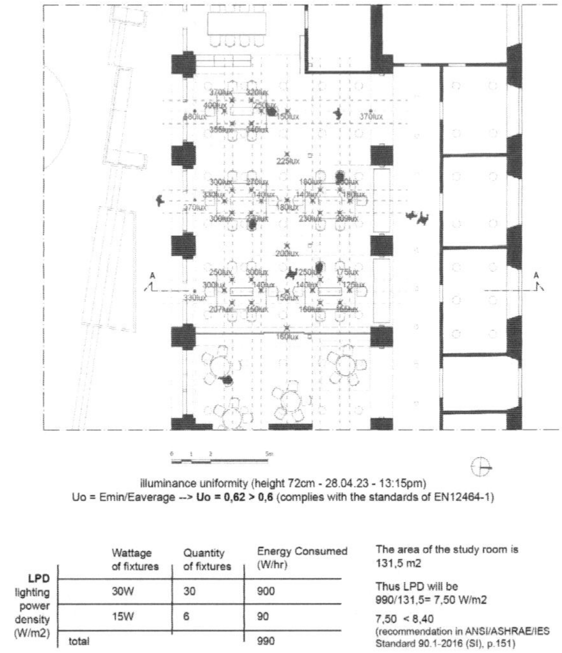

Figure 8. Horizontal illuminance, uniformity and lighting power density measurements [17].

3.3. Literature Review

Three articles related to: a) the effect of light distribution on comfort, concentration, calmness, and energy efficiency [18], b) the comparison between adjustable and static lighting and their impact on human behavior and mood [19], and c) the contribution of light distribution and contrast to the perception of space [20], were analyzed as references for the conceptual design proposal (Table 3).

Table 3. Indication of the main findings through experiments, in the articles of the literature review.

article	van Mil, I. W. et al. 2018. *Light at eye level is a means to create energy savings and space for learning, focus and concentration*	Wessolowski, N., et al. 2014. *The effect of variable light on the fidgetiness and social behavior of pupils in school.*	Wänström Lindh, U. and Billger, M. 2021. *Light Distribution and Perceived Spaciousness: Light Patterns in Scale Models.*
findings through experiment	_"improved performance and reduced noise level when working in light from pendants (non-uniform lighting) compared to evenly distributed light from the ceiling" (p.82). _non-uniform lighting like pendants, can lead to energy savings of up to 68%. _non-uniform lighting leads to a variation in lighting conditions within the same learning space.	_dimmed, warm-white light in the classroom can directly lead to a decrease in fidgetiness and an enhancement in social behavior among students.	"darkness can be an enlarging factor" (p.10-13). _light can serve as a simple method to visually enlarge spaces. _"Variations in light distribution affect the visual experience more than the amount of light" (p.1).

3.4. Conceptual design proposal for the KTH library's study room

All the aforementioned results, along with some comments from the library administration, were used to develop a new conceptual lighting proposal for the KTH library's study room which aims to accommodate the needs of multiple users and introduces light zones with different light distributions and adjustable lighting (Figures 9-10, Tables 4-5).

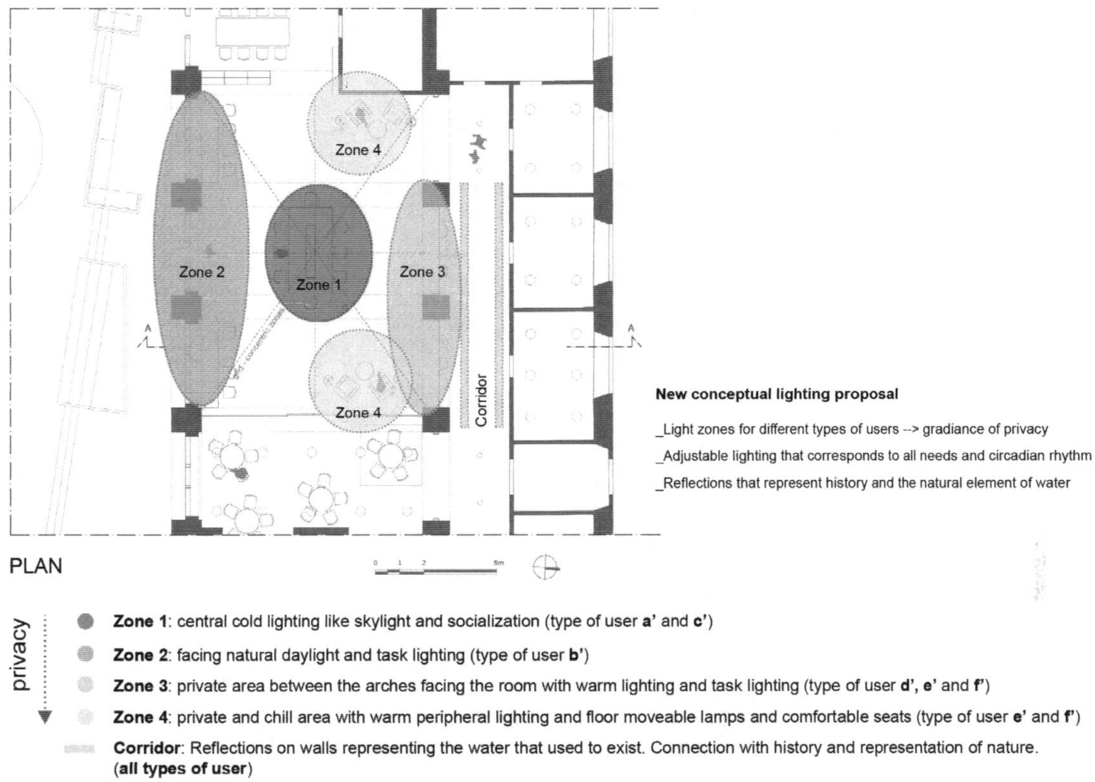

Figure 9. Plan of the new conceptual proposal with different light zones that accommodate the needs of different types of users.

Table 4. Proposed lighting scenarios for the study room in KTH library.

daytime scenario	zones on dimmed 50% - automatic dimming control with daylight sensors | cold CCT in zone 1, 5500K (artificial skylight) | water reflection in the corridor on
evening to night-time scenario	zones on 100% - automatic dimming control with daylight sensors | warmer CCT in zone 1 (3000K) with automatic control following the time of the day | peripheral recessed light in zone 4 and warm wall glazing recessed uplights in the corridor on
closing time scenario	some zones off, with motion sensors (when there is absence of people)
cleaning or event scenario	all zones on 100% | extra downlighting from the ceiling for uniform illumination (when needed), controlled by the library administration

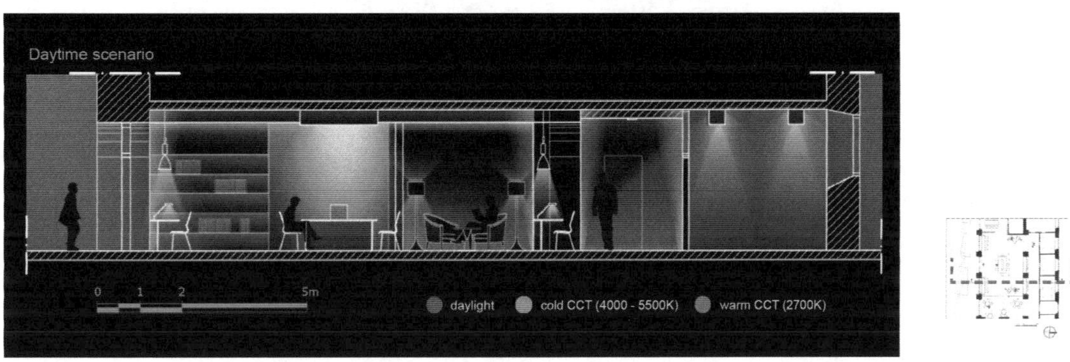

Figure 10. Section and perceived light distribution of the new conceptual proposal with different light zones that accommodate the needs of different types of users (daytime scenario).

Table 5. Representation of the lighting conditions corresponding to users' needs, in each light zone included in the conceptual design proposal.

		Daytime	Nighttime	
Zone 1. Artificial skylight. Cold light during daytime and controlled CCT to follow circadian rhythm.	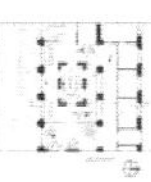			• Compatible with needs of types of user a' and c' that need bright and cold light to concentrate. • Zone with automatically controlled CCT (tunable white lighting from 5500K to 3000K) following the time of the day and circadian rhythm. • Social area. "Light seems almost to act like a glue for all the people sitting around the table" (Alexander, 1977).
Zone 2. Zone affected by natural daylight. Individual task lighting. Warm downlighting during nighttime or if needed (automatic dimming control with daylight sensors).	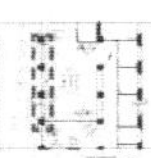	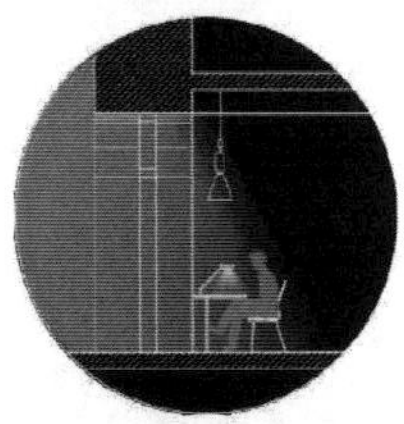		• Compatible with needs of type of user b', that feels comfortable when the light feels natural and is not to bright not too dimmed. • Daylight can reach this zone and affect the natural feeling. • Facing the main hall so it is compatible with the need of "feeling accompanied" while studying.
Zone 3. Private area between the arches. Task lighting (possibillity of no artificial lighting). Slightly affected by daylight during the day.	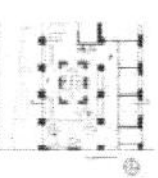		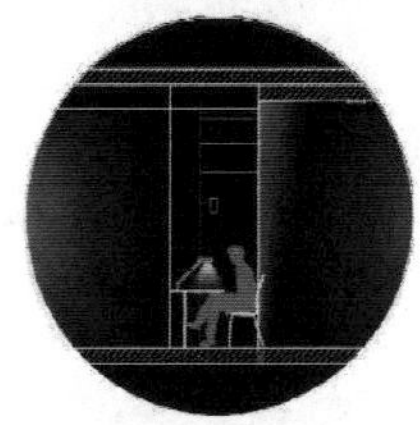	• Compatible with needs of types of user d', e' and f' that appreciate privacy, warm light and task lighting. • These types also commented appreciation of natural light, of reflections of natural elements and of dimmed lighting. • The zone is close to the corridor with water reflections, has some contact with daylight and when the task light is off it can be even more dimmed affected only by its adjacent light zones.
Zone 4. Private and chill area with comfortable seats and moveable floor lamps (that can be turned on or off).	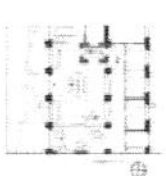			• Compatible with needs of types of user e' and f' that need privacy, warm and dimmed light to feel comfortable and a cozy environment with adjustable lighting to study concentrated. • CCT of 2700K. • The floor lamps can be turned off-on or be moved depending the needs. • Peripheral lighting on the back wall during nighttime. The experience of privacy can be reinforced with non-uniform and peripheral lighting" (Wänström et al, 2021, p.5).
Corridor Reflections of water as representation of nature and of the history of space. During night time wall glazing recessed uplights between arches.	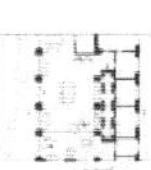	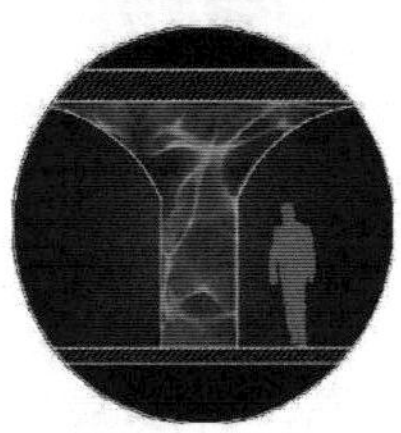		• Compatible with needs of all types of users that appreciate nature. • Representation of water will be a memory of the history of the space that used to be a water gutter. Projections and dynamic lighting help the mind to escape when needed. • Repetitive uplighting of columns highlights the corridor and its architecture.
Peripheral lighting (left) Recessed uplighting and wall glazing effect. **Circulation (right)** The circulation in between different light zones is dimmed.		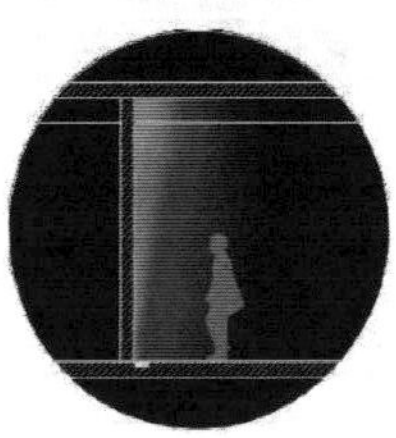 Next to wall	 Circulation	• "It is found that uplighting and wallwashing are more energy efficient than downlighting" (Cuttle, 2013, p.33). • "The function of the intermediate shadow-zones, which limits and defines the light-zones, is thus circulation" (Madsen, 2007, p.55).

The design proposal tries to incorporate the needs of all user types identified and creates divided light zones that collectively form an inspiring, comfortable, and human-centered environment that is compatible with users' circadian rhythm and well-being, addresses their need for privacy or socialization [21], establishes a connection with daylight and the perception of time, and fosters a sense of orientation within the space.

4. Discussion

The research identified different types of users with varying lighting needs in the four case study areas. The qualitative research showed that users' lighting preferences generally matched the provided lighting atmosphere in the study rooms, except for the KTH library's study room. For instance, users in the annex of Kungliga biblioteket, with abundant daylight, preferred bright environments for concentration, ranked privacy as the least important factor, and all of the participants in the questionnaire chose the bright and uniform lighting scenario. Conversely, users at Uppsala Ekonomikum, evaluated as an introverted space, preferred dimmed and warm environments with adjustable lighting and various light zones, considered privacy and comfort as top priorities, and all of them preferred the warm and varied lighting scenario in the questionnaire (Figure 5). While lighting conditions were not the primary factor influencing users' choice of study space, the extreme preferences in these two cases suggest a strong compatibility between user type and space type, indicating that users may select study areas that meet their needs or become accustomed to the provided lighting over time. However, educating users that are not familiar with the effects of illumination on their health and well-being, including the impact of correlated color temperature (CCT) on circadian rhythms, is essential.

The Uppsala Ekonomikum study room was evaluated most positively by its users, and that indicates an appreciation of the zoned lighting (Figure 6). Light zones are "(spatial) groupings of the lighting variables (intensity, direction, distribution and color)" [22] that, while interacting, can affect significantly the perception of space. Moreover, uneven and varied lighting leads to a variation in lighting conditions within the same space, which can consequently lead to an inspiring and versatile spatial environment with multiple lighting choices that can accommodate different needs. Interestingly, Uppsala Ekonomikum was at the same time the most disconnected study room from daylight, which was a factor much appreciated by participants in the questionnaires in all the four study rooms. Upssala Ekonomikum's example shows that careful lighting design that provides a pluralistic light atmosphere and adjustable lighting that can be controlled by users, reduces the discontent due to lack of daylight, at least to an extent.

It is supported by researchers and lighting professionals that contemporary public spaces, like study rooms in libraries, should be equipped with adjustable lighting, not only to allow individual use and control by users, but also to allow automated control (advanced dimming with daylight sensors and tunable white OLEDs) in order to be human-centric, follow the needs of users concerning circadian rhythm and emotional well-being, and be energy efficient and sustainable. According to Myriam Aries, "during the day, bright light (1000-2000lux) in a combination of daylight and electric lighting improves mood and motivation levels" [23]. When there is not much daylight, artificial dynamic lighting could abstractly represent daylight and its visual effects, in order to reduce the results of its insufficiency and affect emotionally the users that could feel more connected to nature (e.g. projections of shadows of trees or water).

Concerning energy efficiency, controllable zoned lighting that is used when it's needed together with task lighting, uplighting, and wallwashing, can save energy comparing to the uniform downlighting of the study room in KTH library. The area was found to be adequate to standards in terms of lighting power density, however it was also found to have poor lighting that doesn't fulfill the wishes of users and could be improved. Standards (like EN 12464-1 [24]) are the only recognized tool for evaluating light quality. However, they still rely on a simplified understanding of how vision functions, which can lead to the installation of excessive lighting systems that can have negative impacts on both well-being and energy consumption [25]. Careful reinterpretation of standards and zoned lighting with control systems and adjustability can be the means to create sustainable study room environments that satisfy their users.

Finally, as was mentioned in the introduction, although uniformity is not commonly found in nature, it was established together with an effort of overlighting, as a regular way of illuminating offices and public spaces like schools and libraries, at least during the last 60 years. Maybe we still broadly create uniform and overlit environments because it's an easy and quick method in terms of design and installation, or maybe because we are used to it. However, it is important to realize that less brightly and evenly lit environments could broaden the palette of perception and enhance the overall spatial experience.

4.1. Conclusion

From all the conducted research and since uniformity can accommodate the needs of only one category of users, uneven lighting seems more appropriate than uniform lighting for public study spaces that have the responsibility to satisfy multiple types of users and their individual needs concerning concentration, alertness and comfort. However, uniform lighting can be beneficial as a controlled scenario for certain situations like cleaning or workshops. Moreover, high-quality study environments should consider the diverse needs and preferences of users, avoiding a one-size-fits-all lighting approach. Prioritizing the well-being of all types of users in study rooms involves providing varied and uneven lighting atmospheres that facilitate both privacy and socializing, giving prominence to existing natural light, and implementing automated control lighting systems that align with the circadian rhythm and the daily passing of time. The conceptual proposal in this research took into account all of these insights and integrated them in the design in order to ameliorate problems with the existing study room in KTH library, with a primary objective of creating spaces that support physical and mental health through lighting, while corresponding to UN Sustainable Development Goals and the vision of a more sustainable future. Further research with added scopes should be conducted in the future concerning users' lighting preferences in study rooms and other public spaces, as well as real-scale testing to investigate the combination and cooperation of different light zones in space, as a whole.

References

[1] Allan AC, Garcia-Hansen V, Isoardi G and Smith SS 2019 Subjective assessments of lighting quality: a measurement review. *LEUKOS - Journal of Illuminating Engineering Society of North America* **15**(2-3) p. 115.

[2] Lowings L 2021 Perception and experience of daylight - Reflections of an architect. *Multimodality & Society* **1**(2) p. 176.

[3] Veitch JA 2001 Psychological processes influencing lighting quality. *Journal of the Illuminating. Engineering Society* **30**(1) pp. 124-140.

[4] Wänström Lindh U and Jägerbrand AK 2021 Perceived Lighting Uniformity on Pedestrian Roads: From an Architectural Perspective. *Energies* **14**(12) 3647, p. 5.

[5] Wänström Lindh U, Billger M and Aries M 2020 Experience of Spaciousness and Enclosure: Distribution of Light in Spatial Complexity, *SDAR* Journal of Sustainable Design & Applied Research* **8**(1), Article 5, p. 35.

[6] Cuttle C 2013 A new direction for general lighting practice. *Lighting Research & Technology* **45**(1) p. 22.

[7] Hiller C, Boork M, Enger J and Wendin K 2023 User-Centric Measures of the Perceived Light Qualities of Lighting Products. *Emerging Science Journal* **7**(2) p. 610.

[8] Tiller DK and Veitch JA 1995 Perceived room brightness: Pilot study on the effect of luminance distribution. *International Journal of Lighting Research and Technology.* **27**(2) p. 100.

[9] Andersson O 1988 *Rum och ljus* (Sweden: ARKUS) p. 27.

[10] Pallasmaa J 2005 *The eyes of the skin* (West Sussex: John Wiley & Sons Ltd) pp. 46-49.

[11] Barazawa N and Hanyu K 2013 Effects of nonuniform lighting on the evaluation of spaces: A comparison between the situations with and without conversation. *Japanese Psychological Research* **55** (3) p. 274.

[12] Keszei B, Halász B, Losonczi A and Dúll A 2019 Space Syntax's Relation to Seating Choices

from an Evolutionary Approach. *Period. Polytech. Arch.* **50**(2) p. 116.

[13] Moezzi M 1996 *Social Meanings of Electric Light: A Different History of the United States* (Washington: American Council for an Energy-Efficient Economy) p. 124.

[14] Liljefors A 1999 *Lighting Visually and Physically: V/P Lighting Theory Basic Lighting Knowledge* (Stockholm: Lighting Department School of Architecture, KTH) p. 7.

[15] Wänström Lindh U 2012 *Light Shapes Spaces: Experience of Distribution of Light and Visual Spatial Boundaries* (Gothenburg: ArtMonitor. School of Design and Crafts, University of Gothenburg) p. 22.

[16] The 2030 Agenda for Sustainable Development. Sustainable development goals, United Nations, Department of Economics and Social Affairs, URL: https://sdgs. un.org/2030agenda.

[17] ANSI/ASHRAE/IES 90.1-2016: *Energy Standard For Sites And Buildings Except Low-Rise Residential Buildings* (IS Edition).

[18] van Mil I W, Iversen A, Osterhaus W, Garcia Alvarez M, Petersen S and Jeong C-H 2018 *Light at eye level is a means to create energy savings and space for learning, focus and concentration: Research Report for the Danish Electricity Research Fund (ElForsk)* Elforsk.

[19] Wessolowski N, Koenig H, Schulte-Markwort M and Barkmann C 2014 The effect of variable light on the fidgetiness and social behavior of pupils in school. *Journal of Environmental Psychology* **39** pp. 101-108.

[20] Wänström Lindh U and Billger M 2021 Light Distribution and Perceived Spaciousness: Light Patterns in Scale Models. *Sustainability* **13** 12424.

[21] Alexander C 1977 *A pattern language: Towns, buildings, construction* (New York: Oxford university press)

[22] Madsen M 2007 Light-zone(s): As Concept and Tool. *ARCC Journal for Architectural Research* **4**(1) pp. 51-55.

[23] Aries M 2005 *Human lighting demands: healthy lighting in an office environment* (Eindhoven: Technische Universiteit Eindhoven) p. 19.

[24] Guide to DIN EN 12464-1 2019 *Lighting of work places - Part 1: Indoor workplaces* (Brussels: CEN)

[25] Enger J 2018 Visual Concepts and Sensory Experience of Light Environments. *Conf. The 17th Nordic Workshop in Sensory Science: Making Sense.* URL:https://www.researchgate.net/publication/325106260_Visual_Concepts_and_Sensory_Experience_of_Light_Environments.

Acknowledgments

I would like to thank Jörg Frank Seemann, Federico Favero and Foteini Kyriakidou for their guidance and support.

IOP Conf. Series: Earth and Environmental Science 1320 (2024) 012008 doi:10.1088/1755-1315/1320/1/012008

The Use of Immersive Light-Based Art for Well-Being

L Godley[1], R Frasso[2], T Igoe[3], Y Erel [4], J LeNoir[5]

[1] School of Design and Engineering, Jefferson Center of Immersive Arts for Health, Thomas Jefferson University, 4201 Henry Avenue, Philadelphia, PA, United States of America

[2] JCPH – Population Health and SKMC – Center for Research in Medical Education and Health Care, Thomas Jefferson University, 1020 Walnut Street, Philadelphia, PA, United States of America

[3] Interactive Telecommunications Program, New York University, 370 Jay Street, Brooklyn, NY, United States of America

[4] School of Architecture, Rensselaer Polytechnic Institute, 110 8th Street, Troy, NY, United States of America

[5] Population Health, Thomas Jefferson University, 1020 Walnut Street, Philadelphia, PA, United States of America

Corresponding author's email: lyn.godley@jefferson.edu

Abstract. This paper presents the findings of 'Waiting Room - Immersive Art for Well-being,' an exhibition of various dynamic light art installations that investigated the potential impact of this work on viewers' sense of well-being in a mock healthcare setting. Research shows that exposure to art impacts human health, improving overall healthcare experiences, resulting in shortened hospital stays, improved recovery time, and reduced need for pain management. The purpose of this study was to determine if exposure to dynamic light art has an impact on overall well-being. Qualitative data were collected using electronic questionnaires associated with 11 different dynamic light art installations. Open-ended questions explored how participants perceived and valued the presence of and engagement with dynamic light art installations and captured their views on the potential benefit of exposure. Visitors took part in questionnaires to obtain feedback on user experience, the length of exposure/engagement, and the effect the experience had on them. Overwhelmingly, users reported a sense of "calm" and "peacefulness" after spending time with the dynamic light art. One hundred ninety-five responses were collected. This study allowed us to gain a better understanding of how dynamic light art may be used to ameliorate stress and anxiety in spaces where users are confined for periods, as well as identifying key areas for future research, such as particular aspects of the installations or sub-populations that may particularly benefit from this type of intervention. This investigation leads to further studies exploring how exposure to dynamic light art may impact patients, visitors, and family members in various healthcare settings to determine if the perceptions and experiences vary by type of installation. Additionally, this work will inform future explorations of adaptations to dynamic light art, i.e., dynamic and interactive technologies on viewer experience.

1. Introduction

Multiple studies have shown that art can positively impact physiological and psychological well-being, particularly in healthcare environments [1-15]. Most of this research focuses on traditional art forms: painting, prints, or sculpture – work that is 'static'. Other approaches have involved certain types of music, comedy, and especially nature or nature-based imagery [16]. More recently, with the advancements in technology and new media art, virtual or augmented reality approaches such as Distraction Therapy (DT) have been used as well, with promising results [17-20]. One of the main features of the DT is the creation of an immersive experience. Immersive refers to environments where animated media and sound command most of the participant's sensory attention. In the studies cited, immersion is created by having the participant wear a virtual reality (VR) or augmented reality (AR) headset.

It takes time for immersive environments to have an effect. The participant is usually engaged with the headset for minutes to hours at a time. During that time, the stimuli to which they're exposed in the headset have an effect. While effective, immersion in a headset removes the patient from the everyday reality of the space in which they are being treated. However, for many, there are other drawbacks, including physical barriers to wearing a headset, lack of comfort with the technology, and fear of falling.

Based on the immersive quality of dynamic light art, we believe similar outcomes in anxiety management could be achieved using dynamic light-art to create an immersive experience without using headsets. The term dynamic light art, for our purposes, refers to art that incorporates lighting that changes noticeably during the time in which the viewer is observing it. Patients and family members are often kept waiting for extended periods in waiting rooms, recovery rooms, and other clinical spaces. In experiencing dynamic light art in these spaces, we believe that patients could benefit from the effects of immersion, like distraction therapy, without having to block out the rest of the environment. They can still engage with healthcare providers and others in the room while benefiting from the sensory effects of the artwork. In this way, we hope to improve the overall healthcare experience.

An immersive experience creates an effect like the experience of awe or wonder. When people experience awe, their prosocial behavior increases. Perceived social connectedness increases, and feelings of loneliness decrease. Awe evokes a "need for accommodation," shifting our normal understanding and provoking a modification of the mental constructs that we use to understand the world, resulting in less need for cognitive control and being more comfortable with uncertainty [21]. All of which would be beneficial in a healthcare environment.

Digital media allows the creation of dynamic light art that changes during viewing. Based on prior work, we believe that work in this medium can evoke awe and produce a profound engagement that may free the brain from the dominance of reality and may thus reduce anxiety and stress [22].

Figure 1: Exhibition Promotional Card

2. Research Aim: Our study aimed to explore how dynamic light-art might be received by people in a waiting room setting and what impact it might have on their experience.

'Waiting Room - Immersive Art for Well-Being' was an exhibition of various dynamic light art installations arranged as individual waiting areas in mock clinical settings with seating for the viewers to engage in the artwork for an undefined amount of time (see Figure 1).

The entries in this exhibition all used light as a dynamic component of the artwork to create a sense of animation or movement without a set narrative. This was done through various means: programming sequences of LEDs, dynamic reflection and shadows off moving objects, or projection mapping. The installations ran pre-programmed sequences of changes; they were not interactive.

Exhibit visitors were invited to respond to qualitative questions via a link to the QR (Quick Response) code next to each installation. Questions were designed to explore the impact of engaging with each dynamic light art installation on the well-being of the viewers. This paper summarizes findings from analyzing participant responses to the questions linked to each installation. Questionnaires were completed while participants were at the exhibition.

3. Literature Review

The literature review included studies on the effects of art and immersive experience and its potential benefits in healthcare settings. For our purposes, we are focusing on benefits in well-being gained through passive viewing of visual art, as opposed to benefits gained through either participating in making art or engaging with interactive art, where the work changes in response to the participant's actions.

3.1. Methods:
We conducted a review of academic databases, including PubMed, PsycINFO, National Library of Medicine, and Google Scholar, using keywords such as "visual art," "paintings," "art appreciation," "well-being," and related terms. We included studies published in peer-reviewed journals from the past decade. A total of 27 studies were selected for this review.

3.2. Findings:
The studies we reviewed showed that viewing art has demonstrably affected well-being. They associated art museum visits with decreased cortisol levels [1,4,7] and increased relaxation and positive affect [8]. The studies we reviewed further indicated that the arts can promote health and psychological well-being and offer a therapeutic tool for many, e.g., adolescents, elderly, and vulnerable individuals [2,5,6], reducing stress, and enhancing overall psychological well-being [3,9]. The viewing experience can provide opportunities for relaxation, contemplation, and a sense of connection to something greater, eliciting feelings of pleasure, awe, and inspiration [10-12]. These positive emotional responses increase subjective well-being, mood improvement, and overall life satisfaction [13].

Art viewing can prompt introspection and personal meaning-making. Individuals may reflect on their experiences, values, and beliefs by engaging with artworks. Research suggests that art can catalyze self-reflection and exploration of personal narratives, fostering a sense of identity and purpose and improving psychological resilience [14-15].

3.3. Effect by Particular Types of Art or Imagery on Health relevant to this study:
3.3.1. Abstract or Ambiguous Art:
The cognitive effort required to decipher and interpret ambiguous elements in artwork can intensify the aesthetic experience and lead to a deeper engagement [23]. Studies have shown that when confronted with ambiguous stimuli, such as abstract or surreal artwork, the brain activates multiple perceptual systems simultaneously, leading to heightened cognitive engagement. Abstract art connects with viewers more than figurative or realistic art in ways that manifest in heightened emotional responses [23]. The more

ambiguous the image, the more it demands the viewer's imagination, allowing for multiple interpretations and deeper engagement [24]. Some viewers describe this as an expressive, transcendent experience [25].

3.3.2. Light-Based Immersive Art

Immersive experiences most commonly involve dynamic light in one form or another. These experiences, characterized by a high level of engagement and presence, often facilitated by technologies like virtual reality (VR), augmented reality (AR), or multisensory environments, have shown promise in various domains, including stress reduction and enhancing well-being [18-19]. VR-based interventions have been effective in reducing symptoms of stress and anxiety in various populations, including individuals with post-traumatic stress disorder (PTSD) and healthcare professionals [17,20].

The reviewed literature highlights the positive impact of visual art (Ambiguous or Dynamic Light-based Immersive) on well-being. Engagement with visual art and immersive environments has shown potential in reducing stress levels, promoting psychological well-being, improving physical health, enhancing social connectedness, enhancing psychological well-being, and fostering social connections. The effects of immersive experiences may vary depending on factors such as the specific technology or design employed, the content of the experience, and individual differences in perception and response. Further research is needed to explore the long-term effects of immersive experiences, the optimal design principles for creating immersive environments, and the ethical considerations associated with immersive technologies.

3.4. Gaps in the literature and the need for qualitative research on dynamic light art in waiting rooms.

While there is a growing evidence base on the impact of visual art on health and well-being, most existing research on the impact of art on the experiences of people in the healthcare setting focuses on either static art (paintings, prints, or sculpture) or, more recently, Distraction Therapy using VR headsets. While there are some studies on dynamic light art as immersive experiences in large public spaces to draw audiences for entertainment purposes, studies on this type of work to test user experience in settings that replicate healthcare waiting rooms still need to be created.

Most immersive experiences that do not rely on a VR headset take place in large venues where any other light sources must be turned off for the projectors or LED screens to be successful. The artworks developed for the Waiting Room Exhibition are dynamic light art installations that can provide viewers with an immersive experience without such an intervention. The works in the exhibit were scaled to reside in healthcare spaces, where the existing lighting cannot be adjusted for viewing, and healthcare delivery activities (for example, filling out paperwork or drawing blood) must still take place.

There are many studies on the health impact of dynamic "circadian lighting," which changes slowly over hours and is not visibly noticeable to the viewer [26]. However, we could not find any studies on the impact of dynamic light art that changes noticeably during the time the viewer observes it.

4. Methodology

4.1. Study design: Qualitative research approach

This qualitative study design followed the Consolidated Criteria for Reporting Qualitative Research (COREQ) [27] and involved a self-administered questionnaire. Data were captured in Qualtrics © and exported to Excel. Responses were coded and analyzed using standard coding techniques. The study design was reviewed with a Jefferson Institutional Review Board member, and it was determined not to require a formal ethical review board assessment as no identifiable health information was gathered from respondents.

The exhibition space included two large rooms, four art installations in each room, and three smaller rooms that housed single installations. Each installation had seating for three to four respondents to engage with the work. Each installation was designed to promote an engaging environment where the viewer could become immersed in a self-reflective or calming experience. The designs could not change the structure of the space, impede movement within, or interfere with any activities that took place in the space. In lighting design terms, the artwork complements the existing interior lighting, not replaces it. All the installations' dynamic light animations were controlled automatically; the viewer could not interact with or change the lighting effects.

4.2. Participant selection and recruitment

People who visited the 'Waiting Room –Immersive Art for Well-Being' exhibition were invited to respond to questions using a QR (Quick Response) code at each of the 11 installations in the exhibition. The QR code directed respondents to a set of Freelisting [28] (a qualitative questioning approach that asks respondents to generate lists in response to a question) and open-ended questions exploring how participants perceived and valued the installations; how they engaged with the work; and what their views were on the potential benefit of exposure to dynamic light art for people spending time in a healthcare setting. Additionally, we attempted to assess whether the impact varies by duration or proximity of engagement. Freelisting responses were cleaned, that is, like terms were collapsed into one representative term (e.g., if one respondent said "Calm" and another said "Calming," the term "Calm" was used to represent both responses). After data cleaning, frequency scores were calculated to generate representative word clouds. Responses to open-ended questions were reviewed and coded into categories based on the prompts.

5. Results

One hundred ninety-five questionnaire responses were completed by visitors at the HOTBED Gallery in Philadelphia, PA, between September and November 2022. All visitors to the exhibit were eligible to participate. While some respondents completed a questionnaire for each installation (n=11), others chose only to respond to one. Review of responses by installation varied too; some thirty-eight people chose to complete the questionnaire for one installation (#11) (see Figure 2), while only five people chose to complete the questionnaire for another installation (#8) (see Figure 3). Still photographs of selected images appear below, and a link with videos of all works in the exhibit can be found on the exhibition website).

Figure 2: Art Installation #11

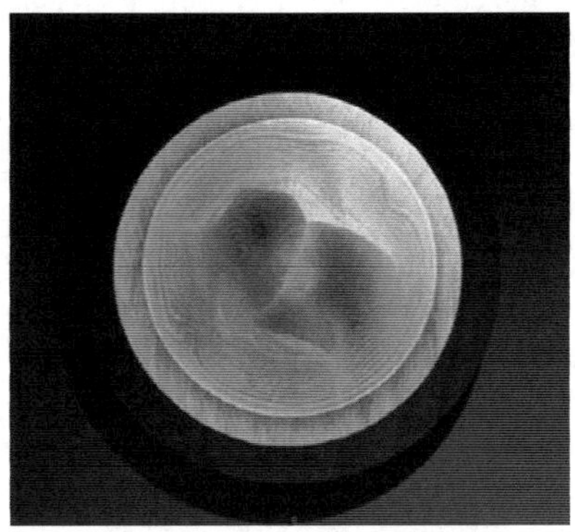

Figure 3: Art Installation #8

Freelisting questions elicited a list of words participants would use to respond to specific questions. After data cleaning, frequency was calculated, and word clouds were generated to represent the most frequent sentiments shared. Word clouds were created with an online generator ("Free Word Cloud Generator," 2023 – see Figures 4+5).

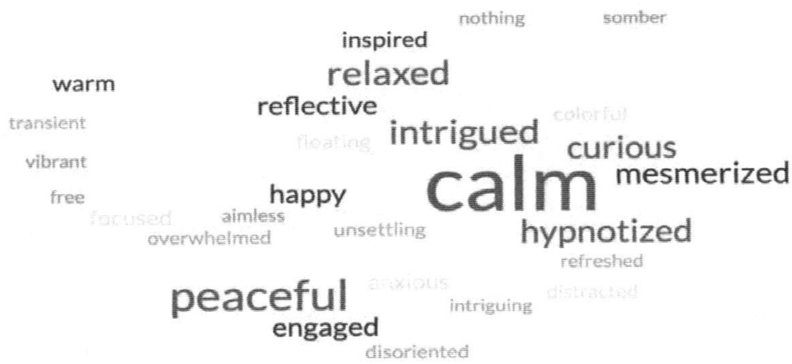

Figure 4: Question:
"How did this installation make you feel? List all of the words that come to mind."

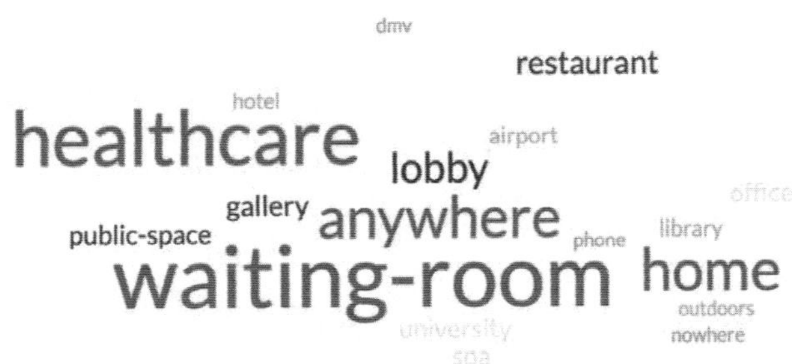

Figure 5: Question:
"Where would you like to see an installation like this?"

Analysis of open-ended question responses revealed a variety of opinions on the different installations of dynamic light art, how they made viewers feel, and if the participants would like to experience a given installation in a doctor's waiting room or other healthcare setting. Respondents reported spending anywhere from 2 minutes to 3 hours with each artwork displayed in the exhibit. They referred to the majority of the installations within the exhibit as calming, peaceful, mesmerizing, beautiful, and intriguing. Each of the 11 installations elicited a calming feeling on at least one of the participants who viewed it. Many respondents reported feeling hypnotized by the art installations and how they distracted them from all other thoughts they were having.

A majority of respondents responded "yes" when asked if they would like to see installations like these in their doctor's waiting room. They gave reasons, including how they were calming, relaxing, distracting, and reflective. Some respondents expressed the artwork as easing the anxiety that was once there, which would be beneficial in a place that causes anxiety for many, like a doctor's waiting room. Art Installation #6 (see Figure 6) had the highest proportion of respondents responding "no" to this question because it was too chaotic or stressful. Some people thought it was unnecessary for this type of artwork to be placed specifically in doctors' waiting rooms but expressed how they would not mind if it were there. Responses to questions at each installation were shared in aggregate, as our goal was not to compare the installations but to collect responses about overall experiences. However, if there were responses unique to a particular installation, that was noted. Below, a selection of verbatim participant responses to open-ended questions are shared:

Question: "How would you describe your experience with the installation to a friend?"

- *Soothing, so mesmerizing with the subtle color changes. Feels like I'm watching the sunrise or sunset.*
- *A zen experience where the movements and changes in light and color relaxes and eases the viewer.*
- *It was illuminating, it made me think about geometric patterns in nature and how there's beauty in that - and all the while I stopped thinking about my hypothetical terminal illness.*
- *A subtle play of light and shadows interfacing happy colors invite the viewer to recall memories or just be in the moment watching the movement.*

Question: "Would you like to see an installation like this in your doctor's waiting room? If yes, why? If no, why not?"

- *Yes, since it would be a good form of distraction while waiting for an anxiety ridden appointment.*
- *Yes, as someone with anxiety especially related to doctors and health this would be very helpful to use as a tool to calm patients down.*
- *Yes, it would preoccupy my mind without things like an obnoxious television screen or other patients' conversations.*
- *Oh yes. So much better than out of date magazines, signs, and stupid daytime tv. It would help to calm me down and prepare for my appointment. Gather my thoughts. Take a deep breath.*
- *Yes- it's calming; it's something to focus your attention on; something to take you outside of whatever else is going on and stop and just imagine something completely different.*

While the vast majority of respondents expressed how the artwork elicited positive feelings of peace and comfort, a few respondents shared that some of the pieces were unsettling, commenting that some of the pieces were dizzying. A few attendees did note that specific installations may not be well suited for a healthcare setting, as viewers might find them overwhelming, dizzying, or unsettling.

- *Not in the waiting room. The light may be overwhelming to some patients who have light sensitivity. The constant movement may be a risk for children to grab it and cause a break.*
- *I wouldn't care? It is very sweet that you're attempting to address lack of access to daylight or views to the outdoors, but also honestly pretty insulting that someone would try to sell plastic and a few lights as an adequate substitute for it.*

6. Discussion

'Waiting Room - Immersive Art for Well-Being' was an exhibition and study designed to allow viewers to view dynamic light art installations in settings that simulated a healthcare waiting room experience. The study allowed us to gather feedback from potential viewers of dynamic light art installations in a mock healthcare setting and capture their advice about where such installations would be beneficial.

Overall, respondents reacted positively to the installations, noting that they would appreciate seeing work of this sort in the healthcare setting. Respondents described the experience of viewing the work using positive words, including "calming" and "peaceful." As visitors to the exhibition are likely to find themselves in healthcare spaces as patients, visitors, or family members of those seeking care, their opinions about the value of the work are important. This study is particularly valuable as a first step in exploring the use of dynamic light art in actual healthcare arenas and allowed us to identify characteristics or aspects of installations that would be ideal or best to avoid as we move toward assessing the utility of this type of art in these spaces where patients and the companions are often stressed or anxious.

The study allowed us to identify the many aspects of art installations that can affect how they are perceived and the feelings they may cause their viewers. These aspects should be considered as we move towards installation in healthcare settings. For example, the brightness or speed of lighting sequences or imagery, the size of the space itself, the seating, the ambiance of the room it is being displayed in, or the resemblance to nature may impact the patient experience.

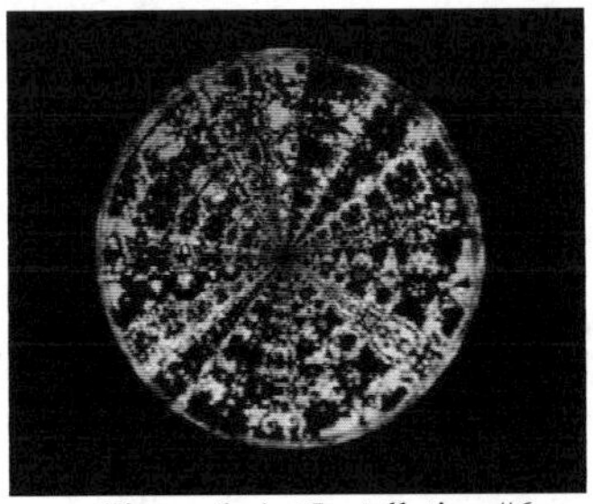
Figure 6: Art Installation #6

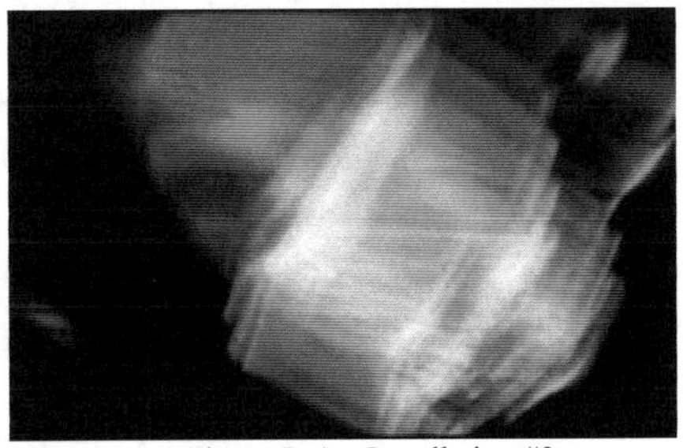
Figure 7: Art Installation #3

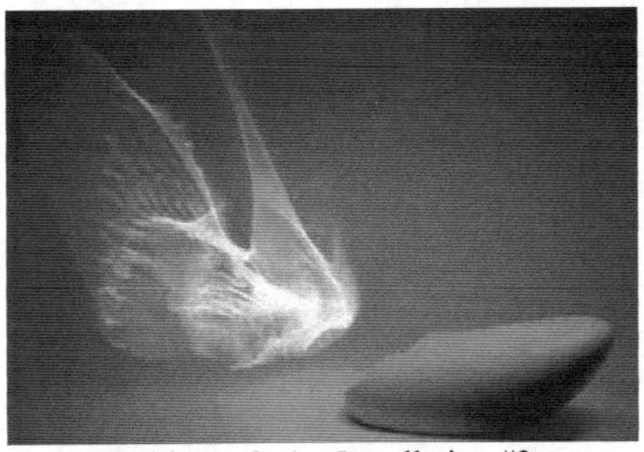
Figure 8: Art Installation #2

For some installations, special attention should be paid to the room size when considering an installation of dynamic light art in healthcare spaces, wherein the goal is to elicit calm and provide patients with a respite from the stress inherent in the experience of seeking care. For example, installation #6 (see Figure 6) elicited responses of feeling unsettled and disoriented, likely due to the constrained viewing space where it was installed, the high visual contrast of black and white imagery created by the bright white lights of the artwork, and the darkened room. This installation also had a faster lighting sequence speed than the other installations. However, installation #3 (see Figure 7) was also in a smaller room with a slower lighting sequence and more ambient light; a respondent described it as "a Zen experience where the movements and changes in light and color relax and eases the viewer." The immersive nature was evidenced by one respondent to installation #2 (see Figure 8), who described it as "A series of sculpted rocks shined intricate lace patterns on the wall, the patterns danced & moved in ways that were entirely not predictable and kept my attention for a long time because of how layered they were." Another installation (#1) (see Figure 9) was a large-scale artwork in one of the larger rooms, with adequate ambient lighting and invoked nature imagery. One respondent said, "It feels like I am staring at a real-life waterfall. The waves draw me in, and the motion is soothing."

This study had several strengths; in particular, it was designed to explore the experiences of community members who are, like all of us, potential patients and visitors to healthcare settings. The study design allowed us to examine the value of the installations in a low-stress mock healthcare setting as we prepared to assess installations in actual patient venues. However, it is important to point out the study's limitations. Throughout this study, there were potential biases. The exhibit's title was "Waiting Room," which could have caused participant bias and affected the responses respondents shared in their questionnaires. They could have been more likely to say that they could see this artwork in waiting rooms because of the known title of the exhibit and the thought that that is the answer researchers were looking for from the participants.

IOP Conf. Series: Earth and Environmental Science 1320 (2024) 012008 doi:10.1088/1755-1315/1320/1/012008

Also, there is potential self-selection bias where people seeking out and understanding artwork and galleries are likelier to participate than others. This bias can lead to results that cannot represent a larger community of people in Philadelphia who would experience this artwork in healthcare settings. In addition, there is an inherent difference between baseline anxiety levels in a gallery versus healthcare environments. The study design did not allow us to measure or assess stress levels.

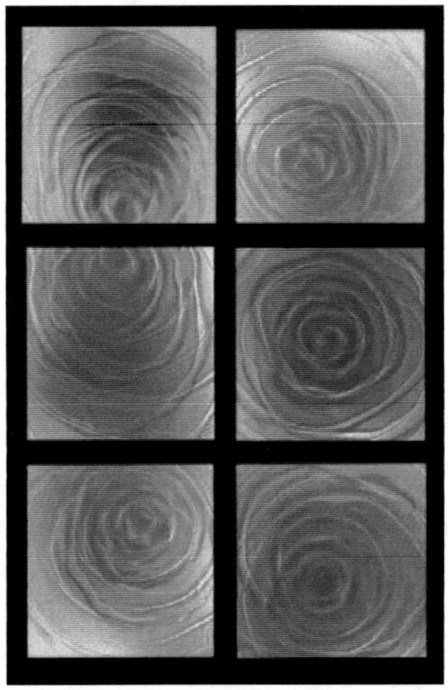

Figure 9: Art Installation #1

We had hoped to assess the level of engagement with the installations to explore if impact varies by duration or proximity of engagement. However, there were not enough responses to accurately make comparisons.

This study was a preliminary exploration to gauge individuals' opinions on dynamic light art and whether it could be a beneficial intervention in waiting rooms or other healthcare settings. Further studies are needed to explore light-based art's effect on those in waiting rooms and healthcare settings and whether it can help decrease anxiety and stress while increasing calming feelings.

7. Conclusion

Our study, 'Waiting Room – Immersive Art for Well-being,' demonstrates that dynamic light art installations in a healthcare setting have the potential to impact the experience of patients positively. As cited in the literature review, most existing research on the impact of art on the experiences of people in the healthcare setting focuses on static art. While there are some studies on dynamic light art as immersive experiences in large public spaces to draw audiences for entertainment purposes, studies on this type of work to test user experience in settings that replicate healthcare waiting rooms still need to be created. The use of technology in art is growing at a rapid pace throughout galleries and museums. The potential to use this new type of artwork to create intentional experiences through dynamic light art to benefit well-being depends on further research.

The dynamic lighting of these art installations was the focal point of the work on display for this study. It has allowed us to gain a better understanding of how dynamic light art may be used to ameliorate stress and anxiety in spaces where users are confined for periods of time and stress levels may be high, as well as identifying key areas for future research, such as sub-populations that may particularly benefit from this type of intervention. With a majority of respondents confirming that their engagement with the installations resulted in an increased sense of calm, peacefulness, reflection, etc., there are research opportunities to investigate the long-term effects of immersive experiences, the optimal design principles for creating immersive environments, and the ethical considerations associated with immersive technologies.

Areas for further investigation into aspects of dynamic light art that could impact overall user experience include on-site studies in healthcare environments to identify demographic considerations; studies focusing on color, dynamic illuminance levels, and spectral power distribution averages of individual installations; the speed of light sequencing; duration or engagement and proximity to installations; as well as the incorporation of interactive technologies in which the participants could, through adaptations of interactive design, choose their level of engagement, from simple observation to co-creation to complete immersion, thus adding a level of agency to the experience, allowing the visitor to have a sense of control over certain aspects of their environment, which is often a source of anxiety during patient visits. The 'Waiting Room – Immersive Art for Well-Being' study documented overall positive community feedback on the potential value of dynamic light art use in the healthcare setting to support patients and reduce anxiety commonly associated with the healthcare experience.

References

1. Treadon CB, Rosal M, Wylder VDT. Opening the doors of art museums for therapeutic processes. The Arts in Psychotherapy. 2006 Jan;33(4):288–301. Available from: https://doi.org/10.1016/j.aip.2006.03.003

2. Daykin N, Byrne E, Soteriou T, O'Connor S. Review: The impact of art, design and environment in mental healthcare: a systematic review of the literature. Journal of the Royal Society for the Promotion of Health. 2008 Mar;128(2):85–94. Available from: https://doi.org/10.1177/1466424007087806.

3. Stuckey HL, Nobel J. The Connection Between Art, Healing, and Public Health: A Review of Current Literature. Am J Public Health. 2010 Feb;100(2):254–63. Available from: https://doi.org/10.2105/AJPH.2008.156497.

4. Ward P. Museums, Health and Wellbeing By Helen Chatterjee and Guy Noble. London: Ashgate Publishing, 2013. 158 pages. Hardcover: $79.95. Curator The Museum Journal. 2014 Oct;57(4):513–9. Available from: https://doi.org/10.1111/cura.12089.

5. Todd C, Camic PM, Lockyer B, Thomson LJM, Chatterjee HJ. Museum-based programs for socially isolated older adults: Understanding what works. Health & Place. 2017 Nov;48:47–55. Available from: https://doi.org/10.1016/j.healthplace.2017.08.005.

6. Thomson LJ, Lockyer B, Camic PM, Chatterjee HJ. Effects of a museum-based social prescription intervention on quantitative measures of psychological wellbeing in older adults. Perspect Public Health. 2018 Jan;138(1):28–38. Available from: https://doi.org/10.1177/1757913917737563.

7. Law M, Karulkar N, Broadbent E. Evidence for the effects of viewing visual artworks on stress outcomes: a scoping review. BMJ Open. 2021 Jun;11(6):e043549. Available from: https://doi.org/10.1136/bmjopen-2020-043549.

8. Cotter K.N., Pawelski J.O. Art museums as institutions for human flourishing. The Journal of Positive Psychology. 2022 Mar 4;17(2):288–302. Available from: https://doi.org/10.1080/17439760.2021.2016911.

9. Hathorn K, Nanda U. A Guide to Evidence-based Art. 2008 Jan. Available from: https://api.semanticscholar.org/CorpusID:73355330

10. Chatterjee A, Vartanian O. Neuroscience of aesthetics: Neuroscience of aesthetics. Ann NY Acad Sci. 2016 Apr;1369(1):172–94. Available from: https://doi.org/10.1111/nyas.13035

11. Cupchik GC, Vartanian O, Crawley A, Mikulis DJ. Viewing artworks: Contributions of cognitive control and perceptual facilitation to aesthetic experience. Brain and Cognition. 2009 Jun;70(1):84–91. Available from: https://doi.org/10.1016/j.bandc.2009.01.003.

12. Fingerhut J, Prinz JJ. Wonder, appreciation, and the value of art. In: Progress in Brain Research [Internet]. Elsevier; 2018 [cited 2024 Feb 9]. p. 107–28. Available from: https://linkinghub.elsevier.com/retrieve/pii/S0079612318300049

13, Fancourt D, Finn S. What is the evidence on the role of the arts in improving health and well-being? A scoping review. World Health Organization. Regional Office for Europe; 2019. (Health Evidence Network synthesis report;67). Available from: https://apps.who.int/iris/handle/10665/329834.

14. Bolwerk A, Mack-Andrick J, Lang FR, Dörfler A, Maihöfner C. How Art Changes Your Brain: Differential Effects of Visual Art Production and Cognitive Art Evaluation on Functional Brain Connectivity. He Y, editor. PLoS ONE. 2014 Jul 1;9(7):e101035. Available from: https://doi.org/10.1371/journal.pone.0101035.

15. Link B. Art Integration in Teacher Education: Aesthetic Tools to Foster Critical Reflection. 2022 [cited 2024 Feb 9]. Available from: https://doi.org/10.26153/TSW/41907.

16. Ulrich R. Effects of interior design on wellness: Theory and recent scientific research. Journal of health care interior design : proceedings from the . Symposium on Health Care Interior Design Symposium on Health Care Interior Design. 1991 Feb;3:97–109. PMID: 10123973.

17. Difede J, Cukor J, Jayasinghe N, Patt I, Jedel S, Spielman L, et al. Virtual reality exposure therapy for the treatment of posttraumatic stress disorder following September 11, 2001. J Clin Psychiatry. 2007 Nov;68(11):1639–47. PMID: 18052556.

18. Chirico A, Cipresso P, Yaden DB, Biassoni F, Riva G, Gaggioli A. Effectiveness of Immersive Videos in Inducing Awe: An Experimental Study. Sci Rep. 2017 Apr 27;7(1):1218. Available from: https://doi.org/10.1038/s41598-017-01242-0.

19. Navarro-Haro MV, Modrego-Alarcón M, Hoffman HG, López-Montoyo A, Navarro-Gil M, Montero-Marin J, et al. Evaluation of a Mindfulness-Based Intervention With and Without Virtual Reality Dialectical Behavior Therapy® Mindfulness Skills Training for the Treatment of Generalized Anxiety Disorder in Primary Care: A Pilot Study. Front Psychol. 2019 Jan 28;10:55. Available from: https://www.frontiersin.org/articles/10.3389/fpsyg.2019.00055.

20. Croghan IT, Hurt RT, Aakre CA, Fokken SC, Fischer KM, Lindeen SA, et al. Virtual Reality for Health Care Professionals During a Pandemic: A Pilot Program. J Prim Care Community Health. 2022 Jan;13:215013192210867. Available from: https://doi.org/10.1177/21501319221086716.

21. Keltner D. Awe: the new science of everyday wonder and how it can transform your life. New York: Penguin Press; 2023. 309 p.

22. Rollins J. 'Purpose-built' Art in Hospitals: Art with Intent [Internet]. Emerald Publishing Limited; 2021 [cited 2024 Feb 9]. Available from: https://doi.org/10.1108/9781839096808.

23. Shimamura AP, Palmer SE. Aesthetic ScienceConnecting Minds, Brains, and Experience [Internet]. Oxford University Press; 2011 [cited 2024 Feb 9]. Available from: https://doi.org/10.1093/acprof:oso/9780199732142.003.0010.

24. Jakesch M, Leder H. Finding meaning in art: Preferred levels of ambiguity in art appreciation. Quarterly Journal of Experimental Psychology. 2009 Nov;62(11):2105–12. Available from: https://doi.org/10.1080/17470210903038974.

25. Kandel E. Reductionism in Art and Brain Science: Bridging the Two Cultures [Internet]. Columbia University Press; 2016 [cited 2024 Feb 9]. Available from: https://doi.org/10.7312/kand17962

26. Brown TM, Brainard GC, Cajochen C, Czeisler CA, Hanifin JP, Lockley SW, et al. Recommendations for daytime, evening, and nighttime indoor light exposure to best support physiology, sleep, and wakefulness in healthy adults. PLoS Biol. 2022 Mar 17;20(3):e3001571. Available from: https://doi.org/10.1371/journal.pbio.3001571

27. Tong A, Sainsbury P, Craig J. Consolidated criteria for reporting qualitative research (COREQ): a 32-item checklist for interviews and focus groups. Int J Qual Health Care. 2007 Dec;19(6):349–57. Available from: https://doi.org/10.1093/intqhc/mzm042

28. Keddem S, Barg FK, Frasso R. Practical Guidance for Studies Using Freelisting Interviews. Prev Chronic Dis. 2021 Jan 14;18:200355. Available from: http://dx.doi.org/10.5888/pcd18.200355.

Acknowledgements:

This work was partially funded by a Faculty Completion Grant from the Kanbar College of Design, Engineering and Commerce at Thomas Jefferson University. We thank all the artists; Jessica Judith Beckwith, Alyson Denny, Aidan Lincoln Fowler, Lyn Godley, Yael Erel, and Philip Hart, who made their work available for the exhibition and accompanying research, the faculty and students at New York University, NY, NY, USA, Parsons School of Design, NY, NY, USA, Jefferson University, Philadelphia, PA USA, KTH Royal Institute of Technology, Stockholm, Sweden, Ariel University, Ariel, Israel, Rensselaer Polytechnic Institute, Troy, NY, USA, Aalborg University, Copenhagen, Denmark, and ITMO University, St. Petersburg, Russia whose work as part of a global student design competition was included in the exhibition and accompanying research, as well as the staff at HOTBED Gallery who helped encourage the participation of visitors with the QR Surveys.

IOP Conf. Series: Earth and Environmental Science 1320 (2024) 012009 doi:10.1088/1755-1315/1320/1/012009

Exploring Simulation Workflows, Tools, and Metrics for Beyond-Vision Effects in Multi-Objective Optimization: A Scoping Review

Alyaá Tabbah[1], Myriam B C Aries[1], Annika Moscati[1], Peter Johansson[1]

[1] Department of Construction Engineering and Lighting Science, School of Engineering, Jönköping University, Jönköping, Sweden
Corresponding author alyaa.tabbah@ju.se

Abstract. Lighting in the built environment affects different aspects, ranging from building performance in terms of costs and energy consumption to human well-being in terms of thermal comfort, visual effects, and beyond-vision effects. Buildings optimised for specific performance metrics rarely consider different aspects simultaneously, leading to sub-optimised, unbalanced, or non-trade-off solutions. Therefore, multi-objective optimisation has commonly been used to overcome conflicting performance objectives. Recently, light effects beyond vision gained more interest in building design but it is unclear if and how they are integrated with other existing building performance metrics and simulation workflows. A scoping review investigated the state-of-art in multi-objective lighting design optimisation regarding building performance and human well-being focusing on performance metrics, simulation workflows, and the overall information flow. Results show that metrics for beyond-vision effects are not integrated with other human well-being and building performance metrics. The simulation workflows included multiple steps and computational tools in multi-objective optimisation. This process has limitations such as a long simulation time, no ubiquitous integrated tool, and a reduced information flow.

1. Introduction

The concept of the built environment encompasses the physical and spatial surroundings that humans create for various activities such as habitation or work. The quality of the built environment profoundly influences human well-being, comfort, and overall quality of life. An essential element of this influence lies in the provision of adequate lighting. To create a sustainable, good-performing lighting environment, a broad spectrum of design variables must be considered, i.e., building envelope properties, building services' systems, occupants' dynamic behaviour, and life-cycle costs. Lighting is one of these variables and must be balanced with many other, sometimes conflicting, performance objectives [1, 2]. Earlier analysis and building performance simulations have shown that optimising building performance for energy and cost provides a great saving potential. However, it is accompanied by a great risk that it contributes to a sub-optimisation of the building's value at the cost of other values, e.g., user comfort and lighting quality [3].

Building optimisation can be performed using different multi-objective/multi-criteria optimisation methods which Yu *et al.* [4] have categorised into five different optimisation levels. Levels M1 and M2 investigate all aspects separately (M2 adds an overall comparison) and make no trade-offs. Level M3 ranks different criteria in a prioritised order, and only solutions that satisfy the first criteria continue for further analyses reducing the solution sample. Level M4 combines different criteria into a single metric.

The highest level M5: multi-objective optimisation (MOO) analyses all criteria simultaneously to balance all objectives and find trade-off solutions. Due to the high complexity of optimising different building design aspects simultaneously, solving this task without integrating computational tools is unfeasible.

Recent studies have increasingly used simulation-based multi-objective optimisation. It is a mathematical optimisation technique that explores solutions simultaneously satisfying multiple potentially conflicting performance objectives [5]. Solutions that consider trade-offs between different objectives, i.e., not improving one objective without worsening the other, are called Pareto-front solutions [4]. It can effectively solve dynamic building design optimisation problems by mimicking its evolution process [6]. The MOO simulation workflows often include a design tool, a visual programming language, and a building performance simulation tool [7]. Using a design tool provides the means to manipulate architectural elements and lighting parameters, while the visual programming language enables the creation of complex algorithms and design processes considering multiple objectives and constraints. Consequently, the information input and digital model complexity increase rapidly [8]. Therefore, standardisation and digital support are required to optimise human-driven, multi-objective lighting design [9].

Optimising the built environment performance is not limited to managing energy consumption and life-cycle costs. It balances visual and non-visual light effects, a methodology now summarised by the International Commission on Illumination (CIE) as 'integrative lighting' [10]. This integration must coexist with other, sometimes contradicting, design criteria. Light affects human physiology and psychology, including vision, circadian rhythm, sleep, melatonin suppression, heart rate, alertness, and (non-) seasonal depression [e.g., 11, 12]. The rods and cones (S, M, L) are the four classical photoreceptors for vision. A fifth type of photoreceptor, the intrinsically photosensitive retinal ganglion cells (ipRGCs), mainly controls the light's non-visual responses based on the photopigment melanopsin [13] and its sensitivity peaks at a shorter wavelength. However, as all five α-opic photoreceptors can contribute to both visual and beyond visual responses to light [14], it is more straightforward to distinguish between light effects related to 'vision' and 'beyond vision'.

Assessment of beyond-vision effects requires appropriate quantification. However, established photometric light measures such as photopic lux, are inadequate [15]. Recently, recommendations were made proposing 'α-opic Equivalent Daylight Illuminance values' (in lux) as metrics to specify the impact of lighting on either of the five α-opic human photoreceptors. Metrics like 'melanopic equivalent daylight illuminance' (m-EDI) are based on the spectral sensitivity of the ipRGCs [14, 16]. The building certification institute WELL [17] recommended to use 'equivalent melanopic lux' (EML) based on the ipRGC spectral sensitivity curves [15]. Other metrics are based on nocturnal suppression of the hormone melatonin such as 'circadian stimulus' (CS) which is a non-linear model based on the quantity and spectrum of light [18] and the 'non-visual direct response' (nvRD), a model that takes a time series of eye-level light stimuli as input to predict human alerting response [19].

Lighting simulation software is pivotal in architectural and lighting design optimisation. Conventional software simplifies the visible spectrum into RGB colour space, but recent spectral simulation tools, like Lark v2.0 [20] and ALFA [21], specialise in modelling effects beyond human vision. These innovative tools divide the visible spectrum into multiple channels, with Lark using nine and ALFA 81 channels and calculate metrics for beyond-vision effects directly. Both tools support static point-in-time simulations. Lark offers annual simulations with the daylight coefficient method for dynamic daylighting [20]. Meanwhile, ALFA uses spectral ray tracing to predict light absorption by α-opic photoreceptors based on location and view direction [21].

In light of the growing awareness of the profound impacts of beyond-vision effects on human health and well-being and according to the review done by Gkaintatzi-Masouti *et al.* [1], new simulation workflows have been introduced to better predict and integrate these effects into the design and optimization of architectural environment by considering luminous and temporal factors of light, specifically light quantity, spectrum, directionality, timing, duration and previous light history [22]. These advanced simulation workflows provide practitioners and researchers with a detailed understanding of how lighting affects both the built environment and human occupants. They enable better decision-making and the creation of spaces that prioritize overall well-being.

However, even though multiple studies may have combined, i.e., light-related indoor environmental quality with human comfort, energy efficiency and life-cycle cost factors, only recently, light effects beyond vision gained more interest in building design. Multi-objective optimization allows for the simultaneous consideration of various factors, but it is unclear if and how beyond-vision effects may interact with other building performance and human well-being aspects using the previously mentioned performance metrics, simulation tools and workflows. Therefore, a scoping review of the literature investigated if and how beyond-vision effects are integrated in multi-objective lighting design optimisation in terms of performance metrics and simulation tools and whether the integration of beyond-vision effects influences the simulation workflows.

2. Method

A scoping literature review was performed according to the PRISMA procedure for scoping reviews. The review process started with defining the scope and research question using the mnemonic Population, Concept, Context (PCC) framework [23]. Population was not applicable since there was no focus on a certain group of people (i.e., age, gender). Three search concepts were applied, see Table 1. The search in Scopus and Web of Science databases was performed within title, abstract, and keywords and limited to scientific journal/conference publications after 2002 ((re-)discovery ipRGCs). Only English and available full text were included.

Table 1. Search concepts and search blocks

Search concepts	Synonyms and search blocks (S)
Building lighting environment (C1)	S1 (light OR daylight OR illumina OR luminous) W/3 (environment OR design OR model OR perform OR system OR technolog OR "built environment" OR building)
Simulation, evaluation, optimisation (C2)	S2 (digital OR comput) AND (simulat OR emulate OR analy? OR assess OR evaluat OR validat OR valuat OR valoriz)
Human well-being and building performance metrics (C3)	S3 (human OR visual OR "non-visual" OR "non image forming" OR "beyond vision") W/3 (metric OR indicator OR impact OR performance OR effect OR outcome OR output OR factor OR index OR parameter OR comfort OR health OR well-being)

Papers were excluded if they investigated building types with a specific (short stay) use (e.g., museums, airports), reported subjective evaluations and did not use computational modelling, investigated building performance aspects with less relevance for the thermal-visual-energy-cost balance (e.g., acoustics), or investigated effects on users/occupants with special conditions and needs (e.g., people with Alzheimer's Disease). Additionally, studies were excluded if the multi-objective/multi-criteria optimisation was lower than M5 [4].

The paper selection process is shown in Figure 1. From a total of 662 papers originally retrieved, 21 papers reached the highest optimisation level. The included papers were imported into NVivo 14 software for further (qualitative) analysis.

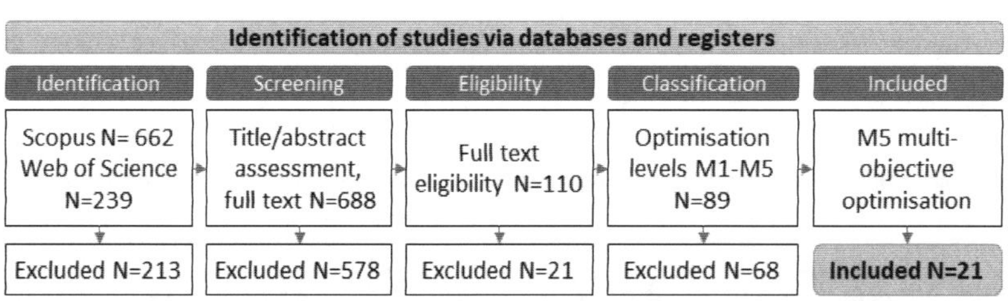

Figure 1. Paper selection process (schematic PRISMA flow diagram).

The information extracted from each paper are presented in Table 2 and included the simulation level (building, floorplan, room), location, if available, and the type of building (office, educational, residential) to define the function of the building. Input properties include information about the case study in each paper and describe the fixed variables of the building studied like simulated light source (daylight, electric light), window, surface, envelope, occupancy, and HVAC properties. Design variables represent parameters that can be adjusted (independent variables) in the built environment or the test object. Different limitations about the aspects, design variables, simulation tools, or any presented columns in this table, were also mentioned.

3. Results

Focused on the issue of MOO, this review investigated if and how beyond-vision effects are integrated in multi-objective lighting design optimisation in terms of performance metrics and simulation tools and whether the integration of beyond-vision effects influences the simulation workflows. The information was categorised according to the building performance and human well-being aspects involved (energy consumption and life-cycle cost, visual effects both comfort and performance, beyond-vision effects, thermal comfort). The specific metrics from each aspect were reviewed, and the combinations of the aspects were investigated. The simulation workflow of multi-objective optimisation and software tools were presented. The main results are shown in Table 2, presenting the results of 21 papers in M5: MOO level.

Table 2. Summary table showing the results of the papers at multi-objective optimisation (MOO) level

Reference	Light source[a]	Simulation level, building type, location[b]	Aspects[c]	Inputs properties[d]	Design variables	Modelling/ simulation/ optimisation tools[e]	Generative system[f]	Main conclusions	Limitations
[8]	☀	Italy	+VP -EC	BD, BO, E, S, W, WWR	concrete shading device		GA	complex concrete shading system outperforms overhang, horizontal louvers	software limitation
[24]	☀	China	+VP +TC	BD, BO, S, WWR, WD	10 building layout position	OpenV EDDy3D	ANN NSGA-II	optimal buildings are located north & east south sites could have higher buildings	long time simulation
[25]	☀	UK	-EC	BD, BO, E, W, WWR, HVAC, B/SH	blind position	, jMetal	GA	prominent energy savings in highest cooling load and clear sky days	long time simulation
[26]	💡		+VP +VC -EC -C	BD, S, L	luminaire type layout	formula	GA-PSO	optimised result is better compared to recommended constraints limits	closed box no window
[27]	☀	Norway	+VP +TC -EC	E, W, HVAC, WD, O, B/SH, WFR	2 ventilation systems	OpenV GenOpt	GA-PSO	EC, TC and VP satisfactions were higher in all-air ventilation system DCV	satisfy all constraints
[28]	☀	Indonesia	+VP +VC	BD, BO, W, S, O, WD	blind type v, h angle material		GA	blinds slat angle is more influential than material specularity on simulated metric	limited objectives
[29]	☀	Australia	+VP +VC -EC	BD, BO, E, W, S, WD, HVAC, O	perforated screens	Galapagos	GA	perforated screen PS1 achieve better VP VC EC compared to no shading case	limited variables
[30]	☀	China	+VP +VC -EC	BO, WWR, S, WD	facade shading ratio		SPEA-II	optimal FSR 0.37 reduces glare and solar radiation compared to no shading	long time simulation
[31]	☀		+VP +TC -EC	E, HVAC, WD, O	PID controlled HVAC system	formula	GA-PSO	control strategy improves comfort and decreases energy consumption	no software tools

[32]		India	+VP +VC -EC	BD, BO, E, W, S, WD, O	motor shading, dimmable	formula	FLC ANFIS	VC TC scheme provides least energy saving but maximum occupant comfort	using data simulated
[33]		Singapore	+VP +TC -EC -C	BD, BO, E, W, WD, O	W type, status, WWR, shade	JEPlus+EA	NSGA-II	design for natural cooling increases energy efficiency and cost effectiveness	fixed shape layouts
[34]		Egypt	+VP +VC +TC -EC	BD, BO, W, WWR, WD, HVAC, O	6 daylighting systems	FlucsPro Apachesim	-	selecting fitting daylighting systems for different BO and site conditions	limited variables
[35]		China	+VP +VC -EC	BD, BO, W, WWR, O, S, HVAC	shading parameters		NSGA-II	integrated shading boards and perforated panel +VP +VC -EC	long time simulation
[36]			+VP +TC -EC	BD, BO, W, WWR, O, HVAC, L	predictive control model	TRNSYS	FLC	MPC achieve high comfort level when the total energy supply is in a shortage	subjective evaluation
[37]		China	+VP +TC -EC	BD, E, O, HVAC, L	window design parameters		SPEA-II	increasing WWR, decreases EC & TC while increase VP, south less EC	long time simulation
[38]		Slovenia	+VP +TC	BD, BO, W, E, WWR, WD,	predictive control model	Dymola/ Modelica	FLC	simulator is faster and process-safe tested compared to real-system experimentation	limited objectives
[39]		Australia	+VP +VC	BD, BO, E, S, W, WWR, WD	shading v, h system angle		NSGA-II	all-purpose polygonal shade showed increase in VP and VC vs. no shade	building conditions
[40]		box Egypt	+VP -EC	BD, BO, E, W, S, HVAC	adaptive facade		CA	adaptive façade outperformed the static configurations	single tool integration
[41]		UK	+VP	BD, BO, S, WD	adapt-façade opening ratio		CA	high grid density, small cell size shows improved visual performance result	no GA, less variables
[42]		Iran	+VP	BD, BO, W, WWR, WD, S	kinetic façade		NSGA-II	Pareto front set solutions fulfil several correlated VP metrics	metrics fitness vary
[43]		China	-EC -C	BD, BO, W, O, HVAC	shading HVAC,	numerical simulation	SDP	controlled blinds and natural ventilation save EC & C, ensure occupant comfort	long time calculation

[a] : daylight, : electric light
[b] : building level, : floorplan level, : room level, : educational building, :office, :residential building
[c] (-): minimise, (+): maximise, EC: energy consumption, C: cost, TC: thermal comfort, VP: visual performance, VC: visual comfort (glare-free), BVE: beyond-vision effects
[d] BD: building dimensions, BO: building orientation, E: envelope, W: window, L: luminaire, S: surfaces, WWR: window-wall-ratio, WFR: window-floor-ratio, DS: daylight systems, B/SH: blinds/shading, WD: weather data, O: occupancy
[e] : Grasshopper, : Rhinoceros3D, : DIVA-for-Rhino, : Radiance, : Ladybug, : Honeybee, : DAYSIM, : ARCHSIM, : DialuxEvo9, : ClimateStudio, : EnergyPlus, : IDAICE, : ECOTECT, : OpenFOAM, : MATLAB, : SIMULINK, : Octopus
[f] GA: Genetic Algorithm, CA: Cellular Automata, SPEA-II: Strength Pareto Evolutionary Algorithm, GA-PSO: Genetic Algorithm-Particle Swarm Optimisation, FLC: fuzzy logic controller, ANFIS: Adaptive Neuro-Fuzzy Inference System, NSGA-II: Non-dominated Sorting Genetic Algorithm, SDP: stochastic dynamic programming, ANN: Artificial Neural Network

3.1 Performance metrics and aspects combination

In Table 2, the aspects column shows which building performance and human well-being aspects are being optimised: energy consumption, life-cycle cost, visual performance, visual comfort, thermal comfort. Interestingly, beyond-vision effects were not investigated by any of the papers. The plus/minus +/- signs indicate if the objective is to maximise or minimise these aspects. In the papers, these aspects are presented in measurable metrics; most used ones are heating, cooling, lighting energy consumption (kWh), operational cost, useful daylight illuminance (UDI), daylight glare probability (DGP), and indoor air temperature, respectively. Visual performance was the most investigated aspect present in all studies. For daylight, there are static metrics like daylight factor (DF) and point-in-time illuminances, and dynamic ones like daylight autonomy (DA), useful daylight illuminance (UDI), or annual sunlight exposure (ASE).

Three papers investigated one aspect (visual performance) in combination with several design variables on adaptive facade design [41], kinetic façade [42] and (energy consumption) for blind position control [25], seven papers investigated a combination of two aspects [8, 24, 28, 38-40, 43], eight papers investigated a combination of three metrics [27, 29-32, 35-37], three papers investigated a combination of four aspects [26, 33, 34], and zero investigated 5 or 6 aspects. The most common aspects combinations are (visual comfort, visual performance, and energy consumption) [29, 30, 32, 35], (visual performance, thermal comfort, and energy consumption) [27, 31, 36, 37]. Most investigated design variables and simulated light sources are related to daylight penetration in buildings. Only four studies investigated both daylight and electric light [32, 33, 36, 38] and only one focussed on electric light sources [26]. Simulations were often done at room level and in office or commercial building types.

Depending on the case study, the different aspects were not always conflicting with each other. For example, Lan *et al.* [33] investigated four aspects (visual performance, thermal comfort, energy consumption, and life-cycle cost) in net-zero energy buildings (NZEB) and found that the social aims do not conflict with the environmental and economical aims. The results show that designing for daylighting and natural cooling contributes to the improvement of energy efficiency and cost effectiveness as well as better visual performance and thermal comfort. In contrary, Rabani *et al.* [27] found that thermal comfort and energy efficiency have a positive correlation while they conflict with visual performance.

3.2 *Simulation workflow*

According to most reviewed studies, the simulation workflow in MOO starts with formulating the problem after preparing the building model and setting the required parameters in the available modelling, simulation and optimisation tools (see Figure 2) [8, 24-31, 33, 35, 37, 39, 42]. Problem formulation includes defining the objective function (e.g., reduce total energy consumption, discomfort glare and thermal discomfort rate), defining the design variable(s), and defining the algorithm parameters or constraints (e.g., illuminance value > 300 lx for visual performance). Design variables can be active variables relying on the use of mechanical and electrical systems, or passive variables relying on natural phenomena to regulate the indoor environmental quality [33].

Next step includes the use of metaheuristics optimisation algorithms which stochastically generate an initial population ('chromosomes') to respond to an optimisation problem where parameters have been identified and encoded. The fitness of each member of the population or chromosome is then assessed using a fitness function or optimal target value for each algorithm parameter. The best fitted parents or chromosomes are then selected to generate a new generation, while poor performers are discarded. This procedure is then repeated using crossover of two or more chromosomes or mutation of random chromosomes and each time fitness values of the new offspring is evaluated until the optimal design solution is reached (see Figure 2).

The selection of an optimal solution could be done by the ranking method or Pareto front according to Sadegh *et al.* [42]. Ranking method typically involve aggregating the objectives into a single score, and then ranking the solutions based on that score, while the Pareto front represents a set of non-dominated solutions that capture the trade-offs between the objectives meaning that no other solution can improve one objective without worsening the other. Sadegh *et al.* [42] investigated the design and development of kinetic façades to optimise visual performance in terms of daylight autonomy (DA), useful daylight illuminance (UDI) using genetic algorithms and both ranking methods and Pareto front. They found that ranking method may provide a limited set of solutions while Pareto front provides a more comprehensive and diverse set of optimal solutions. The ranking method might be beneficial when one objective outweighs other objectives in terms of desired design parameters.

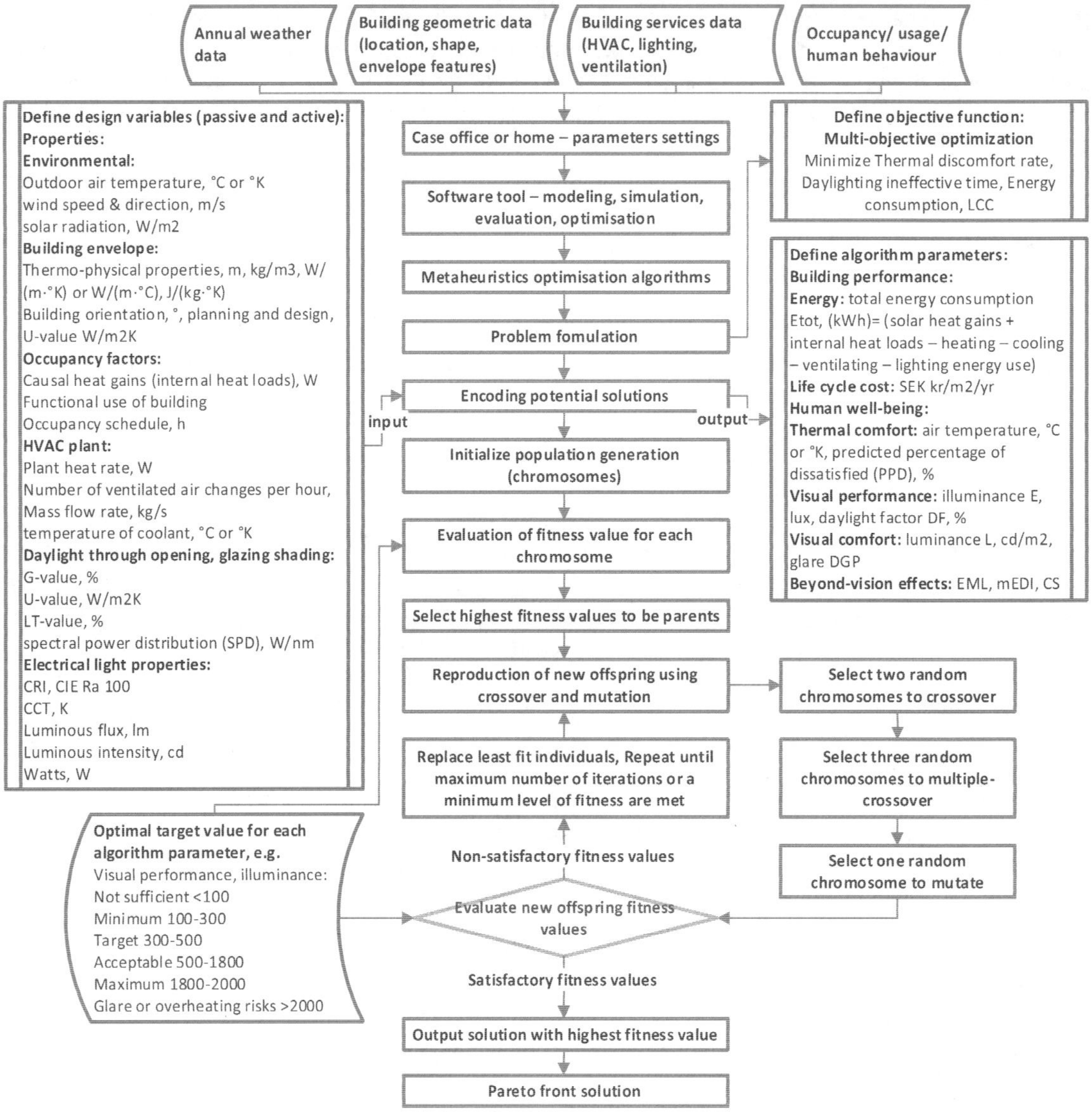

Figure 2. The process of implementing genetic algorithms (GA) (adapted from Harish [31])

Two other studies have used hybrid Genetic Algorithm - Particle Swarm Optimisation (GA-PSO) [27, 31], to optimise controlled ventilation system and energy control system, respectively. Ahmad *et al.* [25] used canonical generational Genetic Algorithm to optimise scheduled operation of window blinds. Kurian *et al.* [32] used ANFIS controller (Adaptive Neuro-Fuzzy Inference System) to optimise window blind controller and Tomažič *et al.* [38] used Fuzzy logic to optimise building-management systems (BMSs). Sun *et al.* [43] used stochastic dynamic programming (DP), rollout technique to optimise integrated shading and ventilation system. Fan *et al.* [30] used image density atlas (IDA) to provide decision-making basis for facade shading ratio (FSR) optimisation.

Two studies by Ayoub [40] and Kim [41] applied cellular automata (CA), which is a mathematical model consisting of a grid of cells that evolve over time according to a set of rules based on the states of neighbouring cells. They aimed to find the applicable cellular automata (CA) parameters and daylighting design criteria by using generative design and genetic algorithm to build an innovative design strategy to search the adaptive building facade. The quantitative evaluation method of CA rules controlled average opacity and perforation rate to ensure an optimal distribution of the natural daylight that enters the internal spaces, and to give an interesting visual appearance.

To ensure the validity of the optimal solution, some studies have tested the same design variables and measured metrics in real-life in comparison to simulation, e.g., [27]. Lavin and Fiorito [29] and Mangkuto *et al.* [28] validated the simulation tool before generating the optimal solutions by comparing daylight factor in measurement and simulation. Others validated the results of one tool or framework with another simulation tool, e.g., [26].

3.3 Simulation tools

The simulation workflow included multiple steps and computational tools in multi-objective optimisation. Many studied cases started with parametric modelling of the building by using the Grasshopper and Rhinoceros 3D software [8, 28, 35, 37, 39-41]. Then multiple simulation tools and plug-ins were used for various purposes (e.g., Ladybug, Honeybee, DIVA). Kim [41], for example, used DIVA to perform daylight analysis on an existing architectural model via integration with Radiance and DAYSIM. Ayoub [40] conducted illuminance and energy computational simulations using Diva-for-Rhino and Archsim plugins that interface Radiance and EnergyPlus simulation engines respectively. Mangkuto *et al.* [28] also used DIVA-for-Rhino that works on validated simulation engines for daylighting (Radiance) and building energy use (EnergyPlus), both of which are open sources. To simulate the effect of adding shading on moderating glare and daylight illuminance, Sorooshnia *et al.* [39] used DIVA plugin and ClimateStudio. In general, Solemma based simulation tools were commonly used in MOO [21].

In the next step, building performance tools are coupled with optimisation tools to enable applying metaheuristic algorithms. For example, Lan *et al.* [33], coupled JEPlus+EA, a multi-objective optimisation tool based on NSGA-II, with EnergyPlus to explore optimised solutions and to seek insights into the design parametric space that leads to zero energy systems. Another example where no parametric modelling was applied was done by Rabani *et al.* [27]. They used the dynamic building energy simulation software IDA-ICE to carry out daylight simulations through the Daylight-tab in IDA-ICE that uses backward raytracing and Radiance as a simulation engine. The software was coupled with GenOpt as the optimisation engine. Furthermore, a detailed thermal and visual comfort analysis of all scenarios were conducted through coupling of IDA-ICE with OpenFOAM, which is open source computational fluid dynamics (CFD) software [27].

In several studies, formulas were used to calculate some metrics in combination with the parametrised model and simulation tools [24, 27, 31-33, 43]. In one of the studies by [43], only formulas in MATLAB were used to optimise integrated control shading and ventilation systems. A lot of statistical models, analysis techniques and mathematical framework were presented in the reviewed studies. For example, statistical models like regression analysis and polynomial functions offer insights into relationships between design variables and objectives. This understanding helps designers make informed decisions [32]. Techniques like lexicographic ordering allow designers to prioritize objectives based on their significance. It can lead to more efficient optimization by focusing on critical objectives first [42]. Techniques like Standardised Rank Regression Coefficients assist in determining appropriate objectives' weights. This ensures that objectives are appropriately balanced in the optimization process [33]. Sensitivity analysis helps identify how variations in input variables affect objective responses. This information aids in designing solutions that are robust against uncertainties [43].

4. Discussion

The study aimed to investigate if and how beyond-vision effects are integrated in multi-objective lighting design optimisation in terms of performance metrics and simulation tools and whether the integration of beyond-vision effects influences the simulation workflows. By analysing the collected data and comparing it with previous research, insights into the broader implications of this work is gained.

The MOO in most studies focused on maximising visual performance and minimising energy consumption. Most studies included three or more aspects [26, 27, 29-37] but mention in the limitations that more aspects with more metrics and design variables must be tested to achieve higher optimisation reliability and obtain a virtual environment that would accurately imitate the real-world conditions. The reviewed studies all lack an essential aspect: optimizing lighting design to incorporate beyond-vision effects alongside building performance and well-being considerations. Notably, these effects were not simulated at the user or product level in any of the studies. This gap highlights the need to comprehensively integrate beyond-vision effects for a more holistic lighting design approach. However, multi-objective optimisation allows adding more objectives and aspects with multiple metrics, meaning that beyond-vision effects can be added to the problem formulation by following the same workflow mentioned before, e.g., the objective could be reducing circadian stimulus at night to benefit beyond-vision effects. Achieving this would entail advanced simulations to capture intricate human perceptions and lighting product attributes [44]. Closing this gap could lead to designs that balance efficiency and occupant well-being more effectively.

Optimisation process often includes aspects with conflicting goals regarding metrics combinations. Many studies found that not all social, environmental, and economic aspects and their metrics are conflicting in results [27, 29, 30, 32-35, 37]. This agrees with Lan *et al.* [33] who found that adaptive shading could reduce glare, solar gains, and thermal temperature, and reduce energy and operational cost for cooling while balancing daylight level and electric lighting. Many studies used lighting quantity metrics like illuminance for evaluating visual comfort in combination with lighting quality metrics like glare and uniformity. Mixing the use of metrics in assessments of the incorrect aspect may lead to reduced quality of the optimised design solution. In addition, established photometric light measures such as illuminance, are inadequate for quantifying beyond-vision effects [15]. Therefore, it is important to distinguish between visual and beyond-vision effects and use appropriate metrics [45, 46].

The MOO simulation workflow included multiple steps and computational tools. Many studied cases started with parametric modelling of the building, then multiple simulation tools and plugins were used for various purposes, which were coupled with optimisation tools and algorithms [8, 24, 27-30, 32, 35, 37, 39-42]. This process has some limitations like long simulation time, no single integrated tool that could investigate all aspects, a reduced information flow, and lost information between tools, software compatibility issues. According to Kebede *et al.* [9], these limitations would lead to double the work of recreating models, and manually inputting information into building models. In general, the information flow between product data or existing multi-disciplinary models and simulation tools are not discussed in these studies. Instead, authors described the built environment and input properties in the presented test case.

Some software packages such as Grasshopper, Rhinoceros3D, DIVA, open-source engines Radiance, EnergyPlus, and other plug-ins for Ladybug and Honeybee, were used commonly by many studies due to their compatibility within the simulation workflow. Other simulation tools that are compatible with these packages can be used for simulating beyond-vision effects like ALFA and LARK. Most simulation tools were developed over the years with having one aspect in focus (e.g., 'energy' or 'daylight') with one (set of) metrics as results. Later in time, researchers started to couple tools in order to calculate multiple (sets of) metrics at the same time [47].

MOO is a complex process combining multiple aspects and metrics, and, therefore, many studies avoided evaluating the existing environment first and jumped directly to optimisation. Some studies used a black-box or white-box model to simplify the environment and only focus on the investigated design variable. Many studies omitted validating the optimisation solutions in real-life afterwards. According to Maile *et al.* [48], omitting validation affects both reliability by optimising the incorrect variable and validity if measured and simulated results show significant deviation. In general, solutions that are optimal in simulations, do not necessarily perform as well in complex and dynamic real-life, therefore validation is an important step to ensure quality of the design solution and validity of the simulation tool.

4.1 Limitations

In this review only the highest level of optimisation M5 (multi-objective optimisation) is investigated. Multi-objective analysis includes multi-disciplinary aspects but, in this project, the author has major focus on light-related aspects and beyond vision effects. Building performance aspects with less relevance for the thermal-visual-energy-cost balance (e.g., acoustics) were not considered. Studies that solely focus on simulations of beyond-vision effects were not investigated in the scoping search as this review focused on multi-objective optimisation integrating beyond vision effects.

An overall limitation is the use of only simulation software to assess visual comfort and well-being. As Yusoff *et al.* [49] and Houser and Esposito [44] showed there are certain spatial aspects like special perception, multi-sensory experience and user engagement that cannot be predicted and evaluated if user-based assessment is missing. Therefore, simulations and on-site assessments complement each other and can be used at different design stages [49].

5. Conclusions

The review shows that none of the found papers combine visual and beyond-vison effects with other building performance aspects, and is, therefore, not integrated in the workflow of a full multi-objective optimisation simulation. Multiple steps and simulation tools are used starting with parametric modelling, building performance simulation tools and optimisation tools for generating algorithms. The simulation workflow is very affected by the compatibility of the software packages that hinders the choice of the most suited lighting metrics. Other factors like building type, light source, design variables are less represented in reviewed papers. Future research could focus on the less represented conditions in MOO to increase the scientific evidence and integrate beyond-vision effects in the MOO workflow.

References

[1] Gkaintatzi-Masouti M, van Duijnhoven J and Aarts M P J 2023 Simulations of non-image-forming effects of light in building design: A literature review *Lighting Research & Technology* **55** 669-89 doi:10.1177/14771535221142812

[2] Jalilzadehazhari E, Vadiee A and Johansson P 2019 Achieving a Trade-Off Construction Solution Using BIM, an Optimization Algorithm, and a Multi-Criteria Decision-Making Method *Buildings (Basel)* **9** 81 doi:10.3390/buildings9040081

[3] Fedorczak-Cisak M, Kotowicz A, Radziszewska-Zielina E, Sroka B, Tatara T and Barnaś K 2020 Multi-Criteria Optimisation of an Experimental Complex of Single-Family Nearly Zero-Energy Buildings *Energies* **13** 1541 doi:10.3390/en13071541

[4] Yu F, Wennersten R and Leng J 2020 A state-of-art review on concepts, criteria, methods and factors for reaching 'thermal-daylighting balance' *Building and environment* **186** 107330 doi:10.1016/j.buildenv.2020.107330

[5] Shahbazi Y, Heydari M and Haghparast F 2019 An early-stage design optimization for office buildings' façade providing high-energy performance and daylight *Indoor and Built Environment* **28** 1350-67 doi:10.1177/1420326X19840761

[6] Deb K, Pratap A, Agarwal S and Meyarivan T 2002 A fast and elitist multiobjective genetic algorithm: NSGA-II *IEEE Transactions on Evolutionary Computation* **6** 182-97 doi:10.1109/4235.996017

[7] Negendahl K 2015 Building performance simulation in the early design stage: An introduction to integrated dynamic models *Automation in construction* **54** 39-53 doi:10.1016/j.autcon.2015.03.002

[8] Zani A, Andaloro M, Deblasio L, Ruttico P and Mainini A G 2017 Computational Design and Parametric Optimization Approach with Genetic Algorithms of an Innovative Concrete Shading Device System *Procedia Engineering* **180** 1473-83 doi:10.1016/j.proeng.2017.04.310

[9] Kebede R, Moscati A and Johansson P 2020 Semantic web for information exchange between the building and product manufacturing industries: a literature review *37th International Conference of CIB W78, Sao Paulo - Online, Brazil* 248-65 doi:10.46421/2706-6568.37.2020.paper018

[10] Schlangen L J M 2019 CIE position Statement on Non-Visual Effects of Light: Recommending Proper Light at the Proper Time. (Vienna: International Commission on Illumination) p 3 Retrieved from: https://cie.co.at/publications/position-statement-non-visual-effects-light-recommending-proper-light-proper-time-2nd

[11] Boyce P R 2014 *Human factors in lighting* (Boca Raton: CRC Press, Taylor & Francis Group) doi:10.1201/b16707
[12] Vetter C et al. 2022 A Review of Human Physiological Responses to Light: Implications for the Development of Integrative Lighting Solutions *Leukos* **18** 387-414 doi:10.1080/15502724.2021.1872383
[13] Hattar S, Liao H W, Takao M, Berson D M and Yau K W 2002 Melanopsin-Containing Retinal Ganglion Cells: Architecture, Projections, and Intrinsic Photosensitivity *Science (American Association for the Advancement of Science)* **295** 1065-70 doi:10.1126/science.1069609
[14] CIE 2018 System for metrology of optical radiation for ipRGC-influenced responses to light. doi:10.25039/S026.2018
[15] Lucas R J et al. 2013 Measuring and using light in the melanopsin age *Trends in neurosciences (Regular ed.)* **37** 1-9 doi:10.1016/j.tins.2013.10.004
[16] Brown T M et al. 2022 Recommendations for daytime, evening, and nighttime indoor light exposure to best support physiology, sleep, and wakefulness in healthy adults *PLOS Biology* **20** e3001571 doi:10.1371/journal.pbio.3001571
[17] International WELL Building Institute 2021 WELL v2 Q4 2021: WELL building standard. Retrieved from: https://v2.wellcertified.com/wellv2/en/overview
[18] Rea M S, Figueiro M G, Bullough J D and Bierman A 2005 A model of phototransduction by the human circadian system *Brain Research Reviews* **50** 213-28 doi:10.1016/j.brainresrev.2007.08.003
[19] Ámundadóttir M L, Andersen M and Lockley S W 2016 *Light-driven Model for Identifying Indicators of Non-visual Health Potential in the Built Environment*: Ecole Polytechnique Fédérale de Lausanne) Retrieved from: https://books.google.se/books?id=J4K4nQAACAAJ
[20] Gkaintatzi-Masouti M, Pierson C, Van Duijnhoven J, Andersen M and Aarts M 2022 A simulation tool for building and lighting design considering ipRGC-influenced light responses *E3S Web of Conferences* **362** 01001 doi:10.1051/e3sconf/202236201001
[21] SOLEMMA ALFA 0.5.0.7 Beta Retrieved from: https://www.solemma.com/alfa
[22] Khademagha P, Aries M B C, Rosemann A L P and van Loenen E J 2016 Implementing non-image-forming effects of light in the built environment: A review on what we need *Building and environment* **108** 263-72 doi:10.1016/j.buildenv.2016.08.035
[23] Booth A, Sutton A, Clowes M and Martyn-St James M 2022 *Systematic approaches to a successful literature review* (Los Angeles: SAGE) Retrieved from: https://www.researchgate.net/publication/235930866_Systematic_Approaches_to_a_Successful_Literature_Review
[24] Wang S, Yi Y K and Liu N 2021 Multi-objective optimization (MOO) for high-rise residential buildings' layout centered on daylight, visual, and outdoor thermal metrics in China *Building and Environment* **205** doi:10.1016/j.buildenv.2021.108263
[25] Ahmad M W, Mourshed M, Hippolyte J L, Rezgui Y and Li H 2015 Optimising the scheduled operation of window blinds to enhance occupant comfort. International Building Performance Simulation Association) pp 2393-400 Retrieved from: https://www.scopus.com/inward/record.uri?eid=2-s2.0-84963762999&partnerID=40&md5=7a50a675fa3418e821d57e48966e24fa
[26] Mandal P, Dey D and Roy B 2021 Optimization of luminaire layout to achieve a visually comfortable and energy efficient indoor general lighting scheme by Particle Swarm Optimization *LEUKOS - Journal of Illuminating Engineering Society of North America* **17** 91-106 doi:10.1080/15502724.2018.1533853
[27] Rabani M, Madessa H B and Nord N 2021 Building retrofitting through coupling of building energy simulation-optimization tool with cfd and daylight programs *Energies* **14** doi:10.3390/en14082180
[28] Mangkuto R A, Dewi D K, Herwandani A A, Koerniawan M D and Faridah 2019 Design optimisation of internal shading device in multiple scenarios: Case study in Bandung, Indonesia *Journal of Building Engineering* **24** doi:10.1016/j.jobe.2019.100745
[29] Lavin C and Fiorito F 2017 Optimization of an External Perforated Screen for Improved Daylighting and Thermal Performance of an Office Space. ed L Ding, *et al.*: Elsevier Ltd) pp 571-81 doi:10.1016/j.proeng.2017.04.216
[30] Fan Z, Liu M and Tang S 2022 A multi-objective optimization design method for gymnasium facade shading ratio integrating energy load and daylight comfort *Building and environment* **207** 108527 doi:10.1016/j.buildenv.2021.108527

[31] Harish V S K V and Kumar A 2020 Simulation based energy control and comfort management in buildings using multi-objective optimization routine *International Journal of Mathematical, Engineering and Management Sciences* **5** 1324-32 doi:10.33889/IJMEMS.2020.5.6.098
[32] Kurian C P, Aithal R S, Bhat J and George V I 2008 Robust control and optimization of energy consumption in daylight-artificial light integrated schemes *Lighting Research and Technology* **40** 7-24 doi:10.1177/1477153507079511
[33] Lan L, Wood K L and Yuen C 2019 Sustainable design of residential net-zero energy buildings: A multi-phase and multi-objective optimization approach. American Society of Mechanical Engineers (ASME)) doi:10.1115/DETC2019-97171
[34] Reffat R M and Ahmad R M 2020 Determination of optimal energy-efficient integrated daylighting systems into building windows *Solar Energy* **209** 258-77 doi:10.1016/j.solener.2020.08.086
[35] Wang Y, Yang W and Wang Q 2022 Multi-objective parametric optimization of the composite external shading for the classroom based on lighting, energy consumption, and visual comfort *Energy and Buildings* **275** doi:10.1016/j.enbuild.2022.112441
[36] Xu X, Fu B, Wu Z and Sun G 2021 Predictive control for indoor environment based on thermal adaptation *Science Progress* **104** doi:10.1177/00368504211006971
[37] Zhang A, Huang Q, Sun Y and Zhang Q 2017 Multi-objective window optimization of school buildings for thermal and daylight performance in the cold climate of China. ed L Brotas, *et al.*: NCEUB 2017 - Network for Comfort and Energy Use in Buildings) pp 4461-8 Retrieved from: https://www.scopus.com/inward/record.uri?eid=2-s2.0-85086310317&partnerID=40&md5=0c1da95a7c44de26467a5a2e6d088732
[38] Tomažič S, Logar V, Kristl Ž, Krainer A, Škrjanc I and Košir M 2013 Indoor-environment simulator for control design purposes *Building and Environment* **70** 60-72 doi:10.1016/j.buildenv.2013.08.026
[39] Sorooshnia E, Rashidi M, Rahnamayiezekavat P, Rezaei F and Samali B 2021 Optimum external shading system for counterbalancing glare probability and daylight illuminance in Sydney's residential buildings *Engineering, Construction and Architectural Management* doi:10.1108/ECAM-03-2021-0191
[40] Ayoub M 2018 Integrating illuminance and energy evaluations of cellular automata controlled dynamic shading system using new hourly-based metrics *Solar Energy* **170** 336-51 doi:10.1016/j.solener.2018.05.041
[41] Kim J 2013 Adaptive façade design for the daylighting performance in an office building: The investigation of an opening design strategy with cellular automata *International Journal of Low-Carbon Technologies* **10** 313-20 doi:10.1093/ijlct/ctt015
[42] Sadegh S O, Gasparri E, Brambilla A and Globa A 2022 Kinetic facades: An evolutionary-based performance evaluation framework *Journal of Building Engineering* **53** doi:10.1016/j.jobe.2022.104408
[43] Sun B, Luh P B, Jia Q S, Jiang Z, Wang F and Song C 2013 Building energy management: Integrated control of active and passive heating, cooling, lighting, shading, and ventilation systems *IEEE Transactions on Automation Science and Engineering* **10** 588-602 doi:10.1109/TASE.2012.2205567
[44] Houser K W and Esposito T 2021 Human-Centric Lighting: Foundational Considerations and a Five-Step Design Process *Frontiers in neurology* **12** 630553 doi:10.3389/fneur.2021.630553
[45] Reinhart C F, Mardaljevic J and Rogers Z 2006 Dynamic Daylight Performance Metrics for Sustainable Building Design *LEUKOS* **3** 7-31 doi:10.1582/LEUKOS.2006.03.01.001
[46] Lee J and Boubekri M 2022 Introduction of new daylighting metrics for health, wellbeing, and feasibility: a study of the indoor building environment *Journal of Green Building* **17** 105-26 doi:10.3992/jgb.17.1.105
[47] Pan Y et al. 2023 Building energy simulation and its application for building performance optimization: A review of methods, tools, and case studies *Advances in Applied Energy* **10** 100135 doi:10.1016/j.adapen.2023.100135
[48] Maile T, Bazjanac V and Fischer M 2012 A method to compare simulated and measured data to assess building energy performance *Building and Environment* **56** 241-51 doi:10.1016/j.buildenv.2012.03.012
[49] Yusoff M A M, Ahmad J, Mansor A N, Johari R, Othman K and Hassan N C 2016 Evaluation of School Based Assessment Teacher Training Programme *Creative Education* **07** 627-38 doi:10.4236/ce.2016.74065

Indirect lighting in Contemplative Spaces – an Endless Depth

Juan Carlos Mirson
Licht Kunst Licht AG, Berlin, Germany.

Jcmirson@gmail.com

Abstract. This paper explores the relationship between indirect lighting and contemplative spaces. Through an examination of notable architectural works, the study identifies a consistent preference for hidden light sources in contemplative spaces across various religions and eras. The research begins by defining the fundamental aspects of contemplation and its correlation with architecture. This is followed by a bibliographic survey of indirectly illuminated contemplative spaces. Within this investigation, three approaches to understanding these spaces are considered. Firstly, the conceptual ideas that architects sought to infuse into their creations; secondly, sensory descriptions provided by experts aimed at elucidating the luminous atmospheres of these locations. Lastly, the study delves into these diverse themes through the lens of phenomenology, highlighting perceptual peculiarities such as the predominance of sight, the significance of peripheral stimuli, and hapticity. This offers insights into the relationship between specific attributes of indirect lighting and transcendental architectural experiences. The consistent suggestion of perceiving endless depth in contemplative spaces prompts a brief discussion on the essence of our perception of depth. This study underscores light's significance as a tool to connect with architectural spaces, transforming them into realms of imagination and awe.

1. Introduction

Delving into the observation of contemplative spaces, I noticed the inherent connection between them and indirect lighting solutions. Through the study of various significant architectural works, I discovered numerous instances of churches (e.g. St. Moritz Church, 2013, Germany), chapels (e.g. St Henry's Ecumenical Art Chapel, 2005, Finland), cathedrals (e.g. Gloucester Cathedral, 11th century, England), temples (e.g. Jambukeswarar Temple, 600 BC, India), mosques (e.g. Sancaklar Mosque, 2011, Turkey), and other spiritual spaces from different eras that are illuminated indirectly, concealing the light source. This reveals a consistent trend transcending various religions and periods, implying a timeless preference among architects in designing these types of buildings. In a more contemporary context, as we will explore in subsequent chapters through clear examples, prominent modern architects propose a close relationship between this type of illumination and contemplative spaces. This connection extends beyond the realm of symbolism and religious connotations.

These observations lead me to the main questions that will guide this investigation:

- What are the main characteristics of contemplation, and how do they manifest in architectural experiences?
- How does indirect lighting impact and contribute to the creation of contemplative states in this type of environment?

2. Methodology

To achieve the goal of addressing the questions proposed in this investigation, I will begin with a brief introduction to the idea of contemplation, where I will define certain qualities of it and its connection to architecture. The second chapter will focus on indirect lighting, starting with an approach to the definition of this terminology. I will then identify the categories of illuminated spaces that I intend to explore. This will be followed by an extensive bibliographic survey of these spaces, aiming to understand the ideology of the designers behind them. Simultaneously, aiming to shed some light into a topic that goes beyond quantifiable perceptual qualities, I will gather perceptual descriptions of the main characteristics of these indirectly lit contemplative spaces. The theoretical framework will draw from authors and theorists who have explored architectural interventions through the philosophical prism of phenomenology. Finally, I will widen the scope by discussing the recurrent topic of perception of depth, ending by condensing in a synthetic way the main points of the aforementioned aspects and presenting the conclusions drawn from the investigation.

3. Contemplation

The etymology of the word “contemplate” dates back to Latin, more precisely to the word Templum, which means “a piece of ground consecrated for the taking of auspices, or a building for worship” [1] — an origin that inspires reflection by presenting a direct connection between spiritual and architectural notions.

In a religious context, the practice of contemplation seeks a direct awareness of the divine which transcends the intellect. There are numerous historical references within various religions to such metaphysical experiences, and for the purposes of this paper it is pertinent to list some of the defining characteristics:

Spacelessness, timelessness, selflessness, boundlessness, radiance of our fundamental nature, mystical condition, something greater than oneself, union with God, transcendence, pure consciousness, absence of thought, absence of space, time and body awareness, ecstasy, transcending the subtlest level of mental activity, mental silence, trance state, cessation of mental activity, oblivious to external and internal environmental stimuli, no-end, pleasant, peaceful state of mind, all-oneness, bliss, expanded consciousness, letting go, benevolent disregard, non-thought, non-duality. [2]

And it is in seeking to encompass these concepts that the German theologian and philosopher from the early 20th century, R. Otto, introduced the term "numinous" to describe a profound emotional experience present in all religions. Also describing that darkness, silence, and emptiness are among the methods that evoke the numinous experience. [3, p.228]

Beyond the religious realm, it is interesting to highlight that these types of transcendental experiences have been studied since the beginning of philosophy as we know it. For Plato himself, contemplation was an integral part of his philosophy. He believed that through contemplation, one can access the knowledge of the form of the good or other divine forms. [4]

As I have already proposed in the introduction, the last millennia of human history inspire us with breathtaking architectural projects that encourage these contemplative states, “in which physical characteristics turn into metaphysically charged feelings of transcendental reality and spiritual meanings” [3, p.19] says Architect Juhani Pallasmaa, presenting to us spaces that go beyond purely religious ones, unleashing experiences that “arise from the experimental holiness of life itself and a deep existential recognition of one’s own being”. [5, p.21]

Some modern architects have emphasized the importance of these metaphysical aspects, seeking to define and reference these intangible qualities through their language, philosophy, and work. Louis Kahn (2008, p.14), for example, viewed these qualities as the "unmeasurable" and believed that great works of art are born from this realm. LeCorbusier, inspired by his visit to the Parthenon, described these qualities as the "ineffable", an inexpressible and emotionally sensed experience with boundless depth.[3] And it is by exploring these transcendental aspects, that they have been able to incorporate these qualities into their own work, surpassing the physical and reaching into the metaphysical essence of space.

It is concluded from this chapter that the interpretation of these states of the individual —active subjects of contemplation— may be approached from different angles, whether they are defined as mystical, philosophical, psychological, or artistic. The reaction of the subjects will be imbued with all these elements that are difficult to define but which are evidently present in spaces that transport them beyond their limits.

4. Indirect Lighting in Contemplative Architecture

Because of the historic relevance —of ancestral nature— that contemplative spaces propose, we will focus this chapter on the lighting that has been there through all of the human process of evolution, and that, evidently, is of great importance for our understanding of the world: Daylight.

4.1. Indirect Lighting: a definition

There are different types of indirect lighting. In fact, if we understand it from the strictest sense of the word, every perceived light enters our eyes indirectly; for it to be visible it must first hit, or go through, some form of matter [6]. As we can see, the definition of this lighting tool, that is part of the specific terminology commonly used in the lighting design field, has vague limits. Where the only certainty defined by the Society of Light and Lighting is that indirect lighting is achieved when the light reaches the space "via a number of inter-reflections on the room surfaces". [7]

In studies carried out by Rockcastle and Andersen —through meticulous perceptive analyses of daylight in contemporary architecture— we find a categorization of different types of illumination. The result of their studies is a matrix of seventy-five examples divided into fifteen categories. [8] At the same time, the general aspects of each category are compressed in an abstract way to create a single representative space for each of them. [8, p.37]

Rockcastle and Andersen mention that many of the religious spaces fell in the second half of the matrix under the two categories: "Partially Direct and Partially Indirect" and "Spatial Indirect". This observation leads this study to focus especially on these two categories [8, p.32]. Adding that, because of the number of contemplative buildings that belong to the category "Surface Indirect", it will be also kept as being on the limit of the relevant cases.

Now that we have established the typology of illuminated spaces on which this study is based, we will define the main characteristics of these spaces and the implications of these characteristics on contemplative states.

4.2. The Phenomena

To achieve a deeper understanding of the relationship between indirect light and contemplative spaces, we need to approach the qualities of this type of illumination from a sensitive and perceptual perspective, that can allow us to shed light on aspects that go beyond quantifiable and measurable aspects. To accomplish this, I will conduct a bibliographic survey that will explore three fronts:

- The concepts and ideologies behind the architects responsible for these spaces, seeking to understand the ideas they sought to convey to their works and how they connect with the world of lighting design.
- Written descriptions of human experiences within these spaces, such as those by Professor Henry Plummer, winner of the Daylight Award for his sensitive writings and research on light, as well as other magazines and books dedicated to architecture, lighting and perception.
- Contrasting these aspects with philosophical concepts, more specifically related to the phenomenological current of perception, which will approach the essence of these transcendental experiences.

For a clearer understanding, the bibliographical survey will be divided in subcategories that bring together the most relevant aspects that define these illuminated spaces, and aim to collect the main characteristics of the intersection between Indirect lighting and Contemplative spaces.

4.2.1. No-View. Firstly, through the observation of the works included in these categories, we notice that the openings that let the sunlight in have a particular quality: they do not provide a glance at the exterior.

In modern design where the trend promotes extensive use of windows with views, the absence of external visual access is often considered a flaw. Mexican architect Luis Barragan believes that large plate windows strip buildings of intimacy, shadow, and atmosphere, leading to a loss of privacy and a sense of living a public life away from home [9, p.51]. In the same direction, Pallasmaa adds that the window has lost its role as a mediator between the interior and exterior worlds, boundaries that are crucial as they bridge the external world with our consciousness, shaping our understanding of space [10, p.8]. An important point is evoked by emphasizing the feeling of privacy. This is especially relevant in contemplative spaces, where human beings seek to delve into aspects related to their inner selves, exploring the depths of their psyche and soul; therefore, a sense of comfort and security is required. [3, p.46]

Figure 1. St. Ignatius Chapel, S. Holl.

Buildings like the St. Ignatius Chapel (United States, 1997) (Figure 1), designed by Steven Holl, present a space where light is channelled through a huge puzzle immersed in the structure and slides softly and mysteriously to the inside, "a caressing, breathing, and healing light that connects us with the constantly changing nature of daylight and its cosmic ambience" [5, p.26]. These thoughts are directly linked to the ones of Holl himself. According to him "architecture can define fluid movement by determining daytime and nighttime viscosities of light and shadow... style and form at some point disappear..." [11, p.227]

Another good example with these characteristics is Zumthor's Vals Thermae (Switzerland, 1996), where we see a space that presents channels that surround it, creating high contrast and a flow. A space that feels intimate, accompanied by a feeling of freedom. Zumthor describes it as a medium to wander, a state of mind that has to do less with directing people but more with seducing them. [12, p.60]

4.2.2. Carved Architecture. As we can appreciate in the last examples, this type of lighting solution canalizes the light through architecture, creating carvings that are intended to act as optical instruments that distribute and carry the light through the space.

This is an aspect clearly noticeable in the ideas of Japanese architect Tadao Ando, who says about his own work that "the severely built openings caught the movement of light with precision. Space was carved...by a line of light that pierced the darkness, its appearance altered with each successive moment" [13, p.353]. Making us understand how by this specific type of illumination also darkness can be achieved. These are ideas that have a clear connection with those from architect Louis Kahn, who believed that "even a space intended to be dark should have just enough light from some mysterious opening to tell us how dark it really is." [14].

Figure 2. St. Hallvard Church, Lund & Slaatto

Also in this direction, Plummer notes in reference to the St. Hallvard Church (Norway, 1966) (Figure 2), designed by architects Lund & Slaatto, that through the subtlest touch of light, this dimly enigmatic place defies our grasp, "transcending both sight and comprehension, while projecting a nocturnal, gestative image—a threshold for new emergence." [12, p.231]. Understanding these great architectural spaces almost as living organisms, in which there is a "constant deep breathing of shadow and light; shadow inhales and illumination exhales light" [9, p.51].

4.2.3. Beyond Walls. It is clear at this point that our perception goes beyond the walls that limit the space: something that is created in our mind through a series of signs that our surrounding provides but that does not correspond to a direct association but rather to our capacity for imagination. "Something is never merely enclosed into its borders. It gives off heat like an oven and cold like ice." states the artist zur Lippe. [15, p.144]

In this direction, French philosopher Gaston Bachelard claims that inside and outside are an absolutism, where an object or concept is either in or out, 'here' or 'there'. He argues that both inside and outside depend upon one another and that is the only way they can preserve their identities, where our own imagination proposes a revision of their boundaries [16, p.226-231].

According to Plummer, architect Juha Ilmari Leiviskä highlights certain qualities of natural light within his designs, specifically noting the ephemeral nature of their illumination. The writer adds that this light connects us with time and a reality that exists "beyond Men, towards something infinite, and eternal" [12, p.151]. Descriptions that do not only blur the limits of our body but also give a body to the architectural space, as Pallasmaa says: "Natural Light, with its ethereal variety of change, fundamentally orchestrates the intestines of architecture". [5, p.26]

Figure 3. Church of the Good Shepherd, J. Leiviskä

When studying one of Leiviskä's works, the Church of the Good Shepherd (Finland, 2002) (Figure 3), Plummer notes that within these spaces of profound whiteness, the luminous fragmented elements transcend their fragmented perception. This leads to a harmonious integration with the surrounding space, reestablishing the fundamental connection between humanity and nature [12, p.18]. Adding that these factors transport us to calm moments, settling down our internal rhythms and putting us in a contemplative mood [12, p.68].

Additionally, the author describes that the intervening vertical elements within the space gain varying degrees of illumination, creating multiple levels of emptiness, light, and shadows, evoking an infinite sensation similar to a forest, where the mind can endlessly explore [12, p.144]. These concepts echo Bachelard's ideas about the vastness of the forest, where, he says, one feels the apprehensive sensation of delving deeper into a boundless realm, surpassing the shroud of trees to create a psychologically transcendent enigma. Arguing that there is something more to be expressed besides the objective description of this experience— "what should be expressed is hidden grandeur, depth." [16, p.203].

4.2.4. Directionality. When analysing the descriptions of these spaces, we understand that this type of indirect lighting solution often imparts a sense of directionality to the perceived experience. This directionality emerges, on one hand, as a possibility of movement, defining our behaviour in the space. For instance, in the case of the Kiasma Museum of Contemporary Art (Finland, 1998), its architect, Steven Holl, states that "when moving through space, with a twist and turn of the head, mysteries of gradually unfolding fields of overlapping perspectives are charged with a range of light—from steep shadows of bright sun to the translucence of dusk". [11, p.278]

On another hand, this directionality extends itself as a sensation beyond our physical capabilities. For example, in the case of the Dybaer Church (Denmark, 2005), by Regnbuen Architects, the horizontal channel on the ceiling behind the altar catches a larger amount of light, looking brighter and "marking the holiest point in the church… making a vertical link with the heavens," describes Plummer. [12, p.36].

Figure 4. Bagsvaer Church, J. Utzon

A sensation similar to what occurs in the Bagsvaerd Church (Denmark, 1976) (Figure 4) by Jorn Utzon, where the bright "upcurving vaults lift the eye and appear to hover free of the earth, imparting celestial mood. Emphasizing this levitation is the detachment of vaults by the skylit aisles". [12, p.56] About which Utzon itself states: "the light in the corridors provides almost the same feel as the light experienced in the mountains during a sunny day in winter, making these elongated spaces happy places in which to walk." [17, p.22]

Or, alternatively perceivable, in a different manner, at the Church of Enghoj (Denmark, 1994) designed by Henning Larsen. The walls of concrete, touched by light entering through the ceiling's perimeter, allow us to slide into the immensity of the sky. Simultaneously, they connect us to the force of gravity that keeps us within this never-ending tunnel. [12, p.176]

As we see from these descriptions, these types of luminous solutions can trigger outstanding feelings, but there are not particularly placed in the focal point of the visitor. On the contrary, they spread extensively to the peripheral part of the visual field. And it is precisely this stimulus of the peripheral vision that can transport us to a world that is less rational, somehow less physical, and can grant us a blurred vision of our surroundings, connecting us to it in a less conscious way, enveloping us in the world, blurring depth and distance, and allowing for a more participatory and empathetic gaze. [9]

4.2.5. Diffuse Light. As a characteristic of indirect light, it is noteworthy how, in many of these spaces, the light, upon reflecting off various surfaces before reaching the interior of the space, acquires an especially diffused quality, forming a significant bond with encountered materials and surfaces. A diffused light that can provide both a dark space where, as we have seen, it spreads faintly, as well as white luminous spaces that acquire heavenly characteristics when lighted with soft gradients.

Figure 5. Riola Parish Church, A. Aalto

Architect Alvar Aalto, for example, developed a curvaceous sfumato in which light gradually shades, creating softened outlines and hazy forms. Specifically, in the Riola Parish Church (Italy, 1978) (Figure 5), we can appreciate how Aalto manages to escape physical form by making the space look blurry, where "light is sprayed onto vaults from three scalloped monitors, casting an array of imprecise tones that seem airbrushed on, veiling the church in a dreaminess that is free, in part, of any solid object". [12, p.179]

4.2.6. Thingness of Light. When writing about these spaces where openings have the main task of letting the light in, without allowing a visual of the exterior, Corrodi and Spechtenhauser claim that "materiality could emerge in tangible sensuousness". [18, p.85]

When we look, for example, at the already mentioned St Ignatius chapel (Figure 1), we see how the light passes through a series of carved, indirect figures that diffuse it and grant it characteristics of the surfaces it passes through, paradoxically heightening "both the immateriality and materiality of light". [5, p.26] These characteristics are also evident in the spaces created by Alvar Aalto, where the daylight arrangements frequently reflect light into curved white surfaces. This results in light taking on qualities of being material and malleable, with a heightened sense of presence. Additionally, it acquires distinct characteristics like weight, temperature, and tactile qualities, appearing as a moulded and slowed-down light that feels like matter. [5, p.26]

And in this sense, it's worth mentioning the importance of touch for many philosophers within the phenomenological movement, for whom "the world touches us and we touch the world, primarily through the medium of light." [9, p.45] Ideas that go even a step further in architectural works like Luis Barragan's Capuchin Convent Chapel. (Mexico, 1952) where light turns into a warm, coloured liquid that even suggests sonorous qualities invoking an imaginary humming sound. [19]

4.2.7. Silence. The idea of atmospheric silence is studied, mentioned, and elaborated upon by the vast majority of the architects and philosophers mentioned in this chapter.

Figure 6. Baumschulenweg Crematorium, A. Schultes

As an outstanding example, it is interesting to see how in the Baumschulenweg Crematorium (Germany, 2000) (Figure 6), architects Schultes and Frank set out a monumental forest of concrete columns that join the ceiling through holes that let the light through in an almost mysterious way, providing the space with a silence that grants it a transcendental power. [20, p.196]

A concept that can be further understood by mentioning some of the thoughts of the many architects that have been already mentioned in this research:

- Juhani Pallasmaa: "transcending architecture fundamentally necessitates the assistance of light and silence, phenomena that raise our ontological, existential, and spiritual sensibilities." [5, p.11] In architectonic experiences, silence eliminates surrounding noise, directing our awareness inward to our own mystic existence [9, p.55]
- Peter Zumthor describes this beautiful silence of buildings as something that is not related to its aesthetics or to what it represents, but to its attributes like "composure, self-evidence, durability, presence and integrity". In his mind, it is about a "building that is being itself, being a building, not representing anything, just being". [21, p.32]
- Louis I. Kahn: develops the idea of silence as part of the unmeasurable of architecture, a silence that is capable of providing the space with a particular spirituality, a silence that "is not very, very quiet. It is something that you may say is lightless, darkless." [22, p.20]
- Tadao Ando: claims to believe that architecture should not speak too much. "It should remain silent and let nature in the guise of sunlight and wind speak…They activate space, make us aware of the season, and nurture within us a finer sensitivity." [13, p.196]
- Steven Holl: "A building speaks through the silence of perception orchestrated by light. Luminosity is as integral to its spatial experience as porosity is integral to urban experience." [23]
- In Henry Plummer: the atmospheric silence is achieved when the forms calm down and surfaces and materials are reduced, generating a necessary simplicity for the buildings to transcend their own physical limitations. [20, p.178]

And it is through these ideas that we begin to reflect on contemplation not as something that lasts an instant, but as a serene process where calmness and slow appreciation reign.

In this chapter, I described how these spaces demand a perceptual sensitivity that allows us to define certain qualities of them which are not particularly measurable. An aspect that leaves us considering how, as professionals, this type of description is necessary to encompass certain elements that aren't so specific but add great value to the spaces we design. Next, a brief immersion into the world of art will allow us to further develop these ideas.

5. Discussion: An Endless Depth

Many authors describe how these spaces break the boundaries, take infinite dimensions and evolve beyond their physical limits. Philosopher Merleau-Ponty argues that it is a mistake to conceive depth perception as an issue regarding distances in space; that the look from a God's eye perspective does not belong to us, since Depth can only be experienced from where the individual is standing. He argues that we must rediscover what spatial perception is, moving away from the traditional way of addressing this topic, to be able to elucidate the true relation between subject and object [24, p.299]. Claiming that it is this re-interpretation of depth that is the key that will allow us to understand how we experience space, since it "forces us to reject the preconceived notion of the world and rediscover the primordial experience from which it springs: it is, so to speak, the most existential of all dimensions, because it is not impressed upon the object itself, it quite clearly belongs to the perspective and not to things. Therefore, it cannot either be extracted from, or even put into that perspective by consciousness" [24, p,298]. He proposes to understand depth rather as a "lived distance", that binds us to things which count and exist for us, and links them to each other. A distance that measures the 'scope' of our life at every moment [24, p,333].

In turn, it is interesting to refer to Bachelard's thoughts, who argues that since immensity "is not an object, a phenomenology of immense would refer us directly to our imagining consciousness" [16, p.202]. To further present us the notions on this paradoxical approach to infinity, he quotes a passage of the poet Oscar Milosz, in which the author beautifully describes this transgression of limits:

> "As I stood in contemplation of the garden of the wonders of space, I had the feeling that I was looking into the ultimate depths, the most secret region of my own being.... Away with the boundaries, those enemies of horizons! Let genuine distance appear! .... Everything was bathed in light, gentleness and wisdom; in the unreal air, distance beckoned to distance. My love enveloped the universe." [16, p.207]

6. Conclusion

This paper explores the connection between indirect lighting and contemplative architecture that is observed from an anthropological perspective. The study of this link has led to the identification of certain aspects that, on one hand, define themselves as characteristics of this type of illumination, and on the other, are those that in some way imbue these spaces with transcendental qualities.

Within these spaces, light enters through channels created by cuts in the architecture, becoming a fundamental element of the perceived space. The interior space unfolds allowing the gaze to extend beyond walls, breaking boundaries and transcending its own physical limits. The lack of visual connection with the exterior can create an effect even more impressive than the view itself. It is as if the absence becomes present in our minds, triggering imagination and memories. The ephemeral light entering through these openings provides a glimpse to the outside, connecting us with the passage of time and our own ephemerality. These openings illuminate in a subtle manner, where not everything is immediately comprehensible. They also disperse stimuli in our peripheral sight, enveloping us in a softer, more blurred environment. These luminous elements provide a diffuse light, capable of creating a high contrast with the most mysterious darkness or a gentle spectrum of celestial whiteness, granting these spaces the necessary intimacy for contemplative states without introducing unnecessary elements or noise. On the contrary, they foster tranquillity and silence that transport us to realms resonating more with our subconscious than with physical boundaries. As light moulds and flows through these openings, it adopts the characteristics of the materials it reflects upon. It becomes diffuse, acquiring attributes typical of matter: weight, colour, temperature. Additionally, light gains volume, filling empty spaces in various forms, sometimes even becoming the space itself. Some spaces present a similarity to the endless depths of a forest, and others present endless depths that connect us with the gravitational forces —

spaces that lead us to immerse ourselves in the earth, leap towards the universe, escape towards the horizon, or stretch out like waves towards infinity. As individuals contemplate the vastness of these spaces extending beyond their physical confines, they create a medium for interpretation and metaphors. Ultimately, this can elevate their experiences to extraordinary levels of consciousness, turning awareness inward to their own mystic existence.

And this is how, starting from a specific link in the field of lighting design, fundamental characteristics of spaces have been uncovered, delving into various aspects of this connection. These aspects shown not to be limited solely to the realm of contemplative spaces, but also tug at strings related to our sensitivity, not only as beings who perceive, but also as those who design the spaces we inhabit.

7. Limitations

The intention of this study is not to create a formula for illuminating contemplative spaces, nor to define indirect light as the only means to achieve these states. Instead, it aims to mention intrinsic characteristics of these experiences that are often overlooked in the realm of lighting design, yet they offer insights into issues inherent to our nature and our relationship to architecture and its illumination. It serves as a homage to daylight and the sophisticated management of it within architectural spaces. It involves a descriptive challenge where merely using lighting terminology is insufficient, going beyond the realm of lighting, urging us to step outside our comfort zone and witness the synergies that light and architecture can create when combined.

References

[1] Lewis T C 2017 Definition of Templum. A Latin Dictionary. December 2017. http://latinlexicon.org

[2] Wahbeh H Sagher A Back W Pundhir P and Travis F 2017 *A systematic review of transcendent states across meditations and contemplative traditions. The Journal of Science and Healing.* Institute of Noetic Sciences. Accesesd December 2017 http://www.sciencedirect.com/science/article/pii/S1550830717300460

[3] Barrie T Bermudez J and Crosbie M J 2015 *Transcending Architecture: contemporary views on sacred spaces* Catholic University of America Press (California, USA)

[4] Rosen F 1980 Contemplation and Virtue in Plato. *Religious Studies*, *16*, 85–95. *http://www.jstor.org/stable/20005634*

[5] Pallasmaa J 2015 Light, Silence, And Spirituality in Architecture. In: Bermudez, J. *Transcending Architecture: contemporary views on sacred spaces*, USA: CUA Press

[6] Cuttle C 2015 Lighting Design: a Perception-based approach Routledge (New York, USA)

[7] Society of Light and Lighting 2012 SLL Code for Lighting. CIBSE. Accesesd December 2017 http://app.knovel.com/hotlink/toc/id:kpSLLCL00H/sll-code-lighting/sll-code-lightingAnother reference

[8] Rockcastle and Andersen 2013 Annual Dynamics of Daylight Variability and Contrast A Simulation-Based Approach to Quantifying Visual Effects in Architecture. London: Springer London.

[9] Pallasmaa J 2012 The eyes of the Skin, Architecture and Senses. UK: John Wiley & Sons.

[10] Pallasmaa J 2013 Towards a Neuroscience of Architecture. *Architecture and Neuroscience.* Finland: Tapio Wirkkala Rut Bryk Foundation, Finland. Accessed: December 2017 *http://www.anfarch.org/wp-content/uploads/2014/01/*

[11] Wong J F 2012 The script of viscosity: the phenomenal experience in Steven Holl's museum architecture, *The Journal of Architecture*, Accessed December 2017. http://www.tandfonline.com/doi/abs/10.1080/13602365.2012.678646

[12] Plummer H 2012 Nordic Light – Modern Scandinavian Architecture. UK: Thames and Hudson

[13] Shirazi M 2009 *Architectural Theory and Practice, and the Question of Phenomenology*. A thesis in: Brandenburg Technical University. Germany. Accessed: December 2007. https://opus4.kobv.de/opus4-btu/files/1840/

[14] Schielke T 2013 Light Matters: Louis Kahn and the Power of Shadow. Archdaily *Accessed Aug 2023. https://www.archdaily.com/362554/light-matters-louis-kahn-and-the-power-of-shadow*
[15] Griffero T 2016 *Atmospheres: Aesthetics of Emotional Spaces*. ed.2010 Routledge (London UK)
[16] Bachelard G 1958 *The Poetics of Space* ed 2014 Penguin books (New York, USA)
[17] Weston R 2015 Jorn Utzon: crafter of light. *Lighting Illumination in Architecture*. Volume 47 Issue 1. Jan 2015. Accesesd December 2017. http://digital.lighting.co.uk
[18] Corrodi M Spechtenhauser K 2008 *Illuminating* Birkhauser Verlag AG (Switzerland)
[19] Pallasmaa J 2016 Daylight & Architecture #26 2016. Accessed April 2023. https://www.daylightandarchitecture.com/dwelling-in-light-tactile-emotive-and-life-enhancing-light/
[20] Plummer H 2009 *La Arquitectura de la Luz Natural*. Barcelona, Spain: Blume
[21] Zumthor P 1988 *Thinking Architecture* ed 2010 Birkhauser (Switzerland)
[22] Kahn L I 2008 *Between Silence and Light* Shambhala Publications (London,UK)
[23] Santos S 2016 Steven Holl Wins 2016 Daylight Award, ArchDaily. Accessed December 2017. https://www.archdaily.com/795468/steven-holl-wins-2016-daylight-award-in-architecture
[24] Merleau-Ponty M 1945 *Phenomenology of Perception* ed 2005 Gallimard (Paris, France)

The Influence of Urban Lighting on the Sense of Belonging

Laura de Frutos[1,#], Luca Pellizzari[2,#] and Mette Hvass[3]

[1] Kardorff Ingenieure Lichtplanung GmbH. 10789 Berlin, Germany.
[2] Les Éclaireurs, Lighting Design Agency. 69001 Lyon, France.
[3] Aalborg University Copenhagen. Department of Architecture, Design and Media Technology. 2450 Copenhagen, Denmark.
[#] These authors contributed equally to this work and share first authorship.

Corresponding authors: laura.frutos.llorente@gmail.com; lucapelli97@gmail.com

Abstract. This research aims to gain a better understanding of the influence of artificial lighting on the sense of belonging in urban spaces and to investigate whether lighting is a determining factor in establishing bonds between humans and the city. An exploratory method is employed to give possible answers with the purpose of sparking a discussion in the lighting and urban design practice but without intending to be conclusive. The topic is addressed by an integration of a literature review on related concepts such as place attachment, the connections between urban lighting and atmospheres, the social dimension of urban lightscapes and placemaking, and input from conversations with professional lighting designers. The proposed answer consists of a series of models, design principles and methods for the lighting designer. It is argued that lighting can act as a generator of atmosphere, propitiating shared experiences in the outdoors, or as an object of memory when it acquires the category of a symbol. Both qualities can contribute to building a collective identity and, consequently, a sense of belonging. The lighting design - if unique, meaningful and sustaining social interaction - and the design process employed - if participatory - have a role in strengthening ties within the community.

1. Introduction

We are social beings. We grow up, live and dwell in communities. As described by Abraham Maslow in his 1943 paper *A Theory of Human Motivation*, we human beings hunger for social relations and a place in our group, "*If both the physiological and the safety needs are fairly well gratified, then there will emerge the love and affection and belongingness needs*" [1].

Throughout history, an important part of our social interactions has taken place in the public realm: people exchanged news, made deals, arranged marriages or attended events; goods were sold, artists entertained and processions, celebrations or political acts were held. The city has been the meeting place [2]. But city life takes place at night as well. Since the appearance of electrical light, our economic and social life has become more and more nocturnal, tending to blur the limits between day and night [3]. In recent years, recognised lighting designers such as Roger Narboni have stood by a change in the profession, going from lighting urbanism to nocturnal urbanism, which considers the subjective experience of its inhabitants [4]. Moreover, the latest developments in lighting technology make it possible to create a large variety of atmospheres. Indeed, there has been a change in the paradigm of both research and design practice in lighting, taking an interest in the human and social dimensions of urban lightscapes, but research and evidence are missing [5]. Over decades, lighting research has provided vast quantitative information about light levels and task performance, as well as visibility and perceived safety, but information about other dimensions of the perceived environments is lacking [6, 7]. As part of their work, lighting experts need an understanding of the social aspects of light and how to integrate these into their everyday practice [8]. As Casciani says: "*Because lighting is primarily intended for people, urban lighting should be focused on the human experience with a qualitative social engaging role*" [9]. Among all the possible emotions related to public life, the need to belong and connect with others is universal, can be found across all humans and has a proven impact on our mental and physical well-being [10].

1.1. Objective of the paper

The purpose of this research is to gain a better understanding of the role of artificial lighting concerning the sense of belonging in urban spaces. Can artificial lighting improve mental well-being and avoid alienation by establishing an emotional bond between citizens and urban space? How can we create more inviting neighbourhoods that support social interactions and collective identity? What kind of atmospheres do citizens interpret as belongingness? How can the qualities of "feeling at home" be transferred to the public realm?

The content of this paper is based on the master's thesis "The Influence of Urban Lighting on the Sense of Belonging" realised by the same authors Laura de Frutos and Luca Pellizzari during the year 2022 at Aalborg University. No previous studies or literature that directly links a sense of belonging (henceforth referred to as SOB) and urban lighting have been found. Thus, the main effort of this study will consist of finding the steps to connect the theories about SOB, use of public space, and outdoor lighting, as well as providing some conclusions and models that facilitate its implementation during the design practice. This study aims to open a discussion and give possible answers and solutions but does not intend to be conclusive.

2. Method

The research topic, i.e. the influence of urban lighting on SOB, lacks a clear definition and precise boundaries as no previous related data could be found. Therefore, in order to navigate, investigate and structure it, an exploratory method is employed, aimed at presenting new findings and reflections that can serve as a foundation upon which further research and investigation are drawn [11]. Exploratory research is meant to connect the information coming from different sources of knowledge, so as to get to an understanding of the topic, its nature and its underlying dynamics. It allows a thorough analysis of the problem and it explains how its variables interact with each other [12], namely identifying the steps that connect the theoretical knowledge about SOB, public space and outdoor lighting.

For that purpose, awareness of SOB is achieved thanks to the integration of a literature review and individual conversations with experts in the field. In practice, peer-reviewed research papers and books have been searched in the months from February 2022 to May 2022 in the following online databases: Google Scholar, EBSCOhost and Aalborg University Library. The starting point of the literature research consisted of defining SOB, from which it was then possible to individuate and define related concepts such as place attachment and collective identity. The research perspective was continuously widened by investigating new concepts and keywords that appeared along the way, with a major focus on topics such as urban atmospheres, the human and social dimensions of lighting, placemaking, and community engagement. All the information gathered was critically examined and organised, and further developed with the data collected through conversations with four professional lighting designers: Leni Schwendinger, Luciana Martinez, Lisbeth Skindbjerg Kristensen, and Anna Lykke Thorup (permission for publication was obtained from all interlocutors). They were chosen because they all have experience working with urban lighting but with different backgrounds and approaches. The intended purpose of the conversations was to collect knowledge on how emotional and social approaches can be applied during the design process, as well as reflect on their understanding of urban belongingness and personal experiences and methods when working together with citizens. Topics such as social interaction, collective identity, atmospheres or memories were intentionally addressed as well.

The inputs acquired from the literature review and conversations were synthesised into a theoretical framework. This framework presents a series of models, design principles and methods for the lighting designer, using real projects as examples for a better understanding.

3. Literature review

A literature review summarising existing knowledge on belongingness as a subjective emotion within an urban context and its links with place attachment and social dynamics, as well as the role of atmospheres and outdoor lighting, is hereafter expounded. Comments from the conversations with the aforementioned lighting designers are also included.

3.1. Sense of belonging

Sense of belonging, or belongingness, is "*the human emotional need to be an accepted member of a group to maintain close and safe ties that generate a sense of security, care and affection.*" [13]. It is a subjective feeling, based on an individual's perception of their connection to a chosen group (family, friends, school or work environments, communities, cultural groups, etc.) or physical place. It produces the feeling that "*one is an integral part of their surrounding systems*" [10]. Furthermore, this SOB is dependent on opportunities for interaction with others [10].

"*Belonging and love needs*" are in the middle of Maslow's hierarchy of needs. They cannot be satisfied until physiological and safety needs are met, but they precede esteem and self-actualization [1]. The need to belong is universal and can be found across all humans and cultures, although with cultural and individual differences in intensity, as well as how they satisfy and express this need [14]. Even if it is agreed that almost all people seek to satisfy the human need for belonging, there is no full agreement about how belonging should be measured and what people can do to satisfy it [1]. The benefits of connecting deeply with other human beings and secure places, as well as feeling that one is part of cultural identity, are multiple. These positive effects include more positive social relationships, academic achievement, occupational success, and better physical and mental health. On the other hand, alienation can lead to mental and physical problems, such as an increased risk for mental illness, antisocial behaviour, lowered immune functioning, physical illness, and early mortality [1].

Two kinds of belongingness can be distinguished: state or trait belongingness. State belongingness is related to the situation-specific SOB and it is influenced by several daily life events and stressors. Thus, the perception of one's own belonging can change over the day, as it happens with other emotions. However, trait belongingness, which is belonging as a core psychological need, is relatively stable even if the variability across time and situations depends on the subject and one's awareness of environmental and social context [1]. Since the need for belongingness can be only satisfied by other people, this is from outside the person, there is an important dependence on the environment [15].

When talking about urban belonging, each individual can associate a variety of words with a sentiment of belongingness. The more personal experiences a group of individuals have in common, the easier it will be to find an agreement [16].

3.2. Place attachment

The development of SOB in a public space relies on the development of place attachment. If place attachment describes the direct relationship between people and space, SOB describes the relationship among people within and in relation to the space. Place attachment is "*the symbolic relationship formed by people giving culturally shared emotional and affective meanings to a particular space that provides the basis for the individual's and group's understanding of the environment and its relation to it*" [17]. Place attachment relies on place identity and place dependence [18].

A space becomes a place when it is assigned meaning through personal, group, or cultural processes. Place attachment, for instance, is more developed for settings that evoke personal memories [18]. An undifferentiated space becomes a place when we visit it recurrently and endow it with value [17]. According to Manzo, what can be called *experience-in-place* creates meaning [19]. In the conversation with Anna Lykke Thorup, she further developed this point in relation to lighting: light is sometimes the actual centre of the experience and it is sometimes just something that sets the mood and the scene for other experiences to happen. She mentioned New Year's Eve fireworks as a past memory around a lighting scene that is collectively experienced. In this way, there is a past memory of the lighting. On the other hand, when lighting sets the mood and atmosphere to make it possible for people to enjoy themselves and have a good time with friends and other people, you have instead a memory of those social interactions and events sustained by the lighting.

Kevin Lynch stated that the concept of place attachment comprehends three components: identity (the distinctiveness of a place, the qualities that distinguish it from any other place), structure (the mental representation, spatially outlined) and meaning (subjective feelings linked to physically separate places) [20].

Place attachment is reliant on the social and physical dimensions of the space [18]. The social dimension consists of social ties and familiarity with the other residents of the neighbourhood. Physical attachment is also predicted by the duration of residence and the plans to stay. Certain physical characteristics of the space, such as density, proximity, and the presence of amenities and other social areas, can influence these interactions. Spatial bonds become important because they reflect social bonds and the identity of the group. You are attached to the place because of its distinctiveness from other places, and because it supports the specialness of a certain group [18].

3.3. Urban atmospheres and light

The term "atmosphere" appeared within a meteorological context, but since the eighteenth century, the word has been also used to define a "*certain mood hanging in the air*" [21]. Atmospheres are always something spatial and emotional at the same time. To be more specific, Böhme defines the atmosphere of a city as "*the subjective experience of urban reality which is shared by its people*". An urban atmosphere is constantly produced by the locals throughout their lives but the stranger easily notices it as a characteristic of the city. Since it is continuously produced, this atmosphere can mutate over time [21]. Böhme also argues that atmospheres must be studied from the point of view of the subject, by exposing ourselves to them, but also from the side of the object, by analysing the generators and conditions. Atmospheres are characterised subjectively [21], but they can be shared because even if they do not arouse exactly the same feeling for each individual, people may share the verbal representation of the affective impact [22, 21]. To reach an agreement, previous experiences, knowledge and culture play a key role [22]. Leni Schwendinger also mentioned the interplay between culture and language when defining atmospheres.

Urban lighting (natural and artificial) has a role in generating atmospheres in the city, and it should be therefore studied both quantitatively and qualitatively, as a physical phenomenon and as the result of physical perception. Scholars agree that there is a relationship between lighting, emotions and atmospheres: lighting enables visibility but it also can influence people's mood, well-being and behaviour, impart a certain sense of place and create a certain atmosphere in the space [23, 24]. Lighting, including urban one, influences the emotions of the observer with an effect that is to a large extent subjective, with light that "*transcends the cognitive and moves into the non-representational, the realm of the affective and sensual*" [25]. The affective qualities attributed to a place can be differentiated from the emotions evoked by the lighting itself, with them being "*more related to the experience of the lighting than to the appearance of the place*" [6].

3.4. *The social dimension of urban lightscapes*

Daria Casciani's book *The Human and Social Dimension of Urban Lightscapes* elaborates extensively on the role of emotional and cognitive perception in the lighting and nocturnal design to achieve more accessible, sociable, hospitable and sustainable cities by defining the criteria for a good place [5]. Although not focused on the nocturnal city, Jan Gehl's vision is also especially relevant to understanding the social dynamics in urban spaces [2]. The way a space is perceived has an impact on the emotional and behavioural response, shaping the activities that take place in it. Framed in his comprehension of the city as a meeting point, Gehl distinguishes three kinds of activities: necessary, optional and social activities. While necessary activities happen under all conditions, good physical conditions are essential to facilitate optional activities. Thus, the increase in activity level causes an increase also of social activities. A successful public realm must provide opportunities for multiple degrees of social engagement: from passive (e.g. seeing or hearing others) to active (e.g. chance meetings, children playing), including planned activities (e.g. markets, demonstrations) [2].

Lighting has a role in sustaining social activities. Good lighting conditions enable correct visibility of the surroundings and faces, thus facilitating observation of ongoing events and people. On the other hand, lighting is a determinant of generating inviting and comfortable atmospheres to stand and sit, so active social activities such as events, exchanges and conversations can take place. As emerged in the conversation with Luciana Martinez, design by itself cannot force people to interact, but it can set the conditions for social interaction by helping people feel comfortable in the space. She also states that designers cannot plan spaces targeted for everyone, but it is important that diverse people are still allowed to be in the space without feeling excluded. People with different backgrounds can coexist in the place although, necessarily, with different degrees of ownership or at different day times. This is aligned with the Urban Belonging Project [16], which questions if designing for belonging does not mean making all-inclusive spaces but designing for community and variation in ownership.

3.5. *Placemaking and community engagement*

Traditionally, urban planning has been exclusively placed in the hands of experts. However, since 1960, theorists and urban designers such as Jane Jacobs, William H. Whyte and Jan Gehl, have advocated for bringing the public into the design process [26]. Currently, different organisations and city planners contribute to involving local communities through placemaking approaches. Project for Public Spaces (PPS) defines placemaking as "*the collaborative process by which we can shape our public realm in order to maximise shared value*". It strengthens "*the connection between people and the places they share [...] and facilitates creative patterns of use, paying particular attention to the physical, cultural, and social identities that define a place and support its ongoing evolution.*" [27]. The International Association for Public Participation proposed five increasing levels of community involvement based on the degree to which the participation impacts decision-making: inform, consult, involve, collaborate and empower [26].

Three of the lighting designers who took part in the individual conversations (Leni Schwendinger, Lisbeth Skindbjerg Kristensen, and Luciana Martinez) have broad experience including communities in the design process. The three of them pointed out the necessity of a deeper understanding of necessities, uses of the space, and subjective perceptions when designing for a community. From Martinez's point of view, though, the biggest impact on people developing a shared identity has to do more with the process of how the lighting design is implemented than the lighting design itself, with the community being actively involved in shaping the design of the space they live in. They have a role and they feel listened to. People remember the activities they have contributed to, such as workshops, and by working together with their neighbours they build new relationships and strengthen already existing ones.

IOP Conf. Series: Earth and Environmental Science 1320 (2024) 012011 doi:10.1088/1755-1315/1320/1/012011

4. Designing for Sense of Belonging: an implementation into the lighting practice

Based on the knowledge acquired through the previous literature review, a series of models are developed as an attempt to connect the previous theoretical knowledge and the lighting design practice. These models must be understood as a potential but not exclusive interpretation and, therefore, open to discussion.

4.1. Sense of belonging model

The model shown in Figure 1 allows synthesising a definition of SOB and connecting it to lighting conditions. Social interaction in a space produces shared experiences among people. The more time is spent in the space (recurring time), the more experiences are developed and so a social place attachment is built. These shared experiences are then translated into collective memories. Elements that serve as symbols with a shared meaning among the community and the groups involved, translate also into collective memories. Developed collective memories are then responsible for building a collective identity, bringing SOB to the members of the community.

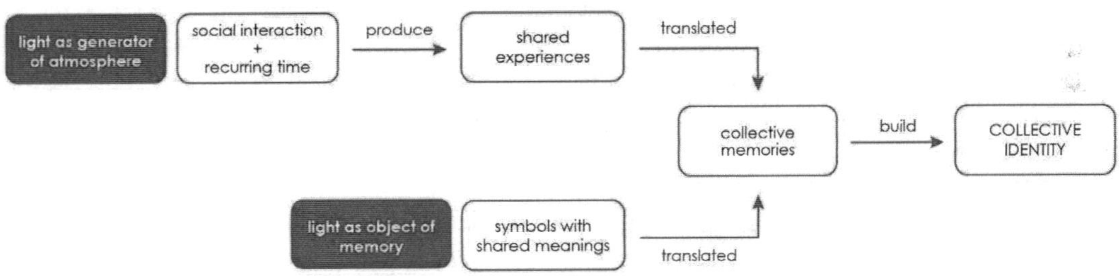

Figure 1. Proposed sense of belonging model

Light can sustain this process when intended in two different approaches:

4.1.1. Light as generator of atmosphere. By providing the proper atmosphere, lighting can set the optimal conditions to encourage social interaction, resulting in a more attractive space for people to return and spend more time (recurring time). The experience in place creates meaning. Consequently, the atmosphere of a place can sustain the development of shared experiences. In this way, new memories of the place and social interactions that took place in it are created over time, sustaining SOB. The design for Exchange Square in London by Speirs Major [28] is a good example of light as a generator of atmosphere. An intimate ambience that encourages easy social interaction is preserved by supporting intuitive wayfinding and low-height lighting.

4.1.2. Light as object of memory. The light itself can have a symbolic value when it has been assigned meaning through personal, group, and cultural processes. When the meaning is shared due to a common cultural background, the symbolic value of light reflects collective memories. These memories can either be past memories evoked by a meaning that is already assigned to the light, or new memories that are built by assigning new meanings over time. For instance, the same glowing star at the top of the Christmas tree which is used for a family year after year acquires an emotional meaning for the members of that family. In this case, we talk about personal memories. *Sankthans* fire that is made for the summer solstice is a type of lighting that everybody knows and has a memory of in Denmark. In this case, light is part of a shared and collective memory, mediated by culture. New memories are developed when the light is so unique and characteristic to be remembered. Meaning can be given when the light starts to be an identifier of the space, becoming symbolic and representative of it, even if initially it does not hold meaning. A good example of that is the lighting design made by

Arup for the University of Sheffield [29]. The bridge of the university is lit by dynamic coloured light, with a different palette provided for each day representing the branding of the student event within the Students Union. With its uniqueness and symbolic value, the lighting becomes a strong identifier of the university and its life. It is assigned meaning and remembered.

4.2. Hierarchy of urban belonging

Inspired by Maslow's hierarchy of needs, a hierarchy of SOB in urban spaces is developed (see Figure 2). The knowledge acquired during the previous Literature Review and mentioned interviews with lighting designers, along with information from the chapter *The Human-Scale Urban Lighting Experience* in Casciani's book [5] and Boyce's journal article *The benefits of light at night* [30], is interpreted, discussed and structured in two pyramids of needs. Pyramid A organises a hierarchy of urban belonging, while Pyramid B translates those needs into lighting requirements and considerations. The hierarchy consists of four main levels and 5 sub-levels. The three first levels are the same as the ones described by Casciani when defining the criteria for a good place: functionality, comfort and sociability. A place can still be well-designed and not necessarily sustain belongingness, therefore a fourth level is added. The lighting pyramid (B) distinguishes between "good lighting" and "lighting that sustains belongingness" as well.

Accessibility. It is identified with the functionality of the space: does the place have the minimum conditions to get into and be used? In terms of outdoor lighting, related factors such as visibility or visual acuity are shown at the bottom of Pyramid B.

Comfort. It is organised into two sub-levels. On the first sub-level, safety and security perception and reassurance are placed. On the second sub-level, an appropriate atmosphere and urban beautification are placed. These aspects are especially connected to emotions and they can motivate visiting or spending more time in a place. To set the appropriate atmosphere for the different activities in an outdoor space and make people feel comfortable, all the qualities of light (colour spectrum, luminance, distribution, etc.) should be balanced. Using different lighting layers and designing from the point of view of the user (human-scale lighting) contribute to a more pleasant atmosphere. The use of smart lighting to excite or adjust the atmosphere depending on the activities, times or seasons can be also helpful.

Sociability. Certain atmospheres can contribute to social interactions, from passive to active ones. Besides facial recognition, setting up a pleasant ambience to make people feel relaxed can contribute to making social interactions easier. Dimmed light levels and warm CCT can lead to a relaxed atmosphere. Not feeling overexposed and providing different conditions, since not everyone feels comfortable in the same situations, must be taken into account. Interactive installations use lighting as a way of communication and motivate people to mingle.

Belongingness. The development of a collective identity gives extra value to a place. There is a great interplay with the previous level since belongingness is deeply dependent on social interactions. However, it is considered a different level since some places can sustain spontaneous and sporadic social interactions and still not strengthen the collective ties among citizens. Time, symbols and shared experiences are determinant factors. A new term, *Belonging Lighting*, has been adopted to define the urban lighting that sustains belongingness. The characteristics of this belonging lighting will be explained in the next section. Nevertheless one might wonder if it can be possible to develop SOB even without covering the foundation levels of the hierarchy. Indeed, it can be argued that it is possible in certain cases. We all can think of deprived areas, which are unsafe or lack an appropriate atmosphere and their inhabitants still feel part of a community. It can be deduced that if the social bonds are solid, they can compensate for the shortages. Similarly, the recurring time factor can supply them. We all probably have grown up using some "ugly" or not idoneous spaces but we still feel a connection with them because they were on the way to our school or because strong memories related to friends or family take place in them.

IOP Conf. Series: Earth and Environmental Science 1320 (2024) 012011 doi:10.1088/1755-1315/1320/1/012011

Environmental issues have not been included in this hierarchy because, although important for human well-being, it has been considered that there is no explicit relationship with SOB.

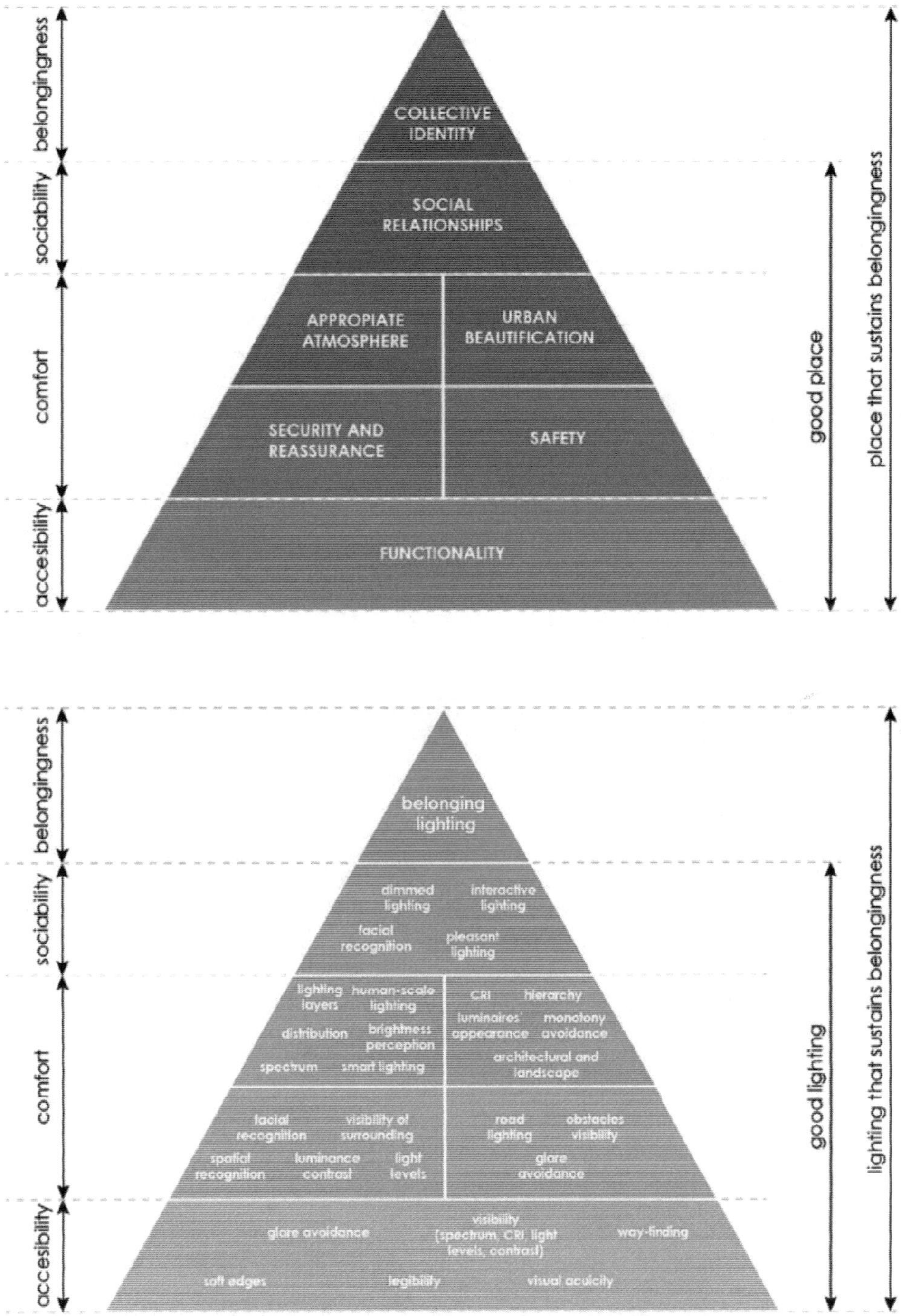

Figure 2. Hierarchy of SOB in urban spaces at the top (A) and its translation into lighting requirements and considerations at the bottom (B).

4.3. Design principles: Belonging Lighting

Three design principles that a lighting design should follow in order to sustain SOB are introduced: unique, meaningful and social. They are developed from the concept expressed in the SOB model and the hierarchy of urban belonging previously presented.

4.3.1. Unique lighting. Lighting is singular, giving the space a distinctive identity. A unique design should consist of innovation, such as the design for Ishøj Station by ÅF Lighting [31], with gobo projections reinterpreting water waves, or the Van Gogh Path by Studio Roosegaarde [32], made of thousands of twinkling stones. If standardised fixtures are planned, these must at least be used in an ingenious way, being perceived as something non-seen before.

4.3.2. Meaningful lighting. Lighting has shared recognizable meanings related to memories and reflects the identity of the community. A lighting solution can be singular and not found anywhere else and still not necessarily related to the identity of a place. A good example of an urban element that can have both qualities at the same time is the Torèt, a standpipe with more than 800 reproductions spread throughout the whole city of Turin in Italy [33]. The dispensing spout has the shape of a bull's head, recalling the animal icon of the city, which is also present on the official city symbols. They have become a symbol and strong identifier of the city and people have developed a sense of affection towards them. They are inextricably linked not only to the satisfaction of needs (functional attachment) but often also to childhood memories and sentimental events (emotional attachment). Also, streetlights as urban elements can be unique and meaningful and become a symbol and identifier of the city, strengthening SOB. In the conversation with Lisbeth Skindbjerg Kristensen, she said that the Copenhagen street luminaire produced by Philips *"makes you recognize the street as being your home and it belongs to your city or your neighbourhood"*. Speirs Major's lighting design project for the Sheikh Zayed Grand Mosque in Abu Dhabi [34] is a good example of a lighting concept with symbolic value. The colour of the façade lighting changes according to the lunar cycle on which the Islamic calendar is based. This way, the design is endowed with a cultural and religious meaning. Another example is catenary lights, which are used in many cultures and, although with different local meanings, they are universally associated with moments of social celebrations.

4.3.3. Social lighting. Lighting sustains the atmosphere for developing social experiences, either by creating the conditions or by making people interact. Social lighting can sustain the atmosphere for developing social experiences by setting the atmosphere and creating optimal conditions to encourage social interaction. The lighting proposal of Light Follows Behaviour for Terry Spinks Place [35] increases the dwell time and activates it through the help of bold geometries and graphics, vibrant colours, shapes and sounds. On the other hand, social lighting can directly encourage people to interact with each other. In most cases, this is achieved with interactive lighting installations as in The Pool by Jan Lewin [36] or BruumRuum! by Artec Studio [37].

4.4. Design process: public participation

As has been seen before, the method is revealed as important as the final design: citizen engagement is a powerful tool to generate SOB but also to educate people about the role of lighting. An understanding of the inhabitants and social dynamics is as crucial as an understanding of the physical space. Public participation strategies enhance the local sentiment in a double way, as Martinez mentioned during the conversation. On the one hand, citizens feel more identified with these places because they have contributed to making them possible and they reflect their needs and decisions. On the other hand, the participation itself helps to strengthen the community bonds, since it gives the opportunity to meet your neighbours.

The tools for community involvement can be divided into two big groups: passive and active. Passive methods are centred on gathering information by observing and collecting data without direct interaction with the users. The information can be registered and systematised by strategies such as counting, behavioural mapping, tracing, tracking or shadowing, looking for traces, keeping a diary, photographing or filming, drawing, or in-situ or digital observations by the resource of the flâneur [5, 38-41]. Among the tools for active participation, it can be listed: night walks, talks, interviews, on-site questionnaires, collaborative maps, personas, workshops, guerrilla lighting, or a combination of the previous methods. Different practices and organisations carry out placemaking activities within the lighting field. These initiatives are sometimes part of a larger design project and sometimes just actions to educate citizens about the culture of light or to engage the community. Among others, can be mentioned: Lighting Detectives, NightSeeing™, Light Follows Behaviour, Configuring Light / Staging the Social, Social Light Movement.

Traditionally, these kinds of methods have been mainly used during the pre-design process to gather data. However, there is potential to introduce them at other design stages by conducting a process of co-design together with citizens. Smart lighting and new control technologies allow also adjusting some parameters after new inputs of the community or changes in the use of the space. The role of municipalities and local associations is indispensable, but the lighting designer, together with sociologists, can plan and guide this participatory process.

5. Limitations and discussion

The study has certain limitations that are hereinafter considered and discussed. Some of them are intrinsic to the method employed in structuring the study since the exploratory research method is not meant to lead to conclusive statements. Its purpose is, in fact, to open a discussion in the lighting and urban design practice by providing a possible answer to what is the influence of urban lighting on SOB. The findings of the work can serve as a foundation upon which further research and investigation are drawn. Another aspect of exploratory research is that it provides qualitative data. Interpretation of such information does not follow a strict cause-and-effect relationship, due to its nature, and it is subject to an inference process that can be biased, also due to a lack of preexisting knowledge on the topic. As a production of knowledge that follows a qualitative method, the results are not directly verifiable and reproducible. The topic itself, SOB, concerns dimensions such as identity, emotions, atmospheres and memories which can hardly be generalised and considered universally valid. Finally, it must also be noted that lighting has limits in relation to the topic. There are many other factors that affect the generation of urban belongingness, and lighting is probably not among the most affecting ones. Given the complexity of the topic and the many disciplines involved, a transdisciplinary perspective in partnership with experts in urbanism, sociology or anthropology, among others, could lead to more significant and comprehensive results.

6. Conclusions

The findings of the study propose that artificial lighting can sustain and strengthen SOB in urban environments. This happens when it has the qualities that characterise Belonging Lighting, as defined in the hierarchy of urban belonging. Belonging Lighting has the potential to generate shared memories that sustain a collective identity in the urban space according to the process delineated in the sense of belonging model. Lighting can in fact act as a generator of atmosphere, propitiating social interactions and shared experiences in the outdoors, or as an object of memory when it acquires the character of a symbol. Both qualities contribute to building a collective identity and SOB as a consequence. Lighting, like other layers of urban design, cannot however force the social interactions by itself but it can contribute to creating the right conditions for them to happen.

Moreover, it can be contended that both the lighting project and the design process employed have a role in sustaining an SOB. A lighting design supports belongingness if it follows specific design principles: it is unique and meaningful and it sustains social interaction in the space. On the other hand, the design process must be participatory by bringing neighbours together and making them connect to have an active role in shaping the design. Group activities strengthen collective ties and having contributed to the design makes people identify more with the space thus strengthening a collective identity.

Acknowledgements
We thank Leni Schwendinger, Luciana Martinez, Lisbeth Skindbjerg Kristensen and Anna Lykke Thorup for sharing their experiences and reflections, which enriched this research. We also acknowledge Mette Hvass for her supervision, advice and encouragement during the completion of our Master's Thesis at Aalborg University as well as the preparation of this article.

References

[1] Maslow, A. H. (1943). 'A theory of human motivation'. *Psychological Review*, 50(4), 370-396.
[2] Gehl, J. (2010). *Cities for people.* Island Press.
[3] Dohna, F. z., Schwendinger, L., & Arup. (2015). *Cities Alive. Rethinking the Shades of Night.* Arup.
[4] Narboni, R. (2016). 'From Light Urbanism to Nocturnal Urbanism'. *Light & Engineering*, 24(4), 19-24.
[5] Casciani, D. (2020). *The Human and Social Dimension of Urban Lightscapes.* Springer International Publishing AG.
[6] Calvillo Cortés, A. B., & Falcón Morales, L. E. (2016). 'Emotions and the Urban Lighting Environment'. *SAGE Open, 6*(1)
[7] Hvass, M., Waltorp, K., & Hansen, E. K. (2022). 'Lights out? Lowering Urban Lighting Levels and Increasing Atmosphere at a Danish Tram Station'. In S. Sumartojo (Ed.), *Lighting Design in Shared Public Spaces* (pp. 152-172). Routledge.
[8] Entwistle, J., Bordonaro, E. & Slater, D. (2018) 'The social study of urban lighting'. In: Davoudian, N, (ed.) *Urban lighting for people: evidence based lighting design for the built environment.* RIBA Publishing, London, UK.
[9] Casciani, D., Rosso, M. (2012) 'ELSE, Experience of Lighting Sustainability in the Environment'. *Cumulus Conf. Northern World Mandate –Towards Open and Participative Cities* 24–26.
[10] Allen, K., Kern, M. L., Rozek, C. S., McInerney, D. M., & Slavich, G. M. (2021). 'Belonging: a review of conceptual issues, an integrative framework, and directions for future research'. *Australian Journal of Psychology*, 73(1), 87-102.
[11] Singh, K. (2007). *Quantitative social research methods (1. publ. ed.).* SAGE.
[12] Stebbins, R. A. (2001). *Exploratory research in the social sciences*, Vol 48. SAGE
[13] Escalera-Reyes, J. (2020). 'Place attachment, feeling of belonging and collective identity in socio-ecological systems: Study case of pegalajar (Andalusia-Spain)'. *Sustainability (Basel, Switzerland)*, 12(8), 3388.
[14] Baumeister, R. F., & Leary, M. R. (1995). 'The need to belong: Desire for interpersonal attachments as a fundamental human motivation'. *Psychological Bulletin,* 117(3), 497–529.
[15] Maslow, A. H. (1968). *Toward a psychology of being (Second ed.).* Van Nostrand Reinhold Company.
[16] Gehl Architects, The Techno Anthropology Lab (Aalborg University), Visual Methodologies Collective (Amsterdam University of Applied Sciences), Service Design Lab (Aalborg University), Center for Digital Welfare & Public Data Lab. (2022). *The Urban Belonging Project.* Available at: https://urbanbelonging.com/ (Accessed: 4 August 2023)
[17] Tuan, Y. (1977). *Space and place, The perspective of experience.* Edward Arnold.
[18] Scannell, L., & Gifford, R. (2010). 'Defining place attachment: A tripartite organizing framework'. *Journal of Environmental Psychology*, 30(1), 1-10.

[19] Manzo, L. C. (2005). 'For better or worse: Exploring multiple dimensions of place meaning'. *Journal of Environmental Psychology*, 25(1), 67-86.
[20] Lynch, K. (1960). *The image of the city.* Technology Press.
[21] Böhme, G. (2017). *The Aesthetics of Atmospheres.* Routledge.
[22] Bille, M. (2019). *Homely Atmospheres and Lighting Technologies in Denmark: Living with Light.* Taylor and Francis.
[23] Bille, M. (2015). Lighting up cosy atmospheres in Denmark. *Emotion, Space and Society*, 15, 56-63.
[24] Seuntiens, P. J. H., & Vogels, Ingrid M. L. C. (2019). *Atmosphere creation: the relation between atmosphere and light characteristics.* Zenodo.
[25] Edensor, T. (2015). 'Light Art, Perception, and Sensation'. *The Senses & Society,* 10(2), 138-157.
[26] Ellery, P. J., & Ellery, J. (2019). 'Strengthening Community Sense of Place through Placemaking'. *Urban Planning*, 4(2), 237-248.
[27] Project for Public Spaces. (2007). *What is Placemaking?*
Available at: https://www.pps.org/article/what-is-placemaking (Accessed: 4 August 2023)
[28] Speirs Major (2022) *Exchange Square.*
Available at: https://www.smlightarchitecture.com/projects/2972/exchange-square (Accessed: 4 August 2023)
[29] Arup UK (2020) *University of Sheffield Concourse.*
Available at: https://www.arup.com/projects/university-of-sheffield-concourse (Accessed: 4 August 2023)
[30] Boyce, P. R. (2019). 'The benefits of light at night'. *Building and Environment*, 151, 356-367.
[31] ÅF Lighting (2015). Ishøj Station.
Available at: http://www.urbanlightscapes.net/ishoj-station-by-af-lighting/ (Accessed: 4 August 2023)
[32] Studio Roosegarde (2018) *Van Gogh Path.*
Available at: https://www.studioroosegaarde.net/project/van-gogh-path (Accessed: 4 August 2023)
[33] Museo Torino (2021) *Torèt.*
Available at: https://www.museotorino.it/view/s/4bead4cf54f744f4831feae465b9f057 (Accessed: 4 August 2023)
[34] Speirs Major (2008) *Sheikh Zayed bin Sultan al Nayhan Mosque*
Available at: https://smlightarchitecture.com/projects/536/sheikh-zayed-bin-sultan-al-nayhan-mosque (Accessed: 26 November 2023)
[35] Light Follows Behaviour (2018) *Terry Spinks Place.*
Available at: https://lightfollowsbehaviour.com/portfolio/terry-spinks-place-canning-town/ (Accessed: 4 August 2023)
[36] Jan Lewin (2022) *The Pool.*
Available at: https://www.jenlewinstudio.com/portfolio/the-pool/ (Accessed: 4 August 2023)
[37] Artec Studio (2014) *BruumRuum!.*
Available at: https://artecstudio.net/BruumRuum/ (Accessed: 4 August 2023)
[38] Gehl, J., & Svarre, B. (2013). *How To Study Public Life.* Island Press/Center for Resource Economics.
[39] Cullen, G. (2012). *The Concise Townscape.* Taylor and Francis
[40] Sumartojo, S., Edensor, T., & Pink, S. (2019). 'Atmospheres in Urban Light'. *Ambiances (En Ligne)*, 5(5)
[41] Pink, S. (2013). *Doing visual ethnography (3. ed. ed.).* SAGE.

IOP Conf. Series: Earth and Environmental Science 1320 (2024) 012012 doi:10.1088/1755-1315/1320/1/012012

Urban Illuminations – Light Art Activating The Public Realm

Yael Erel[1], Foteini Kyriakidou[2], Rodrigo Eduardo Muro Avendano[3]

[1] School of Architecture, Rensselaer Polytechnic Institute, 110 8th St, Troy, NY 12180, USA;

[2,3]Division of Lighting Design, Department of Architecture, KTH Royal Institute of Technology, Osquars backe 5, 114 28 Stockholm, Sweden;

Corresponding author's e-mail: erely2@rpi.edu

Abstract. How can contemporary light art installations impact the sensorial perception of the urban context to modify and reinterpret urban space? In Robert Irwin's words, how can they provide "...an extended way of looking at the world"? In this paper, we examine how temporary light installations may impact the public realm by modifying citizens′ perceptions of moments in the city: a habitual urban passage is interrupted with a temporal experience, changing the way the urban space is experienced. We explore how site-specific light installations create unique intersections of audience and spatial perception through the artists′ intent of creating new urban atmospheres through an experiential layer added onto the urban fabric during light festivals. To do so we analyze and discuss two light installations that we designed and executed, which were shortlisted for the 2022 [d]arc awards. Both projects took part in larger urban light festivals and were located within urban passages. "Reflecting on Troy" [Author 1 - Troy Glow Light Festival] created a temporal light graffiti in an urban alley, while "Riddle 102" [Authors 2,3 - Nobel Week Lights], a dynamic lighting installation in a 231m pedestrian tunnel, created an immersive environment that alters the perception of the tunnel's physical dimensions. Through reflections, as well as visitor and curator reviews, we unpack the intent and impact of the works on the perception of the urban context. We discuss and analyze the similarities and differences between the two urban settings, identify the design processes and principles pertinent to site-specific light installations [materials, lighting fixtures, testing] and address the impact darkness and over-lit spaces have over spatial perception. Lastly we discuss the impact a temporary light installation may have on the collective memory of the city, perceived in person or online.

1. INTRODUCTION

How can contemporary light art installations impact the sensorial perception of the urban context to modify and reinterpret urban space? In Robert Irwin's words, how can they provide "...an extended way of looking at the world" (1)?

Cities are places for heightened cultural production. Light art installations (LAI) use the city fabric itself as a canvas for temporary art to unfold, producing art that is truly public. One does not choose to interact with the art; it temporarily occupies the public realm by modifying citizens′ perceptions of moments in the city: A habitual urban passage is interrupted with a temporal encounter, changing the way the urban space is experienced. We explore two different site-specific LAIs taking place in two light festivals: "Reflecting on Troy" (project 1) and "Riddle 102" (project 2). We analyze and compare

IOP Conf. Series: Earth and Environmental Science 1320 (2024) 012012 doi:10.1088/1755-1315/1320/1/012012

the installations and see how each installation creates a unique intersection of audience and spatial perception via the artist's intent of creating new urban atmospheres through an experiential layer added onto the urban fabric.

2. BACKGROUND

2.1. Light Festivals

Light festivals have become popular throughout the globe in the past few decades. This development is attributed largely to the development of new lighting technologies such as 3D projection mapping, LED light sources, controls, and automation (2). They are increasingly globalized; at times itinerant works go from one festival to the next, which makes them less site specific (e.g., from Nobel Week Lights 2021, to Nordic Bridges 2022, to Allt Ljus på Uppsala 2022). This approach enhances the potential that light art has as a desirable public object in the urban realm. It is difficult to precisely assess how many light festivals are active around the globe, although it is clear they are present in all five continents. Light festivals may mark a specific event in the urban space and allow for a celebratory transformation of the city in different durations. They also serve as an economic driver to boost tourism, support commerce and establish a reputation for the city (2). This commercialization and globalization has also, as Tim Edensor surveys (3), drawn criticism: fear of replacing cultural substance with spectacle (4), increasing inequities (5), and billing the urban festival experience as another product for consumption (6). This criticism is apt. Yet as Edensor observes, light festivals also have the potential to stimulate critical awareness of perception and sense of space, as well as recall forgotten histories (3). Therefore, some light festivals organizers show interest to expand the experience beyond boundaries; light festival artists' briefs ask for LAI that possess the capacity to revitalize spaces, transform the perceptions of citizens and visitors, and foster a positive influence within communities. Whether in one city or another, LAI could be works that grace landscapes internationally (7). They are ground for experimentation and for innovation (3). These last thoughts align with the idea and importance of producing site specific LAI.

2.1.1. Troy Glow Light Festival

The Troy Glow Light Festival (Troy Glow) was a five-week public art light festival in downtown Troy, brightening its streets during the darkest time of year through unique and thoughtful public art. The city of Troy is a small post-industrial city (population 50,000) located on the Hudson River, New York. Troy Glow was established in 2022 to invite people back to downtown Troy after two years of pandemic precautions by using art as a way to begin rebuilding the collective space of the city (8). The festival featured six light-based, outdoor art installations created by regional artists, as well as eight partner organizations' unique light displays. The pieces were located within walking distance of one another. Throughout the festival run, there were guided tours as well as additional events to draw people to the festival. Free to all, the festival has also become part of the Troy urban fabric, and was observed by visitors, residents and passers-by. The festival was organized by the Arts Center of the Capital Region, Troy Cultural Alliance and Rensselaer County Chamber of Commerce, and financed by a grant from Empire State Development and I LOVE NY/New York State's Division of Tourism through the Regional Economic Development Council initiative.

2.1.2. Nobel Week Lights

Established in 2020, Nobel Week lights [NWL] is an annual light festival at Stockholm city center, Sweden during the Nobel Week [beginning of December]. During a span of ten days, local and international artists and designers share their artistic light installations with approximately 350,000 visitors each year. On 2021, twenty installations were opened to the public at a portion of Stockholm city centre. LAI were encountered as citizens roamed the city or accompanied guided tours. The festival invites citizens and visitors to explore the city during the Nordic dark season by encouraging new insights about Nobel Laureates and their research, lives and discoveries. Nobel Week Lights is presented by the Nobel Prize Museum with support from the City of Stockholm, the Erling-Persson Foundation, the Knut och Alice Wallenberg's foundation and Stockholm's Chamber of Commerce, as well as numerous other partners and artistic lighting companies (9).

2.2. The role of light in creating an atmosphere

A light installation is an art form that allows for a significant and palpable effect on a space with minimal material means. As Canepa describes, "atmosphere is a complex phenomenon because it is invisible, intangible, and elusive, without physical limits, unstable, instinctive, highly subjective, and often described through metaphors." (10). Light is one component of an atmosphere, and it is a key ingredient in humans' spatial perception. As Robert Irwin explains, an artist may extend the perception of his viewers to see the world in a new way (1). Thus immersive light art is a unique art form, using light as a temporal medium that affects how our surroundings are perceived. To see, experience and be immersed in these public light installations is an exercise of embodiment, as Maurice Merleau-Ponty argues in reference to art: "I do not see it according to its exterior envelope; I live it from the inside, I am immersed in it. After all, the world is around me, not in front of me." (11).

2.3. Temporality and Time

Temporality and time play a significant role in LAI. Light festivals are temporary moments in the time of the city. While Troy Glow was a five-week celebration throughout the Christmas and New Year's holiday season, NWL is a ten-day long light festival demarcating and celebrating Nobel. Both festivals took place during the winter, a time of year that has short days and extended night hours. Troy Glow, located in Troy, NY experienced approximately 9 hours of daylight, and 15 hours of darkness, while NWL, located in Stockholm, had approximately 6.5 hours of daylight, and long nights of 17.5 hours. The installations were programmed for specific hours, 16.00-23.00.

Another aspect of temporality in these events were the duration of the visitor stay in each LAI. The location of each work within the flow of the city provides for different uses and interactions of the site. Project 1 in Troy Glow is an urban intersection where the passer-by may choose to interact, stay or pass an open, unbounded public space. Conversely, Project 2 in NWL is an urban, enclosed linear passage – an experience that contains not only a duration of time required to traverse it but also features an entrance and an exit (see Figure 1).

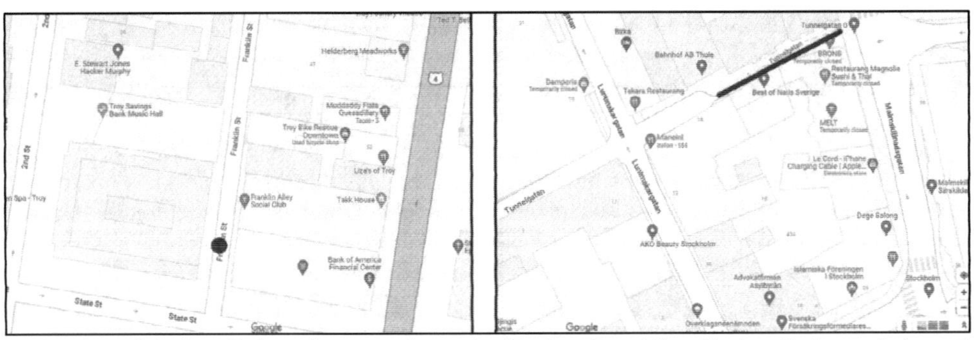

Figure 1-The sites of the installations demarcated as point (Project 1) and line (Project 2) due to their spatial qualities.

Temporality is also understood as the duration of impact the light installation has on the citizens, not only the time they were immersed in the installation, but also how long they carry the memory of the experience with them. As Juhani Pallasmaa explains; "all experience implies the acts of recollecting, remembering and comparing. An embodied memory has an essential role as the basis of remembering a space or a place," and that "we identify ourselves in space as space becomes part of our very existence" (12).

2.4. Activating the Public Realm

"Activating the public realm" refers to the process of making public spaces more vibrant, engaging, and versatile. This activation involves changes to the physical environment, as well as events or activities that encourage more diverse use of the space. An artistic light intervention could potentially promote this diversity of use; through it the public space is activated, providing a temporal variety from daytime to night-time for an array of needs and interests (13). The regular programming of lighting festivals can help attract people to a public space and encourage them to move along the different interventions throughout the city, hence adding to its activities. Engaging the public realm calls for a sustainable and responsible approach–an approach which could manifest through universal access to the space of the intervention, welcoming all different cultural and social groups. One where the experience of the space should feel safe and comfortable, which can involve considerations of lighting, sightlines, and availability to move safely in space. And where the community engaged in the design by involving local artists who help ensure the space meets the needs and desires of the people who use it. In our view, light festivals, and, in particular, site specific LAI can be a tool for the creation of Tactical Urbanism for the benefit of the citizens and the activation of the public realm. Both presented cases use Tactical Urbanism - enacting low-cost temporary change on the built environment - to generate a particular impact in the public space, with the ultimate goal of inspiring a long-term evolution of the public space that could change perceptions, behaviours, and the uses of space.

Activation of the public realm may not be connected only to organized activities, but to the experience itself. As Edensor proposed, the light installation may ‘defamiliarize’ a viewer because “light possesses a particularly rich capacity to defamiliarize familiar places, transforming what is well known into an uncanny realm and thereby suggesting that place may be apprehended otherwise.” (3). The visitor thus is allowed to experience a known space in a new way, extending the experience and memory of the city. Kevin Lynch´s concept of imageability (14) of the physical public realm hinges on qualities or attributes that assist in forming a mental representation of the city, creating a memorable imprint. The lighting scheme provided by the LAI supports certain legibility and visibility. It can render an environment distinct and noteworthy, captivating both the eye (as seen in project 1 & 2) and ear (as in project 2), promoting heightened awareness and engagement. Through the temporary LAI, citizens experience a novel sensory impact, differing from their daily encounters. As Lynch suggests, imageability is a mutual process between the observer and the observed. The city's image can be enhanced through symbolic devices or light effects, which can concurrently influence perceptions of the surroundings.

3. METHODOLOGY

The objective is to understand how interventions in the urban landscape influence the behavior and perceptions of urban dwellers by adding experiential layers to city fabric.

1. Project Overview: Constraints Analysis: Evaluate initial limitations set upon the intervention. Site Selection Review: Understand the criteria behind site choices. Design Process Examination: Delve into the design process and key decisions. Intentional Analysis: Discuss the primary objectives and motivations of the designers.
2. Data Collection: First-Person Observations: Document spatial and atmospheric properties of the intervention spaces. Record behaviors and reactions of spectators within the environment. Third-Person Feedback: Gather testimonies from spectators about their experiences and impressions. Collect insights from curators regarding the impact of the intervention on the urban setting.
3. Data Assessment: Assess changes in duration of stay, activities undertaken, and engagement levels within the intervention spaces. Movement Metrics: Observe if there's a change in the speed of movement through passages or intersections. Note any altered patterns or behaviors due to the intervention.
4. Synthesis and Interpretation: Urban Experience Transformation: Evaluate how the intervention modifies the typical urban experience by adding layers of meaning and activity. Behavioral Impact Assessment: Determine if and how viewers' interactions with space change once the intervention is introduced.

Synthesis of the different aspects of the methodology provided insights into whether interventions significantly activate the public realm and influence urban dweller behavior and perceptions. Specifically, discern if there's an evident change in speed, overall behavior, and attention due to the intervention.

3.1. Project descriptions Brief and Designer approach
The two light installations that are analysed in this investigation were designed and executed by the authors and were shortlisted for the 2022 [d]arc awards- art category/ low budget. Both projects took part in larger urban events [Troy Glow and NWL] and were located within urban passages.

3.1.1. "Reflecting on Troy" [Project 1]
"Reflecting on Troy" (Project 1- in Troy Glow) was a light installation that was selected, commissioned, produced, and funded by the Arts Center of the Capital Region through the inaugural Troy Glow Light Festival 2022 (see Figure 2). The curators of the Troy Glow Light Festival selected the presented art works through a two stage competition that referred to the City of Troy Master Plan and to the civic importance of public art. The projects were asked to celebrate Troy, creating site-specific, high-level art work that would interact with its environment and celebrate this time in the city. The initial prompt called for a concept that would be flexible to adapt to multiple site typologies. "Reflecting on Troy"'s response to the brief was to take the principle of projecting data from Troy's industrial past and translating it into light, adding a new, unexpected layer onto the city. In its final iteration "Reflecting on Troy" cast an analogue animation in reflective light onto the back facade of one of Troy, New York's most beloved historic buildings. The animation is created by shining a high-powered, directional light onto two rotating reflective disks on the other side of an urban alley. The disk's surfaces are etched with Troy-inspired artifacts (drawings of the Burden Iron Works water wheel and Troy's topography) to create spatial light drawings onto the textures of the urban fabric. At the end of the project all materials and equipment were either recycled or re-used in different contexts thus maintaining a sustainable material use. The project was awarded the second place [d]arc award in the art-low budget category.

Figure 2 - "Reflecting on Troy". Installation on site (a); Installation with urban context (b) and the urban context without installation (c).

3.1.2. "Riddle 102" [Project 2]
"Riddle 102" (Project 2) was an installation in NWL, created by two of the authors, who are lighting designers and educators. The project was located in Brunkeberg tunnel, a 231-meters-long enclosed interconnection of two of the most central and busy districts of Stockholm [Stureplan and Hötorget]. The designers were invited to participate with the additional value that their students would engage in the mounting of the on-site installation on a voluntary basis. The choice of place was paramount to the installation: pedestrian- and cyclist-only tunnel is frequently used and represented a relevant space within Stockholm's urban fabric. With a carte blanche as a project brief, the designers used Alfred Nobel and the Nobel Prize-awarded discoveries and Nobel Laureates as inspiration. This initiative was adopted

by the subsequent editions of the festival and became the theme for the artists and designers participating at the following NWL.

Conceptually, "Riddle 102" captured the binomial character of Alfred Nobel. On one hand, his scientific knowledge and creation of numerous inventions such as Patent 102 (Dynamite), and on the other, his artistic expressions of writing poems, such as "The Riddle." It also captured the two different construction tunnelling techniques [freezing and blasting] used at the construction of Brunkeberg tunnel back in 1866. The existing static lighting scheme was adapted to fit the theme and the concept of the project (see Figure 3). It was complemented with layers of dynamic lighting, coloured gels, fabric material and sound. The designers used Richard Kelly´s three light tenets so that the installation divided the tunnel into three segments/ scenes. The two main sections with blue and red ambient light [freezing and blasting technique respectively] were separated by the white illuminated core of the tunnel (see Figure 4). There, the diffused and the reflected light of the existing fixtures created a spatial riddle, a moment/pause preparing the citizen to immerse into the next portion of the tunnel. Lastly, an added DMX controlled dynamic iColorFlex with a 1min 50sec sequence, looped endlessly along all segments, enhancing their perceptual connection. The journey through the tunnel from one end to the other is approximately a three-minute walk or one-minute cycle ride, allowing different speeds and experiences. Thus, the commonly perceived yellow tunnel was transformed into the blue and red experience – an immersive environment that altered the perception of the tunnel's physical dimensions and time perception. The project was funded by the Nobel Week Lights organization, the Royal Institute of Technology and technically supported by Signify and Stockholm lighting. The installation of complementary lighting fixtures and materials was supported by KTH students in lighting design as part of their educational experience. After dismantling, most of the materials were kept as students' lab materials. It was awarded the fifth place [d]arc award in the art-low budget category.

Figure 3 - "Riddle 102" - Views of Brunkeberg tunnel without and with installations [(a)+(b), (c)+(d)].

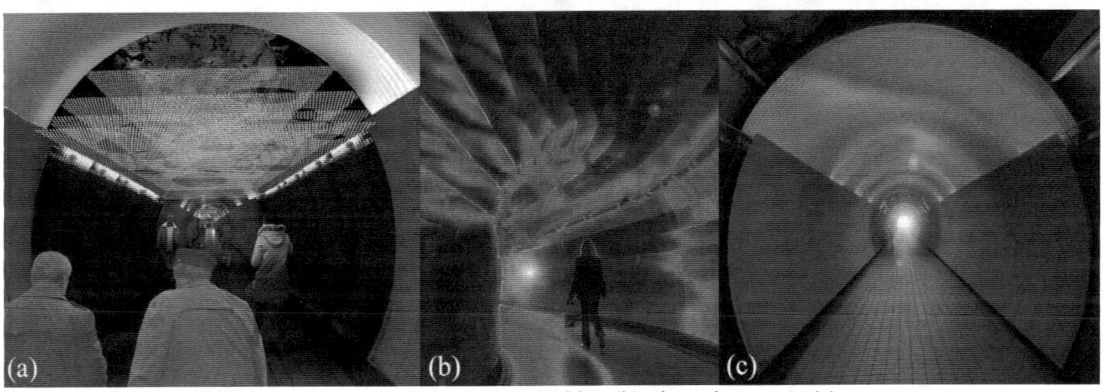

Figure 4 - "Riddle 102" concept images: the core (a); the blue (b); the red segment (c).

3.2. Site and process:

Both sites were located in an urban passage, but experienced very differently: while one is experienced as a point at an urban intersection, the other is perceived as a linear experience (see Figure 1).

There was a difference in approach on how the organizers of both festivals requested the artists to tackle the site-specific light installation. The Troy Glow selection committee was interested in site flexibility during the siting process and asked that the different artists submit flexible proposals that could be adapted to provide multiple settings. "Reflecting on Troy" was developed through a series of site alternatives, from an interior back projection, to a wall projection, to a projection in an alley utilizing both alley walls (See Figure 5). Once the artists were selected, they worked together with the festival curators in an intense testing period to develop the project through: identifying and securing the best site, specifying a light source that would provide a crisp light drawing and could compete with the site's existing urban lighting, and finally installing the project on site (see Figure 6).

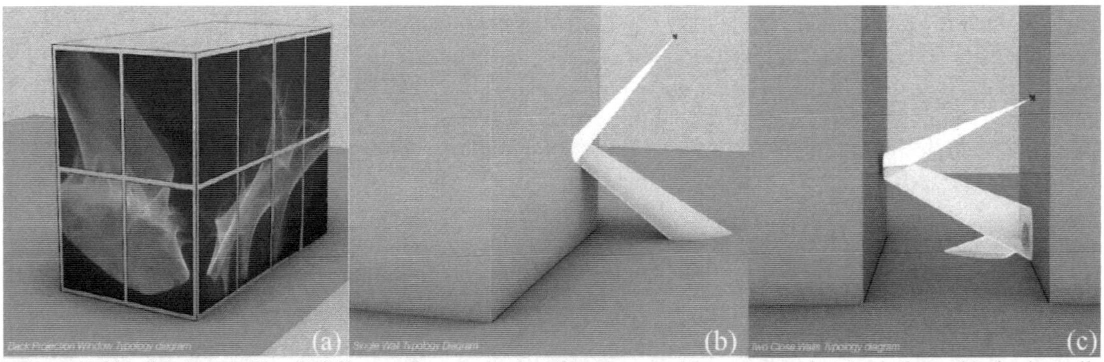

Figure 5 - "Reflecting on Troy" different site typologies, from competition proposal: (a) window, (b) single wall, (c) two close walls.

Figure 6 - "Reflecting on Troy" tests of different sites, light sources and installation.

The two designers for "Riddle 102" used an extensive creative process involving site selection, theme definition, site analysis, concept, and lighting schedule. Additionally, equipment selection and definition of the control strategy and scenarios were discussed together with the programmer collaborator. Their architectural lighting design master students were of great support in mounting the installation. This installation served not only as an educational activity, but also provided experience in a real, functional project as well. In parallel, another collaborative creative process involved two sound specialists, who together with their students at Konstfack [University of Arts, Crafts and Design], developed the installation's soundscape using only the schematic control scheme with time and rhythm of light. The layering of both dimensions, light and sound, were in synchronicity, confirming that the result was bigger than the sum of its parts (see Figure 7).

Figure 7 - "Riddle 102" Installation Photos.

4. RESULTS

4.1. "Reflecting on Troy" [Project 1]

The final site for "Reflecting on Troy" was at the intersection of Franklin Alley and State Street in downtown Troy: reflectors were attached to rear windows on one side of the alley and projected textures onto the Troy Music Hall back facades on the other side (see section Figure 8). The light source was placed within the Troy Music Hall backstage area. The location at an urban passageway was one that people would frequent as a destination within the festival or as part of their urban routines. As such, it did not inherently solicit a pause or a stay. One programmatic exception was that this site was used as a habitual smoking corner for the neighbouring "Franklin Alley Social Club"'s patrons who spent their smoking breaks viewing an unexpected light art element. Such a break provides a unique opportunity to view a public art work, with the desired duration of smoking a cigarette.

Activating the public realm could be seen in subtle behaviour changes in the way one spends their smoking break or experiences a known urban space. A walk home is changed through an event that one passes through, either intentionally or accidentally finding and stopping to observe and decipher.

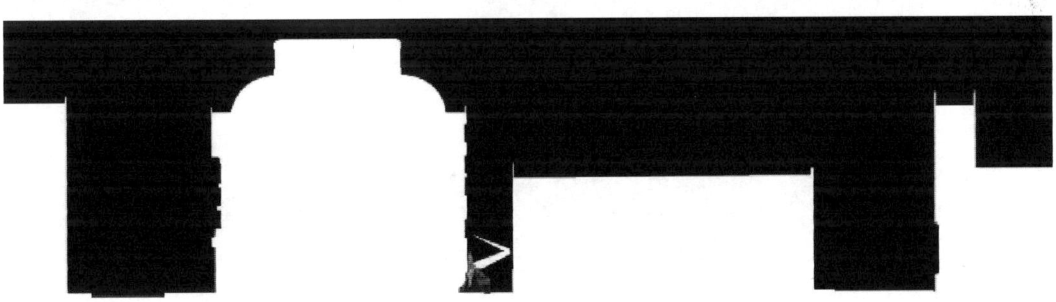

Figure 8 - "Reflecting on Troy" section.

This is evident in three different testimonies: "I slowed down my pace as I approached the alleyway. I made a conscious decision to stay and pause, something I rarely do while walking alone in the dark through downtown. However, this pause felt safe, even though there was no one else around me. Being able to be immersed in the way the brick and cement facades of the building changed, noticing subtleties of literal brick and mortar interacting with moving light became an immersive meditative experience." (Interview Anonymous Troy Citizen testimony August 7, 2023).

"The installation "Reflecting on Troy" reminded me how important public art is by presenting my family with an urgent and interactive set of fundamental questions that stimulated and caused us to marvel. Such as: WHAT is this? WHERE is it coming from? And, HOW was it made? Unlike art that is framed and housed in traditional ways, "Reflecting on Troy" 'met' us during a spontaneous dark-night stroll, and provoked both intense curiosity, and an invitation to a reckoning of the abstract, both uniquely engaging feelings not encountered nearly enough! It was both an exquisite and whimsical encounter that we revisited numerous times, and is still etched in the imagination of the brick alley where it once danced." (Elizabeth Woodberry Cassius testimony June 20, 2023).

"The installation insisted viewers explore the urban environment in a novel way. A grungy alleyway was transformed by ephemeral, dancing light. The installation was surprisingly simple in concept and construction, but created an otherworldly effect. Unlike a static projection the instability of the image created the illusion of animation, bringing the organic shapes to life. Curious viewers were rewarded by discovering the mechanics of the projection, a slowly rotating, dimpled surface reflecting a single reflector" (Troy Glow Curator Judie Gilmore testimony Aug 3, 2023).

In addition to planned tours and festival events, the space was also used for other spontaneous activities such as a photoshoot backdrop (see figure10). Both planned and unplanned uses created an activity that does not typically take place in this location. The testimonies above speak of the viewer's engagement through the puzzle of deciphering the installation. Its simultaneous simplicity of structure and system, versus its complex spatial effect generate a moment of viewer engagement. The project added light to existing light levels, but in a relatively dark passage created a shimmering experience where no event existed before.

4.2. "Riddle 102" [Project 2]

As mentioned previously, "Riddle 102" was located in a pedestrian/cyclists tunnel, a 231m long vault under the central neighbourhood of Norrmalm in Stockholm. The tunnel's existing material and construction details impacted the results and hence the user's spatial perception. The existing materials consisted of exposed natural rock covered with corrugated white matte metal on one half of the tunnel [red segment], while the other half was covered with highly reflective specular metallic sheets [blue segment] (see Figure 9).

These materials affected the light conditions differently, contributing to the perceived difference in brightness levels between the two segments as "the blue light area is perceived of as much darker and harder to see in than the red light area despite similar lux measurements." (15). Similarly, Lara Szabo Greisman, NWL curator, mentioned at her interview that "[...] some found that the blue section was too dark and that they felt there were some security concerns." (NWL curator Lara Szabo Greisman testimony June 4, 2023)

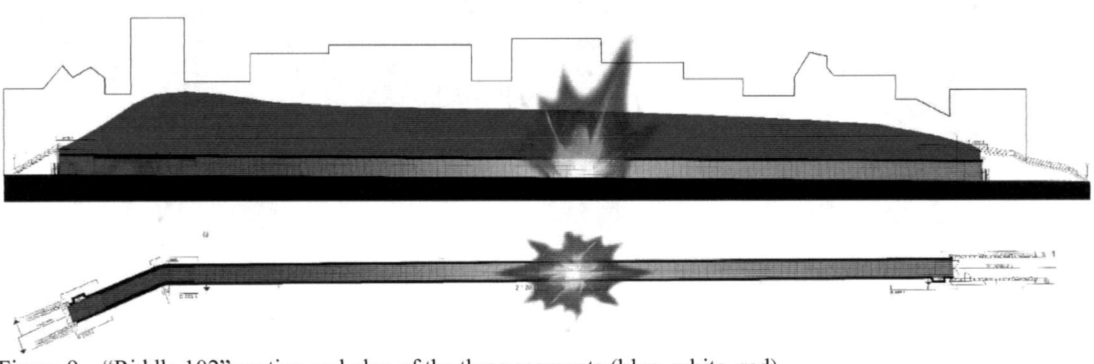

Figure 9 – "Riddle 102" section and plan of the three segments (blue, white, red).

The walls of the tunnel are fully covered with corrugated yellow metal sheets, which was taken into consideration when selecting filters/hue and avoiding the yellow tone. The existing yellow metal cladding was interrupted in some parts in the middle of the tunnel, where the natural rock was left exposed, only protected by a metallic open grid. These exposed segments were illuminated in white light and constituted the central core segment, creating a site specific division of the tunnel. It is interesting to note that the dimensions of the two long segments [blue and red] were interpreted differently according to the survey conducted on-site by the students. It was found that there was a tendency "to perceive the first area (red or blue) crossed as the longest one" (16) despite the objective length, light colour or perceived brightness. According to this, the core segment/white part worked as a focus point. "[...] the brightness of white light, which, in the middle of the tunnel, acted as a visual objective for people walking through." (16). The students considered Michel Lou's visual objective definition "not only direct paths of movement but relays significant distance clues for awareness of walking or driving times" (17). As the students articulated, this finding suggests that "people with a bright visual element on their path were more focused on the walk that they [were] taking, while on the contrary, once they passed the 'attraction point', they possibly walked with lower attention paid to the surroundings." (16).

Apart from the festival visitors, the majority of the users were citizens during their everyday commute – by foot or bicycle – and had no other choice but to be immersed in this installation. The non-stop linear flow of people set the principles for the creation of three segments/characteristic atmospheres, where dynamic light and an expressive soundscape accompanied the traveller, making this installation a multisensory experience [light-sound-movement]. This experience potentially transforms the relationship to site, how it is remembered and discussed, as a testimony noted "The installation made me relate to the tunnel in a different way - it was not an anonymous space anymore, as it gained some context for me, something to remember and to relate to. Afterwards in my podcast, for example, that tunnel came up a few times, which made me realise that there is even more value in temporary projects like this - they become a topic for discussion." (Interview Anonymous Stockholm Citizen testimony August 14, 2023).

5. DISCUSSION

5.1. Can LAI have a sustainable impact on the public Realm?

Art installations in the public realm have a social responsibility as they have the capacity to create a fundamental traction between the citizen and the urban setting. This social component aids to sustain the ability of the citizens to embody the city and promote social cohesion. This responsibility relies on ensuring the use of the public space for future generations. Using the three pillars of sustainable development, both light installations aimed to address the social component on a higher level while still incorporating aspects of Ecology and Economy. Ecological and economical considerations are placed into the aforementioned temporality, where minimal material resources are used without removing or discarding any existing installation, as well as not procuring new equipment as the used one has been on loan. Likewise, the use of energy and running costs of a temporary installation are, by nature, finite, with no burden to municipality and taxpayers. Light pollution and light trespass are not an issue: "Reflecting on Troy" offers a controlled lighting effect on carefully studied surfaces, while "Riddle 102" is an enclosed space. The social human aspect of the light installations reside in the collective memory of those who have experienced the installations. The identification or even rejection of the temporal light installation promotes discussion among the public of the current values or missing aspects in their own city (as is evident in the above testimony). Either by transporting the citizen to an imaginarium or by highlighting architectural features and effects, the spatial configuration of both sites have been transformed through light. The power of light, as intangible as it is, has the ability to change the expression and atmosphere of a place: from daytime, where the true undressed space is visible with no lighting effects but natural daylight, to a nocturnal space, where light dresses up to accompany the citizen.

5.2. How does activating the public realm transform the city in collective memory?

A city is a complex organism that provides both physical and social organization of material, people and events. As seen from recollected testimonies, the installations have left a mark on their visitors that last beyond the experienced durations; once an event is activating the public realm it remains in the collective memory of the visitors. These moments spent in an atmospheric event layered on urban space are adding to how the city spaces may be recalled. The installation may be missed or added to the catalogue of what makes a city an intersection of life and culture.

5.3. What is the impact of the online urban presence on the perception of the public realm?

For millennia people have used mnemonic devices to recall the past (18). There are certain aspects of a city that we can only experience physically. However, because of our increased online presence, an alternate, virtual image of our experiences can be shared beyond the local community (see Figure 10, Figure 11). As both projects were shortlisted to the [d]arc award, they were exposed to global viewing and voting. This reach (19) makes one wonder what are the limits of how a public realm is experienced. This virtual way of experiencing the public realm goes through a process of image selection, editing and selected photographic methods. For example, the projects can be "experienced" differently with a drone than in a moving, breathing, sensing body. Our cities extend, through the light of our photographs, allowing us to communicate a spatial experience remotely to our viewers and peers.

Figure 10 - "Reflecting on Troy" visitor images on Instagram.

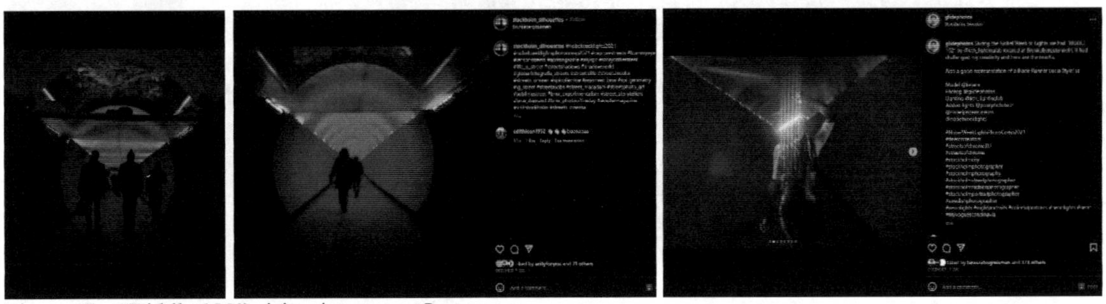

Figure 11 - "Riddle 102" visitor images on Instagram.

6. CONCLUSION

As designers and educators, we find it is important to explore how light may activate our cities and impact how they are perceived and remembered. We do not design only for "now" but for the collective memories of the city as well. Our actions in physical space impact not only our corporeal environment, but also our virtual one. Since light festivals are temporary and repeating events that act as a testing ground in the body of the city, each city can be imagined and reimagined anew adding to the palimpsest of how a city is recalled. Likened to Freud's subconscious unfolded through the unedited physical history of Rome (20). A city may be seen as a multi-layered entity. Thus, one city can also have multiple faces. As Calvino imagines Venice as numerous cities in invisible cities (21). Similarly, different installations, at the same place may, if successful, exist simultaneously in the collective memory and add layers to the identity and perception of a city.

Acknowledgements

We would like to thank the Troy Glow and NWL festival teams as well as our universities and all the students who helped to install the projects. Special thanks to Dr. Lorelei Wagner for her precise and thoughtful editing.

References

1. Feinstein L, Irwin R. Robert Irwin the Beauty of Questions. 1997.
2. Zielinska-Dabkowska. Night in a big city. Light festivals as a creative medium used at night and their impact on the authority, significance and prestige of a city. Lodz: University of Lodz, Department of International Marketing and Retailing; 2016.
3. Edensor T. From Light to Dark : Daylight, Illumination, and Gloom Minneapolis; London: University of Minnesota Press; 2017.
4. Quinn B. Arts festivals, urban tourism and cultural policy. Journal of Policy Research in Tourism Leisure and Events. 2010 November; 2(3): 264-279.
5. Gotham KF. Theorizing urban spectacles. City Analysis of Urban Change, Theory, Action. 2005 July; 9(2): 225-246.
6. Jakob D. The eventification of place: Urban development and experience consumption in Berlin and New York City. European Urban and Regional Studies. 2012 October; 20(4): 447-459.
7. Collection LA. Light Art Collection. [Online].; 2023 [cited 2023 October 27. Available from: https://lightart-collection.com/.
8. Manno M. Times Union. [Online].; 2022 [cited 2023 08 01. Available from: https://www.timesunion.com/news/article/Troy-Glow-light-festival-to-illuminate-city-17552896.php.
9. Team NWL. Nobel Week Lights. [Online].; 2022 [cited 2023 08 06. Available from: https://nobelweeklights.se/about-nobel-week-lights/?lang=en.
10. Canepa E, Scelsi V, Fassio A, Avanzino L, Lagravinese G, Chiorri C. Atmospheres: feeling architecture by emotions. preliminary neuroscientific insights on atmospheric perception in architecture. Ambiances. Environnement sensible, architecture et espace urbain. 2019 December; 5(1): 1-30.
11. Merleau-Ponty M. Eye and mind. In Theatre and Performance Design.: Routledge; 2012. p. 243-245.
12. Pallasmaa J. An Architecture of the Seven Senses. In Holl S, Pallasmaa J, Perez-Gomez A. Questions of Perception: Phenomenology of Architecture.: William Stout Publishers; 2014. p. 27-37.
13. Jacobs J. The death and life of great American cities New York: Random House; 1961.
14. Lynch K. The Image of the City Cambridge: MIT Press; 1960.
15. Leijonhufvud JC. Examining the impact of white, red, and blue light on. Student report. Stockholm: Royal Institute of Technology, Architecture; 2021.
16. Marin M. The emotional impact of colors of light and of different sequences of them. student report. Stockholm: Royal Institute of Technology, Architecture; 2021.
17. Michel L. Light: The shape of space: Designing with space and light: John Wiley & Sons; 1995.
18. Yates FA. The art of memory Chicago: University of Chicago Press; 1966.
19. [d]arc awards t. [d]arc awards. [Online].; 2023 [cited 2023 08 01. Available from: https://darcawards.com/art-low-budget-2022/.
20. Freud S. Civilization and Its Discontents. In. New York: Norton; 2010. p. 17-19.
21. Calvino I. Invisible cities. 1st ed. New York: Harcourt Brace Jovanovich; 1978.

IOP Conf. Series: Earth and Environmental Science 1320 (2024) 012013 doi:10.1088/1755-1315/1320/1/012013

People's conceptual environmental appraisals of urban space in daylight and electric light

M Johansson[1], **G Tsiakiris**[1] **and J Rahm**[1]

[1]Environmental Psychology, Dept of Architecture and Built Environment, Lund University, Box 118, SE-221 00 Lund, Sweden

maria.johansson@abm.lth.se

Abstract. Urban spaces should be resilient, inclusive and safe for all residents including children, elderly and vulnerable groups. This implies that urban planning, management and design should provide public spaces that support human needs of active mobility, social interaction and psychological restoration across seasonal and diurnal variation in natural light. Environmental psychology postulates that an understanding of how people experience or appraise an environment is a critical component in the relationship between urban design and human behaviour. In this conceptual paper we propose and discuss overarching (visual) impressions of the environment (environmental appraisals) that could be utilized to deepen the understanding of how people experience urban green spaces under different lighting conditions. In a certain setting, people respond to specific components of the environment; such perceived molecular environmental qualities make up the overall impression of the setting, and shape critical human-environment transactions fundamental to human well-being. The perceived stimuli are considered to act together to form the perceptual basis for higher order, molar aspects, so called conceptual environmental appraisals. We introduce Küller's human-environment interaction model and Brunswik's lens model to describe conceptual environmental appraisals. Recognising urban forests as important settings for physical active mobility, social interaction, and psychological restoration, we outline the theoretical underpinnings of three conceptual environmental appraisals: prospect – refuge pertaining to the perceived safety of the environment, environmental preferences, and the perceived restorative potential of the environment. The appraisals have been shown to be relevant in studies on electric lighting in urban space, and possible to assess by means of observer-based assessments also among children, elderly and vulnerable groups. Assessment outcomes may inform discussions on conflicting goals for sustainable development with regard to people's health and well-being, energy use, light pollution and the welfare of other species in urban planning, management and design.

1. Introduction

Urban spaces should be resilient, inclusive and safe for all residents, including children, elderly and vulnerable groups [1]. This implies that urban planning and design should provide public urban spaces that support people's needs across seasonal and diurnal variation in natural light. There is ample support that people's experiences of urban spaces differ between daylight and electric lighting, as well as different lighting applications [2-5]. This is amongst others evident in urban green spaces [6, 7]. Historically, the introduction of electric lighting profoundly changed the human way of life, allowing for a broad range of activities previously confined to daylight conditions [8]. Today, a growing body of research shows that benefits of outdoor electric lighting also come with conflicting goals for a sustainable development, including immense costs associated with energy use (SDG 7) and light pollution, negatively impacting ecological systems, wildlife (SDG 14 & 15), and human physical and psychological well-being (SDG 3) [1, 9-11]. Hence, electric lighting must carefully balance the needs of light and darkness with consideration for human and non-human welfare [12].

So far, in the literature on urban planning, people's experiences of urban spaces during dark hours are often restricted to the existence/non-existence of electric lighting, while the field of lighting research is focused on the technical characteristics of the electric lighting [13]. Moreover, lighting is primarily discussed in association with pedestrians' experience of safety (or reassurance) and security [14]. This conceptual paper aims to propose and discuss a broader set of overarching (visual) impressions of the environment (environmental appraisals) that could be utilized to deepen the understanding of how people experience urban green spaces under different forms of electric lighting, which would benefit the planning, management and design of such environments. We first describe human-environment interaction and conceptual environmental appraisals and stress the importance of using observer-based environmental assessment for management and evaluation of inclusive urban space [15]. We introduce three different appraisals established in environmental psychological theory on human-environment interaction and shown to be important across gender, age and levels of vulnerability. These appraisals are discussed in relation to green spaces in residential areas, so-called urban forests (i.e. a forest in a metropolitan area, with self-organizing vegetation and a dynamic system that entails different species and their exchange [16]). Urban forests may be valuable assets for multiple use providing for active mobility, social interaction and psychological restoration. Urban forests have been shown to be important settings for recreation, physical activity [17] that support mental well-being [18], and development of connectedness to nature [19], all aspects relevant to the SDG 3.

2. Evaluating people's experiences of the lit environment

Systematic evaluation of outdoor electric lighting preferably should rely on both technical environmental assessment (TEA) and observer-based environmental assessment (OBEA). OBEA draws on people's self-report through which they express perceptions, observations, and environmental appraisals [20, 21]. In other words, by OBEA human perception is used for assessing environmental quality [22]. OBEAs should involve activities and perspectives that pertain to broad groups of urban residents and visitors (see for example [23] for an overivew of children and outdoor lighting, and [24, 25], for discussions on visually impaired and outdoor lighting). Moreover, OBEAs need to cater for a diversity of settings amongst others urban forests, include a variety of specific places and their associated activities. Current literature largely focuses upon how light supports or hinders active mobility to different destinations, in particular walking and cycling [24, 26-28] and discusses visual orientation, obstacle detection and facial recognition as important pedestrian tasks [29-31]. A few studies address activities such as visiting public (green) spaces for different recreational activities and psychological restoration [32-35], and social interaction [36, 37]. While OBEAs of the perceived qualities of outdoor electric light per se can give important information with regard to perceived visual accessibility, perceived safety, and behavioural intentions [38, 39], environmental appraisals should be considered as outcomes of how the light interacts with the surroundings [30, 40-43].

3. Human – environment interaction and conceptual environmental appraisal

In environmental psychology it has been shown that assessments of people's experience of various environmental features often are better predictors of well-being outcomes than are assessments of the exposure to these features [44, 45]. Accordingly, an understanding of how people experience or appraise an environment is a critical component in the relationship between environmental design and human behaviour. The Human – Environment Interaction model (HEI) [46] presents a theoretical framework that describes why we should distinguish between what people are exposed to in a certain setting and how they experience the environment in this setting. The HEI model posits that a basic emotional process is at the heart of the interaction. According to the HEI model, the basic emotional process is affected by the physical and the social environment. The experience of the environment is seen as an active process involving both external factors (environment) and internal ones (e.g. the individual's emotional state), sensitive to the individual's personal characteristics, experiences, values and attitudes. Moreover, people may experience one and the same setting differently depending on the activities they are engaged in. In a certain setting, people respond to specific components of the physical environment, perceived environmental qualities, which make up the overall impression of the setting, and shape critical human-environment transactions (psychological processes) fundamental to human well-being. The HEI model thereby differentiates between molecular aspects (micro-level infrastructure elements such as surface materials and their quality, landmarks, light sources, etc.) and molar aspects that relate to the way in which individual molecular aspects coexist within the overall environmental setting [25, 47]). Rahm et al. [7] further developed the notion of molecular and molar aspects of urban environments drawing on Brunswik's lens model [48, 49]. Brunswik differentiates between distal stimuli (objective descriptors of the setting, i.e. what we are exposed to), and proximal stimuli (imperfect sensory information serving as cues that represent the distal stimuli). The proximal stimuli are to a varying degree utilized by the individual in his/her perception of specific molecular aspects of the environment. The specific perceived stimuli act together [50], and as such form the perceptual basis for higher order, conceptual environmental appraisals in which perceptions of specific stimuli are integrated [51]. The conceptual environmental appraisals in turn shape the affective appraisal of the environment – which influences intentions to visit the environment [7, 52].

Mehta [53] proposed that public space should be meaningful in its design and the activities it supports, provide a sense of safety, physical and environmental comfort and convenience, a sense of control, and sensory pleasure. The programming and the design of the space and the implementation of light design all contribute to achieve these goals during dark hours [2, 37]. These qualities may also apply to urban forests as a public space. However, urban forests are situated on the fringe of built environments and become a zone of transition between programmed space and more self-organized and dynamic natural systems. Below we describe conceptual environmental appraisals that would have relevance for urban space subject to relatively more or less programming, urban forests included: appraisals of prospect and refuge considered as fundamental to the perceived safety of an environment, appraisals that uncover a more general satisfaction and preference for an environment, and finally appraisals that address the psychologically restorative potential of the environment. Figure 1 gives an overview of the theories.

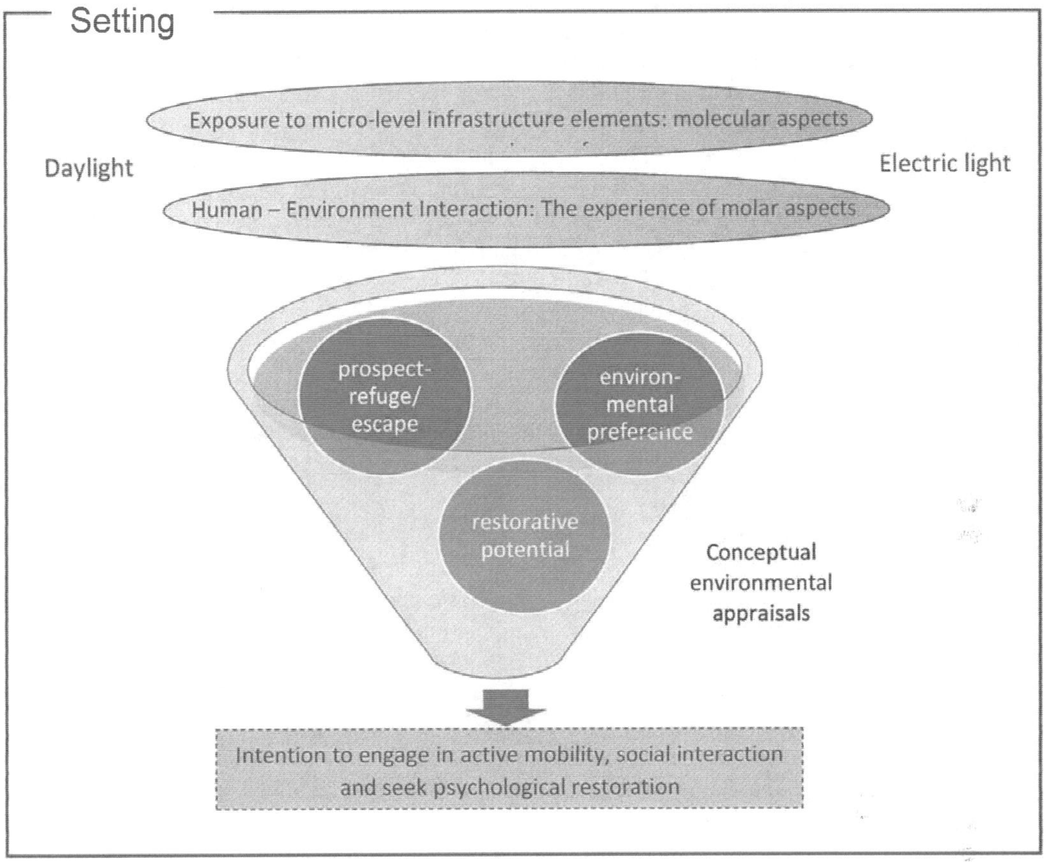

Figure 1. Theoretical overview of conceptual environmental appraisals.

3.1. Prospect –refuge

Prospect-refuge theory [54] posits that there is an evolutionary basis for people's environmental preference and that it is dependent on the extent to which it is possible to get an overview of an environment (prospect) while at the same time not being seen (refuge). In a meta-review of applications of the prospect – refuge theory, Dosen and Ostwald [55] found general support for the importance of prospect whereas refuge was found to be mostly relevant in natural settings. Appraisals of prospect and refuge are associated with the inherent perceived safety of a place. Fisher and Nasar [56] proposed that for urban environments refuge should rather be considered in terms of potential hiding places for offenders, and found that prospect, refuge and perceived possibilities to escape an unsafe situation (escape) were important factors for the perceived safety of the environment. Variations of the prospect-refuge theory have been applied to assess the impact of lighting on the experience of prospect, refuge and escape in urban night time environments [57-59]. Blöbaum and Hunecke [57] found escape to be the most influential aspect, whereas Haans and de Kort [59] found prospect to be of greater importance for perceived safety. In fact, van Rijswijk and Haans [60], in a correlational study drawing on prospect –refuge theory, concluded that assessments of environmental appraisals of prospect, concealment, and entrapment were relatively more important than the assessment of light quality to perceived safety.

3.2. Environmental preference

Küller [46] proposed that all settings can be described by certain visual criteria. These criteria and how they play together in the visual experience of a place could say something about to what extent a person can make sense of the visual environment and find it interesting. By employing the statistical technique of factor-analysis of semantic scales, Küller found that the visual experience of the environments could be divided into eight dimensions associated with environmental preference, including perceived pleasantness, complexity, unity, enclosedness, potency, social status, affection, and originality (further described in Table 1). These dimensions have been employed to describe a variety of built and the natural landscapes in daylight [61, 62]. In the SED the balance between unity (how well different objects in an environment are perceived to fit together) and complexity (how much variation the environment offers and how strong the intensity, contrast, and abundance are in the environment) could be considered a higher-order variable associated with the perceived pleasantness, and a key characteristic of preferred environments [63].

Table 1. Overview of the eight dimensions of Semantic Environmental Description (SED).

Dimension	Description
Pleasantness	The pleasantness, beauty and security which the individual experiences in the environment
Complexity	The environment's liveliness and complexity
Unity	How well the various components in the environment fit and function together
Enclosedness	The closedness and degree of demarcation
Potency	The intrinsic, potential force in the environment
Social status	The evaluation of the environment in economic and social terms
Affection	An age aspect of the environment as well as a feeling of recognition for the old and genuine
Originality	The unusual and surprising in the environment

Kaplan and Kaplan [64], in a similar vein, presented an environmental preference model specifically adapted to natural settings. This model is limited to four dimensions and covers 1) coherence - how well the environment hangs together and can be seen as a whole), 2) legibility - how easy it is for the viewer to understand or to categorize the environmental scene, 3) complexity, defined in terms of how many different visual elements that are present in the environment, and 4) mystery – to what extent the environment draws attention and triggers a person's imagination and gives the impression that there is more interesting information if they walk further into the scene. These four dimensions may act together, and environments assessed as high in coherence and legibility are likely to be preferred. Moreover, in line with the dimensions proposed by Küller [46], coherence (unity) and complexity interact. An environment assessed as low in both coherence and complexity gives few visual impressions, whereas an environment high in coherence and low in complexity is likely to be experienced as monotonous and boring. However, an environment that is assessed high in both coherence and in complexity is visually rich and well-organized, and likely to be preferred.

The dimensions of visual experiences and environmental preferences developed by Küller, and Kaplan and Kaplan respectively, have been extensively researched in preference for urban and natural settings in daylight conditions. Less is known about how these visual experiences may shift between light and dark conditions. A review showed that one reason may be that the assessed environments often lacks systematic objective descriptions that would allow comparison between studies [65]. In an experimental study, Stamps [66] followed up the systematic evaluation and showed, by manipulating photos of different settings, that daylight and dusk were assessed differently on the dimension mystery, where dark conditions was rated higher. Lis et al [67] also employed photos to systematically explore the effects of different lighting solutions (lighting of horizontal and vertical elements respectively lighting the foreground and background) along a path in an urban park on appraisals of amongst others legibility, mystery, and preference. This approach showed that lighting of vertical space surrounding the paths increased the appraisal of mystery and in turn preference, and lighting of the background resulted in the landscape being appraised as more legible and mysterious, with an increased sense of safety. Patching et al. [68], in a field study located in an urban park, applied Küller's [46] overarching SED dimension of pleasantness and found various outdoor lighting solutions to significantly differentiate. The solution perceived to provide the most comfortable light was also assessed as the most pleasant one.

3.3. *Restorative potential.*

Stress recovery theory (SRT; [69]) and Attention restoration theory (ART; [64, 70]) describe the restorative potential of an environment, i.e. to what extent it may support recovery from stress. SRT assumes that, through evolution, humans became innately attuned to features with adaptive significance, and that people today still show biologically prepared responses to such features. A setting would support restoration in the absence of threats and presence of natural contents, like vegetation and water, that signaled possibilities for survival, and visual scene characteristics such as moderate complexity and depth. Such environmental features would block negative affect and thoughts in a stressed person, and promote physiological deactivation. ART describes components of environmental experience that support recovery from directed attention fatigue. Recovery can occur when a person's attention can go without effort to what they find interesting (i.e., fascination). It also requires a change away from routine mental contents, a sense of being away from these routines. Further, the situation should involve compatibility with the person's activities in a setting, and a sense of extent. Rahm et al. [7] assessed the ART dimensions along an urban path and found daylight conditions to have greater restorative potential than nighttime conditions with electric lighting. However, simulation studies by Nikunen and Korpela [40, 41] suggest that outdoor nighttime environments may be experienced as restorative if the greenery is made salient by focusing light on it.

4. Discussion and conclusion

This paper sheds light on how conceptual environmental appraisals could be used to understand how electric lighting introduced in urban forests may hinder or support human needs. The conceptual environmental appraisals suggest that it may be insufficient to analyse perceptions of molecular aspects represented by micro-level infrastructure elements if we are to understand the extent to which the presence and quality of electric lighting support people's needs. Moreover, the appraisals point to the importance to go beyond people's experiences of the light per se, and as different dimensions of visual experiences interact, the analysis should have a broader scope to also capture how people experience the configuration of the lit environment (see also [7, 67]). Departing from the recognition that urban forests are important settings for active mobility, social interaction, and psychological restoration, we outlined the theoretical underpinnings of three conceptual environmental appraisals: prospect – refuge pertaining to the perceived safety of the environment [54], visual experiences associated with environmental preference [46, 64], and perceived restorative potential of the environment [69, 70]. One major benefit of applying this perspective is that the theoretical and methodological foundations for these three sets of conceptual environmental appraisals are well-established. Another benefit is that they are possible to assess by means of OBEAs with valid and reliable instruments [57, 71-73]. The theories and appraisals rely on the perspectives of laypeople and do not require expert knowledge of urban space or lighting. Further, the instruments can easily be adapted to be applied in OBEAs among children, elderly and vulnerable groups [25, 74] thus facilitating the consideration of the experiences and needs of these groups in attempts to design inclusive urban spaces. However, to compare different settings and to be able to make sense of OBEAs, they need to be systematically described and complemented with detailed and relevant results from technical measures (TEAs) [65]. We introduced the appraisals in the context of urban forests, but the appraisals are relevant for diverse settings across built and natural environments and could likely be applied to any type of urban public space. However, the appraisals are complementary with respect to what experiences are captured and the relative importance may therefore vary depending on the type of setting investigated. For example, appraisals of restorative potential might be more important to obtain in spaces aimed at recreational purposes, whereas a generally visually pleasant space that provides a good overview and opportunities for refuge might be more important along walking paths. It should, however, be recognised that active mobility, social interaction and restoration in practice co-occur in time and space, and a strict division would therefore not be desirable. All of the appraisals discussed have been developed in environmental psychology and have previously been mentioned in the literature on electric lighting of public urban space. However, there are other disciplines that could contribute with additional perspectives, and a systematic review would be necessary for a complete overview, and any conclusions regarding the relative importance of the appraisals in this context require empirical studies applying the appraisals in parallel. The theories proposed are not exhaustive, but selectively chosen from the field of environmental psychology as examples to highlight the benefit of relying on established theories and a methodological approach in systematic evaluations of people's appraisals of urban forests. In the present case, to assess and compare daylight with dark conditions, as well as different designs of electric lighting. OBEAs of daylight conditions could serve as a baseline for how well an urban space could function for different purposes, and a comparison with a dark condition would indicate if and to what extent the function changes in darkness. Assessment outcomes may inform discussions on conflicting goals for sustainable development with regard to people's health and well-being, energy use, light pollution and the welfare of other species in urban planning, management and design.

References

1. United Nations (2015) *Transforming our world: The 2030 Agenda for Sustainable Development.* Available at: https://wedocs.unep.org/20.500.11822/9814 (Accessed: 14 February 2024)
2. Casciani D, 2020 *The Human and Social Dimension of Urban Lightscapes* (Springer Cham)
3. Fotios S and Cheal C 2007 Lighting for subsidiary streets: Investigation of lamps of different SPD. Part 1—Visual performance *Lighting Res Technol* **39** 215-32
4. Fotios S and Cheal C 2011 Predicting lamp spectrum effects at mesopic levels. Part 2: Preferred appearance and visual acuity *Lighting Res Technol* **43** 159-72
5. Lin Y and Fotios S 2015 Investigating methods for measuring face recognition under lamps of different spectral power distribution *Lighting Res Technol* **47** 221-35
6. Rahm J, Sternudd C and Johansson M 2021 "In the evening, I don't walk in the park": The interplay between street lighting and greenery in perceived safety *Urban Design International* **26** 42–52
7. Rahm J, Niska A and Johansson M 2024 The environmental experience during daylight and dark conditions: A conceptual model for pedestrians and cyclists *PsyEcology* accepted
8. Garnert J 2016 *Ut ur Mörkret:Ljusets och Belysningens Kulturhistoria* (Lund: Historiska media)
9. Jägerbrand A K and Spoelstra K 2023 Effects of anthropogenic light on species and ecosystems *Science* **380** 1125-30
10. Tavares P, Ingi D, Araújo L, Pinho P and Bhusal P 2021 Reviewing the role of outdoor lighting in achieving sustainable development goals *Sustainability* **13** 12657
11. Wang T, Kaida N and Kaida K 2023 Effects of outdoor artificial light at night on human health and behavior: A literature review *Environmental Pollution* **323** 121321
12. Falchi F, Cinzano P, Elvidge C D, Keith D M and Haim A 2011 Limiting the impact of light pollution on human health, environment and stellar visibility *Journal of Environmental Management* **92** 2714-22
13. Rahm J and Johansson M. *Walking After Dark – A Systematic Literature Review of Pedestrians' Response to Outdoor Lighting* (Lund: Dept. of Architecture and Built Environment)
14. Fotios S, Unwin J and Farrall S 2015 Road lighting and pedestrian reassurance after dark: A review *Lighting Res Technol* **47** 449-69

15. Sundevall E P and Jansson M 2020 Inclusive parks across ages: Multifunction and urban open space management for children, adolescents, and the elderly *International Journal of Environmental Research and Public Health* **17** 9357
16. Roman L A, Pearsall H, Eisenman T S, Conway T M, Fahey R T, Landry S, et al. 2018 Human and biophysical legacies shape contemporary urban forests: A literature synthesis *Urban Forestry & Urban Greening* **31** 157-68
17. Lehto C, Hedblom M, Öckinger E and Ranius T 2022 Landscape usage by recreationists is shaped by availability: Insights from a national PPGIS survey in Sweden *Landscape Urban Plan* **227** 104519
18. Knez I, Butler A, Ode Sang Å, Ångman E, Sarlöv-Herlin I and Åkerskog A 2018 Before and after a natural disaster: Disruption in emotion component of place-identity and wellbeing *Journal of Environmental Psychology* **55** 11-7
19. Elbakidze M, Dawson L, Milberg P, Mikusiński G, Hedblom M, Kruhlov I, et al. 2022 Multiple factors shape the interaction of people with urban greenspace: Sweden as a case study *Urban Forestry & Urban Greening* **74** 127672
20. Craik K H and Feimer N R 1987 Environmental Assessment *Handbook of Environmental Psychology* ed D Stokols and I Altman (New York: John Wiley & Sons)
21. Gifford R 2007 Environmental psychology and sustainable development: Expansion, maturation and challenges *Journal of Social Issues* **63** 199-212
22. Bonaiuto M, Fornara F and Bonnes M 2006 Perceived residential environment quality in middle- and low-extension Italian cities *Revue Européenne de Psychologie Appliquée/European Review of Applied Psychology* **56** 23-34
23. Litsmark A, Rahm J, Mattsson P and Johansson M 2023 Children's independent mobility during dark hours: A scoping review *Frontiers in Public Health* **11**
24. Johansson M, Rosén M and Küller R 2011 Individual factors influencing the assessment of the outdoor lighting of an urban footpath *Lighting Res Technol* **43** 31-43
25. Mattsson P, Johansson M, Almén M, Laike T, Marcheschi E and Ståhl A 2020 Improved usability of pedestrian environments after dark for people with vision impairment: An intervention study *Sustainability* **12**
26. Rahm J 2019 Urban outdoor lighting: Pedestrian perception, evaluation and behaviour in the lit environment *PhD Thesis* Lund University, Lund
27. Uttley J, Fotios S and Lovelace R 2020 Road lighting density and brightness linked with increased cycling rates after-dark *PLOS ONE* **15** e0233105
28. Fotios S, Uttley J and Fox S 2019 A whole-year approach showing that ambient light level influences walking and cycling *Lighting Res Technol* **51** 55
29. Fotios S and Johansson M 2019 Appraising the intention of other people: Ecological validity and procedures for investigating effects of lighting for pedestrians *Lighting Res Technol* **51** 111
30. Rahm J and Johansson M 2021 Assessment of outdoor lighting: Methods for capturing the pedestrian experience in the field *Energies* **14**
31. Fotios S and Uttley J 2018 Illuminance required to detect a pavement obstacle of critical size *Lighting Res Technol* **50** 390-404
32. Nikunen H J, Puolakka M, Rantakallio A, Korpela K M and Halonen L 2014 Perceived restorativeness and walkway lighting in near-home environments *Lighting Res Technol* **46** 308-28
33. Stanhope J, Liddicoat C and Weinstein P 2021 Outdoor artificial light at night: A forgotten factor in green space and health research *Environmental Research* **197** 111012
34. Hu B and Zhao J 2022 Factors promoting nature-based outdoor recreation during the daytime and evening *Journal of Outdoor Recreation and Tourism* **40** 100572
35. Zhao J, Liu G and Li X 2023 Comparison of the restorative quality of green spaces between the evening and daytime *Proceedings of the Institution of Civil Engineers - Urban Design and Planning* **176** 65-76

36. Hennig V K R, Fotios S, Gentile N, Sternudd C and Johansson M 2022 Social interaction in local public squares after dark *IOP Conference Series: Earth and Environmental Science* **1099** 012008
37. Hennig V, Gentile N, Fotios S, Sternudd C and Johansson M 2023 User behaviour in public squares after dark *Lighting Res Technol* **55** 621-42
38. Johansson M, Pedersen E, Maleetipwan-Mattsson P, Kuhn L and Laike T 2014 Perceived outdoor lighting quality (POLQ): A lighting assessment tool *Journal of Environmental Psychology* **39** 14-21
39. Markvica K, Richter G and Lenz G 2019 Impact of urban street lighting on road users' perception of public space and mobility behavior *Building and Environment* **154** 32-43
40. Nikunen H J and Korpela K M 2009 Restorative lighting environments: Does the focus of light have an effect on restorative experiences? *J Light Vis Env* **33** 37-45
41. Nikunen H J and Korpela K M 2012 The effects of scene contents and focus of light on perceived restorativeness, fear and preference in nightscapes *Journal of Environmental Planning and Management* **55** 453-68
42. Nasar J L and Bokharaei S 2016 Impressions of lighting in public squares after dark *Environment and Behavior* **49** 227-254
43. Nasar J L and Bokharaei S 2017 Lighting modes and their effects on impressions of public squares *Journal of Environmental Psychology* **49** 96-105
44. Nasar J L 2008 Assessing perceptions of environments for active living *Am Journal of Preventive Medicine* **34** 357-63
45. Bratman G N, Anderson C B, Berman M G, Cochran B, de Vries S, Flanders J, et al. 2019 Nature and mental health: An ecosystem service perspective *Science Advances* **5** eaax0903
46. Küller R 1991 Environmental assessment from a neuropsychological perspective *Environment, Cognition and Action: An Integrated Approach* ed T Gärling and G W Evans (New York: Oxford University Press) p 111-47
47. Marcheschi E, Ståhl A, Almén M and Johansson M 2020 A theoretical model for urban walking among people with disabilities *Frontiers in Psychology* **11**
48. Brunswik E 1952 *The conceptual framework of psychology* (Chicago: Univ. of Chicago Press)
49. Cooksey R W 2001 Brunswik's The conceptual framework of psychology: Then and now *The essential Brunswik* ed K R Hammond (New York: Oxford University Press) p 225-37
50. Cupchik G C and Berlyne D E 1979 The perception of collative properties in visual stimuli *Scandinavian Journal of Psychology* **20** 93-104
51. Johansson M, Pedersen E and Weisner S 2019 Assessing cultural ecosystem services as individuals' place-based appraisals *Urban Forestry & Urban Greening* **39** 79-88
52. Johansson M, Sternudd C and Kärrholm M 2016 Perceived urban design qualities and affective experiences of walking *Journal of Urban Design* **21** 256-75
53. Mehta V 2014 Evaluating public space *Journal of Urban Design* **19** 53-88
54. Appleton J 1975 *The experience of landscape* (London: Wiley)
55. Dosen A and Ostwald M 2016 Evidence for prospect-refuge theory: A meta-analysis of the findings of environmental preference research *City, Territory & Architecture* **3**
56. Fisher B S and Nasar J L 1992 Fear of crime in relation to three exterior site features: Prospect, refuge, and escape *Environment and Behavior* **24** 35-65
57. Blöbaum A and Hunecke M 2005 Perceived danger in urban public space. The impacts of physical features and personal factors *Environment and Behavior* **37** 465-86
58. Boomsma C and Steg L 2014 Feeling safe in the dark: Examining the effect of entrapment, lighting levels, and gender on feelings of safety and lighting policy acceptability *Environment and Behavior* **46** 193-212
59. Haans A and de Kort Y A W 2012 Light distribution in dynamic street lighting: Two experimental studies on its effects on perceived safety, prospect, concealment, and escape *Journal of Environmental Psychology* **32** 342-52

60. van Rijswijk L and Haans A 2018 Illuminating for safety: Investigating the role of lighting appraisals on the perception of safety in the urban environment *Environment and Behavior* **50** 889-912
61. Johansson M and Laike T 2007 Intention to respond to local wind turbines: The role of attitudes and visual perception *Wind Energy* **10** 435-51
62. Bengtsson A, Hägerhäll C, Englund J-E and Grahn P 2015 Outdoor environments at three nursing homes: Semantic environmental descriptions *Journal of Housing For the Elderly* **29** 53-76
63. Laike T 1997 The impact of daycare environments on children's mood and behavior *Scandinavian Journal of Psychologyl* **28** 209-19
64. Kaplan R and Kaplan S 1989 *The experience of nature: A psychological perspective* (Cambridge: Cambridge Univ. Pr.)
65. Stamps A E, 3rd 2004 Mystery, complexity, legibility and coherence: A meta-analysis *Journal of Environmental Psychology* **24** 1-16
66. Stamps A E, 3rd 2007 Entropy and environmental mystery *Percept Mot Skills* **104** 691-701
67. Lis A, Zienowicz M, Kukowska D, Zalewska K, Iwankowski P and Shestak V 2023 How to light up the night? The impact of city park lighting on visitors' sense of safety and preferences *Urban Forestry & Urban Greening* **89** 128124
68. Patching G R, Rahm J, Jansson M and Johansson M 2017 A new method of random environmental walking for assessing behavioral preferences for different lighting applications *Frontiers in Psychology* **8** 1-13
69. Ulrich R S, Simons R F, Losito B D, Fiorito E, Miles M A and Zelson M 1991 Stress recovery during exposure to natural and urban environments *Journal of Environmental Psychology* **11** 201-30
70. Kaplan S 1995 The restorative benefits of nature: Toward an integrative framework *Journal of Environmental Psychology* **15** 169-82
71. Hartig T, Korpela K, Evans G W and Gärling T 1997 A measure of restorative quality in environments *Scandinavian Housing and Planning Research* **14** 175-94
72. Küller R 1975 *SMB manual* (Stockholm)
73. Küller R 1972 *A semantic model for describing perceived environment* (Stockholm: National Swedish Institute for Building Research)
74. Litsmark A Children's experiences of electric outdoor lighting (in preparation)

IOP Conf. Series: Earth and Environmental Science 1320 (2024) 012014 doi:10.1088/1755-1315/1320/1/012014

Spatial planning for dark hours in Sweden – conditions and proposal for action

C Frænkel[1], M Tastare[2] and M Billger[3]

[1] White arkitekter, Östgötagatan 100, 116 92 Stockholm, Sweden
[2] Sweco architects, Gjörwellsgatan 22, 112 60 Stockholm, Sweden
[3] Dept of Architecture & Civil Engineering, Chalmers, 412 96 Göteborg, Sweden

Corresponding author's email: clara.fraenkel@white.se

Abstract. The purpose of this study was to investigate how the absence of daylight are handled within spatial planning in Sweden, and to propose measures for action. The study includes a review of spatial planning documents, interviews with planning officers, and an overview of how outdoor lighting is included within the Swedish Planning and Building Act (PBL). The result shows that planning documents according to the PBL generally present few specific analyses and proposals linking to the design and the use of outdoor environments during dark hours, such as lighting. Instead, outdoor lighting is often managed within lighting masterplans, that lack a common methodology and a formal planning status. Possible consequences include that spatial planning provides limited support for how different public interests should be weighed against each other in a nocturnal context, and that citizens have little formal influence over strategic decisions on outdoor lighting. As a proposal, we suggest developing spatial planning for dark hours ('nattstadsplanering' in Swedish), that integrates the use, the design, and the outdoor lighting within the same planning process, using the PBL as a framework. The proposal includes clarification of lighting regulations, methodology development, and knowledge dissemination, which requires further research and investigations.

1. Introduction

When night falls and daylight disappears, the use and the character of the outdoor environment change in a profound way. Most living species are biologically adapted to the regular light and dark patterns, that emerge from the rotating earth, and are either diurnal or nocturnal. As humans are diurnal, we generally spend most of our awake time in daylight, but ever since the invention of electric lighting, society has undergone a temporal revolution and is now active 24-hours-a-day [1]. Outdoor lighting has made public spaces accessible and populated during dark hours and has large benefits for the individual and for the society [2]. The extensive use of outdoor lighting [3], also has multiple negative impacts, including astronomical and ecological light pollution and impacts on humans, in terms of health and obtrusive light [4]. Societies located in high latitudes, that are characterized by strong variation in day-length during different seasons of the year, such as Sweden, are especially depending on outdoor lighting. In December, the number of dark hours per day (sunset-sunrise) last around seventeen hours in the south of Sweden (Malmö) and twenty-four hours in the north (Kiruna). The light conditions at high latitudes dissolve the connection between the astronomical definition of 'night', when the sun is below the horizon, and the temporal definition of 'night', for example between 22-06 [5]. The lack of daylight affects the use of public spaces. For example, in subarctic areas, darkness is seen as one of the most important climatic barriers for walking outside [6], and due to unsafety, many women avoid certain areas and means of transportation during dark hours [7]. Furthermore, the experience of the townscape and the landscape alters completely, when the light from the sun and the bright sky is replaced by the dark vault of the night sky, with its subtle moonlight and starlight, and the layer of electric lighting. Through the quality and the quantity of light, in relation to the surroundings, outdoor lighting changes the character and the ambiance of the nocturnal city [8]. The illumination of the city could be used to underline the interplay and the relationality between light and darkness, and to embrace them both [9].

1.1 Spatial planning for dark hours in Sweden

The background to this study, is that we, in our professions within architecture, urban planning and lighting design, have reflected on the idea that the daily shift from day to night should be considered more within spatial planning, to ensure that public spaces are well-designed and sustainable throughout the day and night. Within this study, we reflect on the concept of 'spatial planning for dark hours' ('nattstadsplanering' in Swedish), that should be seen as a perspective on urban spatial planning that considers the specific conditions that occur when there is no daylight. According to our definition, spatial planning for dark hours includes both the use, the design, and the lighting of the outdoor environment. In this context, 'night', 'night-time', 'nocturnal' and 'dark hours', refer to the astronomical night, i.e., the absence of daylight, regardless the time of day. A similar approach to urban planning, that focuses on a certain temporal or seasonal change, can be found within the concept of winter urban planning [10, 11]. In this study, we have used Sweden as a case study, to examine how the absence of daylight is handled within urban spatial planning, and if the national planning framework within the Swedish planning and building legislation (PBL) [12], could form the basis for spatial planning for dark hours in Sweden.

1.2. Research questions

The following research questions have been addressed:

- In what way are the special conditions during dark hours, including outdoor lighting, managed within spatial planning according to the PBL, and within lighting masterplans?
- Does spatial planning according to the PBL, and lighting masterplanning, consider relevant public interests for dark hours? Does the planning process provide citizens, authorities, and stakeholders the possibility to influence the night character of the outdoor environment?
- How is outdoor lighting included within the PBL legislation? Could existing tools and processes within PBL form the basis for spatial planning for dark hours?

1.3. Limitations

This study has been carried out with the ambition to identify tendencies and possibilities as a basis for further research development. Even if there is other relevant legislation, this study focuses on legislation within the PBL. The research material, in forms of interviews and planning documents, is dated before 2017, when a new ordinance on building permission for light source facilities was adopted.

2. Relevant tools and processes within the PBL

To identify relevant tools and processes for spatial planning for dark hours, we have reviewed the PBL legislation with a focus on outdoor lighting. The review has included the Swedish Planning and Building Act and the Planning and Building Ordinance [12], relevant mandatory provisions and general recommendations from the Swedish National Board of Housing, Building and Planning [13,14,15,16], including relevant guidance on the legislation [17,18,19].

2.1. Public interests

The PBL [12] describes public interests that should be considered within all spatial planning. Regarding outdoor lighting, it follows from Chapter 2, Section 6, that light source facilities must be designed and placed in a manner that is suitable with regard to certain public interests. In addition, it follows from

IOP Conf. Series: Earth and Environmental Science 1320 (2024) 012014 doi:10.1088/1755-1315/1320/1/012014

Chapter 2, Section 9, that outdoor lighting may not be located, placed, or designed so that the intended use could result in the impairment of the surroundings that would entail a danger to people's health and safety, or cause significant negative impact in any other way. Pursuing to the PBL [12], conflicts between different public interests must be consciously weighed against each other in an open and democratic planning process, considering the rights of individuals. In this study, a selection of public interests that are specifically affected by the presence or absence of daylight is summarized (figure 1).

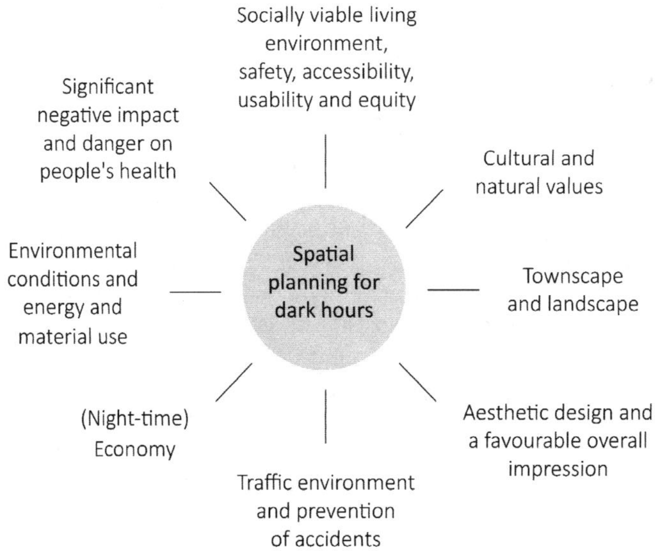

Figure 1. Selection of public interests, specifically relevant for spatial planning for dark hours. In comparison to the PBL, some public interests have been reformulated and grouped by the authors.

2.2. *Planning system*

Pursuant to the PBL, the Swedish planning system, consists of spatial plans of different kinds (figure 2) [12,19]. All planning documents within the PBL are generally subject for consultation, exhibition and review by the public, relevant authorities, and other stakeholders.

In the comprehensive plan, the municipality presents how the built environment should be used, developed, and preserved, including public interest considerations. The comprehensive plan is not legally binding but gives guidance for detailed planning and building permits. The comprehensive plan may be amended regarding a particular part of the municipality (amendment) or a particular public interest (addendum). On a regional level, the County Administrative Boards are responsible for planning material and supervision of national interests. Within this study, we have not found any guidance on how dark hours, including outdoor lighting, could be managed within comprehensive planning.

Within detailed development plans, the municipality regulates the use and the design of the built environment within certain geographical areas. The detailed development plan may not be more detailed than needed to fulfil the purpose of the plan. It follows from the PBL, Chapter 4, Section 12, that the detailed development plan can be used to determine maximum permissible values for disturbance through light, if there are special reasons [12]. Within a detailed development plan, the municipality may also decide on exemptions from building permit requirements for light source facilities. Within valuable environments, building permit requirements for outdoor lighting may also be extended, using provisions in detailed development plans or area regulations [12,17,18]. However, the guidance on provisions from the Swedish National Board of Housing, Building and Planning, gives limited support regarding outdoor lighting, bringing up the possibility without providing concrete examples [18].

Regarding building permits, The PBL legislation valid until 2017 stated that light source facilities require a building permit within areas covered by a detailed development plan [20]. Legal practice from earlier legislation has stated that normal street lighting does not require a building permit [17]. In 2017, an amended ordinance clarified that a building permit is required for light source facilities within an area covered by a detailed development plan if i) 'the intended use may have a significant impact on the surrounding', or ii) 'it is placed on or adjacent to a building or a development area that is particularly valuable from a historical, cultural, environmental, or artistic point of view' [12,17]. According to the Swedish National Board of Housing, Building and Planning, there are uncertainties in how to interpret the legislation [17]. Light source facilities that require a building permit must demonstrate a good effect of design, colour, and material and may not be distorting [12].

Planning material and programmes, such as aesthetic design programmes, can be annexes to all spatial plans according to the PBL, and can be used to provide more detailed guidelines, for example regarding design principles. The municipalities can also develop other types of planning documents, such as policies, strategies, programmes, and guidance on different topics, that are not connected to the PBL spatial planning system. Lighting masterplans are generally developed within this category.

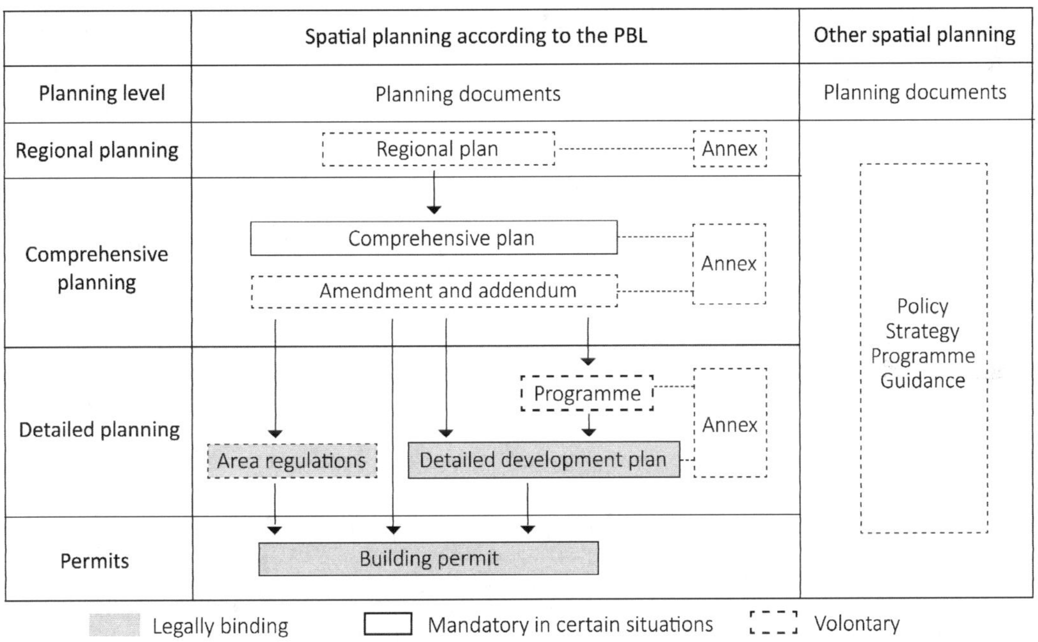

Figure 2. Overview of regional and municipal planning in Sweden.

3. Methods

To analyze how the special conditions during dark hours, including outdoor lighting, are managed within spatial planning, we have interviewed urban planners and building permit officers in different municipalities. We have also reviewed a selection of planning documents according to the PBL, as well as a selection of lighting masterplans (table 1). Qualitative research methods have been used. The selection of municipalities has been made with the ambition to cover a spread of geographical locations and population sizes. The municipalities in the interview series were partly selected based on our knowledge that they had a certain interest or experience in planning for dark hours in some ways. The three most populated municipalities in Sweden have been excluded from the study, since we judged that they have special planning conditions. The identity of the municipalities, the respondents, the spatial plans, and the lighting masterplans are withheld for confidentiality reasons.

Table 1. Municipalities and research material included in the study

Research method	Municipalities Population size < 20 000	20 000 - 70 000	> 70 000	Interviews	Planning documents
Interview series	2	4	3	10	
Document review Spatial plans, annexes and building permits	0	4	2		3 comprehensive plans 1 programme 5 detailed development plans including 3 aesthetic design programmes (as annexes) 3 building permits
Document review Lighting masterplans	1	5	6		9 lighting masterplans (municipal scale) 7 lighting masterplans (district scale)

3.1. *Interview series with urban planners and building permit officers*

The interview series consisted of separate, individual phone interviews, that were carried out during 2015-2016. The ten respondents worked as urban planners or building permit officers at urban planning departments in nine different municipalities (table 1). The interview questions focused on how they plan for the special situations that occur during dark hours (table 2). In the interview, we asked for the respondents' opinions and professional experience within the municipality where they were based. The interviews were semi-structured, using yes and no questions, that were followed up with questions where the respondent could answer in an explanatory descriptive way, as a basis for conversation. The interviews were documented by notes.

Table 2. Summary of interview questions.

1	If and how the special conditions during dark hours, including lighting, is managed within the municipal spatial planning practice in general.
2	If and how the municipality manages the 'dark hour perspective', including lighting, within plans and documents that are not in accordance with the PBL's spatial planning system.
3	If and how the municipality manages the 'dark hour perspective', including lighting, within spatial plans and building permits in accordance with the PBL, and if lighting is subject for regulations.
4	If planning for the special conditions during dark hours, including lighting, involved internal or external participation, and if so, in which stage of the planning process and with whom.
5	The respondent's interest in methodological support within spatial planning for dark hours.

3.2. *Document review of planning documents according to the PBL*

The document review included a selection of spatial plans, annexes and building permits (table 1), covering all municipal spatial planning levels according to the PBL, except for area regulations (figure 1). In this study, these are all referred to as 'planning documents according to the PBL'. The planning documents were selected and provided by the respondents in the interview series. The respondents had an explicit ambition to share material that they judged relevant for the study. Respondents from three of the municipalities chose not to share any planning documents, since they judged that they had no relevant spatial plans or building permits within their municipality. The documents were asked to have become legally binding between 2010–2015. Two of the detailed development plans in the study were adopted before 2010 but were important to review because they regulated the building permits included in the review. The review of the detailed development plans included the legal documents of regulation and recommendation (plan map and planning description), as well as aesthetic design programmes, that were carried out as annexes. As a research method, we have screened the spatial plans for selected words related to the 'dark hour perspective', including outdoor lighting. The selection has been made based on relevant words and expressions used within spatial planning. such as *'dark'*, *'night'*, *'evening'*, *'light'*, *'lighting'*, *'lux'*, *'lumen'*, *'luminaire'*, *'pole'*, *'sign'*, *'safe/unsafe'*, *'secure/unsecure'*, *'pollution'*, *'obtrusive'*, *'glare'*, *'energy' 'health'*, *'experience'*, *'architecture'*, *'design'*, *'character'*, *'landscape'*, *'environment'*, *'cultural heritage'*, *'accessibility'*, *'disability'*, *'vision'* (in Swedish). The context has been subjectively analyzed to examine how the special conditions during dark hours were managed within the documents.

3.3. *Document review of lighting masterplans*

The document review included lighting masterplans on both municipal and city district scale, from twelve municipalities (table 1). The selection of lighting masterplans has been based on internet search on the word for *'lighting program/plan/strategy/policy/guidelines'* (in Swedish). In this paper, all these documents are referred to as lighting masterplans. The plans were dated between 2009–2015 and covered a spread of different consultant firms that had been engaged in the document development. Consultant firms in which we had been employed ourselves, at the time of the document review, were excluded. The content of the lighting masterplans was divided into five categories (table 3). The site-specific or specific lighting guidelines were divided into four detail levels (table 4). To review in what way the lighting masterplans considered public interests, we analyzed how a selection of related themes, such as *'natural values'*, *'light pollution'*, *'obtrusive light'*, *'biodiversity'*, *'cultural values/heritage'*, *'façade lighting'*, *'safety'*, *'accessibility'*, *'traffic environment'*, *'energy use' and 'economy'*, were managed within the plans. To analyze whether there was a common terminology and methodology, the document's titles and contents were compared. We also reviewed which representative bodies that had adopted the lighting masterplans and if the lighting masterplans had been subject for consultation.

Table 3. Content categories within the lighting masterplans.

1	General lighting knowledge and non-specific guidelines that we judged as general practice or inspirational.
2	Site-specific planning material.
3	Site-specific and specific lighting guidelines (including all quantitative lighting guidelines).
4	References to municipal spatial planning and national legislation, standards, and guidance.
5	Descriptions of lighting planning methods.

Table 4. Detail levels of guidelines and proposals within the lighting masterplans.

1	Guidelines that describe that something specific should, or should not, be illuminated, without any specific guidance on how.
2	Guidelines using qualitative characteristics to describe how something specific should be illuminated.
3	Guidelines using quantitative characteristics to describe how something should be illuminated.
4	Guidelines that present a clear lighting principle of how something specific should be illuminated.

4. Results

4.1. Spatial planning according to the PBL

In the interview series, the urban planners and the building permit officers explained that they consider the special conditions that occur during dark hours in their spatial planning in general, especially regarding safety and design. One example that was brought up in the interviews was that buildings were designed and located so that paths would be active and populated throughout the day and night.

Most of the respondents thought that it is relevant and possible to manage the special conditions that occur during dark hours within spatial planning according to the PBL. Aesthetic design programmes, that are annexes to the spatial plans, were evaluated as the most adequate planning form to describe lighting principles within spatial planning according to the PBL. Several respondents considered that it, in some situations, could be relevant to regulate outdoor lighting in detailed development plans and building permits, but expressed an uncertainty about how the PBL should be interpreted regarding lighting regulations. The interviews clearly showed that the respondents interpreted the legal support for lighting regulations within the PBL in different ways. Many of the respondents requested methodology support on how to express lighting guidelines and regulations in spatial plans and building permits. They also requested clear guidance on how to interpret the PBL, as well as the Swedish Environmental Code [21], regarding lighting regulations.

The review of the planning documents according to the PBL showed that when the 'dark hour perspective' was taken into consideration, it was primarily on a general level, and rarely in connection to specific proposals. Proposals for outdoor lighting were more often described in the annexes.

4.2. Lighting masterplanning

In the interview series, the municipalities expressed a general need to plan outdoor lighting for larger areas. Almost all respondents had developed, or were developing, a lighting masterplan for either the entire municipality or for specific districts or sites. Most of the lighting masterplans that were reviewed within this study had been developed with the support of external consultants.

The review of lighting masterplans revealed that the plans lacked a common terminology, methodology and detail level. The content of the lighting masterplans that covered an entire municipality were often similar and on a general level. The lighting masterplans that covered one or several city districts generally contained more site-specific planning materials and guidelines, which had a wide spread of detail levels. The differences in content and detail level did not correspond to differences in document titles. Almost all lighting masterplans focused on the municipal's main city.

Only two out of sixteen lighting masterplans mentioned that certain lighting principles (façade lighting) could require a building permit.

4.3. Public interests

The review of the planning documents according to the PBL showed that when the 'dark hour perspective', including outdoor lighting, was considered in relation to public interests, it was most often highlighted in connection to safety. Some isolated proposals on how to reduce negative environmental impact from outdoor lighting were found within the planning documents.

The review of the lighting masterplans showed that traffic environment, safety, cultural heritage, façade lighting and energy use, were central themes within most of the plans. The lighting masterplans on a city district scale generally considered the themes in a more site-specific way, including more analysis and proposals, in comparison to the lighting masterplans on a municipal scale. Negative impacts from outdoor lighting were brought up on a general level by almost all lighting masterplans on a municipal scale, but by only half of the lighting masterplans on a city district scale. Site-specific or quantitative guidelines on how to limit negative effects from outdoor lighting were rarely included in any of the plans.

4.4. *Democratic influence*

The PBL provides citizens, authorities, and stakeholders opportunities to gain transparency and influence throughout the spatial planning process. Usually, it is also clear which representative body within the municipality is responsible for decision-making. The review of planning documents according to the PBL indicated that lighting proposals were rarely presented. The review of lighting masterplans showed that only one of the plans stated that it had been subject for consultation. The review also showed that the lighting masterplans had been adopted by various representative bodies.

5. Discussion

The result indicates that municipalities consider the special conditions that occur during dark hours within their spatial planning in general, but that the 'dark hour perspective', including lighting proposals, is rarely presented within planning documents according to the PBL. Instead, strategic lighting planning is often managed within lighting masterplans that lack a common methodology and a formal status, not linked to the PBL. Other studies confirm a lack of established common terms, definitions, and methods for lighting masterplanning [22,23]. Especially, we have found limited guidance on lighting planning as an integrated part of urban spatial planning. For example, CIE [24] points out the importance of stakeholder inclusion and suggests lighting masterplans to be an essential part of the wider urban plan. However, they provide limited support on relevant processes. A reason could be that the planning and building legislation differs between countries, which underlines the importance of national guidance.

This study also indicates that spatial planning generally provides limited support for how different public interests should be weighed against each other in a nocturnal context. Since lighting masterplanning generally focuses on the municipality's main city, the lack of guidance is most significant for smaller villages, as well as rural and natural areas. Negative aspects of outdoor lighting, such as ecological and astronomical light pollution, were generally addressed to a lower extent than, for example, safety and traffic environments. However, it is important to note that both research and general awareness regarding light pollution and the need of darkness have increased since the planning documents within this study were produced. The increased awareness can be noticed within more recent Swedish lighting masterplans, that address light pollution and darkness to a higher extent. Another example is that several detailed development plans in Sweden have been appealed due to the risk that the planned buildings could harm habitats for bats, that are nocturnal and legally protected [25].

Since lighting proposals are rarely included in the planning documents according to the PBL, and lighting masterplans are rarely subject for public consultation, the result of this study indicates that citizens generally have a little formal influence over strategic decision on outdoor lighting.

This study also shows that the PBL offers many existing tools and processes relevant for spatial planning for dark hours. Several benefits of using the PBL as a framework for spatial planning for dark hours could be identified. To plan the use, the design, and the lighting of the built environment within the same process, and to treat lighting as one architectural tool among others, would support important public interests, such as aesthetic design and a favourable overall impression. In addition, the PBL provides the possibility to limit negative effects from outdoor lighting. The selection of public interests relevant for spatial planning for dark hours (figure 1), has several correlations with key aspects of sustainable urban lighting, identified by Zielinska-Dabkowska [26, p.25–26]. Also, the PBL planning process would ensure that public interests are considered and weighed against each other, and that citizens, authorities and stakeholders would be given the opportunity to influence during the planning process. Another benefit is that building on and implementing established tools and processes, would gain an equivalent management, and support interdisciplinary cooperations.

However, there is no established methodology or guidance for how the 'dark hour perspective', including outdoor lighting, could be addressed within spatial planning according to the PBL. An important obstacle is the uncertainty of how the legislation concerning lighting regulations should be interpreted and implemented. Even if legislation for building permits for light source facilities have been updated since the research material in this study was conducted, uncertainties remain [17]. Different interpretations of the legislation may lead to an undesirable diversity between municipalities.

6. Proposals

To promote well-designed and sustainable outdoor environments throughout the day and night, we propose to develop spatial planning for dark hours ('nattstadsplanering' in Swedish), focusing on public interests, the use, and the experience of nocturnal outdoor environments. Our proposal is that spatial planning for dark hours should be based on tools and processes within the PBL. When possible and relevant, the planning for dark hours should be integrated within the regular spatial planning, as a perspective on the overall planning situation. For a successful development and implementation, we propose the following measures of action, that include further research and investigations.

6.1. Clarification of the legal basis for lighting regulation

To be practically applicable, we propose that the legal basis for lighting regulations through building permits and provisions should be evaluated and clarified, both within and outside valuable areas. It should be clear which kind of outdoor lighting that may require a building permit since it 'may have a significant impact on the surroundings', or, since 'it is placed on or adjacent to a building or a development area that is particularly valuable'. Guidance is also needed regarding which planning material within comprehensive plans, detailed development plans and programmes, could support the municipality's examination of permits, as well as provisions in the plan map. As a response to the important sustainability challenge of light pollution, we propose to evaluate if and how building permits for light source facilities could form the basis for national and local regulations against obtrusive light and to protect the cultural and natural values of the night sky. Further research and investigations are needed to define which kind of lighting 'may have a significant impact on the surroundings' in this regard. As reference, the guide on the limitation of obtrusive light from the CIE [27] and the French legislation for the limitation of light nuisances [28] include different quantitative thresholds for outdoor lighting for different types of area.

6.2. Methodology development

We propose that methods for spatial planning for dark hours should be developed for all planning levels according to the PBL (figure 2). The different planning levels ensure a process from comprehensive to detailed spatial planning for dark hours and allow municipalities to adapt their planning to local conditions. The methodology development needs to include relevant definitions to support a common terminology. Tools for analysis and proposals should be developed with adequate detail levels, following the formal status of both guiding and legally binding spatial plans.

The development of spatial planning for dark hours within comprehensive planning should focus on long-term guidance that provides the basis for more detailed planning, including building permits. The 'dark hour perspective' should, when relevant, be included within strategies for sustainable development and public interest considerations. We suggest that comprehensive planning could be used to describe the intended use and the desired characteristics of the townscape and the landscape during dark hours. This could include strategic guidance on the quality and quantity of outdoor lighting, including guidance on the limitation on obtrusive light within different areas. Comprehensive planning could also be used to give guidance on within which situations lighting regulation could be used to protect valuable environments. On a regional level, the County Administrative Boards should support the municipalities with planning material regarding the development and the protection of cultural and natural values connected to dark hours. Regional collaboration for comprehensive spatial planning for dark hours could be relevant for municipalities with similar nocturnal conditions.

The development of spatial planning for dark hours within detailed planning should focus on descriptions of how the area is intended to be used and experienced during dark hours. As a perspective on the overall design, the design of outdoor lighting should be included, in relation to public interests such as safety, townscape, cultural and natural values, and the limitations of obtrusive light. The methodology development should provide guidance on how lighting proposals can be expressed within the plan, including how provisions for the lighting can be used to ensure important values. The methodology needs to consider that detailed development plans are valid for a long period of time, while lighting technologies and knowledge evolve continuously. The methods also need to consider that the detailed development plan may not be more detailed than necessary.

We conclude that aesthetic design programmes, that are annexes to spatial plans according to the PBL, are an adequate planning form to describe more detailed design proposals, including lighting. We propose that lighting masterplans could be developed as annexes to spatial plans according to the PBL, giving them the same status as aesthetic design programmes.

Methods for spatial planning for dark hours should also be developed for situations where there is no ongoing planning according to the PBL. To meet the definition of spatial planning for dark hours, these plans (spatial plans for dark hours and lighting masterplans) would need to consider relevant planning material, site-specific public interests and be subject for a democratic planning process.

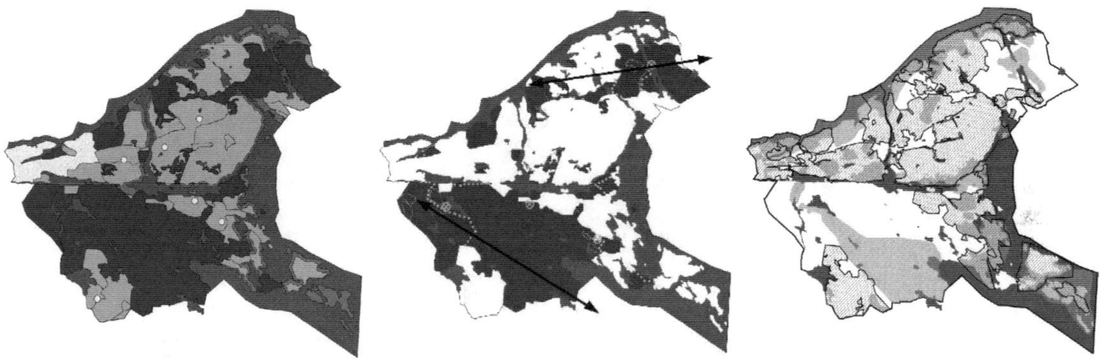

(a) (b) (c)

Figure 3. Examples of relevant planning tools on a municipal scale: (a) shows guidance on obtrusive light [29], including environmental lighting zones according to the CIE [27], (b) shows guidance on development of dark infrastructure [adapted from 29], c) shows guidance on building permits for lighting, describing areas with high cultural values and areas covered by a detailed development plan. Illustrations by C Frænkel. Published by courtesy of Nacka kommun and White arkitekter.

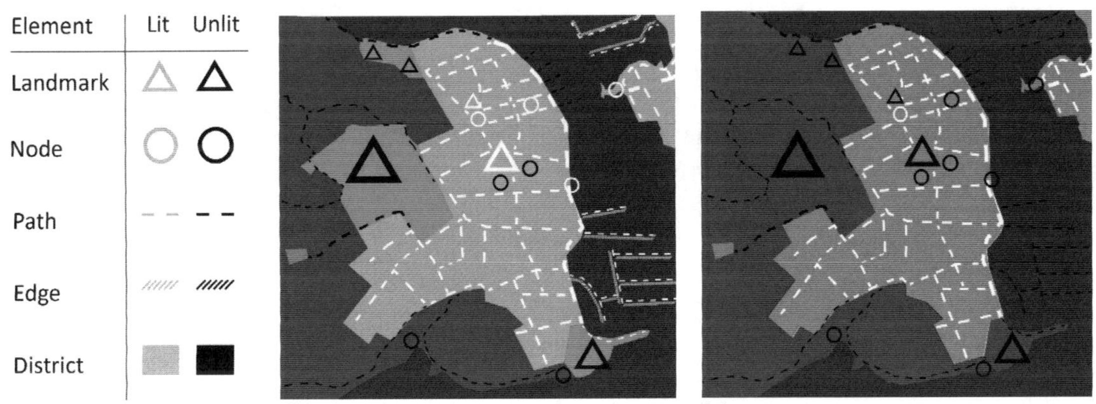

(a) (b) (c)

Figure 4. Example of relevant planning tools on a city district scale: (a) shows Lynch's five elements for analyzing a city [30] with a 'light and dark perspective' proposed by the authors, (b-c) show plans describing the structure of lighting and darkness on different times (for example evening/night or summer/winter). Illustrations by C Frænkel.

6.3. Dissemination of knowledge

When relevant lighting legislation has been clarified and adequate methodologies for spatial planning for dark hours have been developed, we propose that a broad knowledge dissemination should be conducted. The knowledge dissemination should include a national guidance, providing basic knowledge about light and the role of outdoor lighting in relation to public interests, as well as a summary of relevant legislation, adequate methodologies, and good practice for spatial planning for dark hours and lighting masterplanning. We propose that the Swedish National Board of Housing, Building and Planning should be responsible for national guidance and coordination. Together with regional support from the County Administrative Boards, national guidance would relieve each municipality to compile general lighting knowledge and practice, and free up resources for site-specific planning.

7. Conclusion

Within this study, the concept of spatial planning for dark hours ('nattstadsplanering' in Swedish) has been elaborated, with Sweden as a case study. According to our definition, spatial planning for dark hours includes both the use, the design, and the lighting of the built environment, and weighs relevant public interests against each other within a democratic and transparent process. After interviewing urban planners and reviewing planning documents and national guidance on planning and building legislation, we can conclude that there is no current, adequate methodology for how the 'dark hour perspective', including outdoor lighting, could be managed within urban spatial planning. However, we have identified many relevant tools and processes within the PBL that could be developed and applied. For a successful development, we highlight three key proposals, including legal clarifications on lighting regulations, methodology development, and knowledge dissemination. We believe that these steps will support sustainable and well-designed outdoor environments for nocturnal situations.

Acknowledgments

The research for this paper has been funded by Bertil & Britt Svensson Foundation for Lighting Technology.

References

[1] Moore-Ede M 1984 The body's inner clocks *Health & Medical Horizons* ed J Elliott, R Hantula et al (New York: Macmillan Educational Company) pp 167–75
[2] Boyce P R 2019 The benefits of light at night *Build. Environ.* **151** pp 356–67
[3] Falchi F, Cinzano P, Duriscoe D, Kyba C C M, Elvidge C D, Baugh K et al 2016 The new world atlas of artificial night sky brightness *Sci. Adv.* **2** 6 1600377
[4] Jägerbrand A K, Nilsson Tegelin M and Durmus D 2022 An overview of the adverse effects of outdoor light at night and the research methods used in different areas *Proc. of the 14th European Lighting Conf. Lux Europa 2022* 20-22 September 2022 Prague (Prague: Czech Lighting Society/Slovak Lighting Society) pp 130–7
[5] *Swedish Working Hours Act (SFS 1982:673)* (Stockholm: Arbetsmarknadsdepartementet)
[6] Chapman D, Nilsson K, Larsson A and Rizzo A 2017 Climatic barriers to soft mobility in winter: Luleå, Sweden as a case study *Sustain. Cities Soc.* **35** pp 574–80
[7] Andersson B 2001 *Rädslans Rum – Trygghetens Rum – ett Forskningsprojekt om Kvinnors Vistelse i Trafikrummet* (Stockholm: Vinnova – Verket for Innovationssystem/The Swedish agency of innovation) Vinnova Report 2001:32
[8] Wänström Lind U and Billger M 2021 Light topography and spaciousness in the urban environment *Nord. J. Archit. Res.* **33** 1 pp 103–34
[9] Edensor T 2013 The gloomy city: Rethinking the relationship bewteen light and dark *Urban stud.* **52** 3 pp 103–34
[10] Chapman D, Nilsson K, Rizzo A and Larsson 2019 A Winter city urbanism: Enabling all year

connectivity for soft mobility *Int. J. Environ. Res. Public Health.* **16** 10 1820
[11] Costamagna F, Lind R and Stjernström O 2018 Livability of urban public spaces in northern Swedish cities: The case of Umeå *Plan. Pract. Res.* **34** 2 pp 1–18
[12] *Plan- och Bygglagen (SFS 2010:900) Plan- och Byggförordningen (SFS 2011:338)* (Stockholm: Landsbygds- och infrastrukturdepartementet)
In english: *Legislation Planning and Building Act (2010:900) Planning and Building Ordinance (2011:338) 3rd ed.* 2018 (Karlskrona: Swedish National Board of Housing, Building and Planning)
[13] *Boverkets Föreskrifter om Detaljplan (BFS 2020:5)* (Karlskrona: Boverket)
[14] *Boverkets Föreskrifter och Allmänna Råd om Tillgänglighet och Användbarhet för Personer med Nedsatt Rörelse- eller Orienteringsförmåga på Allmänna platser och inom Områden för Andra Anläggningar än Byggnader (BFS 2011:5)* (Karlskrona: Boverket)
[15] *Boverkets Byggregler (BFS 2011:6)* (Karlskrona: Boverket)
[16] *Boverkets Föreskrifter och Allmänna Råd om Avhjälpande av Enkelt Avhjälpta Hinder till och i Lokaler dit Allmänheten har Tillträde och på Allmänna platser (BFS 2011:13)* (Karlskrona: Boverket
[17] Boverket 2017 *Bygglov för Ljusanordningar* [Internet] Available from: https://www.boverket.se /sv/PBL-kunskapsbanken/lov--byggande/anmalningsplikt/bygglov-for-ljusanordningar/ (accessed on 10 September 2023)
[18] Boverket 2020 *Planbestämmelser för Detaljplan 1 augusti 2018 - 30 september 2020* (Karlskrona: Boverket)
[19] Boverket 2023 *Planning Process* [Internet] Available from: https://www.boverket.se /en/start/building-in-sweden/swedish-market/laws-and-regulations/planning-process/ (accessed on 10 September 2023)
[20] Boverket 2014 *Uppdrag att Utreda Undantag från Bygglovsplikt för Skyltar och Ljusanordningar* (Karlskrona: Boverket) Rapport 2014:5 p 7
[21] *Miljöbalken (1998:808)* (Stockholm: Klimat- och näringslivsdepartementet)
[22] Jägerbrand A K 2021 Development of an indicator system for local governments to plan and evaluate sustainable outdoor lighting *Sustainability* **13** 1506
[23] Zielinska-Dabrowska K 2019 Urban lighting masterplan – Origins, Definitions, Methodologies and Collaborations *Urban Lighting for People-evidence-based Lighting Design for the Built Environment* ed N Davoudian (London: Riba Publishing) pp 18–41
[24] Commission Internationale de l'Eclarage 2019 *Guide on Urban Lighting Masterplanning 2nd ed.* (Vienna: CIE) CIE 234:2019
[25] Westin S 2023 Så blev fladdermössen avgörande i planeringen *Arkitekten* **4** pp 34–39
[26] Zielinska-Dabrowska K 2022 Rethinking sustainable cities at night: Paradigm shift in urban design and city lighting *Sustainability* **14** 10 6062
[27] Commission Internationale de l'Eclarage 2017 *Guide on the Limitations of the Effects of Obtrusive Light from Outdoor Lighting Installations 2nd Ed* (Vienna: CIE) CIE 150:2017
[28] *Arrêté du 27 décembre 2018 Relatif à la Prévention, à la Réduction et à la Limitation des Nuisances Lumineuses* (Paris: Ministère de la Transition écologique et de la Cohésion des territoires/Ministère de la Transition énergétique)
[29] Nacka kommun 2017 *Riktlinjer och Förhållningssätt för Offentlig Belysning i Nacka 1st ed.* (Nacka: Nacka kommun) pp 21, 57–8
[30] Lynch K 1964 *The Image of the City* (Cambrigde: MIT Press)

Darker Cities due to the Energy Crisis and the need for context-specific urban lighting

M, Hvass

Assistant Professor, PhD, Aalborg University, A.C. Meyers Vænge 15, 2450 Copenhagen, Denmark

Cooresponding author's email: mh@create.aau.dk

Abstract. Outdoor lighting has been dimmed or turned off in European municipalities due to the energy crisis. Dimming of outdoor lighting levels led to energy savings, advantages for biodiversity and less light pollution. But the dimming also triggers the basic human assumption that bright light is connected to safety and darkness is connected to fear. To discuss this basic assumption, human experience of dimmed lighting and perceived safety is explored through three case studies. One study explores human experiences of dimmed street lighting during energy crises, the second study investigates dimmed lighting levels around public transportation, and the third study is an example of a context-specific adaptation of street lighting in a residential area. The case studies reveal that dimmed lighting levels, minimized contrasts, and context-specific lighting can lead to an increased connectedness to the spatial and social urban context and an increased feeling of perceived safety. Thus, more light doesn't always lead to more perceived safety. With a dimmed general lighting level, it is possible to introduce additional subtle lighting layers, structured in a lighting hierarchy to further enhance spatial and social qualities in outdoor contexts in the dark hours. in constructing both.

1. Introduction

In August 2022, energy prices in Europe rose to ten times the average from previous years [1]. European municipalities therefore had to make decisions about how to lower energy consumption. Some municipalities lowered their energy consumption by dimming outdoor lighting, turning it off for selected periods of time or turning it off completely. However, dimming of outdoor lighting has both positive and negative consequences. The positive consequences of dimming lighting levels are not only energy savings, but also a reduction in light pollution [2], less circadian disruption for humans during nighttime [3], and better conditions for biodiversity in the dark hours [4]. Furthermore, when the general lighting is dimmed, the lighting designer gets an opportunity to add subtle layers of lighting, which adds a spatial composition to the urban context which is related to social needs in the dark hours.

The negative consequences of dimming the outdoor lighting levels are mostly related to concerns in relation to crime, personal security, and traffic safety among local citizens [1]. These concerns are often related to the basic assumption that bright light will reduce the risk of crime and increase feelings of perceived safety, while darkness will increase a feeling of danger. This basic assumption is based on culture and history, where darkness is often perceived as something negative [5]. While the benefits of lighting are many, as Peter Boyce describes, we do need lighting to create perceived safety for pedestrians and traffic [6] [7]. However, the need for lighting varies from place to place and general lighting regulations should be evaluated in relation to context-specific spatial and social needs.

Studies about perceived safety among women show that a higher lighting level is not always related to an increased feeling of safety [8]. Likewise, other studies show that when lighting levels are lowered and balanced between lit zones and contrasts thereby are eliminated, surroundings and people become visible, and this leads to a feeling of safety [9] [10]. To fully understand the complex connection between dimmed lighting and human experience of perceived safety, there is a need to further investigate the human and social dimension of lighting in the specific context that is being lit. To achieve more accessible, sociable, and sustainable cities, there is a need to investigate the nocturnal urban experience in terms of people's emotional, cognitive, and motivational perceptions [11].

The following three cases describe people's experience of dimmed and context-specific outdoor lighting in relation to three different types of outdoor lighting: 1) public roads, 2) public transportation and 3) local roads in a residential area. The three cases are chosen to discuss three types of areas where dimmed outdoor lighting affects the experience of outdoor areas. The basic assumption about dimmed lighting being related to an unsafe feeling is evaluated in the three cases to obtain a nuanced perspective on the positive and negative consequences of dimming the general lighting level in outdoor environments.

Case study one describes studies of people's experience of dimmed lighting in Copenhagen Municipality during the winter of 2022/23 [12] [13]. Case study two describes people's experience of dimmed lighting around public transportation at a tram station in Aarhus, Denmark [10]. Finally, case study three, describes how residents took an initiative to change the street lighting in their residential area of Humleby in Copenhagen. These three cases have been chosen because they concern dimming of urban lighting, and in two of the cases lighting has been adapted to context-specific conditions. The three case studies relate to a sustainable agenda, especially the UN Sustainable Development Goal 11. The dimmed and context-specific lighting can lead to inclusive, safe, and sustainable urbanization. Furthermore, case study three is an example of how a lighting project can be carried out, with participatory planning and management, to reduce environmental impact and secure an acceptance of the new design among locals in the residential area [14]. Recent research shows that to ensure sustainable solutions in urban contexts it is necessary to co-design with local communities and ensure local participation in the process [15]. By engaging local citizens when lighting is planned for their community, an ownership is developed in relation to the lighting design and to the neighbourhood [16] [17]. Furthermore, by engaging locals, lighting can be carefully planned to live up to the Sustainable Development Goals in the dark hours and create urban environments that meet the UN target: 'Leave no one behind'.

2. Background

The general lighting level in lit urban zones is essential for how the urban space is experienced. If the lighting level is too high and contrasts are too large, our experience of the urban space can be destroyed in the dark hours. This can influence our feeling of safety. Contrasts in lighting levels can be between a lit object and the lit background [18] or contrast between two lit urban zones in the urban space [10]. If contrasts in lighting levels are balanced and the general urban lighting level is dimmed, it becomes possible for the lighting designer to add subtle layers of lighting, planned in a lighting hierarchy according to context-specific spatial and social qualities in the specific urban space which is being lit.

2.1. Contrast in lighting levels between urban lit zones

To minimize contrast and create balanced brightness levels, it is crucial to create comfortable and safe urban spaces because the human eye is sensitive to light exposure. Therefore, when examining human experience of lighting levels in the urban environment, it is necessary to consider how human night vision functions. The human eye is sensitive to changes and requires time to adapt from one level of light to another. As an example, the human eye requires approximately 20-30 minutes to adjust from sunlight to darkness. The process of adjusting can be divided into three stages of sensitivity: photopic, mesopic, and scotopic vision. Photopic vision occurs when there is a luminance greater than 5 cd/m2. This allows for fine, detailed perception as well as colour perception. Scotopic vision occurs when luminance levels drop below 0,005 cd/ m2 and is characterized by an inability to perceive colours and a reduced ability to resolve details. The stage of sensitivity between photopic and scotopic vision is called mesopic vision, which occurs when luminance levels range between 0,005 and 5 cd/m2. Mesopic vision enables detail and colour perception when luminance levels are close to photopic vision; this ability is reduced as the luminance levels drop to the range close to scotopic vision [19] [20].

As the eye is very sensitive to changes in lighting levels, large contrasts in lighting levels should be avoided or designed with care. If, for example, a brightly lit commercial sign is placed in a dark spot, it will make the surroundings appear even darker. The contrast makes details, textures and shapes disappear. Bright light in shop windows or lighting on construction sites can also make large contrasts and make the surroundings look darker. Contrasts can also appear between lit urban zones with different activities or if one zone is brightly lit because the area is perceived as unsafe. But contrasts can make the space appear unsafe because the surroundings become invisible when the contrast is too high. Brighter is not always better, as Yang et al. claim when studying how women experience perceived safety in urban lit contexts [8]. Lighting and darkness need to be balanced between lit zones [10]; when a general balanced layer of light is established, other layers of lighting can be added in a lighting hierarchy. To make the lighting hierarchy function optimally, it must be developed based on context-specific needs and spatial and social qualities [21].

2.2. Layers of Light

From the theatre, we know that lighting can produce any atmosphere needed to enhance the drama of the play, to direct the attention of the viewer or to provoke certain emotions [22]. In architectural lighting design, some of the same principals of lighting are used, but there are limitations in practice that are not limitations in the theatre. The lighting designer Hervé Descottes refers to the 'instruments of lighting' when describing lighting characteristics, and he divides them into six principles of lighting: 1) illuminance, 2) luminance, 3) colour and temperature, 4) height, 5) density and 6) direction and distribution [23]. These principles of lighting can be seen as different tools to create a lighting setting and need to be balanced to be able to distinguish the one from the other with the human eye. Together the principles represent layers of light in a lighting hierarchy.

IOP Conf. Series: Earth and Environmental Science 1320 (2024) 012015 doi:10.1088/1755-1315/1320/1/012015

The lighting designer Richard Kelly worked with a lighting structure in his lighting design projects. He defined three layers of lighting needed to create a successful lighting design, where each layer of light has a role in the spatial composition. Kelly distinguishes between three kinds of lighting layers: 1) ambient luminescence, 2) focal glow and 3) play of brilliants. Through an interplay, the three layers need to be balanced to provide a perception of visual beauty [24]. The ambient luminescence provides the general illumination of a space: it "makes surroundings safe and reassuring". The focal glow "makes it easier to see", the light is focused and conveys information about an object or a space to attract the viewers' attention. Finally, the play of brilliants "stimulates the spirit". Here, the light gives information or an experience. It "excites the optic nerves, and in turn stimulates the body and spirit, quickens the appetite, awakens curiosity, sharpens the wit. It is distracting or entertaining". These three layers of lighting need to be balanced and be a part of a holistic lighting design concept [24]. For example, if the ambient layer of light is too high, it will not be possible for the human eye to see the play of brilliance.

2.3. Aim

Our night vision is sensitive to bright lighting levels and contrasts, as explained above. Therefore, it is important to lower and balance general lighting levels and add a variation of layers of light composed in a lighting hierarchy.

The aim of this paper is to explore, through three case studies:

- if humans experience an urban space as safe in a lowered lighting setting,
- if changes in contrast in lighting levels between urban lit zones make a difference in the perceived safety of those zones,
- if an adaptation of the lighting to context-specific conditions can influence the human experience of an urban space.

3. Case studies

The outline of each case study is described in the following chapter: the methods used to explore the human experience of the changes in lighting levels and the result of each study are presented.

3.1 Case study one - Copenhagen municipality: human experience of dimmed lighting

In august 2022, the lighting department in Copenhagen Municipality planned a strategy for dimming urban lighting to ensure energy savings [13]. The ideas for where and how to dim the lighting were discussed with the local police to make sure that lighting was not turned off in areas with high crime rates. Street lighting was dimmed 10%, effect lighting was turned off in most places (figure 1), and lighting was switched off between the hours of midnight and 5am in 25 parks. The Copenhagen Municipality lighting department only received a few complaints from citizens due to the changes in the outdoor lighting levels, and no complaints at all from the police.

Since 2009, the Copenhagen Municipality has each year made a 'Perceived Safety Investigation' to examine if Copenhagen citizens experience the city as secure and safe. Since 2020, the citizens' satisfaction with the lighting has been a part of these investigations, and in 2023 there has been a special focus on human experience of dimmed lighting levels [12]. The company Epinion conducted this investigation for the Copenhagen Municipality and used both quantitative and qualitative methods to test how the savings on lighting in 2022/23 have affected Copenhagen citizens' perceived safety.

Figure 1. Lowered street lighting levels due to the energy crisis winter 2022/23 (Photo: Hvass, 2023)

The analyses in the report are based on responses from the last four years' perceived safety investigations in Copenhagen municipality (2020 to 2023). The database therefore consists of more than 13,000 questionnaire responses in the period spanning the past four years. Quantitative data sources from 2022-2023 consist of 6,000 questionnaire responses about citizens' experience of the lighting in their neighbourhood and their satisfaction with the lighting. This information is compiled with geographical data on the lighting in Copenhagen, including the city's light lines, light zones, and lighting systems, as well as data on the population composition and crime committed in the Copenhagen area. Qualitative data sources consist of a mobile ethnographic study among 22 female citizens, spread over six city districts. Over a period of two weeks, citizens have answered thematic tasks about their experience of safety and lighting in their district [12].

The results showed that most citizens found the lighting sufficient, although satisfaction varied from city district to city district. The lighting was experienced as good when it created a cozy atmosphere from different sources and levels. When there were too few light sources, too high contrast or the light became too warm in colour, the lighting was perceived as insufficient. When it came to the question of whether the savings on lighting in 2023 affected the security of the citizens, the result was that none of the citizens questioned had noticed the dimming of the street lighting. The dimming of the lighting level didn´t make the citizens less satisfied with the lighting and didn't lower the perceived safety of the citizens in the evening and night hours [12].

3.2. Case study two – Aarhus tram station: human experience of dimmed and balanced lighting.
Case study two represents an example of human experience of dimmed lighting at a tram station in Aarhus, Denmark. Studies were performed at tram stations to explore human perception of dimmed and balanced lighting in an urban social situation where people share an outdoor space and interact

[10] [11]. The intention of the studies was to explore how test participants experienced the tram station, the surroundings, people in the urban space, and the waiting activity at the station in two different lighting scenarios with different lighting levels.

The methods used were both quantitative and qualitative. Measurements were performed to define the lighting levels, and ethnographic methods were used to perform interviews. Experiences of lighting and darkness can be difficult for humans to describe as a vocabulary is lacking [25]. Both measurements and interviews were performed first in the existing bright lighting setting, with a high contrast between the lighting level at the station and the lighting level in the surroundings. Secondly, the test participants experienced an approx. 80% dimmed lighting setting, the lighting level now being balanced in relation to the lighting level in the surroundings.

The results indicated that a dimmed lighting setting created a more relaxed atmosphere at the station compared to the existing bright lighting setting. Furthermore, as it was possible to see people in the surroundings from the waiting area at the station in the dimmed lighting setting, the feeling of perceived safety and connectedness increased [10].

Figure 2. Bright lighting setting (Photo: Hvass, 2020)

Figure 3. Dimmed lighting setting (Photo: Hvass, 2020)

The photos reveal the lighting settings and the influence the contrast in the lighting levels have in relation to the visibility of the surroundings. When the lighting level in the sheltered waiting area is dimmed, and the lighting doesn't reflect in the vertical glass surfaces, the surroundings then become more visible (figure 3) than in the bright lighting setting (figure 2). What are also revealed in the dimmed lighting setting are the commercials at the station and their very bright surfaces, which produce a high level of contrast.

3.3. Case study three – Humleby, context-specific street lighting in a residential area

Case study three is an example of how important context-specific lighting can be for a local community. Humleby is a residential area built in 1885 in Copenhagen. The residential area has a special character, and the residents were not satisfied with the glary street lighting that was installed by the municipality. Therefore, they spent two years planning how to change the street lighting into a lighting setting more suited to the context-specific character of the place. The lighting design project was carried out in collaboration with the lighting architect Gunvor Hansen and installation was financed by residents in Humleby and Copenhagen Municipality.

The problem with the street lighting was that the fixtures were positioned too high. This caused the lighting to come through the residents' windows. The lighting was too bright and had a cold colour temperature. In the new lighting solution carried out by lighting architect Gunvor Hansen, fixtures were positioned lower than before. A grid was positioned in the fixture to adapt the distribution of light to the design of the individual streets. Thereby, no spill-light came through the windows of the houses in the area.

The project is included in this paper as an example of a lighting project that was initiated and partly financed by residents in a housing area. A lighting designer was involved to ensure lighting quality and that all context-specific qualities were taken into consideration in the final lighting solution. It is an example of participatory design and how a local community can strengthen their sense of belonging in the local neighbourhood with the right lighting design for the dark hours.

Figure 4. Before changes in street lighting, Humleby (Photo: Henrik Ledet, 2004)

Figure 5. The street lighting today, Humleby (Photo: Henrik Ledet, 2023)

The photos illustrate how the street lighting lights up the facades and the spill-light coming through the windows before the position of the fixtures were lowered (figure 4). Subsequently, they illustrate how the street lighting only lights up the street, when positioned lower and a grid is installed in the fixture to direct the light (figure 5).

4. Discussions and findings

Before the energy crisis, legal requirements for lighting have been decisive, but now it has become necessary to circumvent this legislation, due to the need for energy savings. This change can give us valuable knowledge about how people experience dimmed lighting levels, and it can create the basis for new perspectives to the assumption that more light provides more perceived safety, and that darkness is experienced as unsafe.

The findings from case study one display that the test participants who were interviewed about the dimmed lighting in Copenhagen Municipality had not noticed the 10% dimming of the street lighting, and that the dimming had not lowered the feeling of safety.

In case study two, the dimming of the lighting was more significant. An approx. 80% dimming of the lighting at a tram station in Aarhus, generated an unsafe feeling among 20% of the test participants. The remaining interviews showed that test participants had a positive experience of the dimmed lighting setting. They experienced that, when the lighting levels were balanced in the specific context and the contrast between lit zones disappeared, it was possible to see the surrounding space and people. This led to an increased feeling of connectedness to the place and an increased feeling of safety.

Case study three is an example from praxis, an example of how a few changes in the design of a lighting solution can have a great impact. The amount of lighting was reduced, and the improved lighting setting added quality to the experience of the neighbourhood in the dark hours instead of a lighting solution that caused contrast and inconvenience to the residents living in the area.

It is important to recognize that there is a need for using both qualitative and quantitative methods to research human experience of outdoor lighting. Legal requirements are primarily based on a tradition of quantitative research [8]; however, there is a need to create awareness among decision makers about the spatial and social potentials of lighting in urban environments. There is also a need to discuss lighting levels in relation to the risks that we as humans experience in outdoor environments: including the risks of being involved in an accident and the risks of being a victim of violence or threats. It has been found that for pedestrians, the risk of being a victim of a crime is what people fear the most [26].

During the field experiment at Nørreport tram station in Aarhus, even though the tram passes very close to where one waits for the tram, only a few test participants mentioned the fear of being involved in an accident [10]. During the design process, the fear of accidents and the fact that the tram driver could be glared by the lighting on the station and therefore not be able to see the people waiting, was the parameter that was discussed the most, according to Holscher Design, and not the fear of crime.

Studies from Melbourne University [8] concerning the relationship between urban lighting and perceived safety also show that brighter lighting is not always related to a higher degree of perceived safety. Likewise, their results reveal that when contrast-ratio is unbalanced, pedestrians tend to feel unsafe. These studies also investigated other lighting parameters, such as colour temperature, colour rendering and uniformity in urban lit contexts to look for connections between perceived safety and lighting parameters.

The dimming of lighting due to the energy crises has revealed that in many cases lighting can be dimmed without reducing perceived safety. When the general lighting levels are dimmed, this leaves room for introducing subtle layers of lighting in outdoor environments, which can create inviting and safe places for urban citizens.

5. Conclusion

The three case studies show that it is possible to dim the general lighting and create more value. It is not in all places that it is necessary to add extra layers of light to highlight the character of the place, sometimes it is 'just' the right distribution and balance between lighting and darkness, which can make a difference and lead to better lighting solutions. *"Too much light wipes away the sense of place"*, as the architect Juhani Pallasmaa describes [27]. The sense of specific places, i.e., their context-specific spatial and social qualities, are necessary to take into consideration when a lighting solution is planned in the urban context.

The results from the three case studies have led to three concluding findings, which can be considered in future urban lighting design solutions:

A. Lighting Hierarchy: dimming of general lighting levels and supplementing with layers of light in a lighting hierarchy will make the space inviting, safe and sustainable and ensure that context-specific spatial and social qualities are enhanced.
   All three case studies reveal that it is possible to lower lighting levels and still establish safe and inviting outdoor environments. Case study one, lowered street lighting, case study two

lowered lighting around public transportation and case study three lowered lighting levels in housing areas. The lowered general lighting levels allow for subtle layers of lighting in outdoor environment. A lighting hierarchy can be established which enhances social interaction, draws attention to context-specific qualities or activities can be established.

B. Balanced Brightness Levels: consider balanced brightness levels in the general lighting. With a balanced brightness level between lit zones, the local space and the surroundings, a holistic lighting design can be established.
Case study two demonstrates how high contrasts in lighting levels in an outdoor environment can create the experience of dark areas. While when lowering and balancing lighting levels a safe and inviting outdoor environment can be established.

C. Context-Specific Lighting: enhance spatial and social qualities. Context-specific lighting should be developed and co-designed with the citizens living in the area, hereby ensuring social and sustainable solutions. Case study three is an example of how a local community took the initiative to change the street lighting in their neighbourhood. The new context-specific lighting in Humleby is an example of how a few adjustments can create a solution where lighting is only where it is needed.

Acknowledgement

Thanks to Copenhagen Municipality, Team Sikker By, Epinion and to Humleby Housing area for contacting me when the results of my PhD were published in autumn of 2022. Thanks for sharing background knowledge and discussing your experiences of dimming lighting levels and introducing context-specific lighting. Thanks to Ellen Kathrine Hansen, Karen Waltorp, Holscher Design, Schréder and Aarhus Tram for making it possible to perform the field experiment at Nørreport tram station, Aarhus, Denmark.

References

[1] Balázs L, Braun F and Lengyel J 2023 Energy Saving Potential of Traffic-Regulated Street Lighting. *Sustainability.* **15** 6750
[2] Kyba C C, Altıntaş Y Ö, Walker C E and Newhouse M 2023 Citizen scientists report global rapid reductions in the visibility of stars from 2011 to 2022. *Science.* **379** 265-268
[3] Dolsen E A, Wyatt J K and Harvey A G 2019 Sleep, circadian rhythms, and risk across health domains in adolescents with an evening circadian preference. *Journal of Clinical Child & Adolescent Psychology.* **48** 480-490
[4] Pérez Vega C, Zielinska-Dabkowska K M, Schroer S, Jechow A and Hölker F 2022 A systematic review for establishing relevant environmental parameters for urban lighting: Translating research into practice. *Sustainability.* **14** 1107
[5] Dunn N and Edensor T 2020 *Rethinking Darkness: Cultures, Histories, Practices* (New York: Routledge)
[6] Boyce P R 2019 The benefits of light at night. *Building and Environment.* **151** 356-367
[7] Boyce P R, Eklund, N H, Hamilton, B J and Bruno, L D 2000 Perceptions of safety at night in different lighting conditions. *Lighting Research and Technology* **32** 79–91
[8] Yang H, Berry J and Kalms N 2022 Perceptions of safety in cities after dark. In *Lighting Design in Shared Public Spaces* edited by Shanti Sumartojo (New York: Routledge)
[9] Hvass M, Wymelenberg K V D, Boring S & Hansen E K 2021 Intensity and ratios of light affecting perception of space, co-presence and surrounding context, a lab experiment. *Building*

and Environment **194** 107680
[10] Hvass M, Waltorp K & Hansen E K 2022 Lights out? Lowering Urban Lighting Levels and Increasing Atmosphere at a Danish Tram Station. In *Lighting Design in Shared Public Spaces* edited by Shanti Sumartojo (New York: Routledge)
[11] Casciani D 2020 *The Human and Social Dimension of Urban Lightscapes* (Cham: Springer)
[12] Epinion 2023. Link to 'Perceived Safety Investigation' in Copenhagen https://www.kk.dk/politik/politikker-og-indsatser/sikker-by/tryghedsundersoegelsen (accessed 20/06/2023)
[13] Copenhagen Municipality 2023. Link to Copenhagen Municipality homepage: https://www.kk.dk/sites/default/files/2023-06/Dybdeg%C3%A5ende%20Tryghedsunders%C3%B8gelse%20af%20belysning%20og%20tryghed%20i%20aften-%20og%20nattetimerne%202023.pdf (accessed 20/06/2023)
[14] United Nations 2015. Transforming Our World: The 2030 Agenda for Sustainable Development A/RES/70/1; New York, USA: UN General Assembly.
[15] Musango J K, Currie P, Smit S and Kovacic Z 2020 Urban metabolism of the informal city: Probing and measuring the 'unmeasurable'to monitor Sustainable Development Goal 11 indicators. *Ecological Indicators* **119** 106746
[16] Slater D, Sloane, M and Entwistle, J 2015 Configuring light: Staging the Social. http:// www.configuringlight.org/ (Accessed 20/06/2023)
[17] Martinez L and Bordonaro E 2022 Lighting inequality in an urban context: design approach and case studies *IOP Conf. Ser.: Earth and Environ. Sci.* **1099** 012006
[18] Davoudian N 2017 Background lighting clutters: how do they affect visual saliency of urban objects? *International Journal of Design Creativity and Innovation* **5** 95-103
[19] Boyce P R 2014 *Human Factors in Lighting* (CRC Press; Taylor & Francis Ltd.)
[20] Tregenza P and Loe D 2014 *The Design of Lighting*. (New York: Taylor & Francis)
[21] Hvass M 2022 *Balanced Brightness Levels: Exploring human sensory experience of lighting and darkness in urban public contexts*. Ph.d. Thesis, Aalborg University Press. ISSN nr. 2446-1628
[22] Böhme G 2013 'The art of the stage set as a paradigm for an aesthetics of atmospheres' *Ambiances*
[23] Descottes H and Ramos C E 2011 *Architectural Lighting: Designing with Light and Space* (New York: Princeton Architectural Press)
[24] Kelly R 1952 Lighting as an Integral Part of Architecture. In *College Art Journal* **12** 24-30
[25] Brandston H 2008 *Learning to See - A Matter of Light*. Illuminating Engineering Society of North America.
[26] Fyhri A, Hof T, Simonova Z and De Yong M 2010 The Influence of Perceived Safety and Security on Walking; Getting Communities Back on Their Feet. Cost Den Haag, Netherlands
[27] Pallasmaa J 2014 Space, Place and Atmosphere: Emotional and Peripheral Perception in Architectural Experience. *Lebenswelt* **4** 230-245.

Embedding sustainability into the design of street lighting

H I Wåseth[1], J Safi[1], T K Nielsen[4], U Bayr[2], N Johansen[2], E Fjeldaas[3], T Christensen[1]

[1]University of South-Eastern Norway, Hasbergsvei 36, 3616 Kongsberg, Norway
[2]NIBIO, Postboks 115, NO-1431 Ås, Norway
[3]Statens vegvesen, Anton Jenssens gate 5, 3125 Tønsberg, Norway
[4]Lysteam AS, Grorudveien 24, 0962 Oslo, Norway

*Correspondence: helga.i.waseth@usn.no

Abstract. Light pollution is an increasing problem and Artificial Light at Night (ALAN) affects nature in many ways. The Norwegian Road Authorities (SVV) initiated a collaboration with the Norwegian Institute for Bioeconomy (NIBIO) and the University of South-Eastern Norway (USN) to find measures that can help reduce the negative impacts of road lighting on the environment and nature. While USN has looked more generally at sustainable lighting design and ways to reduce the negative effects of road lighting, NIBIO has focused on how road lighting affects insects. We study how the light from roads contributes to light pollution, with the final aim to know more about what measures most effectively reduce the negative effects of road lighting. In this pilot study of light levels in two transects beside a road, the results show that the light levels decreased relatively fast as the distance to the light source increased. Thus, the direct light from the road lighting might not be the most prominent addition to the light pollution in the open landscape. On the other hand, the measurements showed some light influence from the road up to around 160 meters from the road, although the levels were low. While the light was below the detection limit, the light source was visible as a point source at a very large distance. This observation addresses the question of whether organisms are influenced by the average light energy per area, or the intensity of a source as observed from a distance (irradiance vs. radiance). Additional sources of light pollution may affect the result, such as unshielded private light sources and sky glow. This project aims to find measures that can help reduce the negative impacts of road lighting on the environment and nature. We study how the light from roads contributes to different components of light pollution, like skyglow, light trespass and luminance from the light source.

1. Introduction

Road lighting contributes to light pollution in large areas of Norway, especially because this is one of the few countries that uses road lighting on highways. As research shows how light pollution harms nature (1), the need to look at solutions to reduce light pollution has emerged. Between 1992 and 2017, light pollution increased by 49% worldwide (2,3). However, artificial light at night affects (ALAN) nature in many ways, and most species are affected, but in different ways. ALAN can affect foraging, reproduction, habitats and migration. (1,4,5). There are ways to reduce the negative effects of this type of pollution on nature (6), like limiting the light levels, installing timers, managing the spectral power distribution, and shielding the luminaire.

This project was initiated by the Norwegian Road Authorities (SVV) in 2021, and the collaboration includes the Norwegian Institute for Bioeconomy (NIBIO) and the University of South-Eastern Norway (USN). The aim is to find the most appropriate and effective measures that can help reduce the impacts of road lighting on the environment and to be able to evaluate the zone of influence on insects from road lighting. USN is responsible for the part considering sustainable lighting design and ways to reduce the effects of road lighting in Norway. NIBIO focuses on how road lighting affects insects. In the initial part of the study, pilot studies of existing light levels in the side area of the road were performed. This specific work concentrates on the distribution of street lighting into the landscape, and the different components of light pollution, such as skyglow, light trespass and direct light from the luminaire.

Figure 1 Illustration showing some of the considerations that must be taken when designing sustainable street lighting, where insects are the main focus in this project, other species are also affected, and so are the view of the natural starry sky and northern lights. Ill. Jamal Safi.

2. Method

To gain insight into the distribution of light beside roads, a pilot study was conducted on Haldenveien in Frogner, close to Oslo, the 25.10.2022 between 19.00 and 21.00. Light measurements were done on two transects beside the road. One in the forest and one on a partly open field. On the field, the measurements had to be taken on the side of the field, to not destroy the crops. The road lighting was the most prominent light source in the area, but some farms with luminaires were present at a distance of a few hundred meters. The transects were on opposite sides of the road. The forest transect had the light pole closer, and the light distributed from the back of the luminaire, while the transect on the other side of the road was further away from the light pole, and the light distributed from the front of the luminaire. The measurement was taken horizontally at 1.5 meters. First measurement was taken beside the road and then repeated every 10 meters, increasing the distance to the road, until we reached a place where the light meter did not indicate any further decrease. The weather conditions during the measurements were temperature around 5 degrees C, overcast and cloudy (approx. 80% cloud cover). The light measurement was carried out by Lysteam AS with the light meter GL Spectis 1.0 Touch + Flicker (GL Optics, Poland). The instrument can give full spectra and thus, the measured illuminance can be re-calculated to express irradiance or spectral irradiance. In the presentation of data in this article, photometric units are used since the weighting spectra for ecological effects are mostly unknown. Images were taken to show the situation in the field we were analysing. The luminaire at the road was a Philips BGP764 LED 360/740 DM32 SR with an output of 205 W and a stated colour temperature of 4000 K, measured colour temperature was 4062 K.

IOP Conf. Series: Earth and Environmental Science 1320 (2024) 012016 doi:10.1088/1755-1315/1320/1/012016

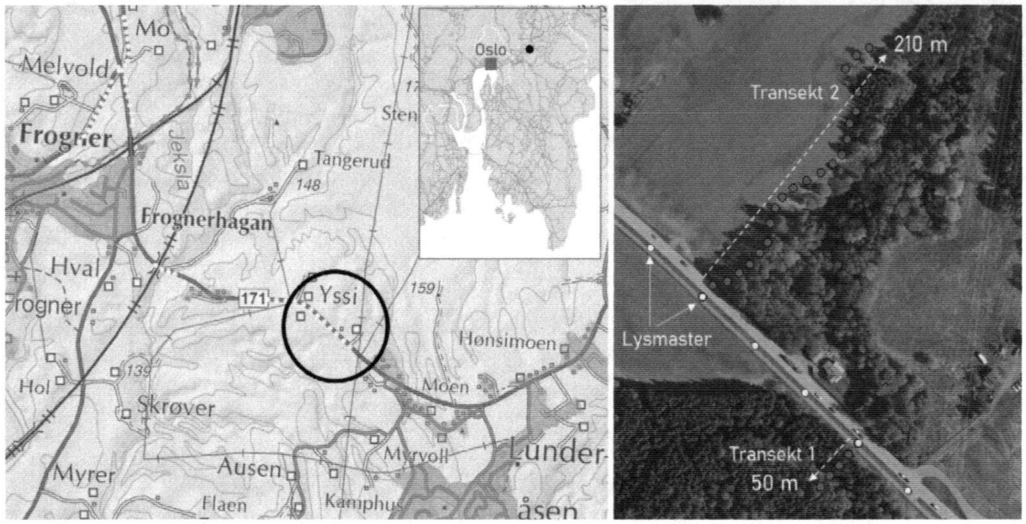

Figure 2The two transects and the location of the measurement area.

3. Results and discussion

1. Transect 1, forest: The results from the light measurements showed that the light levels dropped from 8.2 lx by the road, to 0.25 lx 10 m from the road. At 50 m from the road, the road light did not influence the measured light level, due to a natural shielding of trees and bushes. On the other hand, light reflected from the cloud-covered sky could be observed when the light meter was directed toward the zenith. Also, at some positions, you could see the high luminance point from the luminaire.

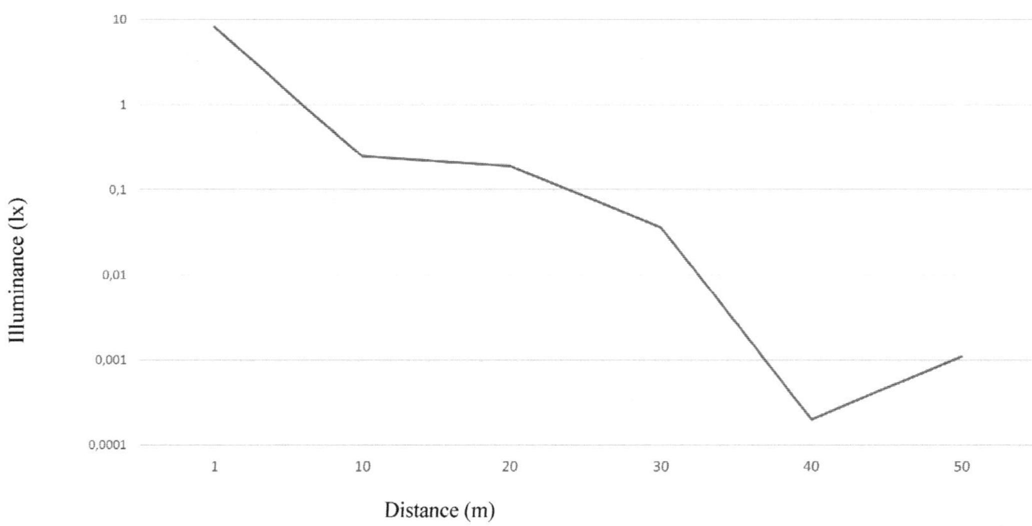

Figure 3 Illuminance in distance from the road, forest-transect.

2. Transect 2, partly open field: Under the fixture the light level was 27.5 lux and dropped to 4.4 lux 10 meters from the light source and further to 0.9 lux 20 meters from the light source. The light level remained elevated at a greater distance from the road than in the forest. At 80 meters from the road, minimal illumination was measured (below 0.1 lux), and after 160 meters the lowest light level was measured to 0.04, and the light level stayed around this low value even at a longer distance.

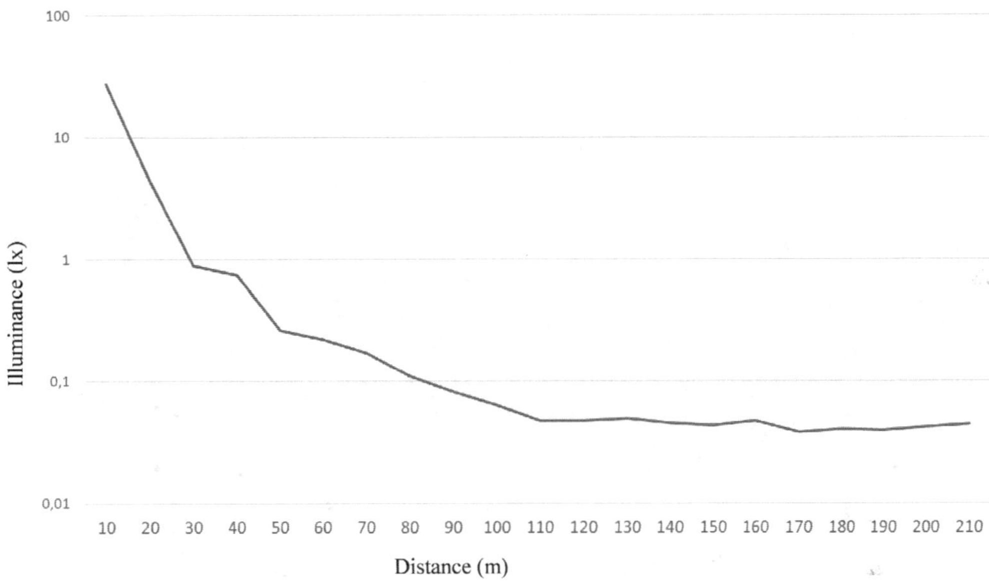

Figure 4 Illuminance in distance from the road, partly open transect.

The results showed that due to the road optics, designed to avoid stray light, the light levels decreased relatively fast as the distance to the light source increased, especially in the forest, where, not surprisingly, the trees shielded the lights. The direct light from the road lighting might not have been the most prominent addition to the light pollution in the open landscape, however, the measurements did show some light influence from the road up to around 160 meters from the road, although the levels were below 0.1 lx. We also know that moonlight (with light levels up to 0.1 lux) affects many species, like rodents, who forage during new moon (7). The reflected light from the asphalt contributes to light pollution and affects greater distances. The strong luminance directly from the light source might also be affecting insects and animals. This light is seen at very long distances, while not being measured on an illuminance meter.

There were several additional sources of light pollution in the area that may have affected the result, such as an unshielded light source on a nearby farm, sky glow from both Oslo/Skedsmokorset and Oslo Airport Gardermoen. At transect 2, 160 meters away from the road, the measured light towards the unshielded light source of the farm (seen in the picture, Figure 5) and towards the airport was greater than when the light meters were directed towards the road. In transect 1, 50 meters away from the road, the light levels were slightly higher measured up towards the cloudy sky (0.002) than measured towards to road (0.001).

Figure 5 Image from approx. 160 meters from the road, with light from streetlamps, private outdoor lighting and sky-glow.

4. Conclusion and further work

The most important result of the light measurements was that the observations provided a basis for ideas and discussion about the focus for further work. 1. Although the light level was below the detection limit, the light source was visible as a point at a very large distance from the road. This observation addresses the question of whether organisms are influenced by the average light energy per area, or the intensity of a source as observed from a distance (irradiance vs. radiance). 2. The preliminary measurements gave some indications of how far from the road the streetlight affects the light environment, but this needs to be verified with more measurements under different types of weather, with greater control of other factors that affect light pollution, different types of road fixtures (lamp types on the road, car headlights), and light pollution from sources other than the road. Here, for example, cubic light measurements in combination with luminance photos can be useful. This can give insights into what factors of light pollution affect an area most, the direct component, the reflected component from the ground, or the sky-glow, which is highly dependent on the particles in the air.

5. Acknowledgements

Thanks to the National Road Authorities og Norway for funding the project.

6. References

1. Sanders D, Frago E, Kehoe R, Patterson C, Gaston KJ. A meta-analysis of biological impacts of artificial light at night. Nat Ecol Evol. 2. november 2020;1–8.

2. Kyba CCM, Kuester T, Miguel AS de, Baugh K, Jechow A, Hölker F, mfl. Artificially lit surface of Earth at night increasing in radiance and extent. Science Advances. 1. november 2017;3(11):e1701528.

3. Sánchez de Miguel A, Bennie J, Rosenfeld E, Dzurjak S, Gaston KJ. First Estimation of Global Trends in Nocturnal Power Emissions Reveals Acceleration of Light Pollution. Remote Sensing. januar 2021;13(16):3311.

4. Follestad A. NINA-rapport 1081 Effekter av kunstig nattbelysning på naturmangfoldet -en litteraturstudie [Internett]. NINA - Norsk institutt for naturforskning; 2014. Tilgjengelig på: https://www.nina.no/archive/nina/PppBasePdf/rapport/2014/1081.pdf

5. Jägerbrand AK, Spoelstra K. Effects of anthropogenic light on species and ecosystems. Science. 16. juni 2023;380(6650):1125–30.

6. Jägerbrand A, Bouroussis C. Ecological Impact of Artificial Light at Night: Effective Strategies and Measures to Deal with Protected Species and Habitats. Sustainability. 26. mai 2021;13:5991.

7. Shier DM, Bird AK, Wang TB. Effects of artificial light at night on the foraging behavior of an endangered nocturnal mammal. Environ Pollut. august 2020;263:114566.

A Framework for Optimizing Lighting in Animal Shelters for Domestic Cats

Kendall Baertlein, Jennifer Scheib and Kyri Baker

Department of Civil, Environmental, and Architectural Engineering, University of Colorado Boulder, Boulder, CO 80309, USA

kendall.baertlein@colorado.edu

Abstract. In this project, we aim to enable designers to optimize lighting conditions for domestic cats in animal shelters. Specifically, we developed an optimization framework that controls the spectral power distribution (SPD) of light sources and returns the effect on feline visual perception. In our model, we used standard light source SPDs and predicted light reflected from different surfaces within an animal shelter. The reflected SPDs were combined with the known brightness sensitivity of domestic cats across the visible light spectrum to develop an understanding of cats' visual experience in the shelter under different lighting conditions. The optimization framework then minimizes the difference between a desired lighting effect (here, daylight) and a modeled effect using a set of light source SPDs. Initial results demonstrate that by using slightly more advanced lighting design and control, shelters can potentially improve the visual experience for cats.

1. Introduction

Much work has been performed in designing favorable lighting conditions for humans. However, less attention has been given to how animals' wellbeing is affected by indoor lighting. The work that has been performed indicates that disruption of the light/dark cycle, specific light intensities, wavelength, and other light-related variables can have significant physiological impacts on animals, as discussed by Emmer et. al [1]. Coupled with an environment such as a research laboratory or an animal shelter, these impacts can be compounded with many additional biological and psychological stressors.

Animal shelters are uniquely stressful environments for many animals. As noted in the 2022 Association of Shelter Veterinarians document *The Guidelines for Standards of Care in Animal Shelters*, "Facilities should be designed to offer as much natural light as possible…When natural lighting is not available and artificial light is used, it should approximate natural light in duration and intensity to support circadian rhythms" [2]. While some work by Wilson [3] has studied the impact of indoor lighting on domestic dogs in animal shelters, the impact on domestic cats, who perceive light similarly but distinctly from dogs, is more underexplored. Domestic cat behavior is sensitive to environment; for example, behavior has been shown to change when indoor and outdoor cats were compared by Parker et. al [4], and cats in shelters face unique challenges to their wellbeing as discussed by Vojtkovská et. al [5]. Stella et. al [6] show that failing lighting control timeclocks (in addition to other unusual external events; lighting effects were not isolated in the study) have the potential to cause gastrointestinal and urinary tract issues as well as skin and behavioral problems. It is well known that daylight and natural

IOP Conf. Series: Earth and Environmental Science 1320 (2024) 012017 doi:10.1088/1755-1315/1320/1/012017

light/dark cycles can have positive impacts on humans as well as animals, but unfortunately not all shelters can provide daylight to their residents.

To contribute to improving the wellbeing of cats in shelters, we developed an optimization framework specifically designed to optimize the spectral power distribution (SPD) (e.g., radiant power as a function of wavelength) of various light sources to mimic a standard description of daylight as experienced by cats inside an animal shelter. In contrast to existing works utilizing optimization frameworks for tuning lighting spectral outputs (e.g., for preserving artwork as discussed by Durmus et. al [7], or conserving energy as explained by Durmus et. al [8]), our work specifically focuses on domestic cats in an indoor environment and develops an optimization framework that does not use heuristic or genetic algorithms. Our framework is in the form of a mixed-integer linear program which can be solved to optimality and easily incorporate different input data (sensitivity curves, objectives, and various SPDs). Future work can include additional objectives such as enhancing color discrimination or targeting the appearance of the cats to potential adopters, whose perception of shelter animals can be improved by lighting adjustments as shown by Heemstra [9].

2. Background on Cats' Visual Perception

To analyze architectural lighting for cats, it was important to consider how each light fixture's radiant power would be perceived by cats. We identified a spectral sensitivity curve for cats from Crocker, et. al [10] to be used as a proxy for spectral brightness efficiency in setting up the optimization framework. For context, the proxy spectral brightness efficiency function for cats is plotted in Figure 1 alongside the CIE spectral luminous efficiency function for human photopic (daytime) vision. Both curves are unitless functions of wavelength that represent the relative ability for narrow bands of radiant power in the visible range to induce equal perception of brightness, as explained in the Lighting Handbook [11]. As demonstrated by Figure 1 below, the proxy brightness efficiency function we used models cats as more sensitive to short wavelengths (can be considered as blue and violet light), while humans are more sensitive to medium wavelengths (green), and have more sensitivity to long wavelengths (red) than cats.

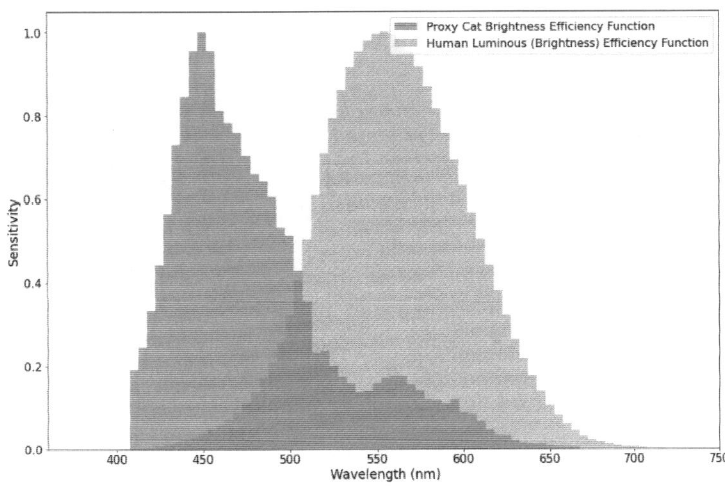

Figure 1: Human and Cat Sensitivities

3. Optimization Framework

The optimization framework was developed in Python and utilized the GUROBI solver within the CVXPY interface [12] to optimize various chosen objectives subject to constraints on the lighting source capabilities. We considered different light sources that a shelter may have access to and their impact on the appearance of light within a space relative to daylight. Specifically, we consider a combination of narrow band (colored) LEDs and white-light sources, including phosphor-based LED and fluorescent, to determine which combination would produce a visual signal in cats closest to that as experienced under a daylight spectrum. We identified a typical SPD for each source from Cree [13] and the Illuminating Engineering Society of North America [14], and through measurements using a Konica Minolta CL-500 illuminance spectrophotometer of each color channel of an ETC Desire D22 fixture. We then scaled the source SPDs to get the same radiant watts for each source at maximum brightness, as can be seen in Figure 2 below. Also, in our optimization, we allowed each source to dim, which required modifier variables. We used the CIE standard illuminant D65 for the daylight spectrum and assumed that the relative SPD was not affected by its transmittance through glazing, as we did not account for the effect of windows on the daylight curve. The SPDs of the sources were modeled with the variables and their modifiers listed below.

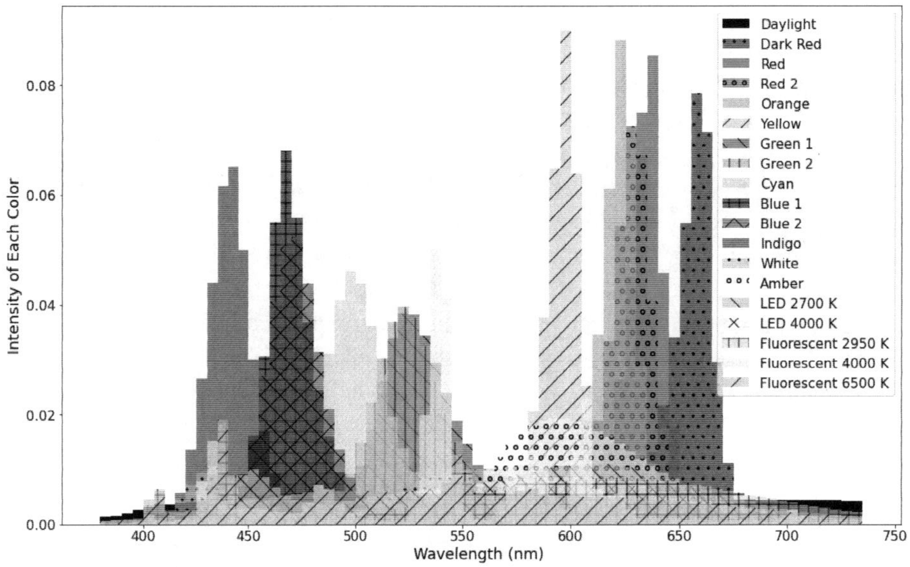

Figure 2: SPD for all Light Sources

daylight	Reference daylight source SPD
blue1, b1	First blue LED source SPD, modifier
green1, g1	First green LED source SPD, modifier
yellow, y	Yellow LED source SPD, modifier
orange, o	Orange LED source SPD, modifier
red1, r1	First red LED source SPD, modifier
dark_red, dr	Dark red LED source SPD, modifier
red2, r2	Second red LED source SPD, modifier
white, w	White LED source SPD, modifier
amber, a	Amber LED source SPD, modifier
green2, g2	Second green LED source SPD, modifier
cyan, c	Cyan LED source SPD, modifier
blue2, b2	Second blue LED source SPD, modifier
indigo, i	Indigo LED source SPD, modifier
LED2700, led2	2700 K LED source SPD, modifier
LED4000, led4	4000 K LED source SPD, modifier
fluorescent2950, flu2	2950 K fluorescent source SPD, modifier

fluorescent4000, flu4	4000 K fluorescent source SPD, modifier
fluorescent6500, flu6	6500 K fluorescent source SPD, modifier

These variables were convolved with a brightness efficiency function representing human or cat vision, $sensitivity$, and the spectral reflectance of a representative wall color, $color$, which can vary, as indicated by Figure 3 below. The resulting values were modified from the initial variable by adding a subscript of c to each variable name. We used these convolved SPDs in our optimization.

We developed several constraints for our optimization problem. We assumed dimming of the light intensity was allowed, using fixed values between 0 and 255. Thus, to constrain our problem, we restricted the modifier variables to be integers from 0-255, as shown in Equation 1 below. Daylight was used as a reference across all objectives, and we assumed it was at its maximum value for all objectives, using a consistent multiplier of 255.

$$0 \leq modifier \leq 255$$

Equation 1: Constraints

We also developed several optimization objectives for our problem. First, we compared several architectural white-light sources to daylight. Next, we iterated through a combination of sources, considering only colored LED sources, only white-light LED sources, and all sources. In our objectives, we considered the convolved SPDs over the wavelengths, λ, 380-730 nm.

In Objective 1, we considered the 2700K LED compared with daylight, attempting to minimize the difference between daylight and the LED source, as shown in Equation 2 below. We squared the difference to penalize higher differences as well as penalize both positive and negative differences between the two curves. Similarly, in Objective 2, we compared the 4000K LED source and daylight, as shown in Equation 3. We compared the 2950K fluorescent source with daylight in Objective 3, the 4000K fluorescent source with daylight in Objective 4, and the 6500K fluorescent source with daylight in Objective 5, as seen in Equations 4-6 below.

$$\min\left(\sum_{\lambda=380}^{730}(daylight_c(\lambda) * 255 - led2700_c(\lambda) * led2)^2\right)$$

Equation 2: Objective 1

$$\min\left(\sum_{\lambda=380}^{730}(daylight_c(\lambda) * 255 - led4000_c(\lambda) * led4)^2\right)$$

Equation 3: Objective 2

$$\min\left(\sum_{\lambda=380}^{730}(daylight_c(\lambda) * 255 - fluorescent2950_c(\lambda) * flu2)^2\right)$$

Equation 4: Objective 3

$$\min\left(\sum_{\lambda=380}^{730}(daylight_c(\lambda) * 255 - fluorescent4000_c(\lambda) * flu4)^2\right)$$

Equation 5: Objective 4

$$\min\left(\sum_{\lambda=380}^{730}(daylight_c * 255 - fluorescent6500_c * flu6)^2\right)$$

Equation 6: Objective 5

In Objectives 6-8, we considered multiple sources. In Objective 6, we considered the colored LED sources compared with daylight. The colored sources consisted of *blue1* and *blue2, green1* and *green2, yellow, orange, red1* and *red2, dark_red, cyan,* and *indigo.* In Objective 7, we compared all LED sources with daylight. Finally, in Objective 8, we compared all sources with daylight. These objectives can be seen in Equations 7-9 below.

$$\min\left(\sum_{\lambda=380}^{730}(daylight_c * 255 - (blue1_c * b1 + green1_c * g1 + yellow_c * y + orange_c * o + red1_c * r1 + dark\_red_c * dr + red2_c * r2 + green2_c * g2 + cyan_c * c + blue2_c * b2 + indigo_c * i))^2\right)$$

Equation 7: Objective 6

$$\min\left(\sum_{\lambda=380}^{730}(daylight_c * 255 - (blue1_c * b1 + green1_c * g1 + yellow_c * y + orange_c * o + red1_c * r1 + dark\_red_c * dr + red2_c * r2 + white_c * w + amber_c * a + green2_c * g2 + cyan_c * c + blue2_c * b2 + indigo_c * i + LED2700_c * led2 + LED4000_c * led4))^2\right)$$

Equation 8: Objective 7

$$\min\left(\sum_{\lambda=380}^{730}(daylight_c * 255 - (blue1_c * b1 + green1_c * g1 + yellow_c * y + orange_c * o + red1_c * r1 + dark\_red_c * dr + red2_c * r2 + white_c * w + amber_c * a + green2_c * g2 + cyan_c * c + blue2_c * b2 + indigo_c * i + LED2700_c * led2 + LED4000_c * led4 + fluorescent2950_c * flu2 + fluorescent4000_c * flu4 + fluorescent6500_c * flu6))^2\right)$$

Equation 9: Objective 8

Figure 3: Animal shelters utilize various light sources, including tubular daylighting devices (left) to help bring in daylight. Most have fluorescent or LED lights (center). Different surface colors (right; all) motivate a dynamic, adjustable approach.

While we squared the objective values to consider absolute differences and penalize larger differences, we wanted to determine the resulting difference, *difference*, between daylight and the light sources for each objective, which we calculated by using the square root of the objective value, *output*, where *n* is the objective number, ranging from 1-8 as detailed above.

$$difference_n = \sqrt{output_n}$$

The resulting objective function thus represents the difference between the sources and the convolved daylight curve, with a higher difference being penalized in our objective. A lower difference indicates the sources better match daylight.

4. Results

Through our optimization objectives, we were able to compare the different lighting scenarios and determine which best matched daylight. In Table 1, we detail the intensity of each light source and the difference between the light sources and daylight, summed across wavelengths. We show both cat and human results based on the corresponding brightness efficiency functions. Human results are included to provide an example of the potentially different lighting needs of cats and humans. In the table, fluorescent sources are denoted by *fluor* and objectives are denoted by *obj.* Moreover, Figure 4 shows plots with the resulting convolved SPDs based on source SPD, cat visual perception, wall color, and dimming of each source. The results below assume a white wall color, though spectral reflectance curves for more saturated room finish colors can be modeled.

Table 1: Summarized Results for Cats and (Humans)

	Obj. 1	*Obj. 2*	*Obj. 3*	*Obj. 4*	*Obj. 5*	*Obj. 6*	*Obj. 7*	*Obj. 8*
Daylight	**255 (255)**	**255 (255)**	**255 (255)**	**255 (255)**	**255 (255)**	**255 (255)**	**255 (255)**	**255 (255)**
Blue 1	0 (0)	0 (0)	0 (0)	0 (0)	0 (0)	**13 (0)**	**31 (0)**	**35 (0)**
Green 1	0 (0)	0 (0)	0 (0)	0 (0)	0 (0)	**0 (97)**	**0 (86)**	**0 (34)**
Yellow	0 (0)	0 (0)	0 (0)	0 (0)	0 (0)	**61 (1)**	0 (0)	0 (0)
Orange	0 (0)	0 (0)	0 (0)	0 (0)	0 (0)	0 (0)	0 (0)	0 (0)
Red 1	0 (0)	0 (0)	0 (0)	0 (0)	0 (0)	0 (0)	0 (0)	0 (0)
Dark Red	0 (0)	0 (0)	0 (0)	0 (0)	0 (0)	**96 (86)**	**54 (76)**	**25 (35)**
Red 2	0 (0)	0 (0)	0 (0)	0 (0)	0 (0)	0 (0)	0 (0)	0 (0)
White	0 (0)	0 (0)	0 (0)	0 (0)	0 (0)	0 (0)	**30 (0)**	0 (0)
Amber	0 (0)	0 (0)	0 (0)	0 (0)	0 (0)	0 (0)	0 (0)	0 (0)
Green 2	0 (0)	0 (0)	0 (0)	0 (0)	0 (0)	0 (0)	0 (0)	**0 (14)**
Cyan	0 (0)	0 (0)	0 (0)	0 (0)	0 (0)	**33 (0)**	**1 (0)**	0 (0)
Blue 2	0 (0)	0 (0)	0 (0)	0 (0)	0 (0)	**0 (39)**	**0 (40)**	**5 (48)**
Indigo	0 (0)	0 (0)	0 (0)	0 (0)	0 (0)	**48 (24)**	**35 (22)**	**18 (3)**
LED 2700 K	**240 (244)**	0 (0)	0 (0)	0 (0)	0 (0)	0 (0)	**100 (23)**	0 (0)
LED 4000 K	0 (0)	**244 (244)**	0 (0)	0 (0)	0 (0)	0 (0)	0 (0)	0 (0)
Fluor. 2950 K	0 (0)	0 (0)	**250 (254)**	0 (0)	0 (0)	0 (0)	0 (0)	**170 (114)**
Fluor. 4000 K	0 (0)	0 (0)	0 (0)	**239 (235)**	0 (0)	0 (0)	0 (0)	0 (0)
Fluor. 6500 K	0 (0)	0 (0)	0 (0)	0 (0)	**253 (249)**	0 (0)	0 (0)	**0 (1)**
Difference	**2351 (3296)**	**640 (888)**	**1621 (2981)**	**1501 (3463)**	**1614 (2868)**	**240 (379)**	**206 (371)**	**126 (311)**

As demonstrated by Table 1, humans and cats have different optimal electric lighting scenarios for mimicking the perception of daylight with common light source technology, with humans needing higher values of dark red and cats needing higher values of indigo, for example. The optimization was better able to match daylight perception for the cats than humans, resulting in lower difference values. Of the single-source objectives, the modeled 4000 K LED source was best able to match the daylight perception as measured through the wavelength-by-wavelength comparison of the convolution of eye-incident SPDs and brightness efficiency functions. Additionally, the optimization with multiple light

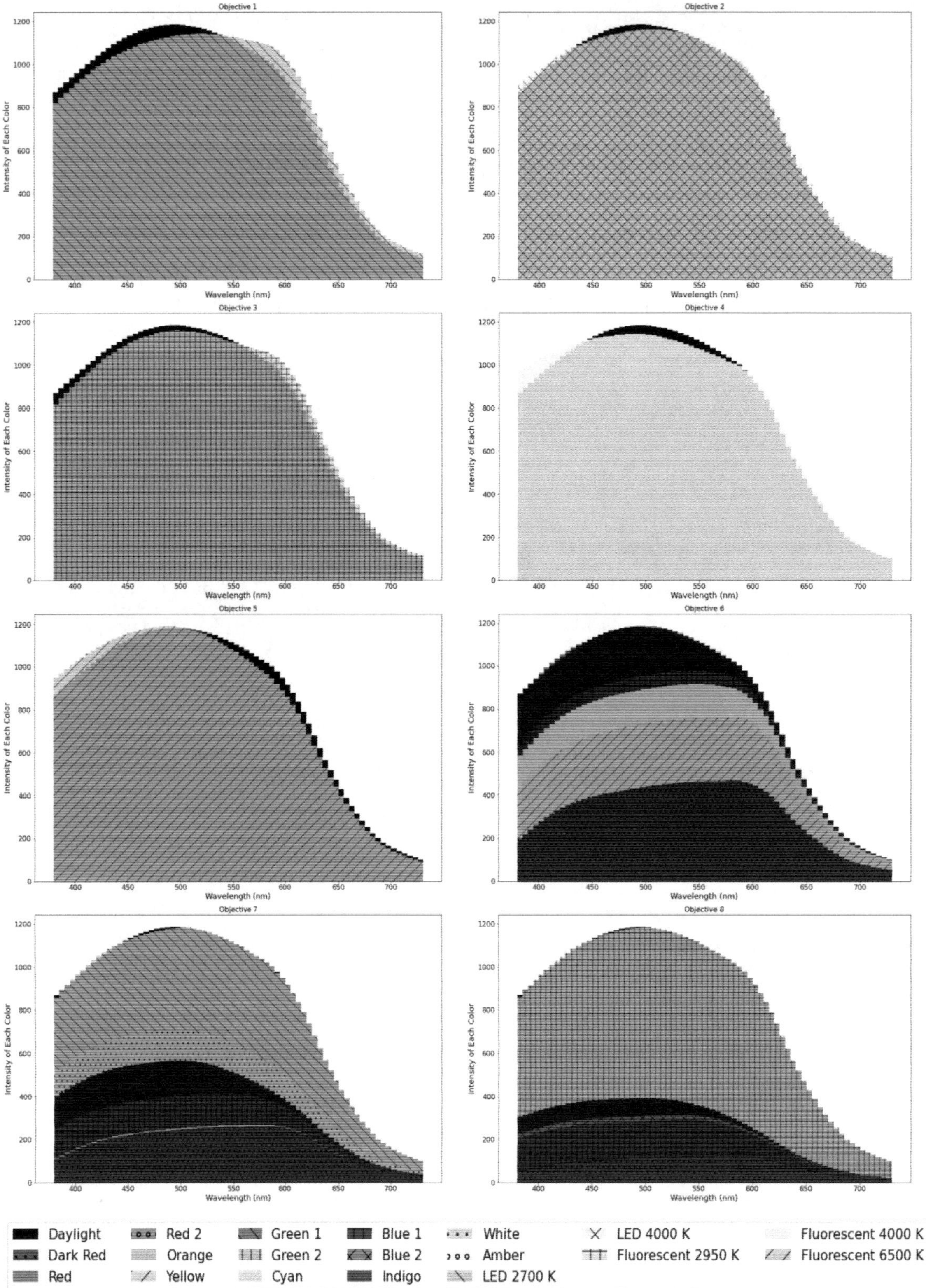

Figure 4: Resulting Convolved Spectral Power Distributions for Each Objective

sources resulted in a lower difference than those with a single light source, with Objective 8 resulting in the lowest difference and Objective 1 resulting in the highest difference. Figure 4 shows the convolved SPDs of the sources in each objective with the proxy brightness sensitivity function for cats compared to the convolved D65 daylight spectrum. Objectives 1-5 had less overlap with the convolved daylight curve compared to Objectives 6-8, which utilized multiple sources.

5. Discussion and Future Work

In this paper, we developed an optimization framework that considers the capabilities of different lighting sources, the spectral sensitivity of cat eyes, the spectral reflectance of walls, and tunes the spectral power distribution of the overall setup to match daylight (as seen by an indoor domestic cat) as close as possible. Animal shelters were chosen as the application due to their potential impacts on cat wellbeing relative to the current practice of lighting for humans in the space. A daylight objective was focused on due to the reasonable assumption that it is an important spectrum for supporting cats' wellbeing. The results indicate that, out of the chosen set of lighting sources, a combination of sources better achieves daylight-matching. Specifically, we found a combination of a 2950 K fluorescent and narrow-band dark red, blue, and indigo LEDs resulted in the best daylight match. The results from this study could be extended to include dynamic light adjustments over time, more complex surface reflectance models, a standard brightness efficiency function for cats, or brightness efficiency functions for other animals, such as dogs. Different light sources could be added into the model, allowing customizable inputs. Daylight spectra beyond D65 could also be considered, incorporating various daylight conditions and considering the shift in the spectrum throughout the day. Additional objectives could be considered as well, such as improving color discrimination for cats perceiving toys, or making shelter cats appear more attractive to potential adopters. For shelters without advanced lighting controls, objectives such as optimizing light timers could also prove beneficial for cats' circadian rhythm, as discussed by Kuwabara et. al [15]. To address circadian entrainment in our optimization, the framework could incorporate spectral sensitivities of the non-image-forming system such as melanopic sensitivity with an objective based on the metric of melanopic equivalent daylight illuminance. Lastly, the optimization framework can include additional decision variables that can improve wellbeing or shelter objectives further, such as temperature control, energy savings, and daylight integration.

References

[1] Emmer K M, Russart K L G, Walker W H, Nelson R J, and Devries A C 2018 Effects of light at night on laboratory animals and research outcomes *Behav. Neurosci.* **132** no 4 302-314

[2] The Association of Shelter Veterinarians 2022 The guidelines for standards of care in animal shelters: second edition *J. Shelter Med. and Comm. Animal Health* **1** no 2 1-76.

[3] Wilson K 2022 The impact of fluorescent light on shelter dog behavior (canis lupus familiaris). *City Univ. of NY Masters thesis*

[4] Parker M, Serra J, Deputte B L, Ract-madoux B, Faustin M and Challet E 2022 Comparison of locomotor and feeding rhythms between indoor and outdoor cats living in captivity *Animals* **12** no 18 2440.

[5] Vojtkovská V, Voslářová E and Večerek V 2020 Methods of assessment of the welfare of shelter cats: a review *Animals (Basel)* **10** no 9 1527.

[6] Stella J L, Lord L K and Buffington C A 2011 Sickness behaviors in response to unusual external events in healthy cats and cats with feline interstitial cystitis *J. Am. Vet. Med. Assoc.* **238** no 1 67-73

[7] Durmus D, Abdalla D, Duis A and Davis W 2020 Spectral optimization to minimize light absorbed by artwork *LEUKOS* **16** no 1

[8] Durmus D and Davis W 2017 Object color naturalness and attractiveness with spectrally optimized illumination *Optics Express* **25** no 11

[9] Heemstra V 2013 Shedding a light on animal shelters: the use of light and its effect on the adoption process *Univ. of Twente Master Thesis*

[10] Crocker R A, Ringo J, Wolbarsht M L and Wagner H G 1980 Cone contributions to cat retinal ganglion cell receptive fields *J Gen Physiol.* **76** no 6 763-85.

[11] Dilaura D L and Illuminating Engineering Society of North America 2011 The lighting handbook: reference and application *IESNA* ed D DiLaura

[12] Diamond S and Boyd S 2016 CVXPY: a python-embedded modelling language for convex optimization. *J. Mach. Learning Res.* **17** no 83

[13] Cree LED 2023 XLamp XP-E LED data sheet 13

[14] Illuminating Engineering Society of North America 2022 Spectral calculator

[15] Kuwabara N, Seki N and Aoki K 1986 Circadian, sleep and brain temperature rhythms in cats under sustained daily light-dark cycles and constant darkness, *Phys. & Behav* **38** no 2

IOP Conf. Series: Earth and Environmental Science 1320 (2024) 012018 doi:10.1088/1755-1315/1320/1/012018

How to illuminate indoor plants sustainably? Tips and tricks to bridge the gap between research and design.

J Hartmann[1] and F Feltrin[2]

[1] Principle and Creative Director lightsphere GmbH, Zürich, Switzerland
[2] Jun. Project Leader lightsphere GmbH, Zürich, Switzerland

Corresponding author's E-mail: hartmann@lightsphere.ch

Abstract. The increasing popularity of biophilic design in architecture and interior design has led to a widespread integration of indoor ornamental plants and green walls. While numerous studies have demonstrated the benefits of such integration for people, only a few have focused on the well-being of the plants themselves. Our latest research project aims to address this gap by exploring suitable lighting conditions for indoor plants, seeking an optimal spectral composition that ensures their well-being, robust growth, and visual appeal, while also optimizing maintenance. Building upon previous applied research, we conducted experiments to analyze the lighting requirements of four species of ornamental plants commonly used indoors. Seven identical room boxes were employed, each illuminated with different light spectra falling within the CCT range of 2800K-5000K. The specific light spectra were meticulously tuned with a focus on the red/blue and red/green ratios, while keeping the photosynthetic photon flux (PPFD) and average illuminance consistent. As a result of this research project, practical guidelines were developed to help lighting designers navigating the intricate world of indoor ornamental plants effectively. By carefully considering the lighting spectrum in the suggested ranges, architects and interior designers can create spaces that not only benefit human occupants but also provide an optimal environment for the well-being and flourishing of indoor plants.

1. Introduction

In recent years, biophilic design has gained significant traction in the fields of architecture and interior design. Rooted in the concept of biophilia, which denotes the innate human affinity for nature, this design approach seeks to integrate natural elements into constructed spaces, thereby enhancing the physical and mental well-being of occupants [1-4]. Architects and designers employ various strategies, such as incorporating natural materials, patterns, colors, optimizing daylighting, creating artificial lighting conducive to the human circadian rhythm, and integrating ornamental plants in the space.

Extensive research has already demonstrated that the implementation of biophilic design principles can substantially improve the overall experience of interior spaces, promoting better health outcomes, reducing the risk of diseases, and alleviating stress [5-13]. Moreover, it has been found to enhance productivity, creativity, and overall user well-being, while aligning with green building strategies and contributing to energy savings [14-16]. The presence of indoor plants has been recognized for its capacity to purify and humidify the air, regulate indoor temperatures, serve as acoustic absorbers, and significantly uplift moods and reduce stress levels, particularly in places where people spend extended periods, such as offices and schools.

In response to the growing significance of indoor plants for the biophilic design approach, the latest version of the WELL Building Standard, a prominent rating system for sustainable and healthy buildings, has integrated many biophilic design principles, including the use of indoor plants. The new standard highlights the importance of direct connections to nature through the presence of plants, water features, daylight, and natural views. In particular, Chapter M09 'Enhanced access to nature' [17] requires a combination of indoor plants and water features in direct sight of at least 75% of workstations and in shared recreational spaces to encourage a stronger relationship between users and natural elements.

Given the increasing popularity and requirements of international building standards for the integration of plants into the built environment, there is an urgent need to research appropriate lighting techniques for ornamental plants and to identify lighting parameters that influence their health and visual appeal. Surprisingly, this area of research is insufficiently investigated, with existing studies focusing mainly on horticultural lighting and neglecting typical decorative ornamental plants (Fig. 1). Furthermore, the few studies conducted so far have been limited by the equipment used or have examined only a limited set of critical parameters.

Figure 1. Difference between visual effects of horticultural lighting (left picture) and lighting for indoor ornamental plants (right picture)

1.1 Lighting for plants versus lighting for human beings

The lighting of ornamental plants in indoor spaces presents a dilemma of its own: how can lighting for plants be combined with the typical lighting for indoor spaces? The impact of light on plant growth and development goes beyond its practical use for human vision. While humans need light primarily for visual perception and circadian regulation, plants have a more complex relationship with light, relying on it for the vital process of photosynthesis [18-23]. The spectrum of light that humans find visually pleasing may not always be the most beneficial for plants, as they require specific wavelengths to facilitate photosynthesis and regulate various physiological responses such as germinating, directional growing and flowering. In order to create an environment that promotes the wellbeing of both humans and plants, it is essential to understand their different lighting needs.

1.1.1. The sensitivity to light differs between plants and humans

Both plants and humans rely on photoreceptors containing specific photosensitive pigments to gather information from their environment through light. Humans are able to perceive wavelengths between 400-700 nm, while ultraviolet (UV) and infrared (IR) wavelengths, which are shorter and longer, respectively, that this range, are not visible to the human eye. Plants also use a specific range of wavelengths for photosynthesis called Photosynthetically Active Radiation (PAR), which corresponds to the 400-700 nm wavelengths. However, plants have five different photoreceptors that can detect light outside the PAR range. Research has shown that UV-A radiation can increase plant vigor and affect leaf pigments, while IR radiation can affect the flowering cycle [24-25]. In contrast, the V-lambda curve shows the different light sensitivities of humans, with our visual system being most sensitive to green-yellow light (Figure 2). On the other hand, the PAR spectrum highlights the preference of plants for blue and red light, which is crucial for optimizing photosynthesis and various physiological processes [26]. While humans thrive in environments rich in green-yellow light, the inclusion of both blue and red light in the PAR spectrum is essential to promote the growth and well-being of ornamental plants. In order to design lighting environments that promote the growth and well-being of ornamental plants, while considering the visual needs and comfort of humans, it is essential to understand the light sensitivities of both plants and humans. By carefully balancing these light sensitivities, we can create harmonious and biophilic spaces that enhance both human comfort and the flourishing of indoor greenery in our built environment.

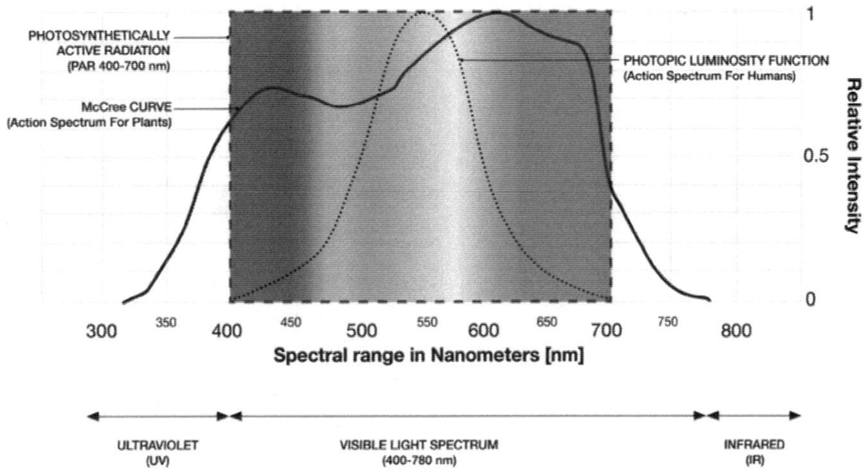

Figure 2. A comparison of action spectra for plants' photosynthesis and humans' visual perception. Source: Zielinska-Dabkowska et al. Sustainability 2019

1.1.2. Plants and human beings refer to different metrics

Since plants and humans respond differently to light, it is essential to employ distinct metrics when designing lighting for each application. Lumens and lux, which are photometric units related to the human visual system, cannot be used for plants' lighting needs. Instead, the Photosynthetic Photon Flux Density (PPFD) and the Daily Light Integral (DLI) serve as the appropriate measurement units for designing lighting for plants (Fig 3). The PPFD quantifies the number of photons reaching a 1 m2 surface per second and is specified in micromoles per square meter per second ($\mu mol \cdot m^{-2} \cdot s^{-1}$), reflecting the instantaneous light intensity received by plants. On the other hand, the DLI measures the number of photosynthetically active photons within the 400-700 nm range delivered to a specific area over a 24-hour period, expressed in ($mol \cdot m^{-2} \cdot d^{-1}$). This integral metric provides valuable information on the total light accumulation that plants receive during the day in their natural environment, helping to determine the cumulative light exposure necessary for their growth and well-being indoors [27]. DLI is crucial for ornamental plants because it directly correlates with their growth, flowering, and overall health. Different species of ornamental plants have varying light requirements. Achieving the ideal DLI for ornamental plants involves considering factors such as the plant's natural habitat, growth stage, light intensity, and duration of exposure. The metrics PPFD and DLI are vital for creating lighting environments that cater to the unique needs of both plants and humans, ensuring optimal growth and well-being in their respective contexts.

IOP Conf. Series: Earth and Environmental Science 1320 (2024) 012018 doi:10.1088/1755-1315/1320/1/012018

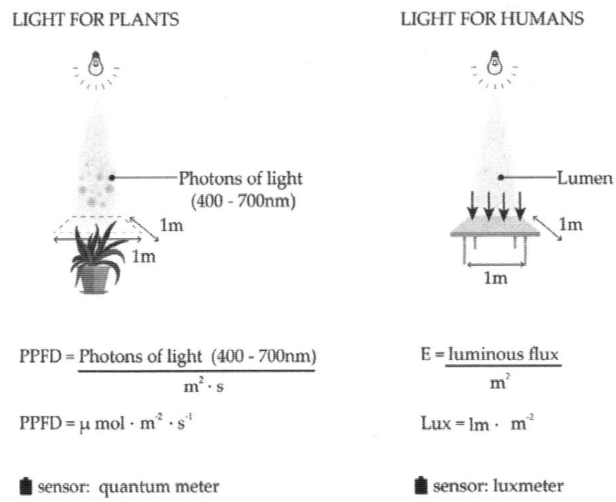

Figure 3. A comparison between metrics used to evaluate light for plant and for humans. Source: Zielinska-Dabkowska et al. Sustainability 2019

2. Research goals

This research project aims to address the disparity between lighting design for people's needs and the botanical requirements of ornamental plants and to offer valuable insights for architects, designers, and lighting specialists to create healthier and more aesthetically pleasing spaces.

The use of standard indoor LED lighting, developed primarily for human visual preferences, falls short in providing the necessary biological health for plants. Previous studies in this area have examined standard light sources with fixed and default CCTs [28-30], but the spectral distribution was not tailored to the specific needs of plants. In order to harmonize lighting for both people and plants, this research aims to demonstrate that typical CCTs used in indoor spaces can be effective for plants if carefully designed for their requirements. The main objective was to identify specific spectra within the typical CCT range of 2800K-5000K and to develop a family of LED chips that can be used in architectural lighting fixtures to achieve a cohesive and a harmonious lighting effect in common spaces. Additionally, considering sustainability, the research aims to investigate the minimum amount of light necessary for plant survival and healthy growth. While previous studies indicate an average illuminance of 2000-2500 lx [28], this project is questioning this reference value and investigating the possibility of ensuring healthy plant development with reduced light levels in line with energy saving requirements. Furthermore, the research aims to contribute to the limited scientific literature on recommended DLI and PPFD-factors for various indoor ornamental plant species, particularly focusing on realistic and energy-efficient possibilities achievable with artificial lighting in indoor spaces.

The research could result in several benefits for the design of indoor environments and the well-being of both space users and plants:

- *Enhanced Biophilic Design* - By developing lighting solutions that meet specific needs of both people and plants, this research can contribute to the advancement of biophilic design principles. Harmonious and biologically supportive lighting environments can create a stronger connection between occupants and nature, leading to improved mental well-being and productivity.
- *Optimized Plant Health.* Identifying spectra within the typical CCT range that promote healthy plant growth and development can lead to better care and maintenance of ornamental

plants in indoor spaces. This knowledge can help architects and designers create conditions that ensure the longevity and vitality of greenery.

- *Energy Efficiency and Reduction of Maintenance*. By questioning the standard reference value of illuminance for plant growth and exploring lower light levels that are still favorable to plants' health, the research can contribute to energy-saving efforts. Reducing the light intensity while maintaining plants' well-being with a natural and compact growth can lead to more sustainable indoor environments with lower energy consumption and lower maintenance costs.
- *Scientific Contribution*. The research's exploration of recommended DLI and PPFD factors for various indoor ornamental plant species fills a gap in scientific literature. This knowledge can serve as a valuable reference for future studies and contribute to a better understanding of optimal lighting conditions for indoor plants.
- *Practical Applications*. The development of a family of LED chips with tailored spectra for plants' needs that can be integrated into standard architectural lighting fixtures, can have practical applications in the lighting industry. Lighting designers can use this technology to create customized and efficient lighting solutions for both the space users and the ornamental plants in indoor spaces.

Overall, the research outcomes have the potential to positively impact indoor environments, plant care practices, sustainability efforts, and scientific knowledge, fostering a more holistic and beneficial approach to lighting design for both humans and plants.

3. Lighting Experiment - Setting

The lighting experiment was conducted between November 2021 and October 2022 to evaluate the formulated theory. The study focused on investigating two fundamental lighting parameters: the Spectral Power Distribution (SPD), gauged through the analysis of the red/blue and red/green ratios, and the Photosynthetic Photon Flux Density (PPFD). These parameters play a vital role in determining the quality and quantity of light received by plants, affecting their growth and overall well-being. In conjunction with these two light parameters, four distinct plant species commonly found in indoor spaces were selected. By studying these representative plant species, the experiment aimed to gain comprehensive insights into their responses to the different lighting conditions, contributing to the advancement of optimized lighting solutions for indoor environments.

Figure 4. Images of the plant's species used for the experiment. From left to right: Codiaeum Petra, Dracaena Janet, Epipremnum and Spathiphyllum

3.1. Ornamental Plants

Four plants species were selected for the experiment: Codiaeum Petra, Dracaena Janet, Epipremnum and Spathiphyllum (Fig. 4). Each species represents a unique set of characteristics relevant to indoor greenery. Among these, Epipremnum and Dracaena Janet are categorized as "full sun" plants, thriving in high light intensity conditions, while Codiaeum Petra and Spathiphyllum belong to the "full sun/half shadow" category, capable of adapting to moderate light exposure. Moreover, Spathiphyllum and Codiaeum Petra have the ability to produce flowers. The selection of these four plant species was done

IOP Conf. Series: Earth and Environmental Science 1320 (2024) 012018 doi:10.1088/1755-1315/1320/1/012018

in collaboration with a botanist and based on their frequent use in indoor applications by landscape architects. To ensure robust experimentation, we employed two plants for each species for every light configuration, thus a total of eight different plants per light box. Additionally, the garden center responsible for providing the plants ensured appropriate soil quality, supplemented with the necessary nutrients, and provided watering guidelines for consistent care throughout the test duration.

3.2. Test Set-up

The experiment was conducted within a controlled laboratory environment. To ensure precise lighting measurements, a physical dark room configuration was established, devoid of any external light sources. The test space was meticulously enclosed with dark cardboard and black opaque fabric, eliminating potential interference from adjacent light. Seven light boxes were arranged in the space, each of them housing one spotlight. The plants were exposed to 12 hours of light and 12 hours of darkness, to simulate the natural light/darkness cycle.

The overall dimensions of the test space measured 4.5m x 2.7m x 3.5m, while each light box measured 0.43m x 0.8m x 1.2m, except for Box B, which measured 0.5m x 0.7m x 1.2m. The spotlights were installed at the top of each box, approximately 60 cm away from the top edge of the plants. Notably, the spotlight in Box B was placed off-center to accommodate specific space requirements. All plant specimens were positioned centrally within the test space, with the "full light" category plants placed directly beneath the spotlight positions (Fig. 5). This setup enabled precise control and measurement of lighting conditions, ensuring accurate and reproducible results throughout the experimentation process.

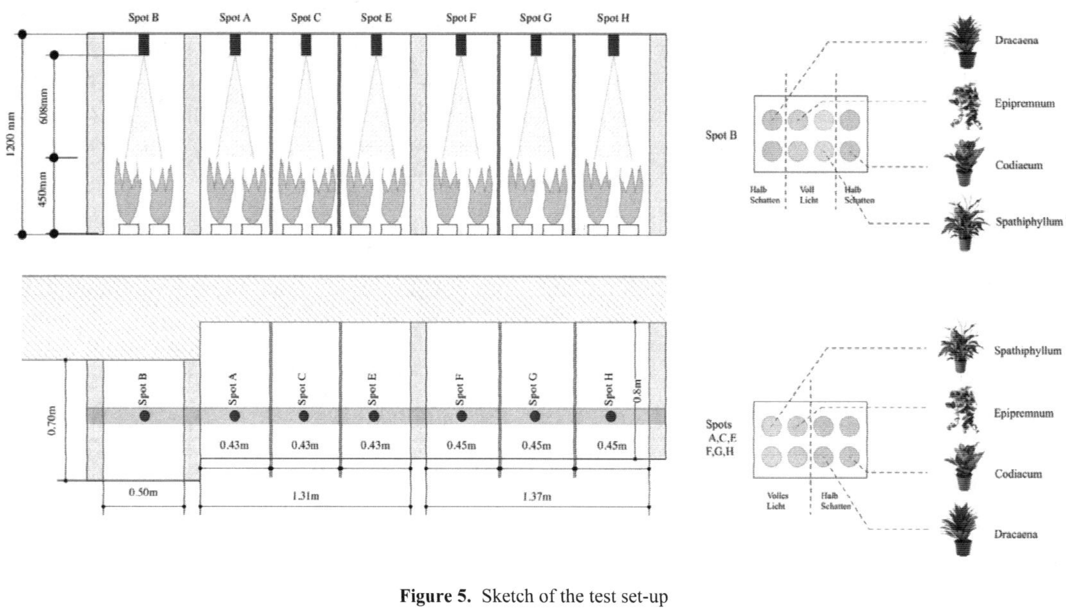

Figure 5. Sketch of the test set-up

3.3. Design of the LED-Chips

The LED chips used for the experiment were custom designed using arrangements of LED modules that were already part of a manufacturer's catalogue for horticultural lighting. Seven distinct LED chip configurations were created to achieve specific values of the investigated lighting parameters.

The design of the LED chips considered the red/blue and red/green ratios, as well as the PPFD values. Each LED chip was composed of 12 LED modules to attain the targeted illuminance levels.

IOP Conf. Series: Earth and Environmental Science 1320 (2024) 012018 doi:10.1088/1755-1315/1320/1/012018

Moreover, individual control of each LED module was enabled through a DMX control system. By independently adjusting each module, the final spectrum of each LED chip was achieved to meet the desired parameter values. This custom-designed LED chip layout allowed for precise control over the light spectrum, facilitating accurate exploration and evaluation of the effects of different lighting conditions on the test plant species.

3.4. Measurements and documentations

The different parameters were assessed at regular intervals of two weeks, specifically on the top surface of the plants, positioned at 43 cm from the floor level and directly beneath the spotlight. The measurements were conducted using a spectrometer[1]. The stability of the spotlights' spectra was examined through the measurement of several key parameters, including PPFD, R/B ratio, R/G ratio, illuminance, and CCT. In addition to the lighting conditions, humidity and temperature were measured inside the test boxes to take into account their possible influence on the plants' health status. Temperature shifted from 27°C in summer to 20°C in winter, whilst the humidity stayed fairly constant throughout the duration of the experiment at values of 40-50%.

Furthermore, visual documentation of the plants' development and growth was carried out. Photographs of each group of plants were taken every two weeks, allowing for a comprehensive record of their progression over the course of the experiment (Fig.6). This combination of quantitative data and visual observations ensures a robust evaluation of the effects of different lighting conditions on the selected ornamental plant species, providing valuable insights for the research analysis.

Figure 6. Pictures of the plants under the different lighting arrangements. From top left to bottom right: Spotlight A, Spotlight B, Spotlight C, Spotlight E, Spotlight F, Spotlight G, Spotlight H.

4. Lighting Experiment – Phase 1

To establish the experiment's parameters, we started with the metrics usually considered by lighting designers and architects for indoor lighting, namely lux levels and CCT. It was decided to explore two

[1] Spectrometer and Flickermeter Lighting Passport Pro Standard, Asensetek, Model ALP-01

IOP Conf. Series: Earth and Environmental Science 1320 (2024) 012018 doi:10.1088/1755-1315/1320/1/012018

illuminance levels, 2,000-2,500 lx and 800-1,000 lx, along with a CCT range of 2800K-5000K, commonly used in various indoor settings.

During the initial phase of the experiment, seven different lighting configurations were designed: four spotlights achieving an illuminance level of 2,000-2,500 lx on the plants' leaves (High output spotlights) and three spotlights achieving an illuminance level of 800-1,000 lx (Low output spotlights). Through multiple tests, it was observed that specific ranges of red/blue (R/B) and red/green (R/G) ratios created a pleasant and natural visual effect on the plants, while maintaining the targeted CCT range. Optimal visual appearance was achieved with R/B ratio < 5.0 and R/G ratio < 2.0, as higher values led to a warm and unnatural visual effect, unsuitable for the desired plants' aesthetics.

To ensure consistent light exposure for all plants, the LED modules were tuned to achieve similar PPFD values for both light levels.

- The spotlights producing 2,000-2,500 lx resulted in approximately 30.0 $\mu mol \cdot m^{-2} \cdot s^{-1}$ PPFD.
- The spotlights producing 800-1,000 lx resulted in approximately 16.0 $\mu mol \cdot m^{-2} \cdot s^{-1}$ PPFD.

The design of the LED chips aimed to maximize spectral variation, incorporating different R/B and R/G values to investigate their impact on plant growth and development.

5. Lighting Experiment – Phase 2

After a 5-month duration of the experiment, significant differences in plant growth became evident between the specimens exposed to light levels of 2,000-2,500 lx and those at 800-1,000 lx. The former group exhibited excessive and too rapid growth for two of the plant's species and a prompting consultation with a botanical expert confirmed that these plants were indeed experiencing unhealthy growth, characterized by elongated and oversized leaves, probably weak roots, and overall abnormal appearance. Therefore, it was concluded that an illuminance level of 2,000-2,500 lx (with a PPFD of 30.0 $\mu mol \cdot m^{-2} \cdot s^{-1}$) was not beneficial to the normal development of these plant species used in the experiment.

Subsequently, the project entered its second phase, where the light output of the High Output spotlights was decreased to 800-1,000 lx while maintaining an unchanged spectrum. Phase 2 then involved comparing the same seven spotlights with similar illuminance level and PPFD values (800-1,000 lx and 16.0 $\mu mol \cdot m^{-2} \cdot s^{-1}$, respectively), while preserving the R/B and R/G ratios determined during Phase 1. Table 1 shows the values of the parameters set for Phase 2. This phase of the experiment aimed to evaluate how the modified lighting conditions influenced the growth trajectory and health of the 4 group of plants (Spot A, B, C and G) that experienced a higher light intensity during Phase 1.

	Spot A	Spot B	Spot C	Spot E	Spot F	Spot G	Spot H
PPFD (μmol.m^-2.s^-1)	15.66	15.8	15.85	15.1	10.9	14.3	12.5
DLI	1.35	1.37	2.25	1.31	0.94	1.24	1.05
R/B	4.5	2.7	2.3	4.5	4.5	2.8	1.6
R/G	1.3	0.9	0.8	1.3	0.8	0.8	0.7
Illuminance (lx)	970	1120	1147	823	695	1050	877
CCT (K)	2985	3344	3682	3460	2901	3540	4514

Table 1. Values of the relevant parameters set during Phase 2 of the experiment.

6. Roots Analysis

After a comprehensive 11-month testing period, an analysis of the root systems was deemed necessary to verify whether the visible health status of the plants corresponded to their non-visible part. A botanical expert was consulted again to carefully extract the plants from the soil, enabling a thorough examination and assessment of the root health.

Each plant underwent a comprehensive evaluation of its roots, leaves, and overall health condition, resulting in a rating system ranging from Very Bad (indicating very poor health status) to Bad, Acceptable, Good, and Very Good (representing excellent health status). Table 2 presents the final evaluations provided by the botanical expert, offering a comprehensive overview of the health status of all plants based on this assessment, whilst Figure 7 shows the diverse roots situation of a few specimens. This critical analysis of the root health provides valuable insights into the overall health and well-being of the ornamental plants, potentially influencing future design and care considerations for indoor greenery.

		Spot A	Spot B	Spot C	Spot E	Spot F	Spot G	Spot H
Codiaeum Petra	Roots Health Status	Bad	Good	Acceptable	Bad	Acceptable	Acceptable	Good
	Foliage Health Status	Very Bad	Good	Acceptable	Acceptable	Acceptable	Good	Good
	Overall Health Status	Bad	Good	Acceptable	Bad	Acceptable	Acceptable	Good
Dracaena Janet	Roots Health Status	Very Good	Very Good	Very Good	Very Good	Very Good	Very Good	Very Good
	Foliage Health Status	Good	Good	Good	Very Good	Very Good	Good	Good
	Overall Health Status	Good	Good	Good	Very Good	Very Good	Good	Good
Epipremnum	Roots Health Status	Good	Good	Acceptable	Acceptable	Very Good	Very Good	Very Good
	Foliage Health Status	Good	Acceptable	Acceptable	Very Good	Very Good	Good	Very Good
	Overall Health Status	Good	Acceptable	Acceptable	Good	Very Good	Good	Very Good
Spathiphyllum	Roots Health Status	Very Bad	Good	Good	Acceptable	Very Bad	Very Good	Acceptable
	Foliage Health Status	Very Bad	Very Good	Good	Acceptable	Very Bad	Good	Good
	Overall Health Status	Very Bad	Good	Good	Acceptable	Very Bad	Good	Acceptable

Table 2. Results of the health assessment conducted in collaboration with the botanical expert.

Figure 7. Examples of roots status for the Spathyphyllum (left side) and the Epripremnum (right side). The first and second images represent respectively roots in good and bad/acceptable conditions.

7. Results discussion

The analysis of the data presented in Table 2 reveals significant variations in the responses of different plant species to the various lighting arrangements. The Dracaena Janet displayed robust growth and development across all lighting configurations, maintaining a healthy pace. Minor leaf signs of burnings or water deficiency were observed, but overall, the plants remained in excellent condition. The Epipremnum demonstrated more variability in response to the lighting conditions, with initial rapid growth followed by favorable development and good health status. In contrast, the Codiaeum Petra exhibited heightened sensitivity to the lighting arrangements, as only two configurations led to good health status, while two had detrimental effects, particularly affecting root development. Similar findings were observed for the Spathiphyllum.

In further examination, the health status of each plant species in relation to each spotlight configuration was analyzed, correlating the root analysis results with the lighting parameters of each spotlight. The objective was to identify lighting configurations that consistently promoted good health status across all plant species. However, no single configuration resulted in excellent status for all four plant species. Nevertheless, four spotlights (Spot B, Spot C, Spot G, and Spot H) demonstrated favorable outcomes, yielding acceptable to very good health status across all four plant species.

Particularly, an interesting trend emerged upon inspecting the spectra of these identified configurations. All four spotlights displayed a R/B ratio lower than 3.0 and a R/G ratio lower than 1.0, indicating that these specific ratios might be crucial in promoting optimal growth and health among the diverse set of ornamental plant species under investigation. Further exploration of these spectral characteristics could offer valuable insights for designing lighting strategies that cater to the health and well-being of indoor greenery in various architectural settings.

8. Conclusions

The primary objective of this project was to establish that light sources with standard CCTs in the 2800K-5000K range can support the healthy development of plants, given careful tuning of their spectrum to provide for plants' specific needs. Through a year-long experiment, it was possible to identify several spectra leading to good health status across all studied plant species while maintaining a pleasing and natural visual appearance. Notably, specific ranges of R/B and R/G ratios indicating favorable impacts on plant health and development were identified.

Furthermore, the research provided evidence that a lower light level of 1,000 lx and a PPFD value of approximately 16.0 $\mu mol \cdot m^{-2} \cdot s^{-1}$ are sufficient to facilitate a healthy growth of the investigated plant species. These findings offer valuable insights into optimizing lighting conditions for indoor ornamental plants, potentially influencing lighting design strategies to promote plant well-being and aesthetics within various architectural spaces.

Once the experiment was completed, it was possible to come up with some guidelines for architects and designers approaching lighting design for indoor plants:

- *Use the right tools*. The quantity and quality of light for plants cannot be assessed through lumens or lux levels using a normal luxmeter, but it must be assessed using a quantum meter able to measure the DLI and the PPFD, ideally in the wavelengths range 300-800nm. CCT is relevant to assess the visual appearance that plants will have together with the surrounding space, but it cannot be used to define the spectrum illuminating the plants. To define the spectrum, it is necessary to refer to the R/B and R/G ratios, along with the PPFD value.
- *Understand your client*. It is important to identify which plants species are going to be used in the space, to understand which needs they have in terms of light, other than water, air and soil. As demonstrated by this research project, the same lighting configuration could be detrimental for some plants species and entirely effective for other species.
- *Understand the context*. Depending on the function of the space in which the plants are integrated, a specific color temperature of light may be preferred to create a harmonious and

visually appealing space. It is possible to use a specific color temperature - in the 2800-5000K range - as long as the spectrum is carefully designed. It must be considered, though, that for the same CCT - and thus overall visual effect - it is possible to design spectra that will have different impacts on plants, by changing the proportion of the radiation that makes up the spectra themselves. Spectra with the same CCT could increase or decrease the growth rate, allow, or not allow flower development or lead to more or less compact growth, just to mention a few. According to on-site needs, a specific spectrum of light will need to be selected.

- *It's possible to be sustainable.* As proven by the research project, plants can grow healthily under lower light intensity, if the light spectrum is carefully tailored on their needs. Using the right light spectrum allows the optimization of the light output and thus the reduction of the light intensity needed for plant subsistence, and consequently enables significant energy savings. In the absence of daylight, an amount of light of 800-1000 lx from the right spectrum can ensure proper plant development. It's important to highlight that this value refers to an arrangement of accent lighting on the plants, in a confined environment where the test spotlights were the only source of light in the room. In a real setting, with spill lighting from the surroundings and possible presence of daylight, this value can be further reduced, and further energy savings can be achieved.

Further investigation and refinement of the identified spectra and light parameters may lead to more tailored lighting solutions for enhancing indoor greenery in the built environment. Illumination of ornamental plants indoors is a complex and interdisciplinary topic, involving daylight planning, building physics, interior design, and a collaboration between all the disciplines is necessary to find the optimal solutions.

References

[1] Kellert SR, Heerwagen JH, Mador M. Biophilic Design: The Theory, Science and Practice of Bringing Buildings to Life. Hoboken: John Wiley & Sons, 2008.
[2] Kellert SR. Nature by Design: The Practice of Biophilic Design. New Haven: Yale University Press, 2018.
[3] Beatle T. Handbook of Biophilic City Planning and Design. Washington DC: Island Press, 2016.
[4] Sharma P. Vertical Gardens: An Innovative Element of Green Building Technology. In: International Conference Gyanodaya 2015. Greater Noida, India. 2015;2–8.
[5] Grinde B, Patil GG. Biophilia: does visual contact with nature impact on health and well- being? Int J Environ Res Publi§c Health. September 2009; 6(9):2332–2343.
[6] Neuwenhuis M, Knight C, Postmes T, Haslam SA. The relative benefits of green versus lean office space: Three field experiments. Journal of Experimental Psychology. 2014;20(3):199-214.
[7] Hosey L. The Shape of Green: Aesthetics, Ecology, and Design. Island Press, 2012.
[8] Bakker I, Van der Voordt DJ. The influence of plants on productivity: A critical assessment of research findings and test methods. Facilities. 2010; 28(9/10):416-39.
[9] Allen JG, Macomber JD. Healthy Buildings How Indoor Spaces Drive Performance and Productivity. Harvard University Press, 2020.
[10] McCormick K. The Business Case for Healthy Buildings. The Urban Land Institute, 2018.
[11] Khan AR, Younis A, Riaz A, Abbas MM. Effect of interior plantscaping on indoor academic environment. Journal of Agriculture Research. 2005;43: 235–42.
[12] Dravigne A, Waliczek TM, Lineberger RD, Zajicek JM. The effects of live plants and window views of green spaces on employee perceptions of job satisfaction, HortScience. 2008; 43: 183–7.
[13] Afroz Nitu M, Grocer O, Wijesooriya N, Vijapur D, Candido C. A Biophilic Design Approach for Improved Energy Performance in Retrofitting Residential Projects. Sustainability. 2022;14(7): 3776.

[14] Shrestha R, Bahadur Bajracharya S, Integrating Energy-Saving Strategies with Biophilic Design Concepts to Increase Thermal Comfort. 12th IOE Graduate Conference, October 2022; V.12.
[15] Wijesooriya N, Brambilla N. Bridging biophilic design and environmentally sustainable design: A critical review. Elsevier. 2021; V.283.
[16] Yin J, Zhu S, MacNaughton P, Allen JG, Spengler, JD. Physiological and cognitive performance of exposure to biophilic indoor environment. Build. Environ. 2018; 132:255-62.
[17] The International Well Building Institute. The WELL Certification Guidebook; 2022 Jun [cited 2023 July 28]. Availabe from: https://resources.wellcertified.com/tools/well-certification-guidebook/
[18] Terashima I, Fujita T, Inoue T, Chow WS, Oguchi, R. Green light drives leaf photosynthesis more efficiently than red light in strong white light: revisiting the enigmatic question of why leaves are green. Plant Cell Physiol. 2009;50(4):684-97.
[19] Hyeon-Hye, K, Wheeler RM, Sager JC, Yorio N, Goin GD. Light-Emitting Diodes as an Illumination Source for Plants: A Review of Research at Kennedy Space Center. Habitation. 2005; 10(2):71–8.
[20] Huner N, Dahal K, Hollis L, Bode R, Rosso D, Krol M, Ivanov A. Chloroplast redox imbalance governs phenotypic plasticity: the "grand design of photosynthesis" revisited. Frontiers in Plant Science. 2012; 3:1-12.
[21] Ouzounis T, Rosenqvist E, Ottosen CO. Spectral Effects of Artificial Light on Plant Physiology and Secondary Metabolism: A Review. HortScience. 2015;50:1128–35.
[22] Mawphlang OI, Kharshiing EV. Photoreceptor Mediated Plant Growth Responses: Implications for Photoreceptor Engineering toward Improved Performance in Crops. Frontiers in Plants Science. 2017.
[23] Kong SG, Okajima K. Diverse photoreceptors, and light responses in plants. J Plant Res. 2016;129: 111–4.
[24] Bugbee B. Toward an optimal spectral quality for plant growth and development: the importance of radiation capture. Plants, Soils, Clim. Fac. Publ. Paper. 2016;763.
[25] Trivellini A, Toscano S, Romano D, Ferrante A. LED Lighting to Produce High-Quality Ornamental Plants, Plants. 2023;12(8):1667.
[26] Zielinska-Dabkowska KM, Hartmann J, Sigillo C. Biophilic Design – the Application of Ornamental Indoor Plants in Sustainable, Green Architecture LED Light Sources and their Complex Set-up for Visually and Biologically Effective Illumination, Sustainability. 2019; 11(9): 2642.
[27] da Costa GJ, Cuello JL. The phytometric system: A new concept of light measurement for plants. Journal of the Illuminating Engineering Society. 2004;33(1):34-42.
[28] Erco. The perfect lighting for green walls. 2022. [cited 2023 July 28]. Available from: https://lightfinder.erco.com/en/contentaccess/page/download/media+whitepaper_green_wall
[29] Lighting Research and Design. Dugar AM. Lighting Green Walls – finding the optimum CCT and SPD of white LED light sources. 2021. [cited 2023 July 28]. Available from: https://www.luxendi.com/content/files/luxendi/news/Lighting%20Green%20Walls,%20Xicato.pdf
[30] Dugar AM, Dileep M, Burhani H. White LED Light Sources – Merging Architectural and Horticultural Lighting Applications within Interior Environments. Int J Hortic Crop Sci Res. 2019; 9(2):83–93.

IOP Conf. Series: Earth and Environmental Science 1320 (2024) 012019 doi:10.1088/1755-1315/1320/1/012019

Skynative – brings daylight into the depths of buildings

Martin Hauer[a], Christian Knoflach[a], Elise Kissling[b], Lutz Mehlhorn[a] Judith Gross[a]

[a] Bartenbach GmbH, research and development, Rinner Straße 14, Aldrans, Austria
[b] Skynative UG, Diakonissenstrasse 62, Speyer, Germany

martin.hauer@bartenbach.com

Abstract. There are various ways of directing daylight into the interior of a building or a room. In cooperation with BASF, Bartenbach has developed a prototype of a horizontal daylight duct integrated into the suspended ceiling that efficiently transports daylight from the outside into the underexposed "core" of a building. "Skynative" is the name of the system, named after BASF spin-off Skynative® UG, which is working on market integration as well as continuous improvement of the system in an open research collaboration with Bartenbach. In this paper, the system concept development, simulation-based application studies as well as the setup of a real-scale mock-up, including results from a long-term performance monitoring at the premises at Bartenbach, are presented. First potential application projects are shown, and the system efficiency is discussed.

1. Introduction

People spend more than 90% of their time indoors, often without direct access to daylight from windows [1]. The problem increases in urban areas due to deeper building floorplans and urban compaction. Furthermore, the EU is driving a drastic increase in the refurbishment rate of existing building stock to help achieve its climate goals of reducing greenhouse gases by 55% by 2035 and reaching climate neutrality by 2050 [2]. In Germany, 64% of the existing building stock was built before 1978 [3]. These buildings usually have a significant lack of quality and quantity of daylight. Revitalization of unused buildings (industry halls, retail centers...) is a vastly growing market, representing 22% of the overall office construction market. And finally, the COVID pandemic has had a massive impact on working environments towards creating inclusive workplaces (e.g. modular seating solutions, desk sharing, less density) and changed therefore the usage concepts for new and existing buildings – also leading to a new focus and requirements on revitalization [4].

At the same time, a growing body of research shows that daylight plays a critical role in human health, wellbeing and productivity [5]–[8]. Lack of daylight negatively influences people. It verifiably causes short-term issues like low concentration, productivity and a disrupted circadian rhythm [9]. Lack of daylight also causes significant long-term health problems such as seasonal affective disorder, fatigue, and stress as well as slow recovery, especially at intensive care units [10]. EU regulatory authorities and independent green building certification agencies are therefore pushing higher daylighting standards (e.g. EN-17037) and supports building certification schemes (e.g. DGNB, WELL, BREEAM, LEED) to promote human health effects through daylight in recent years.

So how to bring daylight effectively and efficiently deep into the interior of a building so that these areas also receive daylight and can manage without electric light during the day? Although the electric lighting industry seeks to position the LED as a daylight-equivalent light source, research reveals that electric light does not provide the same health and productivity benefits as natural sunlight [11]. Consequently, what the market lacks is a daylighting system that enhances daylight levels in the building core, independent of the location and in an optimal balance with electric light via a human-centered control approach.

Especially the so-called "central zones" of a building, which on average account for around one third of the net floor area, are often poorly supplied with daylight and correspondingly dark. Due to normative requirements, hardly any workplaces are found here. Valuable and often cost-intensive areas are thus lost to companies and economic use of the space is hardly possible. Especially in times of the "new work" movement, this is a major aspect, as interior, open spaces in the core area of an office floor plan are increasingly being used as exchange and networking zones across professional and organizational boundaries [12].

The desire for optimal use of space from the occupant's and company's point of view and dealing with a growing urban development density requires a rethinking of daylighting strategies. Systems that direct daylight from the exterior façade into the interior offer an important approach to solving this problem - and therefore the author team developed and tested a daylighting system prototype, referred to under the term Skynative.

A horizontal daylight duct efficiently transports daylight from the façade "invisible" through the suspended ceiling into the interior of buildings. Thus, the Skynative prototype makes it possible to improve the visual indoor quality, promote human health by using daylight, and make numerous rooms usable through daylighting.

Besides new built structures, Skynative can bring benefits especially for the retrofit and revitalization of old buildings, which must be increased drastically within the next years to achieve the climate protection targets. In contrast to most vertical, roof-installed light tubes available on the market (ref. chapter 1.1), Skynative utilizes a horizontal duct system across the façade, which is easier to apply as a retrofit solution. Especially in old industry halls, which are increasingly being revitalized into office buildings, there are often large room heights, which makes a subsequent integration of Skynative very advantageous. However, Skynative not only offers potential benefits in terms of occupant comfort and satisfaction but can also contribute to the building owner's financial interests and help achieve environmental goals.

1.1. State of the art and scientific literature review

On the market, so called "light tubes" (also known as sun tubes) have been established as one variant of daylighting systems for many years and are offered by well-known companies such as Velux, Solatube, Suntek and SunPipe, to name a few. Kingspan Light + Air offers a system of a combined daylighting and air tube solution to enhance natural light and fresh air in buildings within one system. This solution attempts to optimally solve the problem of simultaneous integration of both technologies in the very limited space of a suspended ceiling, which has also proven to be one of the biggest challenges in the design of horizontal light ducts.

But also in research, several works have been published presenting new concepts on daylighting duct concepts. In the following review, a special focus is given to horizontal duct systems.

Just recently, Mesloub et al. [13] published a feasibility study on the potential for visual comfort, overall energy consumption and the economic feasibility of tubular daylighting systems in deep plan offices in the MENA countries. This work also includes a keyword-based bibliometric analysis to extract relevant publications and research done on tubular daylighting in the period of 2010 – April 2022. Thus, most of the research work was focussing on the performance of light pipe design, followed by the energy efficiency and their impact on energy savings achieved.

Numerous works on lighting duct concepts have been done by Obradovic et al. [14]–[16]. In an initial paper, Obradovic et al. invested in a literature study on daylight transport systems spanning from the 80's to 2019 and evaluates their most appropriate application at high latitudes [15]. A clear distinction is made to system applications at equatorial latitudes, as most studies are published from this region. As a conclusion, it is highlighted that the local daylighting conditions (clear vs. overcast conditions, sun angle, façade orientation...) are most essential for deciding about the type of collector, while building structure, room geometry and functionality are decisive factors for choosing the right transport element.

In Obradovic et al. [14], horizontal light pipes in specific are analysed in terms of their daylight efficiency, especially for high sun angles. By attaching light-deflecting panels on the duct entry element, the potential for daylighting performance improvement was analyzed by reducing the number of interreflections in the lighting duct. The result showed that depending on the length of the pipe, an improvement in daylight autonomy (DA_{300}) of an indoor space between 16% and 19% is possible.

As a continuation of the previous mentioned studies, Obradovic et al. investigated the horizontal light pipe concept studied in their previous works in a full-scale test office to analyse illumination levels and lighting energy use [16]. The chosen room, within an office building located in southern Norway, included southwest-oriented windows for natural daylighting via the façade and a horizontal light pipe with daylight entrance facing south. The focus in the case study was to evaluate the improvement of daylight levels. As a result, the reduction of the electric light demand especially in the rear part of the office could be shown because of the daylight provision by the daylight duct. The study included a monitoring of the outdoor and indoor situation and could be used for further theoretical studies. The presented study is indeed a good supplement to the monitoring results gained from the Skynative mock-up in this paper.

Roshan and Barau presented a study on applying an anidolic daylighting system under tropical climate conditions [17]. Mayhoub et al. addressed in their work one of the major challenges to integrate a horizontal lighting duct with other building service ducts like ventilation or air conditioning [18]. They compare the differences and similarities of the different ducting systems and proposes a schematic design on an innovative dual ducting system, which allows the avoidance of conflicts between the building systems and simplifies the installation process. Lu et al. took a more general look and explored multi-dimensional evaluation methods for suitability of applying light duct systems [19].

2. Skynative – Design and function

There are various ways of transporting daylight into the interior of a building or room: In cooperation with BASF, Bartenbach has developed from 2017 to 2019 the concept of a horizontal daylight duct that efficiently guides daylight from the outside into the interior of a building.

In 2019, BASF outsourced the development and founded the start-up Skynative, which has since been marketing and further developing the system together with Bartenbach. "Skynative" means as much as "natural sky". The daylight-directing system can be used where conventional daylighting via the facade reaches its limits.

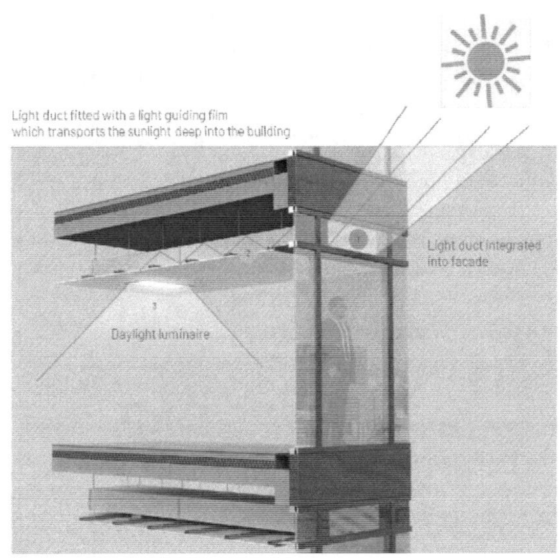

Figure 1: Schematic functionality and components of the Skynative® daylighting system

2.1. *"Skynative" – Functionality and setup*

The light guiding system consists of three components: incoupling element, light duct, and decoupling technology, shown in Figure 1. The daylight entry surface is equipped with a daylight-redirecting film from 3M and is positioned directly above the transparent window of the conventional building façade.

The redirecting film enables that for a broad variety of sun positions, direct sunlight can be transported as effectively as possible through the duct with a reduced number of interreflections. By this, it is possible to transport daylight up to 20 meters into the depth of the building. Depending on the height of the duct, even longer distances could be achieved, but at least 30 centimeters of installation height are required for a proper functionality of the Skynative daylighting system.

3. Mock-up installation and monitoring setup

To test, experience and demonstrate the technology, a real-scale mock-up as system prototype has been installed at the premises of Bartenbach in Aldrans, Austria in 2017 (Figure 2). In addition to providing a real demonstration for future clients, from a research point of view it was the goal to obtain realistic observation data from the mock-up. Therefore, an office room was recreated with a typical furniture and room design.

In Figure 3 a panoramic view of this office room is shown. The external reference ("view outside") is established as a virtual window with an HD screen that mimics the real environment (seen on the right of the image: showing a live video of the external situation in the same viewing direction). Thus, the available daylighting in the room is provided exclusively by the integrated Skynative light ducts in the ceiling.

Figure 2: Installed prototype of the "Skynative" daylighting system at Bartenbach in Aldrans, Austria

Figure 3: Daylit office mock-up through the "Skynative" daylighting system and view out via the virtual window

For research purpose, the two ducts installed were equipped with different reflective materials:

- Duct 1: 3M Specular Film DF2000MA
- Duct 2: Alanod PRIME SILVER 27 DL

The product „PRIME SILVER 27 DL" from Alanod is an aluminium sheet with special coatings to provide high light reflectivity (>98%) and UV-stability. In contrast, the 3M Specular Film DF2000MA is a polymeric film providing specular reflection with high light reflectivity, that is metal free and noncorroding. The purpose of applying two different materials was to test to different application methods (coated aluminium vs. polymeric film) and to obtain comparisons in their optical efficiency, a possible spectral shift and long-term stability. While the 3M-film resulted in a Melanopic Daylight Efficacy Ratio MDER close to 0,9, the Alanod aluminium sheet showed a spectral shift of the reflected light, resulting in a MDER about 0,7.

Table 1: Reached MDER values for the different duct materials

Melanopic Daylight Efficacy Ratio MDER	Outside, vertikal	Duct 2 (3M)	Duct 2 (Alanod)
Clear sky	0.900	0.889	0.700
Overcast sky	0.895	0.878	0.682

The two ducts were sealed on the outside with insulating glazing (Optiwhite) and covered on the inside with a 3M Daylight Redirecting Film. To evaluate the system behaviour in the long term, illuminance levels have been measured indoors and outdoors since October 2018. In addition, luminance cameras were put into operation in January / February 2019. The results of the monitoring at Bartenbach presented in this paper lasted until the end of December 2019.

3.1. Monitoring setup

As part of the monitoring process, illuminance levels are measured at various positions (Figure 4). In addition, fisheye photos are taken, from which luminance images are subsequently created to obtain information about the luminance distributions. The measurement of the illuminances as well as the creation of the luminance images is carried out in the exterior area as well as in the interior area.

The illuminance measurements are implemented with an Ahlborn Almemo809 PC interface and corresponding sensors. Two FLA 613 VLM sensors are used outdoors and four FLA 623 VL sensors indoors. The measured values are stored every minute. The positions of the sensors are shown in Figure 4. The sensors S1 and S2 correspond to those in the outdoor area and are positioned horizontally and vertically. Sensor S3 is placed at the center of the ceiling to create an "integrating space" situation and to estimate the luminous flux. Sensors S4 and S5 are mounted on the standing table (110 cm height) and working table (75 cm height) respectively. In addition, a sensor S6 is mounted on the worktable vertically at eye level. On the standing table and the work table, the sensors are positioned directly under a duct opening: on the standing table under the 2nd opening of the 3M duct and on the work table under the first opening of the Alanod duct.

The positioning and orientation of the sensors is based on following questions, to get a benchmark for the system performance (1-3), possible health effects (4-5), and information to quantify the visual appearance in the room (6-7):

(1) What illuminance levels are achieved at the workplace? (Measurement at the workplace)
(2) Can 500 lx be reached at the workplace? (Measurement at the workplace)
(3) How much light in total enters the room through the system?
(4) What daily light dose can be expected? (Measurement at eye level - vertical)
(5) Occurring luminance in the room? (Luminance camera with fisheye lens)

(6) Sunspots and structures in space? (Luminance camera with fisheye lens)
(7) Dynamics of the light situation and influence of the outside situation? (webcam / L-cameras)

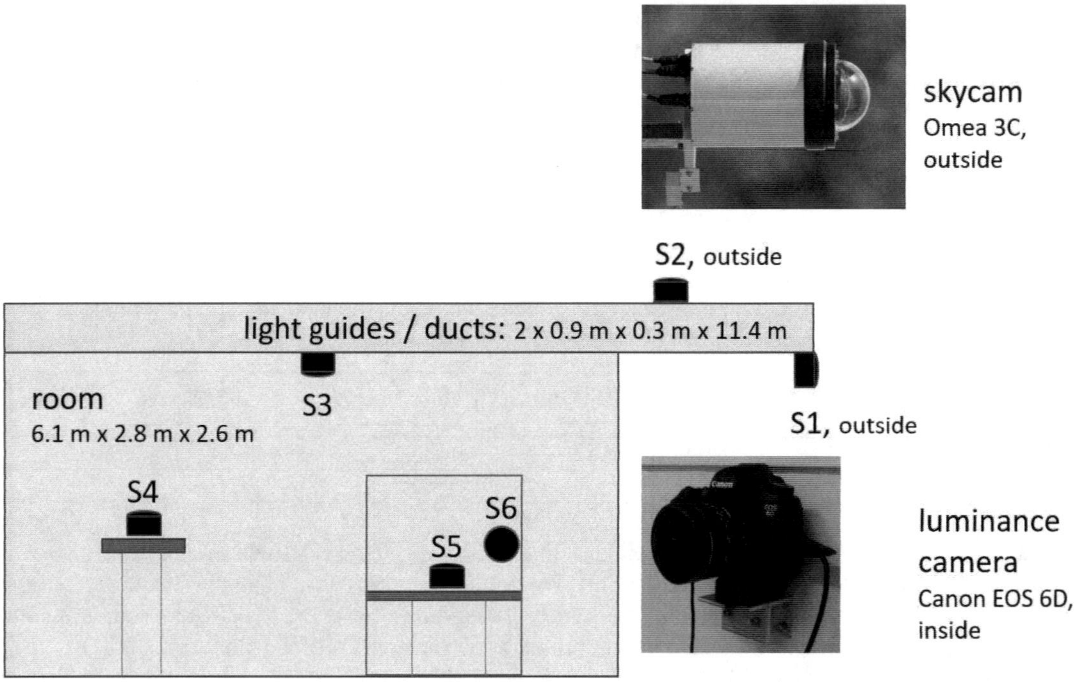

Figure 4: Schematic arrangement of lux sensors (S1 - S6), L-camera and skycam in outdoor and indoor areas

Outdoor measurements are mainly investigated in order to quantify the monitoring boundary conditions as well to serve input data for the simulation models.

This includes:
(1) Horizontal illuminances? (Illuminance levels horizontal – S3, S5)
(2) Illuminance levels directly at the shaft entrance? (Illuminance levels vertical – S1)
(3) Influence of the outdoor situation on the dynamics of the indoor situation? (Skycam)
(4) Proportional contributions (sun, sky, ground) (Skycam)

4. Monitoring results

By monitoring the "Skynative" prototype, its mode of operation and performance can be measured and evaluated. The challenge is to record and map the dynamic processes of the daylighting system, as it is a key aspect, along with the photometric data, that adds value to the solution.

4.1. Daylight availability

For the 2019 measurement period, average continuous daylight autonomy (cDA) at the standing desk was 58.5% based on 300 lx and 46.8% based on 500 lx. While 300 lx is the target minimum for circulation areas in the building, 500 lx is the target minimum for desk lighting. In Figure 5, the monthly values for the achieved continuous daylight autonomy on the standing desk (S4) are shown.

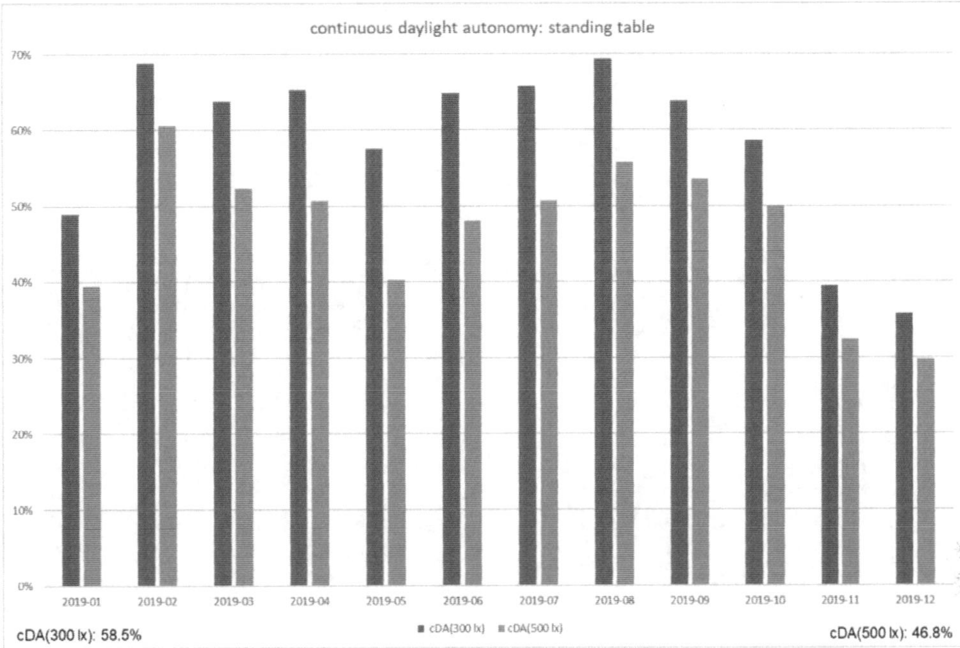

Figure 5: Monthly results for the continuous daylight autonomy at the standing table (S4)

Comparing the collected annual results at the standing desk with the theoretical calculations for different climatic data shows a wide range of variation for the individual months. However, when viewed over the entire course of the year, there is good agreement with the measured values in the monitoring.

4.2. Illuminance levels

In Figure 6, the measured illuminance values on working desk level are shown in one-minute intervals. It can be seen that in the sunny periods with low sun angles (February, October, November) the highest illuminance levels can be achieved.

The higher daylight availability is due both to the generally higher radiation gains in these months on south-facing facades, and because transport efficiency increases due to lower internal reflections in the light channel. The illuminance levels achieved are lowest in the summer months. Nevertheless, the overall level matches the criteria with > 500lx during almost all working hours.

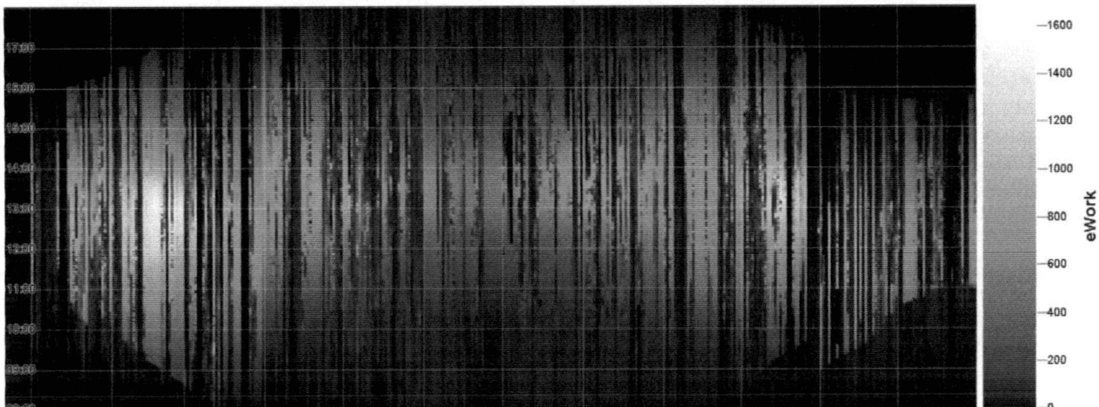

Figure 6: Daylight illuminance levels (in 1-minut interval) on the working desk (S5)

4.3. Glare probability and melanopic equivalent daylight illuminance

While illuminance levels reach peaks especially in months with low sun angles, also highest glare probabilities are measured in this period. Nevertheless, the glare ratings stay within acceptable levels throughout the year (Figure 7). The DGP values determined from the luminance images with values for the DGP_{max} below 0.25 are very moderate. As a comparison, the EN 17037 standard "Daylight in buildings" recommends for high glare protection that a DPG value of 0.35 should not exceed 5% of the usage time.

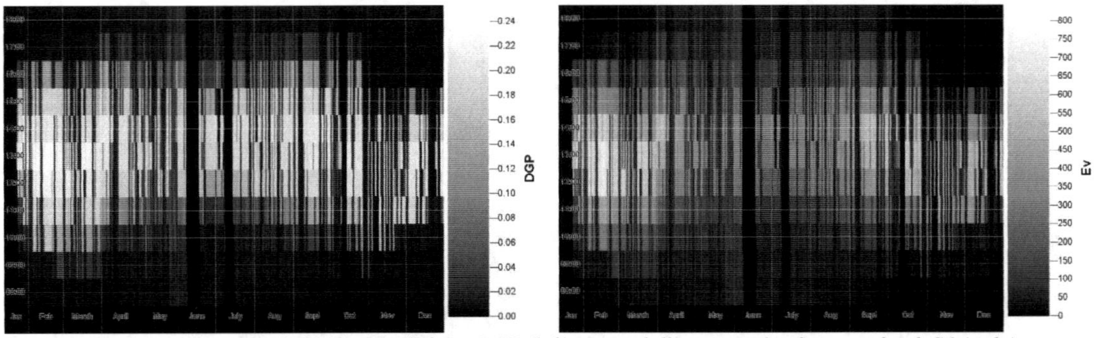

Figure 7: Annual daylight glare probability DGP (left), vertical illuminance levels on eye level, S6 (right)

It should also be noted that the vertical illuminances are often below 380 lx and are thus actually too low for an application of the DPG (compared to the right chart in Figure 7). Also, according to Evalglare, DGP values below 0.2 can underestimate glare. In general, it must also be mentioned at this point that the use of DGP in this context may have some limitations, as this method was originally developed for side windows.

Considering a Melanopic Daylight Efficacy Ratio MDER of 0,889 for the 3M duct according to Table 1, which means comparable to natural sunlight (DER = 0,9) at clear sky conditions, a melanopic equivalent daylight illuminance value (MEDI) of 250 lx vertically at eye level can be achieved for at least four hours at most of the days, especially in the more important winter period, when people working in offices are less exposed to natural sunlight due to the late sunrise and early sunset.

4.4. Daylight dose

As a benchmark for evaluating the potential for non-visual light effects and circadian entrainment of the Skynative daylight system, the daily daylight dose for the working hours of 08:00 to 18:00 was evaluated based on vertical illuminance measured on eye level (S6). A threshold of 5000 lxh is based on recommendations for light therapy [20], [21], where a vertical illuminance of 5000 lx for one hour has a noticeable effect, according to proband studies.
As seen for the horizontal illuminances on work desk level, the highest daylight dose values were reached in the sunny winter period, while the minimum is reached in the summer month due to less radiation gains at the façade as well as higher internal reflections in the light duct (Figure 8). In terms of circadian entrainment, this is an advantageous behavior, as especially in short-day winter months, additional daylight will be beneficial for the daylight supply of office workers, while in long-day summer months, sufficient daylight dose might be reached even outdoors in times before and after working time.

IOP Conf. Series: Earth and Environmental Science 1320 (2024) 012019 doi:10.1088/1755-1315/1320/1/012019

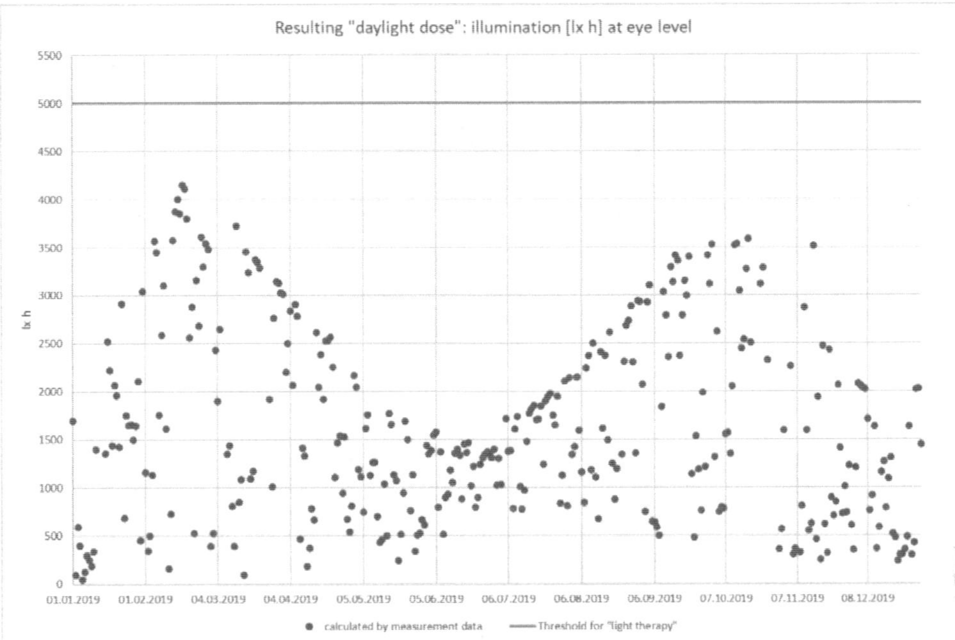

Figure 8: Daily daylight dose levels reached throughout the year of monitoring

5. Design tool and Performance matrix for early-stage decisions

Beside measurements, a simulation model of the duct was setup in Radiance, based on the constructive and optical details of the current mock-up of the prototype. Measured data was used to verify the model in a first step, and to use it for a continuous concept improvement in a second step. The efficiency of the light duct is largely determined by the reflectance of the material. Therefore, the influence of the soiling of the light wells was evaluated in detail by simulations. Likewise, also different coupling variants (with/without light deflecting foil) were simulated.

By further applying the simulation model in performing calculations based on theoretical EnergyPlus Weather Datasets [22], an extended comparison of the simulation results with the gained measurements was possible.

Analysis was performed using an Excel-based design tool, based on pre-calculated simulation data from the simulation model and verified with measurement data from the real mock-up. It is based on a standard room size of 5,4 x 5,4m (double size of the real mock-up) and 4 installed light ducts. The tool estimates, depending on the duct geometry (height and width) and the number of outlets, the maximal possible duct length to achieve a minimum continuous daylight autonomy of 50% on average for the defined room setting. In Figure 9 the outcome is represented in form of a performance matrix, showing the maximal duct length depending on duct width (x-axis) and height (y-axis). The tool allows a fast and easy estimation of the required number of light ducts and the required duct geometry; in addition, the possible length of the light transport can be evaluated. Nevertheless, it cannot replace a detailed engineering of the light duct in a later design stage.

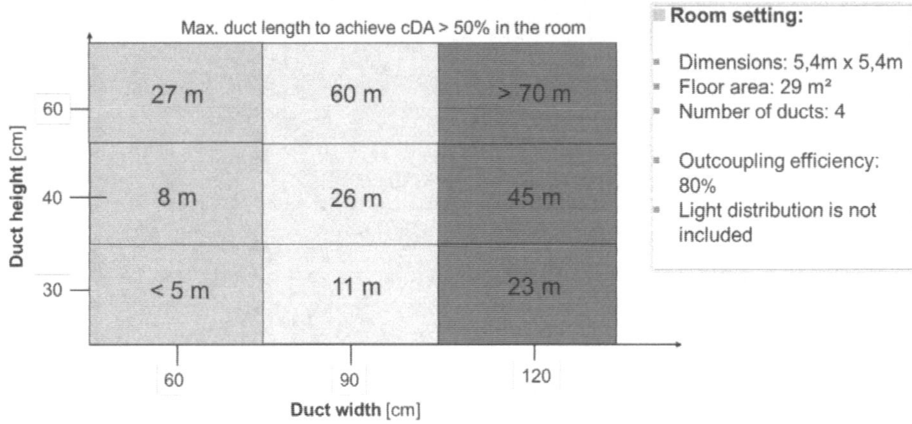

Figure 9: Performance matrix based on the calculation of a reference setting

6. Simulation case study on an open-floor plan

To evaluate the potential and the application of the Skynative daylighting system, especially in combination with a sidelight office setting, a simulation study for a representative floor plan with different room heights (2,75m – 3,75m) was done (Figure 10).

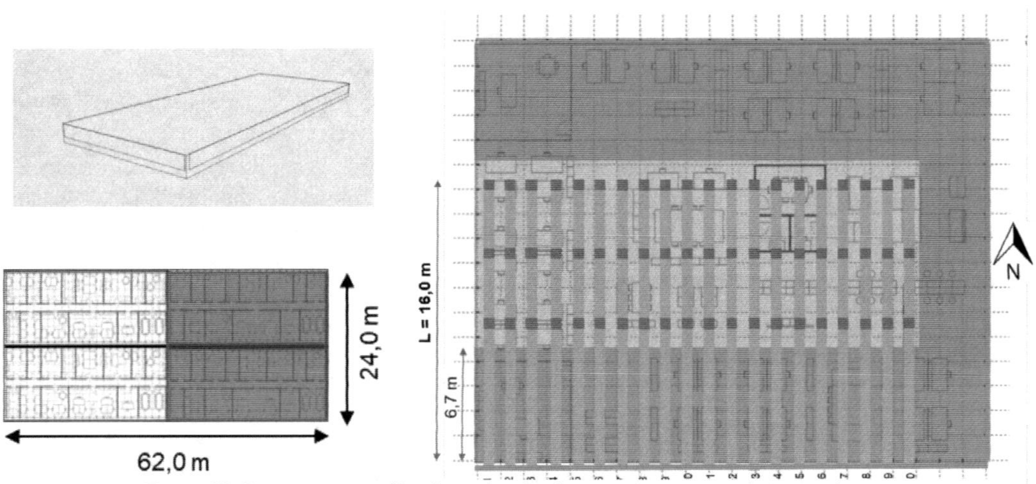

Figure 10: Representative office floor plan for the revitalisation simulation study

For the simulation study. the sidelight windows have been equipped with a shading screen and a daylight-depending control, which activates the shading for an incident solar radiation on the façade of 120W/m2. The façade control was evaluated in two scenarios: an identical operation of the façade in all 4 cardinal directions, and an individual operation.

For the daylight calculation of the floor plan including the sidelight window and the sun shading control, Ladybug tools have been used. The daylight supplementation via the Skynative daylighting ducts have been added on top using the verified simulation model described under chapter 5.

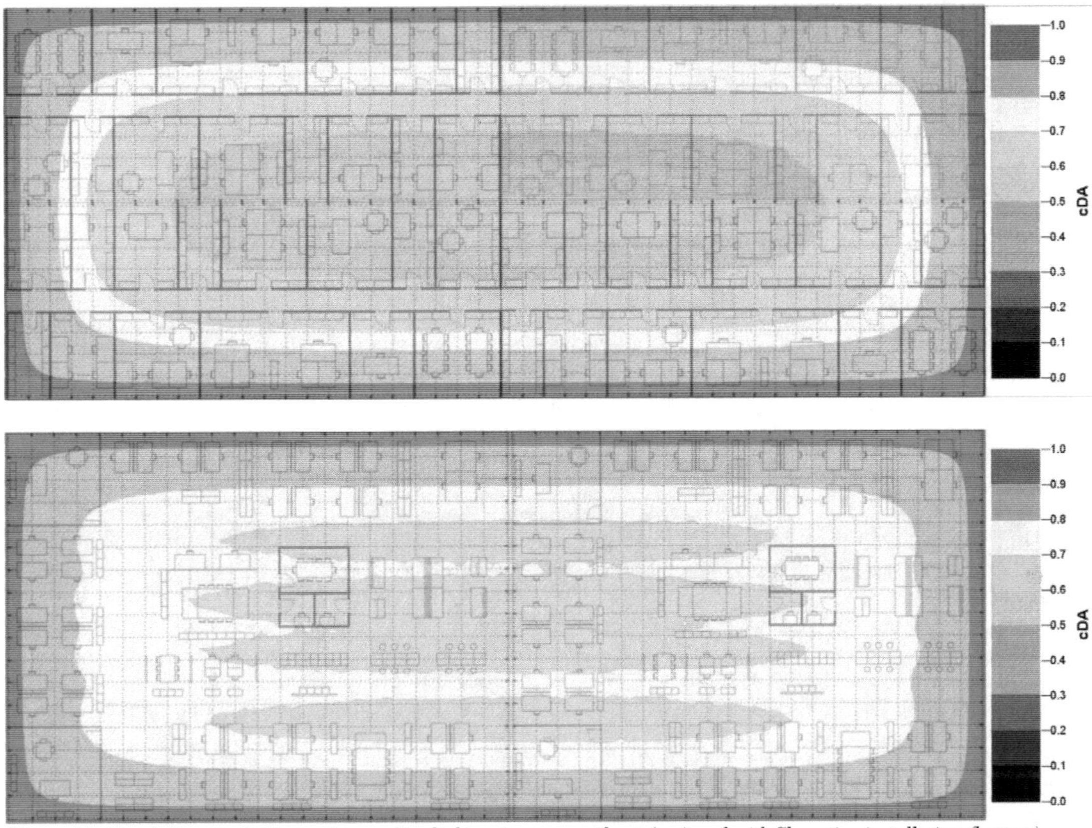

Figure 11: Simulation results in continuous Daylight autonomy: without (top) and with Skynative installation (bottom).

The results for the continuous daylight autonomy in the case of the floor plan daylight simulation without Skynative installation results in an achieved cDA > 50% for 78,1% of the floor area. Especially the core part was clearly below the threshold as seen in the top chart of Figure 11.

By including the supplied additional daylight via the Skynative daylighting ducts, a continuous daylight autonomy of > 60% in a wide range of the core part could be achieved. Thus, a utilization for additional workplaces seems reasonable according to the available daylight levels. The criteria of "view outside" can be still guaranteed by the existing sidelight windows at all four façade directions.

7. Conclusion

This paper shows the concept, the development, and the real-field application of the prototype light duct system Skynative, developed by Bartenbach and BASF and continued by the outsourced start-up with the same name (Skynative). An annual monitoring of a real-scale mock-up has revealed many positive aspects: 'Skynative' reproduces a light spectrum that closely resembles the dynamics of daylight and therefore likely has a similarly activating aspect on people indoors. The fact that daylight dose levels of up to 4,000 lxh per day have been achieved, especially during the winter period, can be evaluated positively. Considering a MDER of 0,9 for the 3M duct, a MEDI-value of 250 lx vertically at eye level can be achieved for at least four hours at most of the days, especially during the winter period. The monitoring results prove that with the use of highly reflective materials with a reflectance > 98.5 % in the light duct, a continuous daylight autonomy of more than 50 % related to 500 lx is possible even in windowless rooms. Orientation and external obstruction play a significant role in the system efficiency, as a maximum number of direct sunlight hours on the duct inlet is required to achieve high performance values as shown in the monitoring as well as the performance matrix in this work.

References

[1] N. E. KLEPEIS *et al.*, "The National Human Activity Pattern Survey (NHAPS): a resource for assessing exposure to environmental pollutants," *J. Expo. Sci. Environ. Epidemiol.*, vol. 11, no. 3, pp. 231–252, Jul. 2001, doi: 10.1038/sj.jea.7500165.

[2] E. K. Velten *et al.*, "State of EU progress to climate neutrality An indicator-based assessment across 13 building blocks for a climate neutral future.," *ECNO Flagsh. Rep.*, 2023, [Online]. Available: https://climateobservatory.eu/sites/default/files/2023-06/ECNO_Flagship report - State of EU progress to climate neutrality_June 2023.pdf

[3] Dena (German Energy Agency), "Dena concise 2018 building report: Energy efficiency in the building stock-statistics and analyses," *ThyssenKrupp techforum*, vol. 2019Januar, no. 1, p. 46, 2018, [Online]. Available: https://www.dena.de/fileadmin/dena/Dokumente/Pdf/9268_dena_concise_2018_building_report.pdf

[4] J. Saunders, C. Brewster, and P. Holland, "The changing nature of work," *Contemp. Work Futur. Employ. Dev. Ctries.*, pp. 1–14, 2020, doi: 10.4324/9781351034906-1.

[5] M. G. Figueiro *et al.*, "Light, entrainment and alertness: A case study in offices," *Light. Res. Technol.*, vol. 52, no. 6, pp. 736–750, 2020, doi: 10.1177/1477153519885157.

[6] M. G. Figueiro, M. Kalsher, B. C. Steverson, J. Heerwagen, K. Kampschroer, and M. S. Rea, "Circadian-effective light and its impact on alertness in office workers," *Light. Res. Technol.*, vol. 51, no. 2, pp. 171–183, 2019, doi: 10.1177/1477153517750006.

[7] R. P. Leslie, "Capturing the daylight dividend in buildings: why and how?," *Build. Environ.*, vol. 38, no. 2, pp. 381–385, Feb. 2003, doi: 10.1016/S0360-1323(02)00118-X.

[8] C. Blume, C. Garbazza, and M. Spitschan, "Effects of light on human circadian rhythms, sleep and mood," *Somnologie*, vol. 23, no. 3, pp. 147–156, Sep. 2019, doi: 10.1007/s11818-019-00215-x.

[9] R. Nagare, M. Woo, P. MacNaughton, B. Plitnick, B. Tinianov, and M. Figueiro, "Access to Daylight at Home Improves Circadian Alignment, Sleep, and Mental Health in Healthy Adults: A Crossover Study," *Int. J. Environ. Res. Public Health*, vol. 18, no. 19, p. 9980, Sep. 2021, doi: 10.3390/ijerph18199980.

[10] M. Engwall, I. Fridh, L. Johansson, I. Bergbom, and B. Lindahl, "Lighting, sleep and circadian rhythm: An intervention study in the intensive care unit," *Intensive Crit. Care Nurs.*, vol. 31, no. 6, pp. 325–335, Dec. 2015, doi: 10.1016/j.iccn.2015.07.001.

[11] M. Canazei, J. Weninger, and W. Pohl, "Human-Centric Lighting (HCL) - eine Zwischenbilanz. ," *LICHT*, vol. 2, 2019.

[12] C. Candido, P. Chakraborty, and D. Tjondronegoro, "The Rise of Office Design in High-Performance, Open-Plan Environments," *Buildings*, vol. 9, no. 4, p. 100, Apr. 2019, doi: 10.3390/buildings9040100.

[13] A. Mesloub *et al.*, "The visual comfort, economic feasibility, and overall energy consumption of tubular daylighting device system configurations in deep plan office buildings in Saudi Arabia," *J. Build. Eng.*, vol. 68, p. 106100, Jun. 2023, doi: 10.1016/j.jobe.2023.106100.

[14] B. Obradovic and B. S. Matusiak, "Daylight autonomy improvement in buildings at high latitudes using horizontal light pipes and light-deflecting panels," *Sol. Energy*, vol. 208, pp. 493–514, Sep. 2020, doi: 10.1016/j.solener.2020.07.074.

[15] B. Obradovic and B. S. Matusiak, "Daylight transport systems for buildings at high latitudes," *Journal of Daylighting*, vol. 6, no. 2. Solarlits, pp. 60–79, Dec. 01, 2019. doi: 10.15627/jd.2019.8.

[16] B. Obradovic and B. S. Matusiak, "Illumination and Lighting Energy Use in an Office Room with a Horizontal Light Pipe: Field Study at a High Latitude," *J. Daylighting*, vol. 9, no. 2, pp. 209–227, Dec. 2022, doi: 10.15627/JD.2022.16.

[17] M. Roshan and A. S. Barau, "Assessing Anidolic Daylighting System for efficient daylight in open plan office in the tropics," *J. Build. Eng.*, vol. 8, pp. 58–69, Dec. 2016, doi: 10.1016/j.jobe.2016.07.002.

[18] M. S. Mayhoub, "Dual ducting: An innovation to increase the use of daylight in buildings," *Light. Res. Technol.*, vol. 47, no. 6, pp. 712–729, 2015, doi: 10.1177/1477153514548472.

[19] S. Lu, Z. Yu, and M. Fan, "Multi-layered and multi-dimensional suitability evaluation of tubular daylight guidance systems," *J. Build. Eng.*, vol. 32, no. May, p. 101820, 2020, doi: 10.1016/j.jobe.2020.101820.

[20] R. G. Lande, L. B. Williams, C. Gragnani, and Albert Tsai, "Effectiveness of light therapy for depression among active duty service members: A nonrandomized controlled pilot trial," *Complement. Ther. Med.*, vol. 19, no. 3, pp. 161–163, Jun. 2011, doi: 10.1016/j.ctim.2011.04.003.

[21] C. Lam and M. H. Chung, "Dose–response effects of light therapy on sleepiness and circadian phase shift in shift workers: a meta-analysis and moderator analysis," *Sci. Rep.*, vol. 11, no. 1, pp. 1–16, 2021, doi: 10.1038/s41598-021-89321-1.

[22] Climate.OneBuilding. All Rights Reserved, "Repository of free climate data for building performance simulation." [Online]. Available: https://climate.onebuilding.org/default.html

Pain Relieving Light – (How) Is it Possible?

Björner Brauer, Hanna, RISE Research Institutes of Sweden, Division Built Environment, SE-41258 Gothenburg, Sweden

Ryberg, Karl, Monocrom, Rörstrandsgatan 40, SE -11340 Stockholm, Sweden.

Wallin, Kjell, Fagerhult Stockholm, Tegelviksgatan 32, SE-11541 Stockholm, Sweden.

Linde, Mattias, Regional Migränmottagning Sahlgrenska, Sahlgrenska Hospital, Blå stråket 5, SE- 41345 Göteborg, Sweden.

Laike, Thorbjörn, Environmental Psychology, Sölvegatan 24 C SE- 223 62 Lund, Sweden.

Karlsson, Bodil S.A., RISE Research Institutes of Sweden, Division Built Environment, SE-41258 Gothenburg, Sweden

Corresponding email: bodil.karlsson@ri.se

Abstract

About 15% of the population suffer from migraines and it is estimated that about 40% of all people with migraines would benefit from preventive treatment, but only 3–13% use it. Migraine is a huge burden for society and individuals. Migraines can be intensified by light, and some patients need to stay in a dark room until the attack is over. People with this type of photosensitivity show a clear preference for light color, specifically green, which has been shown to be more comfortable and even pain relieving in some cases. We will present a feasibility study with the aim of preparing a series of experiments to investigate if regular short-term green-light-exposure can prevent migraines. We will present findings from the literature, developed light equipment, and plans for future testing of migraine friendly light solutions.

1. Migraine

Migraine is a common but painful condition [1], estimated to be affecting around 15% of the world's population [2] Many people with migraine do not seek help even though they could benefit from preventative treatment [3]. Migraines are characterised by powerful and debilitating headaches, often in the form of attacks lasting from hours to several days depending on the individual and situation [4]. Migraines are also associated with other symptoms such as visual aura, nausea, vomiting and anxiety

[5- 6]. Since so many people suffer from this condition, migraines also represent big costs for society, the biggest part being costs of production-loss from sick leave caused by migraine attacks [7]. The annual costs of migraine for society in EU is estimated to be around 50 billion euro [8].

1.1. The relationship between migraine and light

It is well established that there are individual differences between different people's light sensitivity, experiences of light, and light perception. Previous studies have shown that people with migraine are more sensitive to light in comparison to others, both during and between migraine attacks, and that up to 80% of patients are extremely sensitive to light during an attack [9-11]. Many migraine patients therefore prefer to be in a dark room until the attack has passed. Extreme light sensitivity is called photophobia, and there is, as of today, no established way of treating such symptoms [12].
Furthermore, other visual and light related discomfort from, for example, flicker, patterns, and color may also burden patients during migraines [11,13]. Notably, it was found that even when they do not have headache, people with migraine have lower thresholds for detecting flickering light, and people with migraine with aura have the lowest thresholds of all [11]. However, it is argued that light sensitivity for people with migraines does not automatically mean that patients need to or want to avoid all light [13]. Instead, it may be more meaningful and helpful to examine what visual patterns, light frequencies, and colors that affect or discomfort migraine patients more than people without migraines. Migraine also seems to have seasonal variation [5,14], especially in arctic populations, related to the differences in sunlight [15]. A study [15] showed that people suffering from migraine who experienced aura had more attacks in the bright spring and summer, used sunglasses to a greater extent than other patients, and even reported increased light exposure as an activating factor causing the migraine attack. Light was thus concluded to be an important explanation for the seasonal variation of migraine [15].

In a study from 1989, it was suggested that a significant part (30-60%) of migraine attacks were triggered by glare and light [16]. However, this conclusion has since been criticized. The question of whether light itself actually triggers migraine attacks for some patients is difficult to answer with any certainty, as it is difficult to know if the attack has already started when the light discomfort is experienced by the patient, who then may interpret the light as the causing factor.

Regardless of what role light and illumination plays in causing migraine attacks, it is clear that it is an important parameter in many migraine patients' level of discomfort or comfort both during and between attacks [9-11,13]. We therefore argue that more research is needed to understand how light circumstances can be improved to ease the suffering of people with migraines.

1.2. Green Light and Migraines

It is suggested in some studies that people with migraines have altered color vision [17-18].
Furthermore, recent research found that people with migraine have preferences for green colored light [19]. When examining how participating migraine patients responded to green light exposure over ten weeks, results of fewer headache days, less severe migraine attacks and even increased quality of life was observed [19]. Noseda et al. [20] and Nir et al. [21] tested how exposure to lights with different colors affected patients during migraine attacks, and found that all colors but green seemed to intensify pain. Even more noteworthy was that some patients even experienced pain relief when exposed to the green light [20-21]. Burstein et al. [22] suggest that this effect of the green light may be caused by the light reaching the part of the brain that regulates pain, through pathways from the retina that were just recently discovered. Given that light plays an important part for quality of life for many people with migraine, and that recent evidence suggest that green light may be less painful or even relieving for some people during migraine attacks, we argue that more attention needs to be paid to how light environment can be designed and adapted to help people with migraines. Not only does the relationship between green light and less pain need to be further examined and (possibly) verified through medical research, but the

question of where, when and how light can be adjusted to help people with high light sensitivity needs to be addressed.

1.3. Aim

The long-term goal of this research project is to further understand and investigate how people with migraine perceive different lights and to verify the possible effect of green light on patients with migraines thoroughly. Before doing so one may first ask how and when green light should be exposed or administrated to the patient. In addition, for any treatment to be feasible, whether preventive or reactive, it needs to function with the patient's everyday life and needs. We believe that this means that light solutions adapted to the needs of light sensitive people need to be examined from several scientific perspectives including design, environmental psychology, and engineering. This paper reports on a pilot study departing from the above-mentioned work on green light and migraines. The aim of the pilot study was to, based on existing literature on light and migraine, develop green light prototypes for preventive migraine treatments for migraineurs, to be used in future studies with migraine patients. The purpose of this paper is to present the designed and reconfigured prototypes and discuss and suggest pathways for future research.

2. Method

The research team consisted of an interdisciplinary group of researchers and practitioners and a larger reference group to include the competence from a broad spectrum of scientific fields. The total competence involved in the project includes psychology, medicine, optics, light engineering, design, sociology and light researchers.

The prototypes were designed or reconfigured using an iterative work process including several steps. First, a literature review was conducted to synthesize existing knowledge on light and migraine as well as green light specifically. Afterwards, a series of meetings and workshops were held where alternative applications, treatments options and study design were discussed in the research team. After the first meetings and workshops, the designer in collaboration with light counsellor developed several prototypes, meaning different lamps with green light, to be tested. A workshop was held where the research team tried the lamps. After the workshop, a few of the lamps were selected to be developed further, with suggestions of improvements. The lamps were then exhibited at a migraine clinic in Gothenburg, where patients could, if they wanted, try the lamps, and leave a comment. In the final step, suggestions for approaches for future studies and next-step projects were developed by the research team. The overarching phases for the design method is summarized in the figure below.

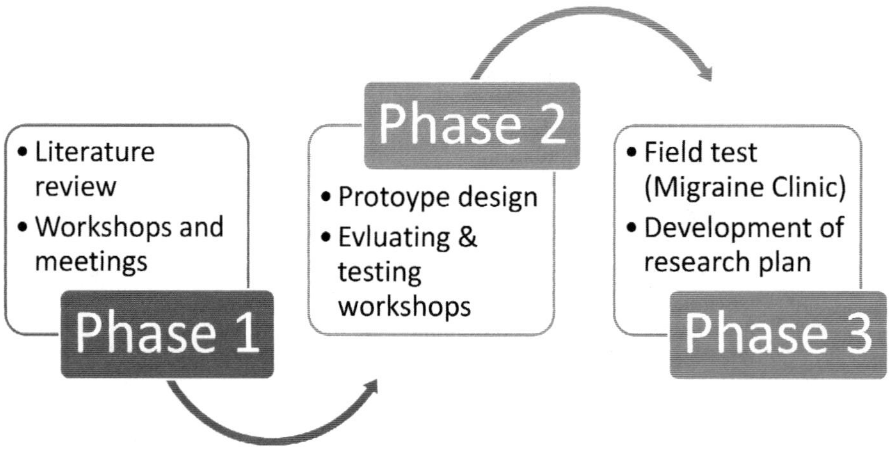

Figure 1. Overview of work- and design process for pilot study.

3. Results

The results of the pilot study will be presented in three parts. First, green light prototypes will be introduced and described briefly. Secondly, the field test at the migraine clinic will be described. Thirdly, ideas for future research and testing will be presented.

3.1 Green Light Prototypes

For green light to be a feasible preventive treatment for migraine patients, illumination needs to be designed with the patient in mind, and how this green illumination can fit into everyday life needs to be considered. A person suffering from reoccurring migraines may not have the option to soak in green light everyday doing nothing for a long time, instead, the lamps may need to be designed to be compatible with performing different tasks or activities. In the design process for this pilot study, we asked how, then, can green light be administrated to migraine patients interested in trying this treatment? Luminaires come in many different shapes and forms, not only in terms of light characteristics such as strength, flicker, color etc, but also what type of lamp it is – a reading lamp, a lamp for ambient light, a ceiling armature and so on. Because of this, the designer of the research team had this in mind when developing different prototypes for preventive treatment. Note that some of the lamps were reconfigured from existing lamps.

3.1.1 The reading lamp

This lamp was configured from a standard IKEA lamp designed for reading. The idea was that a preventive light treatment, that is exposing oneself to green light to prevent migraines, can perhaps be combined with reading or other activities such as washing dishes. A moveable reading light can therefore be a viable option for green light exposure while everyday life of the patient goes on.

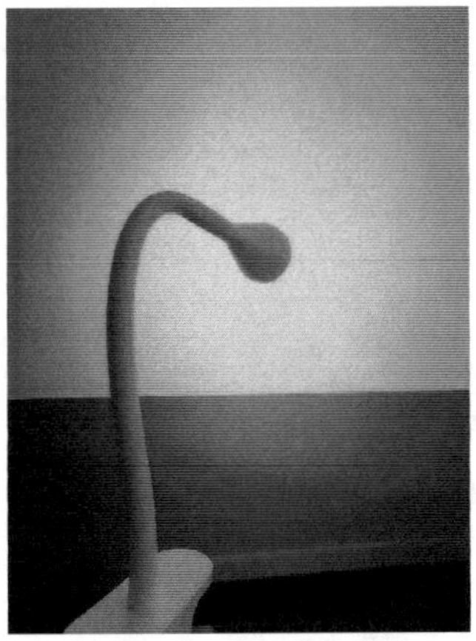

Picture 1. The reading lamp

3.1.2 The ambient globe light

Having a soft, ambient light for moments of relaxation as a form of preventive migraine treatment was also suggested in the form of this globe-like lamp projecting dappled patterns. The idea was that practising relaxation or meditation, serving the purpose of reducing stress, can be combined with green light exposure as a preventive routine. However, the light can also be used when performing other simple activities. The lamp can be placed for example in a bedroom in the privacy of one's own home, but also in a resting room at workplaces as a light option for those with light sensitivity.

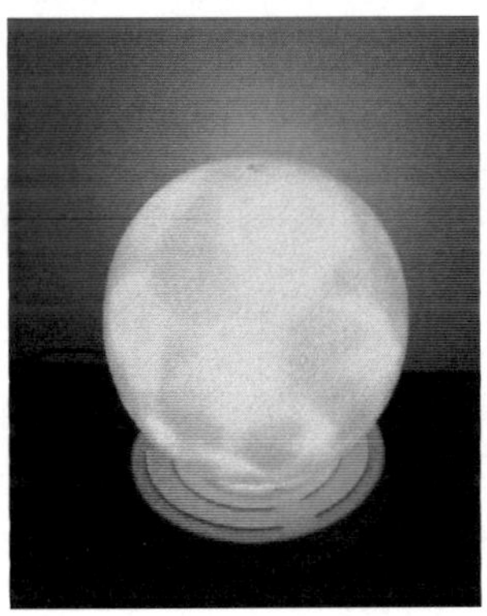

Picture 2. The ambient globe light

3.1.3 The green light glasses

Another option for administrating green light is to wear these green light glasses. The glasses work best when patients decide to relax and do nothing for a while, as the glasses prevent them from seeing anything besides green light. The glasses could be used both preventively and perhaps also reactively during a migraine attack, if the patient finds it comfortable and relieving.

IOP Conf. Series: Earth and Environmental Science 1320 (2024) 012020 doi:10.1088/1755-1315/1320/1/012020

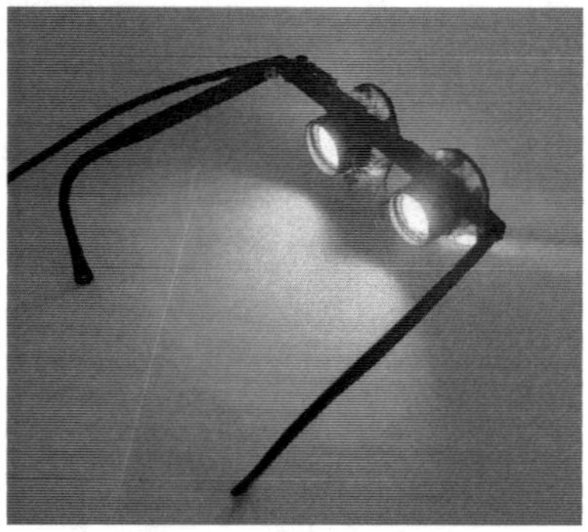

Picture 3. The green light glasses

3.2 Field test at migraine clinic

The three lamps were exhibited for 7 weeks in 2023 at a clinic for migraine and headaches. The lamps were placed in a waiting room that also served as a place for rest for visitors when visiting the clinic. In this room we also put papers informing visitors that these lamps were developed and tested for research on migraine, and visitors were invited to leave written comments on what they thought about the lamps. The authors were not present while the notes were written. The comments were anonymous to the authors and the comments cannot be traced back to identify the commenter, thus no personal information was collected. The purpose of the field test was to get input on these lamps from people who suffer from migraine, to be able to get a first indication of the suitability of the lamps' design as well as ideas for further improvement that could be useful in future studies. As a proper medical study will require extensive ethical approval, significant funding and much planning, this initial field test served as a way to collect some initial response from the target audience used in idea generation for future development of the lamps and ideas for research design. Although our small field test for the feasibility study is not in any way statistically generalisable and the data is limited, it provides some initial insight on opportunities for future studies and design improvement from a qualitative point of view. 13 different comments were collected, and a content analysis provided some common themes and words in the feedback from the field test. Comments were made in Swedish but have been translated by the authors.

3.2.1 Pleasant, calm, relaxing (Behagligt, lugnande, avslappnande)

The comments reveal that some of the visitors found the light comfortable and relaxing. The word 'pleasant' (behagligt) was mentioned nine times in the material, while the word calm (lugn/lugnande) was mentioned six times. Other words used were restful (vilsamt) and relaxing (avslappnande). Most of these comments seemed to refer to the ambient globe light, however some also referred to the other lamps. One comment even described a perceived pain relief from wearing the glasses *"After the lighting and two minutes of wearing the glasses, I felt that the headache and pain in the eye was milder. (...)"* – anonymous commenter.

3.2.2 Powerful, intense & need for adjustments

Other words used to describe the light in a different way were powerful and intense, mainly addressing the reading lamp and the glasses. These descriptions were used both in a positive and a negative sense, for example one comment described how the glasses gave a "powerful and comfortable light", while another stated that the light intensity of the glasses was much too strong and made them take them off immediately. One comment suggested that the glasses were too intense and that the effect would be better if the light was milder and that the light could be more covering. These descriptions indicate a need for the light to be adjustable to meet individual preferences and variation in light sensitivity.

3.2.3 Positive color aftereffect
Two comments described how the green light gave an aftereffect where they saw more pink colors after looking at the lamps. One of them described it as if the globe light had a pink color around it, while the other stated that they had tried the glasses, and saw everything in pink afterwards. Both comments indicated this as a positive experience, for example: *"I wore the glasses and looked at the light (don't know if that was what I was supposed to do), but afterwards everything became pink and it had a calming effect on me"* – anonymous commenter.

3.2.4 Uncertainty in longer effects
Comments also indicated that the commenters were curious about the effects of the lights during migraine attacks, as some preferred darkness, as well as long-term effects from using the light, illustrated by: *"Pleasant light for the eyes. Would be interesting to try it a longer period to see the effect those days when migraine is at it's worst"* – anonymous commenter.

In summary, the comments reflected mainly positive but also some negative reactions to the light, where the light was positively described as pleasant and calming while the negative seemed to be mainly about light intensity and that there was no way to adjust or dim the light strength. The lamps seemed to give certain color aftereffects, but the commenters were unsure how they would react during migraine attacks and if they would use the lights regularly.

3.3 Future plans for research and testing
Throughout the process of developing prototypes and discussing treatment ideas including light therapy, we identified several needs for future research. The effect of green light needs to be further investigated, and the treatment option of light for migraineurs need to be examined from different perspectives. Below, we present three different ideas for future research on green light and migraine. We believe different approaches are needed, and these suggestions should not be seen as replaceable by each other but rather complementary of each other, as they represent different types of research.

3.3.1 Investigation of light color preference in migraine patients
During the work process for developing prototypes, the question of exactly "what green light" is most comfortable for migraine patients came up. It was suggested that there may be individual differences in what green nuance people prefer, as well as other colors. We therefore suggest a small study investigating what light frequency preferences migraine patients have. A suggestion is to combine interviews on photophobia and light experiences with testing preferences through immersion in light. We suggest two alternative light technologies for this test. The first is a light dome, where the patient puts their head in a dome with colored light, providing a field of view without borders, which can be compared to the experience in a floating tank. The second is a light goggle that is simply put over the eyes, an experience similar to using VR-glasses. The patient can decide what frequency they find most comfortable, spend a short amount of time in the dome immersed in the light of their choice, or with the light goggles on, and then describe how they felt. This would give insight into what frequencies are preferred and how migraine patients experience colored light outside of their migraine attacks (since the light therapy is suggested as a preventive treatment).

3.3.2 Conceptual replication: Cross-over study of the effect of green light exposure on migraine patients

The research done on green-light [19-21] suggest that green light is experienced differently than other colors by migraine patients, findings that are both very interesting and important. However, more research is needed to examine and validate what potential there actually is in green light, if it is a relatively comfortable or even pain-relieving light for people with migraine or not. A cross-over study where migraine patients are exposed to preventive green light treatment in comparison with another, established effect such as cardio exercise or a prophylactic medicine, could be a way to thoroughly investigate this supposed effect further. A preventive green light treatment may be, for example, to regularly rest in green ambient light as a form of therapy. However, this may need to be compared to just resting, to verify whether the green light as such has an effect in addition to the possible effect of resting. Furthermore, if there is a positive "green light effect" for people with migraine, the question is whether that effect happens mainly during any of the phases of a migraine attack or perhaps as a preventive effect between attacks. This information is needed to understand how green light can be used in treatment of migraine. This means that another alternative would be to investigate what effect green light has during a migraine attack, using a similar crossover study design. Depending on what type of green light treatment is investigated, we suggest that different luminaires may be suitable for different studies. The green light glasses (see 3.1.3) may be more suitable for studies on acute green light treatment during migraine attacks, while the luminaires such as the ambient globe light (3.1.2) may be more suitable for investigating the effect of a preventive green light treatment.

3.3.3 Diary-based study of preventive and reactive green light treatment in the everyday life of migraineurs

How can light treatment for migraines possibly be successful and reach a high degree of diffusion among patients? We believe that this question is not only a clinical matter but also a question of user preferences and treatment design. We therefore suggest a user study on how green light luminaires can function and blend in the everyday life and practices of migraine patients. The prototypes configured through this pilot study have not yet been tested in real life settings, other than in an exhibition at a migraine clinic. We would like to lend these prototypes to migraine patients and ask them to report through diaries where in their homes they were placed, how they were used, when they were used (e.g. in relation to migraine attacks and everyday routines). We are also interested to know how the use was experienced by the patients, which could be investigated, for example, through interviews after the trial. This would give us an idea of further design and configuration improvements, as well as feedback on how compatible the use of green light is in everyday life of migraineurs.

4. Discussion

Migraine is associated with huge cost for society [8] and causes suffering for vast amounts of people around the world, and can in that sense be considered a major societal challenge. In addition to being painful and unpleasant for patients, migraine is often associated with social isolation [23]. With this background, any action that can ease the burden from migraine is relevant to investigate. Taking this reasoning one step further, it is argued that migraine can be considered a social disease and that designing migraine-friendly workplaces, e.g., in terms of adapted light, is an act of social sustainability [24]. The prototypes in this study may make a positive contribution to building migraine friendly environments, but further testing and research needs to be done. In addition, from a sustainability point of view, it is interesting to note that people with migraine in general seem to prefer less light and lower light intensity which indicates less electricity consumption during usage if people with migraine were to choose the design of light solutions.

The process of developing green light equipment, reviewing migraine literature and workshopping to figure out avenues for future research shed light on interesting questions and challenges. One such challenge is the complexity in practicality regarding treating migraine with light therapy. More design parameters than light frequency (color) must be considered to prevent visual stress among people with migraine [13]. A number of light-oriented design actions in order to create a more migraine friendly work-place may be needed; e.g., avoiding lighting fixtures in which the light source is directly visible, opting for indirect lighting, and equipping workstations with adjustable lighting systems (direction, luminance and light temperature) and avoiding reflective surfaces [24].

The research process of configuring and discussing different options for how green light solutions can be administrated to migraine patients raises questions of where, how, and when migraine patients are able and willing to do light therapy and even what the term light therapy means. Light therapy may be performed in the home environment with specially designed lamps in an everyday context [16]. It may also be conducted in a clinical setting. Furthermore, everyday light solutions, that we normally may not consider therapeutic, may have a therapeutic effect in the sense that they may reduce pain, e.g., by providing individual control of the light settings at work.

Since green light solutions may be perceived odd in a public place, what spatial contexts and activities are suitable to combine with green light? From green light flashlights on accessories such as rings or keychains, to green light settings on smart phones, to visiting special light facilities, the question of what people with migraine would prefer is still open [25]. Tamsin and Bach [25] suggest that medical device design is a complicated and special process, and thus needs a specialized cross-functional team to deal with challenges; and given that light solutions at least to some extent in the future may be considered a medical device, in the sense that it may reduce symptoms, a cross-functional approach is needed. The importance of cross- functional teams are also more important the more technologically advanced the product is and the riskier the market is [26]. A successful established light treatment that can be recommended by physicians of course needs to be medically sound, but it also needs to be compatible with patients' lives and needs, if it is to be a viable treatment option or complement other treatments. This requires both successful design of light technology itself, but perhaps also a wider understanding of everyday practices of people with migraine. We suggest four different contexts are important to consider, and most likely light solutions may be different in these contexts 1) Public environments such as work-places and public buildings 2) Home environments 3) Clinical environments such as waiting rooms 4) Well-being facilities such as spas, gyms and massage-rooms. Yet in each place, the light must fit in and possibly also be integrated in automatic light solutions. E.g., at the workplace a possible place where green light may be a feasible alternative used by migraineurs could be resting rooms. A resting room could be suitable for regular meditation or preventive care to maintain wellness, in this case a moment of green light treatment. For those who do not feel that they benefit from green light, it would be easy to switch to other light colors. Meanwhile, a globe light (3.1.2) or green light glasses (3.1.3) could be accessible for the migraine patient who wants to rest their head for a while during the workday.

Finally, a benefit of a wider, interdisciplinary research approach for studying green light treatment is that it may contribute to understanding of light sensitivity and individual differences in light needs overall, beyond migraine. We suggest that people with migraine may be considered an example of a specifically and sometimes extreme light sensitive group, however, not all light sensitive people suffer from migraine. It is also worth noting that not all people with migraine respond well to green light [18] and preference of light frequency may also vary between individuals. As such, increasing awareness of light perception and how it may affect mental and physical health, can be helpful to a wider audience and a wider category of applications. We suggest that there is a need for a complementary, more person-oriented light approach, where individuals of different needs are considered to a larger degree, to increase well-being and reduce light-related headache and other negative symptoms for example in workplaces.

References

[1] Ashina M, Katsarava Z, Do TP, Buse DC, Pozo-Rosich P, Özge A, Krymchantowski AV, Lebedeva ER, Ravishankar K, Yu S, Sacco S. 2016. Migraine: epidemiology and systems of care. *The Lancet*. **397**(10283):1485-95.

[2]. Stovner, LJ, Hagen, K, Linde, M, & Steiner, T. 2022. The global prevalence of headache: an update with analysis of influence of the methodological factors on prevalence estimates. *The Journal of Headache and pain*. **23**(31), 2-17.

[3] Kernick, D, Kondori, N, Pain, A, Mount, J, Appel, C, Ranopa, M, & Gulati, T. 2024. Preventive treatment patterns in the adult migraine population: an observational UK study over 7 years. *BMC Primary Care*. **25**(1), 34.

[4] Ravisankar, P, Hundia, A, Sindhura, J, Rani, B. S, & Sai, P. 2015. Migraine-A comprehensive review. *Journal of Pharmaceutical Research*. **5**(10).

[5] Dodick, D, & Silberstein, SD. 2016. *Migraine*. (Vol. **91**). Oxford university press.

[6] Karimi L, Wijeratne T, Crewther SG, Evans AE, Ebaid D, Khalil H. 2021. The migraine-anxiety comorbidity among migraineurs: A systematic review. *Frontiers in Neurology*. **18**(11): 613372.

[7] Hjalte, F, Olofsson, S, Persson, U, & Linde, M. 2019. Burden and costs of migraine in a Swedish defined patient population – a questionnaire-based study. *The Journal of Headache and Pain*. **20**(1):65.

[8] Linde, M, Gustavsson, A, Stovner, LJ, Steiner, TJ, Barré, J, Katsarava, Z, Lainez, J. M, Lampl, C, Lantéri-Minet, M, Rastenyte, D, Ruiz de la Torre, E, Tassorelli, C, & Andrée, C. 2012. The cost of headache disorders in Europe: The Eurolight project. *European Journal of Neurology*. **19**(5), 703–711.

[9] Drummond, PD, & Woodhouse, A. 1993. Painful stimulation of the forehead increases photophobia in migraine sufferers. *Cephalalgia*. **13**(5):321–324.

[10] Choi, JY, Oh, K, Kim, BJ, Chung, CS, Koh, SB, Park, KW. 2009. Usefulness of a photophobia questionnaire in patients with migraine. *Cephalalgia*. **29**(9):953–959.

[11] Kowacs, P, Piovesan, E, Werneck, L, Fameli, H, Zani, A, & da Silva, H. 2005. Critical Flicker Frequency in Migraine. A Controlled Study in Patients without Prophylactic Therapy. *Cephalalgia*. **25**(5):339–343.

[12] Digre, KB, & Brennan, KC. Shedding light on photophobia. 2012. *Journal of Neuro-ophthalmology: the Official Journal of the North American Neuro-Ophthalmology Society*. **32**(1): 68.

[13] Wilkins, AJ, Haigh, SM, Mahroo, OA, & Plant, GT. 2021. Photophobia in migraine: a symptom cluster. *Cephalalgia*. **41**(11-12):1240-8.

[14] Poulsen, AH, Younis, S, Thuraiaiyah, J & Ashina, M. 2021. The chronobiology of migraine: a systematic review. The Journal of Headache and Pain. **22**(**1**):1-2.

[15] Alstadhaug, KB, Salvesen, R, & Bekkelund, SI. 2005. Seasonal variation in migraine. *Cephalalgia*. **25**(10):811-816.

[16] Vincent, AJ, Spierings, EL & Messinger, HB. 1989. A controlled study of visual symptoms and eye strain factors in chronic headache. *Headache*. **29**:523–527.

[17] Shepherd, AJ. 2006. Color vision but not visual attention is altered in migraine. *Headache: The Journal of Head and Face Pain*. **46**(4):611-621.

[18] Brighina F, Firpo V, Maccora S, Calabró V, Lombardo F, Cosentino G, Baschi R, Bolognini N, Vallar G, Fierro B. 2015. Color vision and visual cortex excitability are impaired in episodic migraine. Simply coexisting or pathophysiologically related dysfunctions?. *The Journal of Headache and Pain.* **16**(1):1-2.

[19] Martin, LF, Patwardhan, AM, Jain SV, Salloum, MM, Freeman, J, Khanna, R, Gannala, P, Goel, V, Jones-MacFarland, FF, Killgore, WDS, Porrera, F, & Ibrahim, MM. 2021. Evaluation of green light exposure on headache frequency and quality of life in migraine patients: A preliminary one-way cross-over clinical trial. *Cephalalgia.* **41**(2):135-147.

[20] Noseda, R, Bernstein, C, Nir, R-R, Lee, AJ, Fulton, AB, Bertisch, SM, Hovaguimian, A, Cestari, DM, Saavedra-Walker, R, Borsook, D, Doran, B, Buettner, C, & Burstein, R. 2016. Migraine photophobia originating in cone-driven retinal pathways. *Brain.* **139**:1971-1986.

[21] Nir, R-R, Lee, AJ, Huntington, S, Noseda, R, Bernstein, CA, Fulton, AB, Bertisch, SM, Hovaguimian, A, Buettner, C, Borsook, D, & Burstein, R. 2018. Color-Selective Photophobia in Ictal vs. Interictal Migraineurs and in Healthy Controls. *Pain.* **159**(10):2030-2034.

[22] Burstein, R, Noseda, R, & Fulton, AB. 2019. Neurobiology of photophobia. *Journal of Neuroopthtalmology*. **39**(1):94-102.

[23] Abu Bakar, N, Tanprawate, S, Lambru, G, Torkamani, M, Jahanshahi, M, & Matharu, M. 2016. Quality of life in primary headache disorders: a review. *Cephalalgia*. **36**(1):67-91.

[24] Schiavone, E, Sorrentino, A, Merighi, L, de la Torre, ER, & Sandrini, G. 2022. How to Create a Migraine-Friendly Workplace. *HealthManagement.org.* **22**(6).

[25] Tamsin, M, & Bach, C. The design of medical devices. 2014. *International Journal of Innovation and Scientific Research.* **1**(2):127-134.

[26] Gemser, G, & Leenders, MA. 2011. Managing cross-functional cooperation for new product development success. *Long Range Planning*. **44**(1):26-41.

IOP Conf. Series: Earth and Environmental Science 1320 (2024) 012021 doi:10.1088/1755-1315/1320/1/012021

Light and Emotional Architecture: An Embodied Experience

R Muro[1]

[1]: KTH School of Architecture, Royal Institute of Technology, Stockholm, Sweden.

E-mail: rodrigo.muro@arch.kth.se

Abstract. The presented explorative research focuses on the phenomenological relationship and emotional impact of light within architectural spaces, grounded in a pilot study and literature review. Serving as the foundational framework for a Ph.D., this study aspires to decipher the role of light in Emotional Architecture—a term introduced by the German artist Mathias Goeritz in his 1953 architectural manifesto crafted in Mexico. This multidimensional pilot study employs experiential observation and structured interviews, complemented with qualitative and quantitative documentation, all anchored by literature-based insights. Set in the Barragán House, which served as both the studio and residence for the renowned Mexican architect Luis Barragán since 1948, the study endeavours to elucidate how architectural nuances—like form, materials, colours, and arrangements—affect lighting conditions, subsequently shaping human emotions and experiences. Through twenty-two interviews, the research captured perceived atmospheres and the inherent intentionality of affordances in architecture, revealing a strong correlation between spatial impact on observers and light-related experiences. The interplay between these elements suggests a compelling avenue for implementing Emotional Architecture conducive to both individual and societal sustainable development.

1. Introduction

What is light and how do we relate spatially to it? The understanding of light as a physical phenomenon is insufficient to comprehend how light shapes a space and how the observer perceives it. Light has an inherent ambiguity, Nobel laureate physicist Niels Bohr (1912) proposed that a photon is neither a particle nor a wave but acquires one or the other identity when it is observed; light presents a wave-particle duality as it can exist as both simultaneously (1,2) This notion of light emphasizes, on the one hand, the limitations of our human perceptual system at a quantum scale and, on the other, contextualizes the potential for light perception at various scales within the built environment, influencing the observer's experience in an illuminated space. Exploring the deep interplay between light and space is intricate, it involves cognition and addresses concepts like experience and subjectivity (3). I follow what architectural historian and theorist Alberto Gomez-Perez remarks on consciousness, that it is not only embodied but also deeply rooted in place. The physical environment plays a pivotal role in shaping human consciousness, closely tying it to emotional well-being (4). The drive to understand the profound impact space has on our emotions stems from my personal experiences within certain architectural spaces that deeply moved me emotionally, spaces I have noticed as a lighting designer to have compelling lighting qualities. Recognizing that a significant portion of our lives is spent within built environments, the works of Sarah Williams Goldhagen, an American author and architecture critic, provide concrete evidence of the profound impact that built environments have on dwellers' emotions, behaviour, and mental and physical health (5). The overarching goal of my research is to ascertain whether "Emotional Architecture" (EA) —a term introduced by the German artist Mathias Goeritz in his 1953 architectural manifesto crafted in Mexico (6)—can promote well-being, from a light and lighting perspective. To help me gain an understanding of the interplay and relationship between space, light, experience, and their emotional resonance, I conducted an explorative research project preceded by a literature review. I investigated the notion of EA from a phenomenological perspective, employing the concepts of "Atmosphere" and "Affordances" through a qualitative and quantitative pilot study. In this paper, I focus exclusively on first-hand empirical material that resulted from structured interviews, leaving for a subsequent paper the quantitative aspects.

2. Theoretical background

To explore how emotions resonate in tandem with space, the notion of EA is addressed. Merely viewing emotions as bodily sensations, a Cartesian perspective (7,8), is insufficient to understand the intricate relationship between the human mind, body, space, emotions, and the dynamics of light. The theoretical approach in this paper provides knowledge to help bridge the phenomenology and philosophical approach to cognition in Architecture. This framework is built upon two pivotal concepts: Atmosphere and Affordance to facilitate the understanding of architecture's embodiment.

Figure 1: Narrowing of the access corridor enhances depth perception, wooden floor planks mirror the tapering.

Figure 2: A stand-alone tall black wall, evokes an out-of-human-scale impression.

The Eco, Mathias Goeritz (1953), Mexico City. Photos: Author, 2023.

2.1. Emotional Architecture

The notion of EA conceived by Goeritz in a manifesto was not merely a written critique of the dominant functionalist architectural style and the perceived dehumanization of architecture of the era. It also manifested tangibly in the form of built space: the experimental museum “El Eco”, constructed in Mexico City that same year by Goeritz (9) (Figure 1 & Figure 2). It can be said that Goeritz contravened with the precepts of modernism, where standardization, mass production, functionality, formal abstraction, and internationalization are overshadowed by needs, idiosyncrasies of handicraft, figuration, symbols, and regionalism similar to what Williams G. expresses on the design methods used by the Finish architect Alvar Aalto (10). Similarly, Goeritz´s friend and collaborator, Mexican architect Luis Barragán, also followed this path. However, reactions to his work vary. For some, like Luis E. Carranza referencing Kenneth Frampton (11), Barragán's architecture embodies a transition from contemporary modernity to an architecture of resistance, or “critical regionalist architecture”.

Conversely, others critique it as “Corbusian regionalism”, viewing it as a fusion of Barragán’s “tropicalization” of architectural Modernism and his reinterpretation of ecclesiastical design. Barragán's reputation as a Pritzker Prize laureate undoubtedly underscores the values many in the architecture community resonate with. Goeritz and Barragán had in the design of the Satellite Towers of 1958 an emblematic and illustrative work of their collaboration. The Satelite Towers are a significant architectural monument that has since become an identity marker for the region bordering Mexico City and Mexico State. Mexican historian and curator Daniel Garza Usabiaga inscribes the towers into the concept of EA (12) and encapsulates the evident influence Goeritz and Barragán had on each other. Through their years of synergy, both Goeritz and Barragán firmly established themselves as pioneers in the realm of EA. The term EA doesn't have a singular definition. For instance, curator and researcher Leonor Cuahonte describes it as an experimental space that elicits emotions and within which one can move (9). Similarly, Usabiaga characterizes it as an ambiguous space where architecture, conceived as sculpture, engages in a dialogue with spatial considerations and atmosphere (13).

2.2. Atmosphere

The perception of a space's atmosphere plays a pivotal role in how we “feel” architecture by our emotions (14), shaping our emotional responses to it. Michael Arbib, a North American computational neuroscientist, posits that the concept of “atmosphere” in a building context is the interplay between the observer and the structure, where the prevailing tone or mood impacts the observer instantly, prompting or preventing certain actions (15). Juhani Pallasmaa, a Finnish architect and theorist, once asserted, “Light is the strongest conditioner of the atmosphere of place, the most comprehensive criteria of the character of space, place, and setting” (16). This perspective aligns with what Italian architect David Ruzzon believes: light *“is one of the most important layers in which we can feature the architectural indoor space”* (17). Still, the question remains, what is “atmosphere”? According to German philosopher Gernot Böhme, the ontological status of atmospheres remains indeterminate. It is ambiguous whether they belong to the objects or environments emanating from them or to the subjects perceiving them. Essentially, atmosphere underscores the relationship between subject and object, focusing on the aesthetics of the interaction between environmental qualities and human emotions. Böhme, drawing from German philosopher Hermann Schmitz, further describes atmospheres as spatial entities that are “borderless, widespread, yet non-localized”. They manifest as phenomenological entities that carry affective feelings and set the mood of space (18). But as Böhme describes, even if “atmosphere” is a term often employed to describe something vague, it does not imply that the term lacks a clear meaning itself. Pallasmaa and Swiss architect Peter Zumthor contend that the essence of architecture is more about crafting atmospheres than geometries. They believe that since architecture can influence mood and its occupants, it should be attuned to promote well-being (4). Perceptual cues, such as atmospheres, can suggest or guide certain actions or affordances. Conversely, they might deter specific behaviours by not providing cues based on the observed atmosphere. Kathleen Stewart encloses the experience of atmosphere as more than just a passive backdrop; it is a dynamic force field where individuals are immersed. Rather than merely being the outcome or effect of other factors, it is a lived affect: a blend of the capacity to affect and be affected, that moulds the present moment into a unique configuration, with potential and expression. Atmosphere represents a tuning of our senses, efforts, and imaginations to potential experiences. It manifests in the palpable uncertainties of everyday life: the unpredictable pull on our emotions (19).

2.3. Affordances

To understand the intricate interplay of light, space, atmosphere, and emotion—what I term the spatio-sensory relation—I will employ James Gibson's 1979 theory of affordances within an architectural context. This will help decipher how spatial conditions and their lighting conditions either facilitate or inhibit specific actions or behaviours. By placing these actions or behaviours within Gibson's defined environment, we can better comprehend the embodied experience of this relation. My research starts with the notion that objects and environments, through their design, materials, and layout, can evoke certain meanings or associations that influence human interactions and experiences. If a spatial design implies symbolic or metaphorical meanings, shaping how individuals perceive and engage with it, then light-generated atmospheres hold similar sway. Gibson emphasized the significance of surfaces and spatial layouts, outlining nine ecological principles related to them (20). These principles describe specific properties of the environment, with four directly addressing light attributes such as intensity, shading, surface colour through wavelengths, and light reflection or absorption. As Sarah Robinson articulates, *"We could understand surfaces not for what they are, but for what they do"* (21). From a philosophical perspective, I see Gibsonian affordances as abstractions, something one cannot escape from to think. In my view, how an affordance is perceived aligns with English mathematician and philosopher Alfred North Whitehead's idea of "prehensive unity of volume", which, in this context, refers to the architectural experience. Similarly, just as this prehensive unity, a collection of parts, varies based on perspective, the spatial configuration and its associated affordance hinge on the collective impact of architectural elements and their relationship to specific experiences or viewpoints (22).

The connection between the two concepts elucidated above is that the atmospheres present varying affordances based on dynamic experiences within a space, as well as dynamic lighting conditions, contrasting with static perceptions from a singular vantage point and static light. Additionally, Austrian-German philosopher and phenomenologist, Edmund Husserl, talked about optimal perception, the way things offer themselves to us. By relating this to light in space and based on Husserl´s ideas, one can say that space can have diverse appearances depending on the lighting conditions (23). The space offered to us and discoverable through light communicates diverse possibilities of atmospheres and affordances that are dependent on the attunement to those. Recognizing attunement as an interplay between internal projections and sensory environmental perceptions (24), it is understood that observers will resonate differently with spaces based on their personal histories, cultural backgrounds, and physical sensations. These diverse perceptions are not discrepancies in affordance theory but opportunities to explore the myriad emotions a singular space can elicit. It is this emotional spectrum that is of interest to me. When architecture, enhanced by light-driven atmospheres, resonates emotionally with an observer, it embodies the human experience and comprehension of the built environment (25). This theory hinges on the notion of affordances and their relation to the attunement to atmospheres since experiences are intrinsically tied to objects—here, architecture. Atmospheric affordances don't merely elicit a subliminal response; they also communicate potential uses, actions, and functionalities (26).

3. Developing the question

Is it possible to define a clear action-response relationship between user and architecture? Learning from the concept of Don Norman, a North American cognitive engineer, of Emotional Design, adapted to spatial experience instead of human-computer interaction and product design (27), this paper explores a methodology to understand the associated phenomena. Precise emotional responses in the spatiotemporal experiences outside the confinements of a laboratory setting can just be imagined or guessed, but not guaranteed. In this spatio-emotional and phenomenological approach, human nature and the nature of reality are framed by what cannot be determined. Precise causes cannot be affirmed, leading to an inherent unpredictability in this given universe. Therefore I explore the possibility of lighting being a pivotal element in crafting EA and enhancing the perception of affordances, based on how it unveils spaces and shapes atmospheres. If lighting atmospheres—whether crafted by natural or man-made sources—can influence perceptions and interactions by highlighting or concealing certain architectural elements, can such spatial experiences consistently evoke specific emotions and responses?

4. Historical context

To address these questions and deepen the understanding of EA, Atmosphere, and Affordances, my research focuses on Barragán House (BH) in Mexico City (1942), as a case pilot study, adopting a phenomenological approach while examining the residence and studio Barragán designed for himself (Figure 3 & Figure 4). This building is widely recognized as embodying the hallmarks of exemplary architecture that aligns with the principles of EA. Barragán emphasized psychic needs in his designs, underlining the importance of refuge, privacy, and an intuitive connection to architecture, while prioritizing aesthetic appeal. His designs often evoked emotions through atmospheres, employing elements such as low walls, thick walls, and limited window surfaces to ensure the feeling of shelter and protection. In his architecture, he aimed for a seamless continuation of the experience of lighting situations between the interior rooms of the house and the natural environment of the gardens. Barragán's designs frequently played with contrasts, shifting from light to shadow and from constrained to expansive spaces (28). Federica Zanco, in reference to Barragán, underscores that architecture is tangible and exists through elements like space, light, colour, size, and proportion. Its perception is intricately linked with bodily movements through spaces (29), these elements echoed in the phenomenological approach applied to a pilot case study of BH. Barragán's words echo: *"My house is my refuge, an emotional piece of architecture, not a cold piece of convenience"* (30). This emotional resonance is crafted through privacy and intimacy. In Barragán's work, light is a potent tool. He adeptly manoeuvres it, employing contrast, texture, colour, darkness, and strategic placements of windows and planes to conjure varied luminous experiences.

Figure 3: Vestibule ground floor. Mathias Goeritz´s golden "Message".

Figure 4: Iconic floating stairs in the personal library lead to Luis Barragan's office. The window, facing the street, has translucent glass, obscuring a clear view.

Luis Barragán House (1942) Mexico City. Photos: Author, 2023.

5. Methodology

In designing and executing a pilot study, I incorporated a blend of methods to establish a robust multimodal methodology rooted in a phenomenological approach. While first-hand experience is pivotal in interpreting affordances, visual sociologist Luc Pauwels and researcher Dawn Mannay assert that such experiences are often influenced by the prevailing visual culture (31). The study of the perception of illuminated spaces leans heavily on visual considerations. As such, visual research methods are employed. Yet, as these authors caution, visual culture should not be limited to merely an "image culture", nor should visual research be confined to "image-based inquiry" (31). Drawing from their insights, my research takes into consideration the study of the phenomenon at BH. The term "phenomenon" in visual research refers to the subject or topic that is being studied. This pilot study involves examining light and the concepts, atmosphere, and affordance through on-site retrieved data and documentation, and ethnographic research. A quantitative and qualitative feed of light and lighting characteristics as well as visitors' experiences to BH are registered via photometric measurements, photography, video and interviews during July of 2023. The photographs are to be used as evidence of the light distribution and other qualities of space, such as colour, proportions, and arrangement of spaces and objects concerning architecture. The video serves as a record of a first-person perspective experience to provide a viewer´s journey of the events while moving at BH. The interviews are used to analyse the visitor´s experience, impressions and preferences.

Prior to the set-up of the pilot study, my exploration into this topic began with a thorough literature review. I sought to understand the essence of EA and the spaces that fall under this designation. Comprehensive investigations into Goeritz and Barragán were undertaken, drawing from monographs, collected manuscripts, specialized literature, and recorded dialogues with both figures' experts. Further, readings from phenomenology and neuro-architecture theories offered profound insights, on concepts such as atmosphere and spatial affordances. A literature review was followed by the setting up of a pilot study and site visits of spaces designed by Goeritz and Barragán in Mexico City. In particular, this paper and the pilot study focus on BH. The explorative research focuses on the phenomenological relationship and emotional impact of light at BH by documenting qualitative and quantitative data. The collected data is approached in two different ways: A) the Ecological Environment, following Gibson´s etymology (20), and B) the Human Aspect. This distinction allowed for the understanding of the relationship between how light in spaces can influence people's perceptions, experiences, emotions, and sense of atmosphere by revealing or concealing form, material, colour and architectural configuration (Figure 5). This paper discusses the literature review and Human Aspect portion of the pilot study.

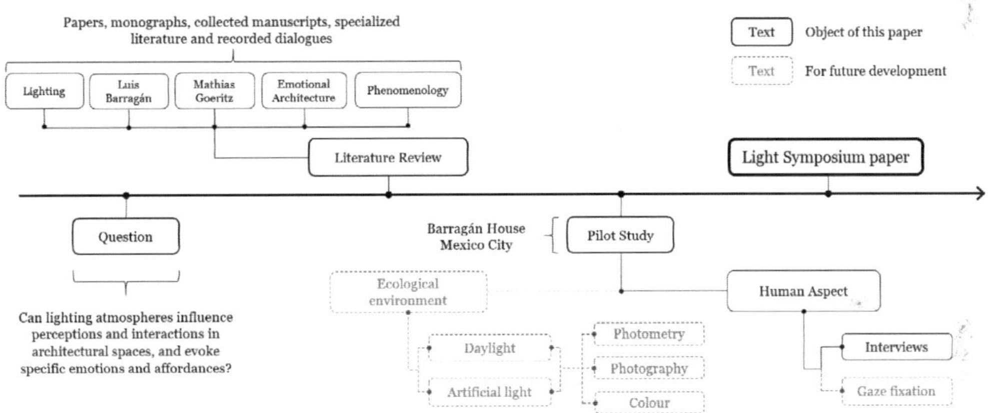

Figure 5: Diagrammatical depiction of information relationships and focus of this paper versus future write-up.

5.1. Designing my study

The Human Aspect pertains to the phenomenological experiences within BH. While this pilot study documents various layers and approaches to the primary question, this paper's results and discussions are concentrated on the Human Aspect. Detailed analyses of the data acquired using the methods described in the previous section, along with the integration step (Figure 5) will be undertaken in future research phases.

My primary method for capturing the Human Aspect involved interviewing visitors to BH. The house offers regular tours from 11:00 to 16:00 hours, Monday to Saturday. Visitors schedule their tours online in advance. Each tour accommodates a maximum of six visitors. Following their tour, visitors were invited for an interview, which they participated in voluntarily, ensuring adherence to the General Data Protection Regulation (GDPR). 22 participants consented to be interviewed and audio-recorded, with the assurance of anonymity and confidentiality. The structured interview contained open-ended questions to ensure consistency but also left room for spontaneous follow-up queries or elaborations. Depending on the interviewee's preference, the interview was conducted in either English or Spanish.

The initial segment of the interview gauged the interviewee's familiarity with BH and Barragán's broader works. It also sought information on their profession, background, and connection (or lack thereof) to architecture and/or lighting. The nationality of participants was asked to assess their familiarity with this architectural style (EA, Mexican, or modernistic) and note their age range to understand the house's visitor demographic better. The interview's second segment addressed concepts discussed throughout this paper and more, such as visit expectations, perceived atmosphere, emotions experienced, spatial perceptions, overall impressions, identified architectural qualities, and familiarity with EA. Concerning the notion of Affordances, which isn't widely recognized, the term was not directly used in the interview questions. Instead, this question was asked: *"Were there specific elements or qualities of the space that seemed to suggest or encourage particular experiences or actions, even if you didn't act on them?"* If further clarification was required, this follow-up question was added: *"Could you elaborate on how these elements influenced your perception or emotions during the visit?."*

6. Results and discussion

As previously mentioned, this paper's results from the pilot study solely address the human aspects of experience. Interviewees had diverse professional backgrounds: 50% were associated with art, design, or architecture, and the remaining had varied professions. The motivation for visiting BH is not predominantly due to an architectural interest or passion for Luis Barragán. According to the BH staff and some respondents, it is a highly recommended tourist attraction both in guidebooks and via word-of-mouth.

Analysis of the interview recordings reveals a dichotomy. On one hand, opinions differ widely; on the other, there's significant consensus in certain areas. The prevailing perceived atmosphere is largely positive. Based on conversations with the interviewees, the atmospheres associated with different spaces in the house include "peaceful," "calm," "tranquillity," "freshness" (thermal perception), "elegant," "introspective," "profound," "positivism," "ease," "inviting," "powerful," "generous," "warm," "soft," "serene," and "a moment to take a pause." I have categorized these semantic descriptions of the perceived atmospheres as positive, negative, and relational. Negative descriptions, validated by interviewees, comprise terms like "trapped", "prison-like", and "very dark". Relational responses tied directly to architectural elements and their relation to perceived ideas, such as "compression & release" (pertaining to transitional spaces), contrasts in light and colour, the "wildness & order" juxtaposition between natural elements like gardens and built forms, "preparation" (an entrance area cueing visitors about what's next), and "perfection," related to terrace features such as volume, geometry, scale, and colour.

Among the atmospheric descriptors, "peaceful" was most common with seven mentions, followed by "calm" with three, and "tranquil" with two. It is noteworthy that many interviewees used similar language for both felt emotions and perceived atmosphere, even though the interview delineated these aspects. Terms like "peaceful", "calm", "tranquil", and "serene" recur in both contexts. This suggests that many find it challenging to articulate the emotional dimension of their experience and lean towards architectural expressions, regardless of their professional background. This convergence between emotional responses and atmospheric perceptions underscores T. Griffero's notion that atmospheres elicit specific feelings when we interact with our environment (26) or, expressed in other words, the relation between atmosphere, emotion and architecture is a very tight one, I could say even inseparable. The interview also helped to elaborate on the impact spaces had on the interviewee (Figure 6). Respondents found that both the terrace and the living room stood out (Figure 7 & Figure 8). Both areas offer strong connections to nature and exude an outdoor feel, aspects recognized by respondents who found these spaces especially memorable. Interestingly, two of the most photographed and presumably well-known areas—the vestibule and the library (Figure 3 & Figure 4)—didn't rank highly in impact. There is a distinct consistency in how Barragán's work appears in media, stemming from a controlled narrative (32). This underscores the primacy of direct experience over preconceived notions, a point further emphasized by interviewees' detailed feedback.

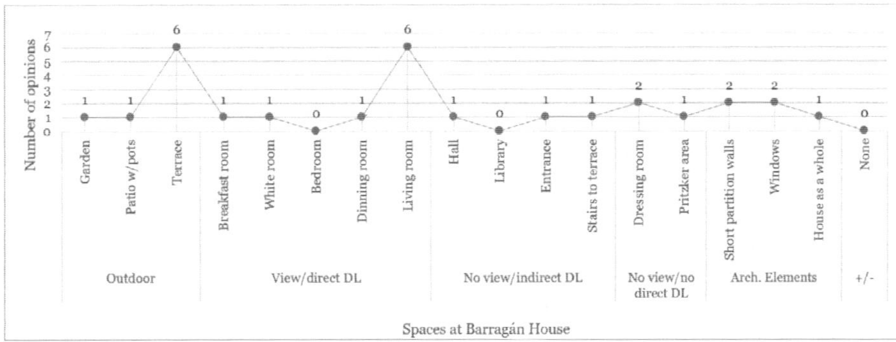

Figure 6: Illustrates the evaluation by participants of the impact of spaces and contained "elements" in them. Correlation to view to the outdoors and daylight (DL) is presented.

Figure 7: Roof-terrace, features the recognizable and characteristic "Mexican Pink" that Barragan used. No view over the city.

Figure 8: Living room, view over the back garden. Right wall, replica of Josef Albers's "Homage to the Square".

Luis Barragán House (1942) Mexico City. Photos: Author, 2023.

The qualities most frequently identified as having the strongest influence on the perceived atmosphere were clear: "light" was mentioned on fourteen occasions, "colour" on six, and "window/view" three times. Beyond these, there were three distinct spatial/architectural aspects, which, while not frequently mentioned, were described in a particularly insightful and analytical manner. First, there is the use of low walls as divisional elements in spaces (Figure 10); these walls, standing between 1.90m to 2.50m tall in rooms that have a 6.00m ceiling height, obstruct views. Second is the non-linear layout of spaces, filled with corners, which lends a "mysterious" and "surprising" quality to the atmosphere (Figure 11). Third are the transitional spaces or doorways, some as low as 1.90m, which create a "compression and release" effect (Figure 9). This effect has been perceived in two ways: positively, as it intensifies the experience and sets the tone for the next space, and negatively, as it can invoke feelings of claustrophobia. Windows, unsurprisingly, were often associated with views, light, or a combination of the two. Emphasis was placed on their role in "framing the view", "bringing the outside in", and fostering a "connection to nature" (Figure 8 & Figure 13). Intriguingly, the few negative atmospheric perceptions were often tied to a lack of view or light. This was particularly evident in the terrace (Figure 7), a space that left dual impressions—both overwhelmingly positive and distinctly negative. The primary source of dissatisfaction in this area stemmed from the high walls that, while offering a visual conduit to the sky, obstructed any panoramic view of the surroundings.

Regarding the concept of affordances, nearly all respondents—twenty-two in total—indicated they had similar intentions or inclinations while in BH. Cross-referencing these with their perceptions of emotion and atmosphere reveals a parallel between the "afforded" actions and space. Atmospheres described as peaceful, calm, and tranquil seemed to suggest an activity that is receptive in action. These included sitting, lying down, or simply relaxing, and they were often accompanied by complementary activities like sipping coffee or wine, reading, contemplating, or listening to music or the rain. The most dynamic engagement was the desire to perform a head-stand, demonstrated by a yoga practitioner, while the most receptive action responses were to "just be present" and to be introspective. Generally, the house, in its varied spaces, positively impacts the emotional states of visitors through the perceived atmosphere, encouraging actions, dynamic or receptive, that resonate with that emotional state. The immersive experience and movement through the house, navigating various lighting zones, enable individuals to position themselves within a space and discern the distinct atmospheres present in the house, mediated by light and affordances.

Figure 9: Transition point from vestibule to living room.

Figure 10: Study area between living room and studio. Low walls that hinder vision but allow natural light over.

Luis Barragán House (1942) Mexico City. Photos: Author, 2023.

As a pilot study, there are a few limitations to be considered and adjusted in a future study. The influence or bias that an interviewee can have due to explanations provided by the tour guide is to be taken into account. The predetermined path and sequence of spaces in a guided tour and the possible influence of other people´s movements and reactions when visiting the house in groups of six also creates the possibility that an individual's consciousness is shaped by their personal and shared experiences. It is important to mention that the visit duration is of relevance, where the approximately 1-hour visit and the reasonably constant weather conditions make no room for the appreciation of daylight dynamics and the experience of the house for extended periods. From my perspective as a researcher and observer who has spent several extended days inside the house, I've noted that the daylight's character and atmosphere remain rather consistent, with its natural gradual variations as dusk approaches. The most notable changes in the Atmosphere are during dusk and night when artificial lighting takes over (Figure 12). The interview process made it clear that the experience at BH was not so consciously lived, but happened almost automatically, just as Goeritz expressed in his EA manifesto.

IOP Conf. Series: Earth and Environmental Science 1320 (2024) 012021 doi:10.1088/1755-1315/1320/1/012021

Figure 11: Nonlinear layout of spaces, filled with corners.

Figure 12: Twilight moment, artificial lighting takes over.

Figure13: Windows that provide view selection and framing

Luis Barragán House (1942) Mexico City. Photos: Author, 2023.

7. Conclusion

In conclusion, the intricate interplay between light, space, and emotion in architecture manifests as a multifaceted equation where architecture is perceived through experience and the assimilation of offered affordances. Visual cues, such as windows, views, and light, play a pivotal role in reinforcing affordances. They have the power to shape individual perceptions, emotions, and behaviours in space, thus enriching the overall architectural experience. Built environments like the BH exemplify the deliberate design of atmospheres buttressed by light and colour, offering visitors an emotive architectural experience. The visitors´ personal experiences are witness to how Barragán's work resonated on an individual level and was interpreted depending on the perceived affordances and atmospheres. Simultaneously these perceived concepts triggered a particular emotional response. In conjunction Affordances, Atmospheres and emotional response are in synergy with light and architecture.

The pilot case study affirms, through its interviews, that architectural design together with light interaction is discerned in ways that significantly influence atmospheres. This is instrumental in comprehending how architecture is "felt" through our emotions, a sentiment articulated by E. Canepa (14). The revelations from this pilot study echo H. Plummer's characterization of Luis Barragán as a luminary in the realm of light, crafting spaces using light and its interplay with shadow. These atmospheric spaces invite dwelling and are capable of evoking emotional responses (33).

The introduction of the Inner Development Goals (IDGs) (34) as an extension of the Sustainable Development Goals from the UN, is an initiative that underscores the significance of prioritizing human inner growth and development. Designing architectural spaces that actively incorporate the dimensions of the IDGs can foster sustainable development, benefiting both individuals and society at large. Dimensions like Being, Thinking, and Acting present opportunities to integrate EA for well-being, emphasizing the creation of atmospheres and cultivating inspiring affordances through light. The forthcoming cross-referencing of measured data, in tandem with the experiential insights from interviews, will aim to deepen our understanding of the relationship between light, space, and the perceived atmospheres outlined in this paper's findings. As EA, BH, far from being an experimental architectural space such as Goeritz´ El Eco, is a building that delivers exactly what is described in the EA manifesto: *"creator or receiver of our time aspires to something more than a nice and adequate house. He/she asks – or will have to ask one day – for architecture and its modern means and materials, a spiritual elevation; simply put an emotion".*

8. Acknowledgements

I am deeply grateful to my PhD supervisors Meike Schalk and Ute Besenecker for their invaluable guidance in shaping this paper. My thanks also extend to Casa Barragan's director, Esguiarte, G., for granting access to Barragan House. Further appreciation goes to Bertil & Britt Svenssons Foundation and the Swedish Energy Agency for their crucial financial support, making this project feasible.

9. References

1. Montwill A, Breslin A. Let there be light. The Story of Light from Atoms to Galaxies London: Imperial College Press; 2008.
2. DiLaura Dea. The Lighting Handbook : reference and application. 10th ed. New York: Illuminating Engineering Society of North America; 2011.
3. Chalmers D. Facing up to the problem of consciousness. Journal of Consciousness Studies. 1995; 2(3): 200-19.
4. Gómez-Pérez A. ALN/CITYX Venice: Attunement [Digital lecture].: Arts Letters & Numbers; 2021 [cited 2023 05 12. Available from: https://www.youtube.com/watch?v=Eiw7YYmNld4.
5. Goldhagen SW. Welcome to Your World: How the Built Environment Shapes Our Lives New York: Harper; 2019.
6. Morais F. Mathias Goeritz Mexico City: Universidad Nacional Autónoma de méxico; 1982.
7. Colombetti G, Evan T. The Feeling Body: Towards an enactive approach to emotion. In M WFOaU, editor. Developmental Perspectives on Embodiment and Consciousness. New York: Lawrence Erlbaum Associates; 2008.
8. Gómez-Pérez A. Timely Meditations: Selected Essays on Architecture Montreal: Rightangle International; 2016.
9. Cuahonte L. El Eco de Mathias Goeritz Mexico City: UNAM Instituto de Investigaciones Estéticas; 2007.
10. Goldhagen SW. Ultraviolet: Alvar Aalto´s Embodied Rationalism. Hardvard design magazine. 2008: p. 38-52.
11. Carranza LE. Barragan Lecture #1: "On Time and Experience" [Video recorded lecture].: Vitra Design Museum; 2023 [cited 2024 02 14. Available from: https://www.youtube.com/watch?v=5YyncSECDXc.
12. Garza Usabiaga D. Las Torres de Satélite: ruina de un proyecto que nunca se concluyó. 2009 January;(94): 127-152.
13. Garza Usabiaga D. Mathias Goeritz y la integración plástica. Artes méxico. 2015;(116): 32-39.
14. Canepa Eea. Atmospheres: Feeling architecture by Emotions. Ambiances. 2019; 5.
15. Arbib MA. When Brains Meet Buildings United Kingdom: Oxford University Press; 2021.
16. Pallasmaa J. Dwelling in light: Tactile, emotive and life enhancing light. Daylight Matter(s). 2016; Autumn(26): 4-13.
17. Ruzzon D. Tuning humans with architecture [Conference]. Matosihnos; 2022 [cited 2024 02 14. Available from: https://vimeo.com/aceanfa/review/739028277/750391dc59.
18. Böhme G. Atmosphere as the Fundamental Concept of a New Aesthetics. Thesis Eleven. 1993;(36): 113-126.
19. Stewart K. Atmospheric Attunements. Environment and Planning D: Society and Space. 2011 June 22; 29: 445-453.
20. Gibson JJ. The Ecological Approach to Visual Perception New York: Psychology Press Classic; 1979.
21. Robinson S. Architecture is a verb New York: Routledge; 2021.
22. Whitehead AN. The Eighteenth Century. In Science and the Modern World. New York: The Free Press; 1997.

23. Husserl E. Analyses Concerning Passive and Active Synthesis. Lectures on Transcendental Logic Bernet R, editor. Dordrecht: Springer Science+Business Media; 2001.
24. Ruzzon D. Neuroscience applied in Architecture [Digital interview].; 2022 [cited 2024 02 14. Available from: https://www.youtube.com/watch?v=_wyZ0VLCzjM.
25. Robinson S, Pallasmaa J. Mind in Architecture: Neuroscience, Embodiment, and the Future of Design London: The MIT Press; 2015.
26. Griffero T. Architectural Affordances: the Atmospheric Authority of Spaces. Architecture and Atmosphere Espoo: Tapio Wirkkala-Rut Bryk Foundation; 2014.
27. Norman DA. Emotional Design: Why We Love (or Hate) Everyday Things New York: Basic Books; 2004.
28. Riggen A. Luis Barragán: Escritos y conversaciones Madrid: El Croquis; 2000.
29. Noelle L. El legado de Luis Barragán, 1902-2002. Cuadernos de arquitectura. 2002.
30. Protoomo O. Asumisen ihanteet Suomessa: tutkimusraportti. ; 2020.
31. Pauwels L, Mannay D. The SAGE Handbook of Visual Research Methods. 2nd ed. London: SAGE Publications; 2020.
32. Josephy M, Pühl M. Conversation about archive at the Barragan Foundation. 2023 May 30. Digital interview.
33. Plummer H. The Risk and Necessity of Poetic Imagination. In Greenlee C, Hinders K, editors. Juhani Pallasmaa 2011 Recipient of the Plym Distributed Professorship.: University of Illinois, School of Architecture; 2011. p. 15.
34. Jordan T, Reams J, Stålne K, Greca S, Björkman T, Dawsin T. Inner Development Goals: Background, method and the IDG framework. Full Report. ; 2021.
35. Griffero T. Architectural Affordances: the atmospheric authotity of spaces. Architecture and atmosphere. 2014;: 15-47.
36. Garza Usabiaga D. Mathias Goeritz y la arquitectura emocional. Una revisón crítica (1952-1968) Mexico City: Vanilla Planifolia; 2012.
37. Rueda Velázquez , de Rentería Cano , Martínez Durán A. Visual constructions: crossed views from architect and photographer Luis Barragán versus Armando Salas Portugal, José Antonio Coderch versus Francesc Català-Roca. Revista 180. 2019 December: p. 52-63.
38. di Pellegrino G, Làdavas E. Peripersonal space in the brain. Neuropsychologia. 2015; 66: 126-133.
39. Hach S, Schütz-Bosbach S. In (or outside of) your neck of the woods: Laterality in spatial body representation. Frontiers in psychology. 2014; 5(123).
40. Partida Muñoz MG. Armando Salas Portugal: Photography: A Collaborative Communication Tool. A: Seminario Internacional de Arquitectura Moderna y Fotografía. 2015.
41. Quiñones. Mi casa es mi refugio: At the Service of Mexican Modernism in Casa Barragán. Avery Review. 2020 June.
42. Wilson R. Evidence of "Doing Nothing": The Utopic Document of Lacaton & Vassal. Architectural Theory Review. 2013: p. 30-45.
43. UNESCO. World Heritage Convention: Luis Barragan House and Studio. [Online].; 2004 [cited 2023 05 05. Available from: https://whc.unesco.org/en/list/1136/.
44. Abramsom DM. Values of Obsolescence. In Paine A, Holden S, Macarthur J, Abramson DM, editors. Valuing Architecture: Heritage and the Economics of Culture. Amsterdam: Valiz; 2020. p. 24-39.
45. Kirtley C. Eye-Tracking as a Method of Visual Research. In Pauwels L, Mannay D, editors. The SAGE Handbook of Visual Research Methods. 2nd ed. London: SAGE Publications; 2020. p. 143-153.
46. Praun S, Stratimirovic A. You say light, I think shadow Stockholm: Art and Theory Publishing; 2015.

A review study of space perception and navigation of people with low vision: is simulated low vision a reliable methodology?

Jingying Dong[1], Caroline Karmann[1]

[1]Laboratory of Architecture and Intelligent Living, Karlsruhe Institute of Technology (KIT), Karlsruhe, Germany

Corresponding author: jingying.dong@kit.edu

Abstract: The inclusion of visually impaired participants in research protocols concerning their perception of space and navigation is essential for the reliability of the results, given the strategies developed by the people concerned in everyday life. However, the diversity of visual impairments, the scarcity of participants and possible safety issues due to obstacles in the physical space induce limitations and prompt researchers to look into alternative methodologies. Simulated low vision is seen as an option. This method involves sighted participants wearing goggles with customized filters or watching processed images in virtual environments. The objective of this study is to investigate the reliability of simulated low vision as a research method to describe the space perception and navigation of people with visual impairment. We conducted a literature review and identified 36 quantitative studies on low vision spatial performance involving multiple user groups. Simulated low vision proved effective in small-scale spatial ability evaluation, such as object detection and distance estimation, but remained challenging regarding large-scale capacity, such as navigation with mobility requirement. Advances in virtual environments suggest that they are a good alternative to goggles and screen displays because of their precision in mimicking ocular problems in simulation settings. Finally, the use of head-mounted-display (HMD) by people with real low vision could open up the possibility of greater testing in safer and controlled conditions, but requires confirmation of the validity of the protocols.

1. Background

Visual perception is important for human interaction with the world, providing crucial information on spatial interaction, distance estimation, and obstacle recognition, enabling effective wayfinding and orientation. Yet, the transition from merely sensing the environment to perceiving it and then successfully navigating through it becomes an even more critical concern when considering individuals with visual impairments. “Low vision” (LV) means having impaired vision that cannot be corrected by any glasses [1]. According to a meta-analysis of population-based datasets, 216.6 million, roughly 2.8% of world population had moderate to severe visual impairment [2] and 60% of them are aged 65 and over. From social level, only 16% of people with visual impairment at working age are employed [3]. The barrier they faced when moving around, navigating in city and building, getting to and from where they work, are major impediments that isolate this group from the rest of the society. As a response, people with LV typically develop strategies to alleviate their struggles and enhance their overall interaction with spaces. Research pertaining to space perception and cognition proved the importance of visual information (albeit damaged and deficient) in navigation and mobility and endeavored to tap into the barriers that exist in the built environment, in particular the complexity and inaccessibility of the environment from the perspective of people with LV.

1.1. Challenges

The challenges inherent in research on space perception and cognition of people with visual impairment stem from three primary sources:

- **Control over studied variables**: As orientation and mobility assume a critical role in navigation and wayfinding, it frequently introduces a dynamic observation point against a certain route, necessitating constant visual updating [4–6]. Various modes of interaction between an individual and the environment (e.g. active and passive exploration) can introduce uncertainty and stochastic components into the final output.
- **Difficulty in recruiting subjects**: the relative scarcity of the people with LV as well as the diversity of types and severity of visual impairments has led to a smaller and/or non-homogeneous sample, which limits the type of analyses and statistical effects targeted [7]
- **Safety concerns**: Peli [8] also highlighted concerns pertaining to the safety of subjects resulting, for example, from the danger of not seeing the obstacles in experimental context when including people with LV in experimental procedures.

As a result, "simulated LV", defined as an artificial alteration of one's visual perception to mimic the visual barriers faced by individuals with LV, has helped to advance scientific knowledge in the field of visual impairment. There are two classical ways to simulate LV:

- by using specialized filters or lenses to reproduce optical distortions. Fully-sighted people will therefore wear filters in the form of blindfolds and glasses to mimic LV sight
- by post-processing images (or videos) and then presenting them to a sighted subject using a computer screen or head-mounted display (HMD).

Using these technologies, many studies resort to simulating visual impairments by involving sighted individuals (details in Table 1), allowing for the expansion of sample size and precision in controlling impairment severity. However, previous practices have also highlighted an adaptation effect in the visually impaired population, demonstrating enhanced spatial perception and cognition through their long experience of interacting with spaces with limited visual ability. The presence of similar effects in the sighted population who suddenly have their vision diminished via a simulation technology is not guaranteed, and the same experimental method can yield different results depending on the specific objectives of the study (for example, obstacle detection tasks are different from navigation tasks). Moreover, the necessity for a dynamic observational perspective also introduces another methodological challenge to LV simulation since neither goggles nor post-processing method can replicate a full field of vision (FOV) covered by the human eye itself, which includes binocular overlap of up to 200 degrees ($\pm$100 to the temporal side), not counting the additive effect of eye movement [9]. The FOV deficiency arising from traditional flat glasses and 2D images presents an opportunity for compensation through the emerging technology of VR glasses or curved screen, which provide an expanded field of view. With the introduction of VR technology, the reliability of easily manipulated virtual environments (VE) used in LV research in determining real-world validity are another contentious topic [10]. The function of VE is particularly pronounced when investigating the influence of environmental design on the visually impaired. The virtual presentation of the environments with post-processing resolved the problems associated with the variables of the physical experiments and strengthened safety measures. However, these experimental protocols do not involve visually impaired persons, which still raise ethical questions [11].

1.2. Study objective

Our objective is to identify and compare existing tools used in assessing the spatial perception of individuals with visual impairment. In particular, we are interested in the reliability of simulated LV.

Figure 1 to 4 depict some existing methods for simulating LV. With this review study, we aim to improve the accuracy, reliability and suitability of future research methods used in the field of LV in architecture.

Figure 1. Typical vision-degrading goggles with blind (left) and blurred foil (right) [31]

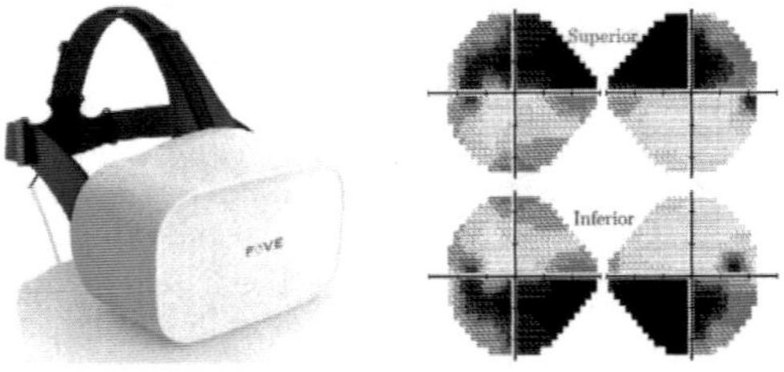

Figure 2. VR headset with integrated eye tracking and the perimetric data used to simulate vision filed loss (VFL) [21]

Figure 3. Simulated visual impairment produced from original photo (left) with blurring and contrast reduction (right) [50]

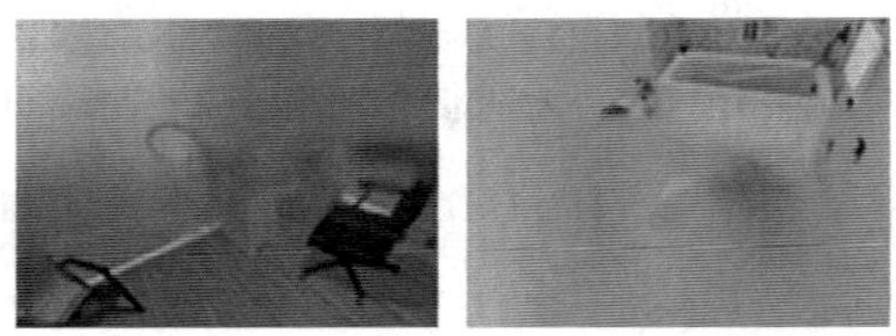

Figure 4. Example of live-screen presented in VR with superior VFL (left) and inferior VFL (right) 21]

2. Methodology

We conducted a literature review on the existing tools and methods used to study the spatial perception of individuals with visual impairment looking at both the methods applicable to LV research, and their reliability for different types of tasks. The preliminary round of screening is based on keyword searches in repositories including Web of Science, Google Scholar, PubMed and SAGE Journals. We looked at the following key words: "visually impaired", "low vision", "field loss of vision", "acuity loss", "visibility", "cataract", "glaucoma", "macular degeneration", "retinitis pigmentosa", "navigation", "wayfinding", "spatial representation", "spatial cognition", etc. Studies concerned with the introduction or validation of technological aids (e.g. apps, intelligent tools) were excluded. We pre-selected papers based on the above keywords with further screening in:

1. Type of publication: papers were written in a peer-reviewed journal, a peer-reviewed conference paper, a Ph.D. thesis, or as a book edited by an established publisher
2. Language: papers were available in English (written or translated) for accurate interpretation.
3. Objective: papers involved human participants and spatial tasks, exclusive of object recognition, reading or typing; or focused on the impact of different environmental factors on the spatial perception, navigation and cognition from LV subjects or on differences between LV and full sighted group
4. Experimental methodology: papers were based on quantitative research method, which involved using numerical data and statistical analysis to draw structured conclusions.

3. Review outcomes and analysis

In total, thirty-six studies published between 1987 - 2023 were judged eligible and were therefore included in this review. Table 1 provides an overview of the paper found, documenting the participants' type of vision (real LV vs. simulated LV vs. fully sighted), the experimental setting (on-site vs. virtual environment), as well as the type of measurements used.

Prior research has explored how different types of visual impairment impact environmental perception through cross-modal and internal comparison setups. Cross-modal comparisons involve comparing sighted individuals with LV groups, including simulated and real LV participants. Internal comparisons focus on visually impaired individuals, considering differences in impairment types and severity (e.g. acuity loss (AL), peripheral vision loss (PFL) with various severity). In this scenario, simulated LV participants are more frequently included to enhance control over variables of visual condition. Most studies incorporating real LV subjects tend to discuss subgroup performance as a whole without delving into within-group categorizations, despite the fact that there are differences existing between subgroups (AL and PFL) [7,12,13]. The adaptation effect, denoting the phenomenon whereby individuals with sustained vision impairment (blind subjects) exhibit heightened perceptual and cognitive abilities in contrast to those with short-term blinded vision [14], stands as a main experimental limitation of artificially induced alterations in vision, which is less addressed in common vision research. Among the reviewed literature, only a handful of articles encompassed both simulated and real LV subjects, aiming to discern disparities between these two groups. In all reviewed papers, this adaptation effect was only mentioned in comparisons of subjects with AL.

Hegarty et al. [15] proposes a differentiation between "small scale" and "large scale" spatial abilities, putting emphasis on the difference between observing an object in space (smaller than the human body) and perceiving the space (in which the human body is surrounded). This implies a possible categorization for the first six target tasks discussed in this review: *small scale spatial tasks* pertain to mentally transforming representations of small shapes or objects, such as blocks, which aligns with dimension estimation and obstacle detection measurement in the context of this paper, while *large scale spatial tasks* entail more intricate tasks, such as mobility, wayfinding and navigation. The discrepancy between space scale was also indicated in an empirical study involving blind individuals, where congenitally blind participants encountered greater difficulty in spatial tasks within a large-scale space than a small-scale space [16]. Figure 5 describes frequency distribution of sample and method types for six different types of tasks: spatial representation (SP), target finding (TF), object detection (OD), gazing behavior (GB), wayfinding and mobility (W&M), navigation and wayfinding (N&W). Among the thirty-six studies, twenty researches incorporated individuals with simulated LV as participants. Eleven opted for virtual environment (screen display or Immersive Virtual Environment (IVE)). The following sections, intended to respond to the reliability of the method, will be organized according to the type of spatial task in relation to the scale considered.

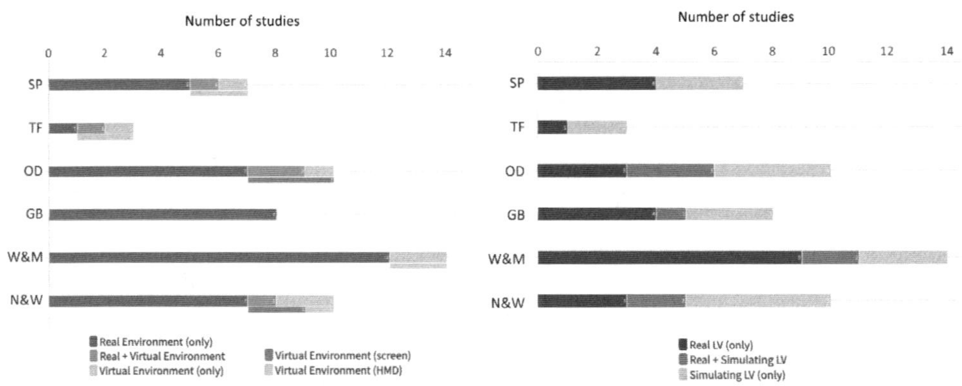

Figure 5. Frequency distribution of selected studies in term of displayed environment and participants type for different spatial tasks (Note: Some studies were double counted due to multiple research objectives.)

Table 1. Summary of 36 studies

Author, date	Participant			Environment			LV Simulating method		Objective measurement							Subjective measurement
	Real LV	Simulated LV	Sighted	On-site	VE (screen)	VE (HMD)	Goggle	Image processing	wayfinding& navigation	Map sketching	Target searching	Dimention estimation	Obstacle detection	Walking task	Gazing and fixation	Psychological questionnaire
Andrew Freedman (2019)	×	$\times_a$	×	×			×		×					×	×	
Aspinall. P. et al. (2014)	×		×		$\times_p$				×							
Barhorst-Cates et al. (2016)		$\times_p$		×			×		×							
Barhorst-Cates (2017)		$\times_a$		×			×		×							
Barhorst-Cates (2019)		$\times_p$		×			×		×							
Barton, R. (2014)		$\times_p$				$\times_s$		×	×							
Black, A.,et al. (1997)	×		×	×									×	×		
Blades, M.(2002)	×		×	×					×	×				×		
Carpenter. B. (2018)		$\times_a$		×	$\times_a$			×					×			
Chow-Wing-Bom, H. (2020)		$\times_p$				$\times_s$		×			×					
Fortenbaugh, F. C. (2007)		$\times_p$				$\times_s$		×				×			×	
Fortenbaugh, F. C., Hicks, J. C., & Turano, K. A. (2008)	$\times_p$			×		$\times_s$						×			×	
Iwata, M., & Kitamoto, H. (2019).	×		×	×										×		×
Jones. P etal (2020)		$\times_p$		×		$\times_s$		×			×				×	
Kalia et al. (2008)	×		×	×	$\times_{a,p}$			×	×	×						
Katemake. P. (2019)	×	$\times_{a,p}$		×			×						×	×		
Kuyk, T. et al. (1996)	×			×									×	×		
Leat, S. J., & Lovie-Kitchin, J. E. (2008)	$\times_{a,p}$			×									×	×		
Legge, G. et al. (2016)	×		×	×								×				
Legge, Gordon E (2010)		$\times_a$		×			×						×			
Liu, S etc. (2021)	×	$\times_a$			$\times_s$			×					×			
Matsuda, Y et al (2019)	$\times_{a,p}$		×	×										×	×	×
Matsuda, Yuji, et al. (2019)	×		×	×										×	×	
Matsuda. Y et al. (2021)	×		×	×										×	×	
Ozlem Beli, (2010)	×			×						×						
Passini, R., Proulx, G., & Rainville, C. (1990)	×		×	×					×							
Peli D. (1987)		$\times_{a,p}$		×			×						×	×		
Rand et al, (2015)	×	$\times_a$		×			×		×							×
Rand. (2019)		$\times_a$		×			×					×				
Rousek, J. B., & Hallbeck, M. S. (2011)		$\times_{a,p}$	×	×			×							×		×
S.E. Hassan et al. (2007)		$\times_p$				$\times_s$		×					×	×		
Sippel, K, et al. (2014)	×		×	×							×					
Tarampi MR, Creem-Regehr SH, & Thompson WB (2010)		$\times_a$	×	×			×					×				
Thompson, Williams B. Shakespeare (2022)	×	$\times_a$			$\times_s$			×					×			
Yamamoto, N., & Philbeck, J. W. (2013)		$\times_p$		×			×					×			×	
Zou, X., & Zhou, Y. (2023).	×			×										×		×

*×a: Real low vision participant with acuity loss only; ×p: real low vision participant with peripheral field loss only
*×s: Virtual environment generated with simulated model ×p: Virtual environment generated by photorealistic picture/video

3.1. Small-scale spatial task: spatial representation and object perception.

Spatial representation refers to the mental or cognitive mapping of physical space [17,18]. It involves the ability to mentally perceive and manipulate the relationships between objects, distance, orientation and self-locations within an environment. Small scale spatial representation allows individuals to create and maintain an internal model of their surroundings, usually confined with a settled space, within which they can conduct subsequent tasks, including obstacle avoidance, distance estimation, object searching, etc. On aggregate, 20 of the 36 papers investigated the issues mentioned above. A common thread across the execution and evaluation of these spatial tasks is that the observed environment remains confined to a small-scale setting (single room or corridor), often restricted to a field of view of 180 to 360 degrees (could involve head movement) and within eye-catching distance, yet devoid of strenuous body-movement or rotation requirement. A relatively high reliability of simulated LV was seen in studies pertaining to such small-scale representation tasks. Legge, Gage, et al., [19] involved simulated LVs (both AL and PFL) via goggles as subjects and reached a conclusion that there was no evident difference between the spatial estimation accuracy of visually impaired individuals and sighted in judging room dimensions, except in cases of severely reduced visual acuity. This finding was further validated in her subsequent study, which included real LV subjects [20]. In several simulated LV studies [21–23], a common finding indicated that individuals with PFL required more time to find target objects compared to sighted individuals with increased eye and head movements as a compensate strategy. Notably, this result aligns with the findings of [24], which involved real PFL, reinforcing the consistency between simulated and real-world PFL cases. Post-processing has been applied more often in obstacle avoidance and discrimination studies, and it has been verified to be with high reliability through direct comparison to real LV. In its initial stages, image processing software employed fundamental techniques such as Gaussian blurring, texture overlays, and tonal adjustments to simulate various eye conditions, serving as a simplified tool for training in accessible design [25]. Designing Visually Accessible Spaces (DeVAS) is another newly-developed tool developed to better access AL perspectives and predict potential hazards in the architectural space [26]. From a photometrically accurate rendering, it uses a parametric post-processing method based on post-editing, combining edge recognition technology and a regional blur effect to calculate an alternative image representing what a person with LV would see. A corresponding predictive value, hazard visibility score, is then calculated from the degree of edge-line continuity to predict the hazard index of the scene for the LV group. In a subsequent validation study that included both simulated LV and real LV subjects, it was confirmed that DeVAS has a good reliability, particularly concerning obstacle discrimination. Liu et al. [27] also utilized both simulated LV, this time with goggles (loaded with diffusive films to accurately reduce acuity to 1.2 logMAR and 1.62 logMAR), and real LV participants to verify the hazard visibility score. The experiment demonstrated the reliability of post-processed images in predicting obstacle detection ability in individuals with visual impairments. For a common target, another system, CatARact [28] applied an intricate pipeline effect successively following acuity and contrast reduction, color shift, etc. to simulate cataracts and has been undergoing validation with real cataract patients.

3.2. Large scale spatial task: mobility, wayfinding and navigation.

Montello [29] proposed mobility as another important layer of differentiation between small-scale and large-scale spatial tasks: whether or not a place can be visually understood from a place without the necessity of movement. In contrast to tasks in static and controlled settings, tasks related to walking, wayfinding, and map-sketching are categorized as large-scale spatial abilities. These activities involve switching viewpoints and require the skill to navigate through extensive environments, such as buildings or outdoor landscapes.

Mobility, referring to the ability to move freely in a space, is a comprehensive capability that builds on the ability to avoid obstacles, and underlying support for navigational behavior. On aggregate, 14 out of 36 papers investigate mobility-related issues. Most of the experiments are conducted on-site and incorporated real LVs as subjects. Black et al. [12] and Leat and Lovie-Kitchin [7] verified that PFL can significantly slow walking speed, whereas AL impaired self-locomotion perception. Matsuda et al. [13] and Kuyk, T. [30] derived the similar conclusion with both AL and PFL patients. For an on-site experiment that included simulated AL and simulated PFL participants, Tarampi [31] indicated spatial updating remains accurate during locomotion under conditions of significantly degraded AL and contrast, albeit with heightened variability compared to non-degraded viewing. Peli [32] showed that neither visual impairment had a clear effect on walking speed unless the impairment was extremely severe. Although the researchers in the latest experiment explain that this result may have been influenced by the exclusion of important hazards in their experimental set-up, the disparity between the results prompts caution about the results obtained when using simulated LV in mobility-related studies. Freeman [33] also tested walking ability incorporating simulated LV with blurred foil. Conversely, he found that the severity of AL affected walking speed, attributing it to longer gaze times and increased pauses during walking. This behavioral change aligns with the adaptive tendencies observed in real LVs during hospital wayfinding, as summarized in Rousek and Hallbeck's [34] experiments. In several studies that both include simulated LV and real LV, Bochsler et al. [35,36] have confirmed that AL alone could lead to an adaptation effect in terms of locomotion. Katemake et al. [37] displayed a consistent output with this claim, showing that chronically visually impaired subjects exhibit better obstacle avoidance performance and higher speed of movement. Fortenbaugh, Hicks and Turano [38] could not detected a similar adaptation effect comparing simulated PFL and real PFL subjects. One possible reason for this is that PFL can be partly compensated for by eye movements with greater amplitude compared to AL, suggesting that the impact of visual impairment on adaptation and related perceptual processes may differ depending on the type and level of visual impairment.

The navigation task appears to be a more comprehensive task including a multi-processing of perception, updating and cognition, and requires the integration of object motion, information, and experience. The results exhibit a diverse bias in studies using different simulation methods. Rand's [39] study, which involved simulated LV, revealed that AL led to increased errors in navigation and biases in orientation, with these effects becoming more pronounced as the severity of the impairment increased. In Barhost's two experiments pertaining to spatial navigation, which adopted the same methodology with simulated PFL [4,40] in two different environments, it was observed that, compared to a simple navigation route in which only a severe PFL would have a negative effect in orientation and map-sketching performance, individuals showed more pronounced effects in navigating through an environment with higher complexity or richer information. Barton, Valtchanov and Ellard [41] derived a similar conclusion via a study in VE with simulated PFL. In a recent navigating and map-sketching experiment, Zou and Zhou [42] incorporated real LV (regardless of type of impairment) who walked and depicted various routes through a hospital. The success of the task was largely influenced by the environmental elements in the route and the patients' individual strategies. Given the constraints imposed by the experimental setup and environmental intricacies, confirming if only simulated low visual acuity leads to bias through cross-modal comparisons remains challenging. In a different experiment involving real LV within on-site environments [43], it was demonstrated that subjective learning efficiency biases originating from distinct environmental tasks (e.g. subjects are informed with a map-sketching or distance estimation task before experiment) also impact navigation performance. Involving both populations in the same environment and protocol for direct comparisons is the only way to understand whether simulation is a reliable method. While navigational efficiency and success may not directly reflect simulated reliability, the observed commonality in behavior exhibited by both simulated LV and real LV subjects warrants further discussion.

Overall, the difference in information load and cognitive processes between "small scale" and "large scale" spatial tasks, as identified by Hegarty et al. [15], was aligned with the findings in this review: various forms of simulation demonstrated high reliability when applied in small scale spatial tasks, while larger scale spatial tasks exhibited more controversy and bias in the results. Gaze behavior is central to both types of tasks, and calls for a more specific examination.

3.3. Gazing behavior

Gazing behavior, including eye movement, fixation, saccades and blinks, is a concomitant activity of environmental observation that helps researchers to understand perceptual and cognitive processes non-invasively. On aggregate, eight out of the 36 papers looked at gazing behavior. Matsuda et al. [13] studied the gazing behavior of a real LV group during street navigation, which revealed that participants with PFL exhibited wider angles in eye-fixation targets compared to participants without visual impairment, while the most significant deviation for participants with AL was a significantly shorter distance and longer duration to their preferred eye fixation targets. The part related to PFL is consistent with several studies, which simulated PFL respectively via blind mask [44] and mixed reality technology with an eye tracking system [21]. Freedman et al. [33] examined gaze behavior during navigation in people with AL, comparing both simulated LV and real LV, and consistently found that gaze patterns were essentially the same between the two groups. Aspinall [45] investigated gaze behavior among individuals with AL using a photorealistic environment displayed on a projected screen and revealed a correlation between differences in pupil diameter and fixation counts, and the participants' visual acuity. Almost all of the gazing behavior studies mentioned above make the point that the subjects generally spent more time looking towards the ground and boundary (between the floor and the wall) in either simulated or real LV perspective. Without compromising the accuracy of the results, applying simulated LV provides the opportunity to conduct separate analysis between a single type of impairment (PFL and AL), which otherwise can be a challenging case in the real LV study.

Despite its efficiency in working procedure, however, the two types of simulation protocol share a common drawback of eye-movement tracking, as most of the occlusion of the visual field or loss of precision is not distributed uniformly across the retina, instead commonly displayed as a region of opacity which remains invariant on the retina [46]. Jones et al. [21] offers a novel way of integrating eye tracking and real-time image processing in AR to enable floating blurring for simulated PFL, indicating that mixed reality could be a potential form of LV research. Although the loss of this trait does not affect some of the objective measurement, the impact on other items, like mobility or navigation in space, is still unknown. There is also a lacking of monocular visual acuity consideration. Amblyopia research has consistently demonstrated that some form of suppression occurs when there is asymmetric visual input. Specifically, individuals tend to ignore the visual input from the worse eye, leading to binocular visual performance equal to the monocular performance of the better eye [47]. Barhorst-Cates, Rand and Creem-Regehr [4,40], conducted subsequent experiments that turned to monocular masks to weaken the side effects of binocular stereo fusion. However, there is also the alternative argument, which suggests that the eye with worse condition still has an impact on people's gazing behavior and spatial performance [22]. Generally, simulated LV and real LV exhibited comparable gaze behavior in various spatial tasks, indicating a certain level of reliability in terms of their influence in spatial observing strategy.

4. Conclusion and discussion: VE in future LV research

This paper provides an evaluation of the feasibility and reliability of various existing simulation tools applied to LV research. While simulated LV showed promising results for studies of small-scale spatial capabilities, such as obstacle detection and distance estimation, large-scale spatial capabilities, such as mobility-related studies, proved simulation LV less reliable. In term of small-scale spatial ability research, goggles and post-processing serve as useful tools for AL simulation; VE/VR integrated with AL or PFL simulations demonstrate substitutability, with HMD setups offering superior immersion compared to screen displays. Particularly, simulations with eye-tracking systems have proven to be more effective than static blur filters and offer promising prospects for enhanced spatial ability testing.

The usage of VE offer greater control in experimental environments, which is a key asset in research. The adaptability of simulated LV or real LV to simulated environments can be considered for effective strategies pertaining to exploring how environmental factors impact spatial performance. VR technology has advanced from VE to Immersive VE (IVE), enabling larger FOV and additional data collection through equipment like eye tracking. This facilitates understanding spatial patterns in the LV group.

Mixed reality technology further integrates simulated LV with the environment, allowing precise control of intrinsic (visual condition) and extrinsic (environmental factors) variables in LV research [21]. Correspondingly, the overlay function of digital information in AR can also be well adapted to LV simulation, integrating with the real environment while addressing the lack of eye tracking in normal goggle simulations [28,48].

Throughout the review of 36 studies, 28 included a physical environment in the experimental setting, with eleven studies utilizing VE or IVE, either as the primary or partial methodology. Among them, there are only four studies involving both real and virtual environments (e.g., photo or rendering), which demonstrated acceptable validity of VE or IVE as experimental settings pertaining to various research objects, including obstacle avoidance [49], dimension estimation, gaze behavior [38], and target finding in space [21]. However, spatial knowledge acquisition in a large-scale task, such as navigation, has been found to be different compared to in real-world environments [50]. This finding was aligned with previous research involving sighted participants, which indicate that there exists bias between orientation and navigation in desktop VE and direct experience in physical environment [51]. Two potential reasons may account for this:

- The disparity in scale between VE and reality. Media-displayed environment may distort the perception of architectural elements, particularly in confined screens. This distortion could significantly affect visually impaired individuals with partial visual field loss, though the potential mitigation through iso-scale simulation in HMDs remains unverified.
- Lack of locomotion in VE. Walking in real environments allows individuals to automatically update their position through physical movement, which is normally lacking in VE [15]. Empirical experiments involving sighted participants suggested that VR manipulated with whole-body rotation and joysticks is sufficient for the navigation task [52] but that outcome remained debated [53–55].

Except for the display media, the image style of VE seems to have additional influence. In terms of small-scale task, such as obstacle detection, Carpenter [49] verified that photorealistic VE and rendering VE can theoretically achieve the same efficacy. For complex environments, Kalia, Legge and Giudice [50] proposed that the presentation and sparsity of information in VE are key factors affecting navigation performance of the simulated LV group. This created the need for corresponding requirements in environment simulations, where high-fidelity and low-fidelity simulations can impact the output. Another potential factor influencing the experimental results is the presence of bias related to the participants' video game experience or their familiarity with VR technology. Experienced users demonstrated better navigation performance than inexperienced among the simulated LV group [50]. The adaptation effect to the VR necessitates the inclusion of more reasonable trial designs or larger sample sizes in VR experiments to mitigate this bias.

Finally, none of the studies fully replicated and verified the navigation scenario in VR for people with LV, despite this approach being considered having a certain predictive value in research involving sighted groups [56–58]. It remains to be seen whether the use of HMD for people with LV is a suitable scientific approach, in other words, whether people with LV identify the same obstacles and have a similar perception of their surrounding in both virtual and physical realities. This would be particularly interesting for large-scale capacity studies (such as navigation) as it would enable more flexible and safer experimental protocols, while guaranteeing the inclusion of visually impaired people in experimental protocols. Lastly, the question of the sound environment in addition to visualization in VR/VE would also be particularly relevant, given the great importance of auditory cues for the spatial perception of visually impaired people [59–61].

References

[1] NIH 2023 low vision
[2] Bourne R R A, Flaxman S R, Braithwaite T, Cicinelli M V, Das A, Jonas J B, et al 2017 Magnitude, temporal trends, and projections of the global prevalence of blindness and distance and near

vision impairment: a systematic review and meta-analysis *The Lancet Global Health* **5** e888–97

[3] Chu H-Y and Chan H-S 2022 Loneliness and Social Support among the Middle-Aged and Elderly People with Visual Impairment *IJERPH* **19** 14600

[4] Barhorst-Cates E M, Rand K M and Creem-Regehr S H 2019 Navigating with peripheral field loss in a museum: learning impairments due to environmental complexity *Cogn. Research* **4** 41

[5] Jacobson R D 1998 Cognitive Mapping Without Sight: four preliminary studies of spatial learning *Journal of Environmental Psychology* **18** 289–305

[6] Li H, Mavros P, Krukar J and Hölscher C 2021 The effect of navigation method and visual display on distance perception in a large-scale virtual building *Cogn Process* **22** 239–59

[7] Leat S J and Lovie-Kitchin J E 2008 Visual Function, Visual Attention, and Mobility Performance in Low Vision *Optometry and Vision Science* **85** 1049–56

[8] Peli E 1995 SIMULATING NORMAL AND LOW VISION *Vision Models for Target Detection and Recognition* (WORLD SCIENTIFIC) pp 63–87

[9] Strasburger H 2020 Seven Myths on Crowding and Peripheral Vision *i-Perception* **3** pp 1–46

[10] Lam A K N, To E, Weinreb R N, Yu M, Mak H, Lai G, Chiu V, Wu K, Zhang X, Cheng T P H, Guo P Y and Leung C K S 2020 Use of Virtual Reality Simulation to Identify Vision-Related Disability in Patients With Glaucoma *JAMA Ophthalmol* **138** 490

[11] Williams J 2008 The Declaration of Helsinki and public health *Bull World Health Organ* **86** 650–1

[12] Black A, Lovie-kitchin J E, Woods R L, Arnold N, Byrnes J and Murrish J 1997 Mobility performance with retinitis pigmentosa *Clinical and Experimental Optometry* **80** 1–12

[13] Matsuda Y, Hara T, Kashiwase M and Nishide K 2019 A study on the eye movement in people with low vision - A case study in an eye clinic *Japan Architectural Review* **2** 588–602

[14] Passini R, Proulx G and Rainville C 1990 The spatio-cognitive abilities of the visually impaired population *Environment and Behavior* **22** 91–118

[15] Hegarty M, Montello D R, Richardson A E, Ishikawa T and Lovelace K 2006 Spatial abilities at different scales: Individual differences in aptitude-test performance and spatial-layout learning *Intelligence* **34** 151–76

[16] Iachini T, Ruggiero G and Ruotolo F 2014 Does blindness affect egocentric and allocentric frames of reference in small and large scale spaces? *Behavioural Brain Research* **273** pp 73–81

[17] Taylor H A and Brunyé T T 2013 The Cognition of Spatial Cognition: Domain-General within Domain-specific *Psychology of Learning and Motivation* vol 58 (Elsevier) pp 77–116

[18] Herweg N A and Kahana M J 2018 Spatial Representations in the Human Brain *Front. Hum. Neurosci.* **12** 297

[19] Legge G E, Gage R, Baek Y and Bochsler T M 2016 Indoor Spatial Updating with Reduced Visual Information ed B Thompson *PLoS ONE* **11** 0150708

[20] Legge G E, Granquist C, Baek Y and Gage R 2016 Indoor Spatial Updating With Impaired Vision *Invest. Ophthalmol. Vis. Sci.* **57** 6757

[21] Jones P R, Somoskeöy T, Chow-Wing-Bom H and Crabb D P 2020 Seeing other perspectives: evaluating the use of virtual and augmented reality to simulate visual impairments (OpenVisSim) *npj Digit. Med.* **3** 32

[22] Chow-Wing-Bom H, Dekker T M and Jones P R 2020 The worse eye revisited: Evaluating the impact of asymmetric peripheral vision loss on everyday function *Vision Research* **169** pp 49–57

[23] Sippel K, Kasneci E, Aehling K, Heister M, Rosenstiel W, Schiefer U and Papageorgiou E 2014 Binocular glaucomatous visual field loss and its impact on visual exploration--a supermarket study. *PLoS One* **9** 106089

[24] Faubert J and Overbury O 2000 Binocular Vision in Older People with Adventitious Visual Impairment: Sometimes One Eye Is Better than Two *Journal of the American Geriatrics Society* **48** 375–80

[25] Banks D and McCrindle R J 2008 Visual eye disease simulator *Proc. 7th ICDVRAT with ArtAbilitation*

[26] Thompson W B, Legge G E, Kersten D J, Shakespeare R A and Lei Q 2017 Simulating visibility under reduced acuity and contrast sensitivity *J. Opt. Soc. Am. A* **34** 583

[27] Liu S, Liu Y, Kersten D J, Shakespeare R A, Thompson W B and Legge G E 2021 Validating a model of architectural hazard visibility with low-vision observers ed G Maiello *PLoS ONE* **16** 0260267

[28] Krosl K, Elvezio C, Luidolt L R, Hurbe M, Karst S, Feiner S and Wimmer M 2020 CatARact: Simulating Cataracts in Augmented Reality *2020 IEEE International Symposium on Mixed and Augmented Reality (ISMAR)* (Recife/Porto de Galinhas, Brazil: IEEE) pp 682–93

[29] Montello D R and Pick H L 1993 Integrating Knowledge of Vertically Aligned Large-Scale Spaces *Environment and Behavior* **25** 457–84

[30] Kuyk T, Elliott J, Biehl J and Fuhr P 1996 Environmental variables and mobility performance in adults with low vision *J Am Optom Assoc* **67** 403–9

[31] Tarampi M R, Creem-Regehr S H and Thompson W B 2010 Intact spatial updating with severely degraded vision *Attention, Perception, & Psychophysics* **72** 23–7

[32] Pelli D G 1987 The Visual Requirements of Mobility *Low Vision* ed G C Woo (New York, NY: Springer New York) pp 134–46

[33] Freedman A, Achtemeier J, Baek Y and Legge G E 2019 Gaze behavior during navigation with reduced acuity *Experimental Eye Research* **183** 20–8

[34] Rousek J B and Hallbeck M S 2011 The use of simulated visual impairment to identify hospital design elements that contribute to wayfinding difficulties *International Journal of Industrial Ergonomics* **41** 447–58

[35] Bochsler T M, Legge G E, Kallie C S and Gage R 2012 Seeing Steps and Ramps with Simulated Low Acuity: Impact of Texture and Locomotion *Optometry and Vision Science* **89**

[36] Bochsler T M, Legge G E, Gage R and Kallie C S 2013 Recognition of Ramps and Steps by People with Low Vision *Invest. Ophthalmol. Vis. Sci.* **54** 288

[37] Katemake P, Radsamrong A, Dinet É, Heng C W, Kuang Y C, Kalavally V and Trémeau A 2019 Influence of LED-based assistive lighting solutions on the autonomous mobility of low vision people *Building and Environment* **157** 172–84

[38] Fortenbaugh F C, Hicks J C and Turano K A 2008 The Effect of Peripheral Visual Field Loss on Representations of Space: Evidence for Distortion and Adaptation *Invest. Ophthalmol. Vis. Sci.* **49** 2765

[39] Rand K M, Creem-Regehr S H and Thompson W B 2015 Spatial learning while navigating with severely degraded viewing: The role of attention and mobility monitoring. *Journal of Experimental Psychology: Human Perception and Performance* **41** 649–64

[40] Barhorst-Cates E M, Rand K M and Creem-Regehr S H 2016 The Effects of Restricted Peripheral Field-of-View on Spatial Learning while Navigating ed A Smith *PLoS ONE* **11** 0163785

[41] Barton K R, Valtchanov D and Ellard C 2014 Seeing Beyond Your Visual Field: The Influence of Spatial Topology and Visual Field on Navigation Performance *Environment and Behavior* **46** 507–29

[42] Zou X and Zhou Y 2023 Spatial Cognition of the Visually Impaired: A Case Study in a Familiar Environment *IJERPH* **20** 1753

[43] Blades M, Lippa Y, Golledge R G, Jacobson R D and Kitchin R M 2002 The Effect of Spatial Tasks on Visually Impaired Peoples' Wayfinding Abilities *Journal of Visual Impairment & Blindness* **96** 407–19

[44] Yamamoto N and Philbeck J W 2013 Peripheral vision benefits spatial learning by guiding eye movements *Mem Cogn* **41** 109–21

[45] Aspinall P A, Borooah S, Al Alouch C, Roe J, Laude A, Gupta R, Gupta M, Montarzino A and Dhillon B 2014 Gaze and pupil changes during navigation in age-related macular degeneration *Br J Ophthalmol* **98** 1393–7
[46] Crabb D P, Smith N D, Glen F C, Burton R and Garway-Heath D F 2013 How Does Glaucoma Look? *Ophthalmology* **120** 1120–6
[47] Pardhan S 1996 A comparison of binocular summation in young and older patients *Current Eye Research* **15** 315–9
[48] Zhao Y, Hu M, Hashash S and Azenkot S 2017 Understanding Low Vision People's Visual Perception on Commercial Augmented Reality Glasses *Proceedings of the 2017 CHI Conference on Human Factors in Computing Systems* CHI '17: CHI Conference on Human Factors in Computing Systems (Denver Colorado USA: ACM) pp 4170–81
[49] Carpenter B 2018 Measuring the detection of objects under simulated visual impairment in 3D rendered scenes [Ph.D. Thesis]. University of Minnesota.
[50] Kalia A A, Legge G E and Giudice N A 2008 Learning Building Layouts with Non-Geometric Visual Information: The Effects of Visual Impairment and Age *Perception* **37** 1677–99
[51] Richardson A E, Montello D R and Hegarty M 1999 Spatial knowledge acquisition from maps and from navigation in real and virtual environments *Memory & Cognition* **27** 741–50
[52] Riecke B E, Bodenheimer B, McNamara T P, Williams B, Peng P and Feuereissen D 2010 Do We Need to Walk for Effective Virtual Reality Navigation? Physical Rotations Alone May Suffice *Spatial Cognition VII* Lecture Notes in Computer Science vol 6222, ed C Hölscher, T F Shipley, M Olivetti Belardinelli, J A Bateman and N S Newcombe (Berlin, Heidelberg: Springer Berlin Heidelberg) pp 234–47
[53] Klatzky R L, Loomis J M, Beall A C, Chance S S and Golledge R G 1998 Spatial Updating of Self-Position and Orientation During Real, Imagined, and Virtual Locomotion *Psychol Sci* **9** 293–8
[54] Ruddle R A and Lessels S 2006 For Efficient Navigational Search, Humans Require Full Physical Movement, but Not a Rich Visual Scene *Psychol Sci* **17** 460–5
[55] Taube J S, Valerio S and Yoder R M 2013 Is Navigation in Virtual Reality with fMRI Really Navigation? *Journal of Cognitive Neuroscience* **25** 1008–19
[56] Ventura M, Shute V, Wright T and Zhao W 2013 An investigation of the validity of the virtual spatial navigation assessment *Front. Psychol.* **4** 852
[57] Coutrot A, Schmidt S, Coutrot L, Pittman J, Hong L, Wiener J M, Hölscher C, Dalton R C, Hornberger M and Spiers H J 2019 Virtual navigation tested on a mobile app is predictive of real-world wayfinding navigation performance ed L Zamarian *PLoS ONE* **14** 0213272
[58] König S U, Keshava A, Clay V, Rittershofer K, Kuske N and König P 2021 Embodied Spatial Knowledge Acquisition in Immersive Virtual Reality: Comparison to Map Exploration *Front. Virtual Real.* **2** 625548
[59] Baus J, Wasinger R, Aslan I, Krüger A, Maier A and Schwartz T 2007 Auditory perceptible landmarks in mobile navigation *Proceedings of the 12th international conference on Intelligent user interfaces* IUI07: 12th International Conference on Intelligent User Interfaces (Honolulu Hawaii USA: ACM) pp 302–4
[60] Afonso-Jaco A and Katz B F G 2022 Spatial Knowledge via Auditory Information for Blind Individuals: Spatial Cognition Studies and the Use of Audio-VR *Sensors* **22** 4794
[61] Massiceti D, Hicks S L and Van Rheede J J 2018 Stereosonic vision: Exploring visual-to-auditory sensory substitution mappings in an immersive virtual reality navigation paradigm ed K Sathian *PLoS ONE* **13** 0199389

IOP Conf. Series: Earth and Environmental Science 1320 (2024) 012023 doi:10.1088/1755-1315/1320/1/012023

Mechanisms of Aesthetics: On the Perception of Materials and Their Properties.

Johannes Weninger[a], Isabel Arnst[a], Maximilian Dick[a], Martina Ascher[a]

[a] Bartenbach GmbH, research and development, Rinner Straße 14, Aldrans, Austria

johannes.weninger@bartenbach.com

Abstract. The visual perception of material surface qualities holds immense importance in our daily lives, serving as the foundation for various environmental interpretations and behavioural adaptations. These encompass critical safety aspects, like assessing floor safety or food freshness, to appreciating intricate sculptural illusions, such as delicately draped fabric in marble statues. Particularly for products with high aesthetic value, perceived properties play a significant role in subjectively attributing material worth. Given the vast array of material surfaces and the pivotal role of lighting in the overall visual perception process, this presents a major challenge for lighting planners and designers. Currently, the achievement of aesthetic effects in architectural spaces heavily relies on the opinions and experiences of professionals, as there are no specific guidelines for combining materials with different lighting concepts. Hence, decoding the connection between physical properties and their subjective interpretations becomes crucial in facilitating the objective planning of materials. In this study, more than 600 materials were systematically examined to explore the interaction between perceived material properties and the resulting aesthetic effects. The resulting perception model establishes a direct link between objectively assigned and subjectively perceived material properties. These findings hold promise in contributing to both a more accurate evaluation and prediction of material appearances in the long term.

1. Introduction

Visual perception of material properties plays a fundamental role in how we perceive and interact with the world around us [1]. From the comfort of a soft fabric to the coolness of smooth marble, our subjective experiences are heavily influenced by visual appearances. As materials are essential components of both the physical environment and products, understanding the relationship between their inherent features and the perceived aesthetic effects is critical to design, architecture, and product development. Moreover, given the increasing availability of computer vision technologies, subjective experiences are currently gaining importance in other application fields, such as process automation [2] and quality control [3], e.g., food inspection [4], which are mostly defined by the same basic problem of a perceptual characterization the visual appearance of materials.

This problem applies in particular to products with high aesthetic value [5, 6], since in these cases the qualitatively perceived properties largely determine the subjectively attributed material value and thus also the price that can be obtained with the materials. Accordingly, objective quantification of material properties requires modeling of subjective human perception. Traditionally, the necessary analyses of visual appearance are performed by trained personnel. However, since this approach provides mainly subjective results and is thus hardly reproducible [7], standardized evaluation methods would have the potential to achieve more reliable quality standards.

In recent years, significant progress has been made in unraveling the intricate connections between material attributes and human perception [8, 9, 10, 11]. Researchers have explored various aspects, including texture [12], roughness [13], color [14], and glossiness [15], as potential predictors of the perceived aesthetic appeal of different materials. However, while numerous studies have focused on individual material types, e.g., stone [16], ceramics [17], parquet [18], and fabric [19], the need for a generalized model that can transcend material boundaries and predict aesthetic effects for various materials remains ever-present.

In addition, lighting situations are largely ignored in current modelling approaches. However, the basic perceptibility of material-specific features [20] is shown to be largely dependent on (a) an intrinsic property of the material (reflectance spectrum), (b) a random condition (illumination), and (c) the response of the sensor (human eye). Therefore, intensities [11], color temperatures [21] and light directions [22] have a potential impact on material perception, which is why lighting-related factors should be given increased importance. From this point of view, it is also surprising that there are currently no concrete guidelines for the use of materials in combination with different lighting concepts and that currently the use still depends exclusively on the experience of lighting planners and designers.

In response to this challenge, this paper presents a comprehensive investigation of the visual perception of materials and their properties. Our primary focus lies in the development and evaluation of a robust regression model capable of predicting perceived aesthetic effects based on perceived intrinsic material properties. The development of the regression model was grounded in an extensive dataset obtained through rigorous experimentation. We systematically collected and analyzed perceptual responses from a diverse group of participants who were presented with various materials representing different levels of homogeneity, roughness, and other essential material attributes. By capturing a broad spectrum of material stimuli, we aimed to enhance the model's versatility and ensure its adaptability to real-world scenarios.

In the following work, we outline the methodology employed to build the regression model, including data collection, feature extraction, and model training. We discuss the experimental design and analysis procedures used to validate the model's accuracy and robustness across different material types. Additionally, we present the results and offer insights into the key material attributes that significantly influence perceived aesthetic effects. The results represent a first step towards the objective predictability of the subjective impression of material surfaces and can serve as a basis to improve future decisions by architects, designers and manufacturers in the material selection and design process. Moreover, this research contributes to the broader field of human perception and cognition, bridging the gap between material science and human psychology.

2. Methods

2.1. Study population

A total of 50 people participated in the study, 33 women and 17 men. The average age of the participants was 30.12 ± 12.50 years, with women (31.66 ± 13.13) being slightly older than men (27.24 ± 11.01). Most of the participants (78.00 %) had a secondary school degree, 28.00 % of them had a university degree (bachelor, master, or doctorate). 40.00 % of the participants stated that they work with different materials in their professional or private lives.

A large proportion of the participants (58.00 %) used visual aids (e.g., glasses, varifocals, contact lenses) at the time of the test. All participants were screened for color vision deficiencies prior to participation in the study using Ishihara color charts. Beyond that, there were no exclusion criteria. All participants gave a written informed consent prior to study participation and received a financial compensation of 250 €.

2.2. *Material selection*

A total of 612 materials of various types were included in the study (Figure 1). With 313 materials (51.14 %), woods represented the largest group of materials studied. Furthermore, the material selection included 154 solid-colored materials (25.16 %), 54 stones (8.82 %), 28 metals (4.58 %), and 18 textiles (2.94 %). The remaining group included a total of 45 materials (7.35 %) with types that could not be clearly assigned to one of the previously listed (e.g., concrete, tiles, or wallpaper).

93 of the investigated materials (15.20 %) were natural materials. The rest were artificially produced materials, which either aimed at a lifelike imitation of real materials (replica) or at a clearly artificial surface design (e.g., solid-colored surfaces). All the materials were standardized to a size of 50 x 50 cm. The thickness of the material samples varied depending on the material type and resistance-related requirements (e.g., fracture strength).

Figure 1: Exemplary presentation of 70 randomly selected materials out of the 612 materials examined.

2.3. *Description of the laboratory setting*

The study was conducted in a controlled laboratory environment in two identical rooms. Both rooms measured 3.65 x 3.90 m and had a ceiling height of 2.80 m. The influence of daylight through a north-facing window opening of 1.5 m^2 each was reduced to a minimum by a shading system (controllable blinds in the space between the double glazing; light transmission < 5 %). The floors were covered with light gray carpet tiles (ρ = 0.35). Both the walls, which were draped with white curtains, and the suspended ceiling of the rooms were kept in neutral white (ρ = 0.85).

Both rooms were equipped with two evaluation stations in the central area of each room, which consisted of two light gray, height-adjustable desks at a height of 1.00 m (Figure 2). The stations were equipped with lighting prototypes for direct illumination of the material samples, which were mounted diagonally across both tables at a height of 82 cm above the desk surface and prevented a view into the respective opposite evaluation station by means of a neutral white back wall (160 x 80 cm, $\rho > 0.80$). Additional up-lights were installed on the light prototypes for both general room lighting and the provision of diffuse light components on the material samples.

Figure 2: Frontal view of an evaluation station with the diagonally arranged lighting prototypes and the applied material positioning.

The material samples were positioned horizontally on the table surface at approx. 10 cm distance normally on the privacy screens in both free table corners. To prevent the possibility of inferring material-specific properties from different thicknesses of the materials, the materials were placed in an approx. 5 cm wide passe-partout frame made of powder-coated aluminum. The coating was in a neutral light gray color.

2.4. Characterization of the light condition

The light condition used in the study was defined visually using various materials and under consideration of both the assessability of material surface properties and the occurrence of gloss effects. After the visual adjustment, the light condition was characterized photometrically. All material samples were illuminated at 5,000 K with a horizontal illuminance of 2,800 lx and CRI 93.

The direct lighting component (Luxeon Z-ES, 2,200-5,700 K) was zoned to the material position and amounted to 2,470 lx. The remaining 330 lx were diffusely generated via the general room illumination (Bridgelux Vesta Series Tunable White 2,700-5,000 K Gen2 29 mm Array). Overall, a min/max luminance ratio of 1/1.8 was achieved on the material samples (Figure 3). The luminance ratio of the material sample to the spatial environment was approx. 1/10. The luminance of the back walls was adapted to the environment and made a negligible contribution to the illuminance on the material sample (< 0.5%).

Figure 3: Luminance image from the position of the viewer defined for material evaluation with clear zoning of the direct portion of the light on the material sample (neutral white).

2.5. Study protocol

The study was conducted as a block randomized trial with repeated measures in the period from 13.12.2021 to 30.05.2022. For this purpose, the 612 materials were divided into 20 blocks of approx. 30 materials each, which were then presented to the participants for evaluation in a randomized order during five participation days (four blocks per day, one day per week for five weeks). Randomization was done with respect to both the order of the blocks and the materials within each block. Daily assessment time averaged approximately 3.5 hours and two individuals could participate simultaneously. Since two time slots were available each day (morning 07:00 - 12:00, afternoon 13:00 - 18:00), 4 participants could be included per day.

During the material evaluation, both participants were always in the same room and occupied different evaluation stations at which two different materials of the same block were presented. Before the evaluation of the material, the participants had about one minute in which they could move freely to look at the materials from different angles to experience any glossy behavior of the surfaces. The participants were not allowed to touch the materials. The subsequent evaluation of the material sample was then carried out from a defined observer position (Figure 3), which was designated by means of a standing stool and was located outside the reflected glare area of the direct lighting. After the evaluation was completed by both participants, they changed the evaluation stations first within the room and repeated the procedure. After both material samples were evaluated, the participants changed to the second laboratory room, which was equipped with two new material samples by the supervising personnel during the preceding evaluation process. The procedure was then repeated until all materials in the daily assigned blocks were evaluated. Both breaks and the consumption of beverages were always permitted.

All participants rated all materials first in terms of their inherent material properties (13 items) and then according to their aesthetic effect (15 items) using bipolar pairs of adjectives (Table 1). All items were presented in German. The scale for each of the 28 items ranged from -100 to +100 with a step size of 1. The survey was conducted using software (LimeSurvey, version 3.24.0+201013 [23]) on two identical tablets and the scales were presented by means of a horizontal slider in a neutral central starting position and without numerical indication of the currently set value. The order of the items was randomized for each material evaluation. The assignment of the submitted ratings to the respective materials was done by means of QR codes, which were attached to the materials, contained referencing identification codes, and had to be scanned by the participants to call up the rating questionnaire. All QR codes were checked for accuracy several times before the start of the study.

Table 1: Bipolar pairs of adjectives used for material evaluation separated according to material-inherent properties and aesthetic material effects. All pairs were evaluated on a scale ranging from -100 to +100 with a step size of 1.

bipolar pairs of adjectives to evaluate material-inherent properties			bipolar pairs of adjectives to evaluate the aesthetic effect of the material		
smooth structure	-	deep structure	attractive	-	unattractive
coloured	-	colourless	modern	-	classic
matte	-	glossy	public	-	private
natural	-	artificial	high-quality	-	cheap
dark	-	bright	dismissive	-	inviting
soft	-	hard	pleasant	-	unpleasant
smooth	-	rough	calming	-	exciting
warm	-	cold	unobtrusive	-	noticeable
regular	-	irregular	usual	-	individual
quiet	-	restless	interesting	-	boring
low contrast	-	high contrast	real	-	fake
plain	-	complex	elegant	-	not elegant
harmonious	-	inharmonious	cozy	-	uncomfortable
			diverse	-	monotonous
			relaxing	-	stressful

2.6. *Data Analysis*
As part of the data analysis, the raw data collected was first adjusted for outliers. Outliers were identified separately for each material and each evaluated item using the 1.5*IQR method. Detected outliers were removed without substitution. Subsequently, material-related characteristic values were derived from the individual scores by calculating the mean, standard deviation, median and inter-quartile range (IQR) for each item. The pre-processed data were used for material classification as well as for the model-based prediction of the aesthetic material effect. To exclude a dependence of the results on the calculated measure of central tendency, all following analyses were carried out for both variants and the results were checked for consistency.

2.6.1. Material classification. The classification of the materials pursued the goal of identifying material groups with similarly rated material-inherent properties. Accordingly, the data analysis focused on the use of the 13 items evaluated in this respect. To reduce the data complexity, the data were normalized in a first step by means of a standard scaling procedure, and then reduced in their dimensionality by using a principal component analysis (PCA) [24]. With an assumed cut-off value of variance to be retained of 80%, a reduction from 26 (13 items for each of two positional measures) to 10 dimensions was achieved.

Subsequently, the classification of the materials was performed in the calculated principal component space by means of a k-means clustering [25]. Since the number of clusters to be created must be specified within the scope of the application of the method, an estimation of the results and the associated derivation of the optimal number of clusters was carried out by means of the elbow method, a graphical representation of the search for the optimal number of clusters using the WCSS (Within-Cluster Sum of Squares), i.e. the sum of the squared distances between points in a cluster and the associated cluster centroid. The cluster assignment resulting from the k-means procedure was finally mapped to the original dataset and the clusters in all items were examined for their distinguishing features and visually verified for coherence.

2.6.2. Aesthetic Predictive Modelling. To build the prediction models of aesthetic material effects, the pre-processed data set was divided into a training and test data set on an 80-20 basis. The data division was randomized. Since cluster-specific correlations between the material-inherent properties and the aesthetic effects could be assumed, the data division was stratified based on the calculated cluster memberships to ensure an equal percentage distribution of clusters in the training and test data set. Modelling was performed separately for each aesthetic item using a linear regression procedure [26]. The generalized applicability of the derived models was determined by means of the determination coefficient R^2, which denotes the proportion of variation in the test data that can be explained by the derived regression model.

2.6.3. Relationship between material-inherent and aesthetic properties. In addition to calculating the determination coefficient R^2 to check the accuracy of the model on the test data, the regression coefficients of the material-inherent input features were calculated to obtain a statement about their impact on the aesthetic material effects. To ensure comparability of the feature-related impact between regression models, individual regression coefficients were normalized with respect to both their percentage contribution to and the explanatory power achieved by each model.

For statistical testing, a one-factor ANOVA was performed with the factor input feature. Because the data violated the equal variance requirement (Levene's test), a Welch correction was applied to the calculation. All post-hoc comparisons to determine pairwise differences between models were performed with Bonferroni-corrected significance levels.

All data analyses were performed using JASP software (version 0.15) with a significance level of 5 % and two-sided testing. Mean (M) and standard deviation (SD) were calculated for descriptive analyses. Effect sizes of ANOVA are shown as partial eta squared values (η^2_p) indicating small ($0.01 \leq \eta^2_p < 0.06$), moderate ($0.06 \leq \eta^2_p < 0.14$), or large effect sizes ($\eta^2_p \geq 0.14$).

3. Results

3.1. Identified Material Classes

The applied clustering procedure resulted in the identification of six distinct material classes. Statistical testing showed significant agreement between the results of the median and mean-based classification (Fisher's exact test, $p < .001$), thus ruling out any dependence of the resulting classification on the applied measure of central tendency.

Clusters 4 to 6 (see Figure 4) proved to be defined by individual material classes, since the position in the principal component space of solid-colored materials (high chromaticity and low naturalness) and metals (low naturalness and high gloss) differed significantly from the rest of the samples examined. Furthermore, there was a clear separation between highly saturated (colored) and neutral (muted) solid-colored materials (clusters 4 and 5).

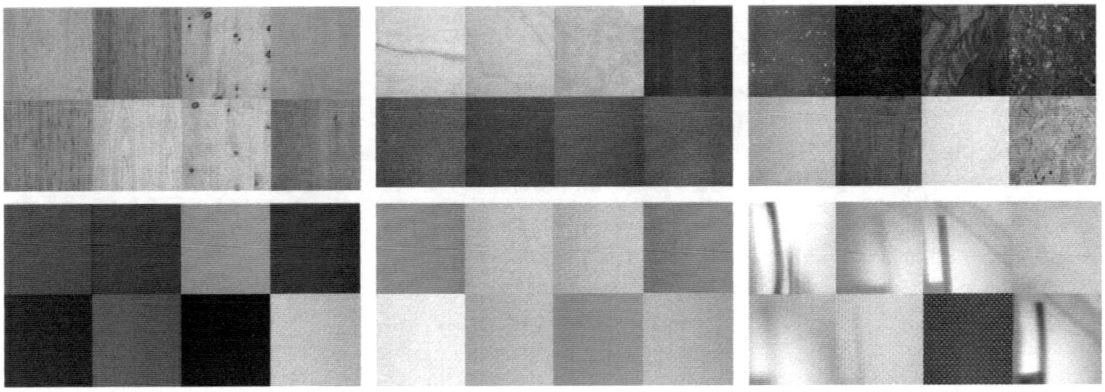

Figure 4: Representation of 8 materials randomly selected from each cluster. The resulting clusters 1 to 3 (top from left to right) are mainly characterized by structural material features, whereas clusters 4 to 6 (bottom from left to right) mainly contain materials of the same type.

The three remaining subdivisions (clusters 1 to 3, see Figure 4), on the other hand, proved to be dependent on structural distinguishing characteristics regarding the materials included, which led to a stronger type-related mixing of the derived material classes. Specifically, the ratings of the items in terms of warmth (mainly warm materials in cluster 1), contrast (neutral light and dark stones and woods in cluster 2), and regularity and depth of structure (strongly patterned materials in cluster 3) proved to be decisive for the resulting classification.

3.2. Prediction accuracy

The 15 specifically derived models for the prediction of the individual aesthetic effects of the materials basically showed a very high quality with 79.86 ± 12.41 % (prediction of the mean values) and 76.49 ± 12.99 % (prediction of the medians). Mean values were predicted significantly better than medians (T-test, paired samples, $p = .001$).

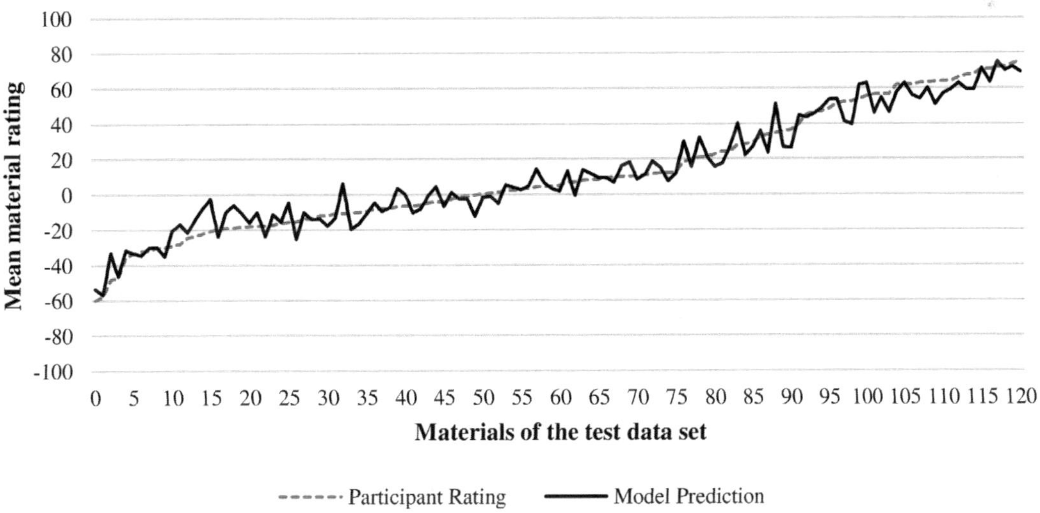

Figure 5: Exemplary representation of the comparison of the real evaluations to the calculated predictions ($R^2 = 0.95$) of the mean values of the aesthetic item diverse (-100) – monotonous (+100) for the test data set of 120 materials.

IOP Conf. Series: Earth and Environmental Science 1320 (2024) 012023 doi:10.1088/1755-1315/1320/1/012023

3.3. *Impact of material-inherent characteristics on aesthetic ratings*

The pairwise comparisons were able to identify three distinct groups of input features (Figure 6) with varying influence on aesthetic ratings. Both the group with low (1.91 ± 2.06 %) and high influence (10.91 ± 8.77 %) had 5 items each. The group with medium influence (5.25 ± 3.84%) accounted for the remaining 3 items. While all groups differed significantly among themselves (all p > .001), the items included within the groups showed no significant differences (all p > .05).

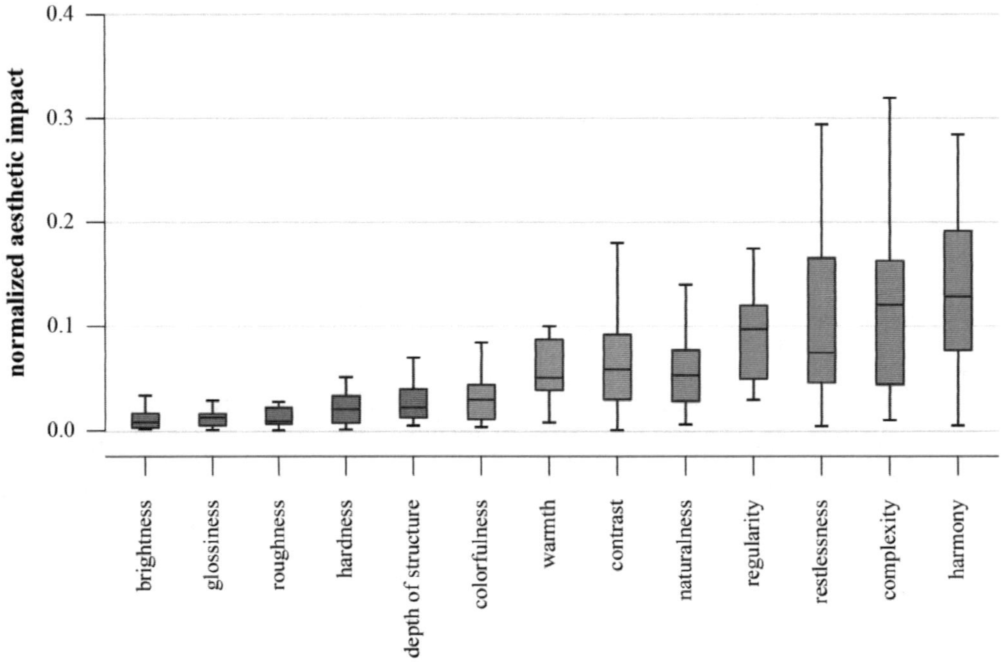

Figure 6: Boxplots of the normalized impacts of the individual material-inherent input features on the aesthetic ratings of the materials, sorted in ascending order based on their respective mean value. The impact groups derived from the analyses are color-coded in the diagram (red: low influence, yellow: medium influence, green: high influence).

4. Discussion

Both the clustering procedure applied in the study and the item selection used proved to be useful for deriving distinct material classes. However, the resulting material groupings only proved to be unambiguously type-related if the underlying material types clearly exhibited strongly pronounced inherent properties that set them apart from other materials (such as the colorfulness of solid-colored materials or the reflective behavior of metallic surfaces). In contrast, for material types, where a clear lack of such properties comes to the fore due to an increased possible variation, superordinate property conglomerates proved to be decisive for class formation, resulting in a clearly type-related mixing of the classes. Therefore, it can be assumed that generally intuitive classifications (e.g., wood or stone) for the classification of materials prove to be insufficiently meaningful and that a perceptually accurate subdivision is far more differentiated. To the best of the authors' knowledge, these results are also largely consistent with existing research [11].

Secondly, the models calculated on the basis of the clustering results for the prediction of the aesthetic material effect by means of perceived material-inherent properties basically showed a very high level of accuracy. The resulting confirmation of a simple and direct connection between the objectively assigned and subjectively perceived qualities of a material surface accordingly suggests the existence of an intrinsic perception model. This would mean that subjectively perceived material qualities are not subject to stochastically intersubjective characteristics but are systematically formed from underlying visually extracted characteristics. However, to complete this model assumption, it is still necessary to find out to what extent these visually extracted characteristics directly represent the material surfaces under consideration or to what extent the influence of both empirical values and interpretation processes contributes to the formation of the objectively assigned material properties. The determination of these reciprocal mechanisms of action was not part of the present investigation.

Finally, it should be mentioned that the accuracies achieved by the simple regression models are limited to the impact of a few material-inherent properties on closer inspection for all aesthetic items. Interestingly, the items with a strong impact on the predicted values are properties that are generally more difficult to define. Concepts such as complexity, restlessness or harmony are explicit material properties, but at the same time require a conceptual interpretation by the viewer, which in turn can be assigned a socio-cultural code. On the other hand, simpler and more direct material properties, such as warmth or brightness, which would also have physically measurable definitions, had only a moderate or low impact on the formation of aesthetic sensations. Although a comprehensive depiction of the material-related perception can be considered necessary to predict all aesthetic effects, the possibility of a greatly simplified consideration of materials opens based on the five decisive influencing factors, which prove to be responsible for almost 68.32 % of the prediction accuracy. Especially when a precise prediction of a specific aesthetic parameter (e.g., attractiveness) is desired. In the long term, this can simplify the applicability of the prediction models and potentially make them easier to put into practice.

4.1. Limitations of the present study

Despite the comprehensive examination of 612 materials, which aimed to ensure broad applicability of the results, it is essential to acknowledge that the materials selected may not entirely represent the entire spectrum of potentially available materials. Especially the inclusion of a larger number of woods in the investigation might introduce certain biases, possibly affecting the accuracy of the prediction models when the dataset is altered.

Moreover, it is crucial to recognize that aesthetic material effects are not inherently fixed but influenced by contemporary trends and socio-cultural backgrounds. Regrettably, these factors were not addressed in the current study, making it difficult to claim universal validity for the derived prediction models.

Additionally, it is worth noting that the modeling process was limited to a single light condition. Given that lighting significantly impacts the recognizability and interpretability of material surfaces, the predictive accuracy of the models might diminish under unexamined lighting situations.

Finally, the validity of the presented results can only be guaranteed in the context of normal-sighted people since all kinds of visual disturbances were excluded. Color blindness, for example, can have a significant impact on the visual appearance of objects or materials. It is also well documented that age-related weaknesses in the visual system, such as age-related macular degeneration, also have a significant impact on both visual acuity [27] and color vision [28, 29]. Appropriate mapping of such groups of people would require an expansion of the existing basic model using experimental investigations.

5. Conclusion

Aesthetic sensations cannot be understood independently of neural systems or physiological sensors, as they are primarily influenced by perceptual and cognitive processes. They arise from an interaction of recognition and evaluation of emotion and meaning [30], which, due to their subjectivity, make an objective evaluation complex and demanding. Since international research is currently still trying to derive a general mathematical model of human aesthetic appreciation [31, 32, 33, 34, 35], the present results offer the potential to make a significant contribution to solving a currently challenging problem.

Due to the complexity of the underlying processing mechanisms and perception processes, the derivation of a generalized model within the framework of a single study proves to be impracticable. The presented models should accordingly be considered as a generalized base model, created under a well-documented lighting situation, making it suitable for multidimensional extension. In the long term, extensions regarding the potential influence of different lighting situations could enable lighting planners and designers to mutually optimize key performance indicators of buildings, such as energy efficiency or melanopic illuminance levels, for the first time, considering the aesthetic effects of the material selection made.

However, this requires further research, which should also take up long-term and intercultural tendencies to rule out a personal exclusion of fringe groups. In general, the presented results confirm the potential that can result from more objective material planning for future-oriented lighting practice.

Acknowledgements
The study presented was funded by the Austrian Research Promotion Agency FFG as part of the project Licht und Material (FFG no. 882556). The authors would also like to thank Chiara Dietz, Tobias Fischer, Claudia Konrad-Soare and Simone Sommerfeld for conducting the study and for their support with data curation.

References

[1] Dant T 2008 The `Pragmatics` of Material Interaction *J Consum Cult* **8** pp. 11–33
[2] Kwon Y 2008 Web-Enabled vision guided robotic tracking within the framework of e-manufacturing *Int J Ind Eng-Theory* **15.3** pp. 323–329
[3] Wang C C and Jiang B C 2001 PCB solder joint defects detection and classification using machine vision *Int J Ind Eng-Theory* **8.4** pp. 359–369
[4] Davies E R 2005 *Machine Vision: Theory, Algorithms, Practicalities* (Burlington: Elsevier Morgan Kaufmann)
[5] Alizadeh A M, Ameri F and Moradian S 2018 A Review on Quantification of Visual Appearance *Journal of Color Science and Technology* **12.2** pp. 135–146
[6] Eugene C 2008 Measurement of `total visual appearance`: a CIE challenge of soft metrology *Proceedings of the 12th IMEKO TC1-TC7 Joint Symposium on Man, Science & Measurement* (Annecy: France)
[7] Boukouvalas C *et al* 1995 Ceramic tile inspection for colour and structural defects *Proceedings of the Advanced Materials and Processing Technologies* (Dublin: Ireland)
[8] Baumgartner E and Gegenfurtner K 2016 Image Statistics and the Representation of Material Properties in the Visual Cortex *Front Psychol* **7** 1185
[9] Jacobs R, Baumgartner E and Gegenfurtner K 2014 The representation of material categories in the brain *Front Psychol* **5** 146
[10] Martín R, Iseringhausen J, Weinmann M and Hullin M B 2015 Multimodal perception of material properties *Proceedings of the ACM SIGGRAPH symposium on applied perception* (Tübingen: Germany) pp. 33–40
[11] Fleming R W 2017 Material perception *Annu Rev Vis Sc* **3** pp. 365–388
[12] Ho Y, Landy M and Maloney L 2008 Conjoint Measurement of Gloss and Surface Texture *Psychol Sci* **19** pp. 196–204
[13] Tiest W M B and Kappers A M 2007 Haptic and visual perception of roughness *Acta psychol* **124.2** pp. 177–189
[14] Sample K L, Hagtvedt H and Brasel S A 2020 Components of visual perception in marketing contexts: A conceptual framework and review *J Acad Market Sci* **48** pp. 405–421
[15] Marlow P and Anderson B 2015 Material properties derived from three-dimensional shape representations *Vision Res* **115** pp. 199–208
[16] Bianconi F, Gonzalez E, Fernandez A and Saetta S A 2012 Automatic classification of granite tiles through colour and texture features *Expert Syst Appl* **39.12** pp. 11212–11218
[17] Kukkonen S, Kalviainen H and Parkkinen J 2001 Color features for quality control in ceramic tile industry *Opt Eng* **40.2** pp. 170–177
[18] Bianconi F, Fernandez A, Gonzalez E and Saetta S A 2013 Performance analysis of colour descriptors for parquet sorting *Expert Syst Appl* **40.5** pp. 1636–1644

[19] Carfagni M, Furferi R and Governi L 2005 A real-time machine vision system for monitoring the textile raising process *Comput Ind* **56.8** pp. 831–842
[20] Wyszecki G and Styles W 2000 *Color Science. Concepts and Methods, Quantitative Data and Formulae* (New York: John Wiley & Sons)
[21] Yu C, Wijntjes M, Eisemann E and Pont S 2022 Effects of inter-reflections on the correlated colour temperature and colour rendition of the light field *Lighting Res Technol* 14771535221126902.
[22] Zhang F, de Ridder H and Pont S 2015 The influence of lighting on visual perception of material qualities *Proc Spie* **9394** pp. 239–248
[23] Schmitz C 2012 *LimeSurvey: An Open Source survey tool* (Hamburg: LimeSurvey GmbH)
[24] Tipping M E and Bishop C M 1999 Probabilistic principal component analysis *J Roy Stat Soc B* **61.3** pp. 611–622
[25] Sinaga K P and Yang M S 2020 Unsupervised K-means clustering algorithm *IEEE access* **8** pp. 80716–80727
[26] Su X, Yan X and Tsai C L 2012 Linear regression *Wires Comput Stat* **4.3** pp. 275–294
[27] Pitts D G 1982 Visual acuity as a function of age *J Am Optom Assoc* **53.2** pp. 117–124
[28] Pinckers A J L G 1980 Color vision and age *Ophthalmologica* **181.1** pp. 23–30
[29] Roy M S, Podgor M J, Collier B and Gunkel R D 1991 Color vision and age in a normal North American population *Graef Arch Clin Exp* **229** pp. 139–144
[30] Chatterjee A and Vartanian O 2014 Neuroaesthetics *Trends Cogn Sci* **18.7** pp. 370–375
[31] Leder H and Nadal M 2014 Ten years of a model of aesthetic appreciation and aesthetic judgments: The aesthetic episode - Developments and challenges in empirical aesthetics *Brit J Psychol* **105.4** pp. 443–446
[32] Chatterjee A 2004 Prospects for a cognitive neuroscience of visual aesthetics *Bulletin of Psychology and the Arts* **4.2** pp. 55–60
[33] Tinio P P L 2013 From artistic creation to aesthetic reception: The mirrormodel of art *Psychol Aesthet Crea* **7.3** pp. 265–275
[34] Koelsch S, Jacobs A M and Menninghaus W 2105 The quartet theory of human emotions: An integrative and neurofunctional model *Phys Life Rev* **13** pp. 1–27
[35] Redies C 2015 Combining universal beauty and cultural context in a unifying model of visual aesthetic experience *Front Hum Neurosci* **9** pp. 1–20

IOP Conf. Series: Earth and Environmental Science 1320 (2024) 012024 doi:10.1088/1755-1315/1320/1/012024

Role of lighting and color in microeconomics: preference and purchase intent

D Durmus[1], T Wang[1], C Gonzalez[1] and R Kore[1]

[1] Department of Architectural Engineering, Pennsylvania State University, University Park, PA, 16802, USA

[*] Correspondence: alp@psu.edu

Abstract. Lighting and color are key factors under the control of lighting designers that can significantly impact shoppers' perceptions and actions in the context of culinary and retail environments. Previous studies have examined the effects of color and lighting on preference and purchase intent, but results are often conflicting due to confounding variables, and a strong connection between these phenomena remains unexplored. To bridge this gap, two visual experiments were conducted, revealing that lighting can indeed influence consumers' perception of products. Participants perceived the changes in objects under different color gamut and illuminance levels. Further research is needed to better understand the mechanisms of purchase intent, particularly across different price and product types.

1. Introduction

In microeconomics, demand refers to the quantity of a good or service that consumers are willing and able to purchase at a particular price and time, while purchase is the act of buying a good or service at a specific point in time. Purchase intent is affected by various factors, such as consumer preferences, income, prices of related goods, advertising, and physical factors in a retails store. Conceptual frameworks of purchase intent indicate seven key factors: cleanliness, music, scent, temperature, display, lighting, and color [1]. Among these seven, lighting and color are aspects that a lighting designer can influence through the rigorous planning and use of electric lighting systems.

Carefully planned lighting can attract and guide customers, help to evaluate merchandise, and initiate purchases [2]. Studies indicate that illuminance and color gamut can significantly impact preference, naturalness, and attractiveness of objects [3-9]. Several studies also explored the influence of correlated color temperature (CCT) on emotional states and product perceptions in culinary and retail contexts [10-27].

1.1 Culinary experience and lighting

The relationship between lighting and food-related has been a focus of consumer behavior and sensory experience studies. For example, Wu et al. investigated the effect of focal vs general, and dim vs bright lighting on customers' intentions toward restaurants using images [10]. Images that present focal lighting (e.g., table lighting) resulted in a boost of the restaurant's perceived attractiveness and visit intention among diners. Steele et al. explore the shelf life of fresh meat products under different lamps (LED vs. fluorescent), and found that LED lighting extends beef *longissimus dorsi*, *semimembranosus* steak and ground beef color life by up to one day [11]. The study also noted that LED lighting has minimal effects on factors that cause meat color to deteriorate. Unfortunately, in these studies, lighting conditions (e.g., spatial and spectral distribution of lighting) were not characterized in detail, preventing the replication of these studies.

Other research papers that focus on the effects of lighting on food perception provided better control of experimental conditions. Utilizing lighting conditions that are less-commonly found in architectural spaces, Hasenbeck et al. found that participants' willingness to consume bell peppers increased the most under yellow lighting and the least under blue, monochromatic lighting [12]. Similarly, Özkul et al. investigate the effect of chromatic lighting (red, blue, green, yellow) on customers' perceptions of service quality and satisfaction in restaurants [13]. Their study highlights that the perception of service quality and the level of satisfaction were higher in red and yellow ambient lighting compared to blue and green ambient lighting.

Several studies also utilized nominal white light sources that are common to architectural spaces, where consumers interact with food (e.g., restaurants, supermarkets). A market analysis of lighting conditions in Slovakia for baked goods and fresh produce found that light levels and CCTs widely varied across retail chains and for different product categories (i.e., between 2530 K and 4450 K; between 200 lx and 1800 lx) [14]. A study by Wu and Wang used nominal white lighting to investigate the effects of CCT and illuminance levels in restaurants on customers' emotional states and spatial impressions [15]. Their study underscores that in a restaurant 2700 K light sources have a greater tendency to "emphasize pleasurable aspects" while 5600 K causes greater "arousal." Chen et al. studies the effect of CCT (2700 K, 3300 K, 3900 K, 4500 K and 5100 K), illuminance (400 lx, 800 lx and 1200 lx), and background (wood vs. white) on baked goods, such as bread and cakes [16]. High illuminance levels promoted the purchase intention for both bread and cakes, while CCT didn't impact results. More recently, Ahmadian et al. studied consumers' preference of vegetables, fruits, and packaged foods under 3000 K at 500 lx to apply the stimulus-organism-response (SOR) [17] framework in the context of retail lighting [18]. The authors created an optimal spectrum for most objects except the can of Pepsi and red cabbage, providing support for the potential for an object-oriented illumination strategy. A tailored-spectra strategy supports previous studies that found similar results [19-21].

1.2. Retail experience and lighting

Understanding how lighting design impacts customer behavior in retail settings is a central theme explored by several research groups around the world. Lombana and Tonello investigated the perceptual and emotional effects of light distribution and object color in retail spaces in Argentina [22]. Their results indicated that overhead direction of lighting improves the appearance of the less favored color conditions. Similarly, Tantanatewin and Inkarojrit studied the effects of chromaticity (warm vs. cool) and light distribution (general vs. accent) on impression and identity using digital images of a bank in a department store in Thailand [23]. Warm color tones significantly enhanced positive impression and space identity, but lighting arrangement was a secondary factor. In a field study conducted in the United States, the lighting (soft vs. bright) in a centrally located retail establishment was varied over a two-month period, and the results indicated that brighter lighting influenced shoppers to examine and handle more merchandise, though sales were not influenced [24].

In a more controlled study conducted in China, color preference and discrimination of blue jeans were the highest at 5000 K [25] Hemalatha et al. extend this discussion to retail apparel stores, examining the effects of lighting conditions (2700 K, 3000 K, 4000 K and 5700 K; 300 lx, 500 lx, 700 lx and 900 lx) on user preferences within the cultural context of India [26]. Both CCT and illuminance influence the spatial impressions of a retail environment, with the most preferred scenes being 5700 K at 500 lx, 5700 K at 300 lx and 5700 K at 700 lx. Meanwhile, Park and Farr explore the impact of lighting on American and Korean consumers' emotions and behavioral intentions in retail environments under 3000 K and 5000 K, and color rendering index (CRI) of 75 vs 95 lighting [27]. The results indicate that 3000 K was more pleasurable, while Koreans found 5000 K to be more "approachable" compared to Americans, highlighting cultural differences across lighting conditions. This body of work underscores the importance of lighting design in creating attractive retail environments and influencing consumers' preferences and behavior.

1.3. Customer experience and lighting

Understanding how lighting influences customer experience (product perception, price fairness, waiting times, purchase intention) in different settings is another key theme in the literature. Hutchison, Thomas, and Elias provide intriguing insights into the world of advertising by investigating the effects of leftward lighting bias in advertisements [28]. Their study reveals that the direction of lighting in advertisements can significantly influence advertisement ratings and purchase intentions, confirming the positive influence of illuminating the objects from the left. Another study exploring the effect of interior colors (blue vs. orange) and lighting (bright vs soft) on price fairness and purchase intent found that blue interiors were associated with more favourable evaluations, higher store patronage intentions, and higher purchase intentions than orange interiors [29]. However, "soft lights" with an orange interior generally nullified the negative effects and produced the highest level of perceived price fairness. Bilgili, Ozkul, and Koc explored the influence of light chromaticity (red, green, blue, yellow, white) on customers' waiting time perceptions, and found a significant impact of lighting on perception of the length of the waiting time, specifically, under green lighting, customers perceived the waiting period to be relatively shorter [30]. Biswas et al. conducted a series of field studies at multiple locations and found that consumers tend to choose less healthy food options when ambient lighting is dim (vs. bright) and connect it to lower mental alertness caused by low light levels [31].

Similar to some of the previously discussed studies, the independent variable (lighting conditions) was not thoroughly reported in these studies. In a better controlled study, Phumchan and Tuaycharoen explored the effects of illuminance (between 100 lx and 900 lx) and CCT (2700 K, 4000 K, and 65000 K) on the emotional state and perception of customers in clothing retail stores [32]. Results indicated that 4000 K lighting and an illuminance of 400 lx were optimal for the emotional state and perception of participants, including price perception and motivation. Despite their limitations, these studies collectively underscore the nuanced relationship between lighting design and customer experience, emphasizing the importance of lighting in shaping how consumers perceive and evaluate products.

1.4. Research questions

Despite the vast literature investigating the effect of lighting and color on subjective evaluations of the environment, the results of these studies are yet to be translated to concrete purchase intent and action. The outcomes of the research studies on purchase intent are at times inconclusive and challenging to compare. For example, Tešić et al. found that CCT can affect the shoppers' perception of the quality of the observed products in the Serbian market, but do not affect the price perception [33]. On the other hand, Liu et al. documented that consumers' purchase intentions followed an inverted "U" curve under the change of luminance and hue (i.e., consumers' purchase intention was stronger under medium luminance and low CCT) [34].

One of the potential explanations for the conflicting results is the lack of control for confounding variables in the experimental protocols. For example, studies often do not acknowledge or account for the limitations of CCT, which is a proxy metric for the chromaticity of light source and do not provide any information about objects or the architectural environment [35]. A complimentary metric D_{uv} (the distance from the Planckian locus) [36] can provide an additional depth to light source characterization. When controlled for CCT and D_{uv}, gamut shape and color shifts in red hue accounts for the visual perception of objects, even more so than color fidelity metrics (i.e., CRI R_a, TM-30 R_f) [37,38]. This understanding leads to two research questions: 1) What is the relationship between preference and purchase intent when lighting is controlled for CCT and D_{uv}? 2) What is the effect of lighting on purchase intent and visual quality of objects in a retail environment under more tightly controlled experimental protocols? Two pilot studies were conducted to close the gap between color preference and purchase intent by addressing these research questions.

2. Methods

Aiming to investigate the influence of lighting on preference, naturalness, vividness, and purchase intent, two psychophysical studies were conducted. These studies examined the impact of illuminance and color gamut on attractiveness, saliency, and purchasing intent in different contexts. Controlled experiments exposed human participants to various color gamut (ANSI/IES TM-30 R_g = 85, 90, 100, 115, 120) and illuminance levels (50 lx, 400 lx, 500 lx, 750 lx). The first experiment investigates the effect of illuminance and color gamut on preference and gaze duration. The second study explores the effect of illuminance and color gamut on purchase intent. A verbal informed consent was collected before each study. Participants were checked for normal color vision.

2.1. Study 1

A vision experiment was conducted by using eye-tracking technology to detect gaze distribution on four natural objects (pear, green apple, orange, red apple). Illuminance levels and tunable LED spectra were controlled using MATLAB®. Six lighting conditions were shown to six American participants (3 males, 3 females) three times (18 trials per participant) in a randomized order, as shown in Fig. 1 (left). The lighting conditions varied in terms of illuminance (nominally 50 lx and 400 lx), color fidelity (nominally ANSI/IES TM-30 R_f = 70, 80, 100), and color gamut (nominally ANSI/IES TM-30 R_g = 85, 100, 115), as shown in Table 1 and Fig. 1. The conditions were generated to represent desaturating, neutral, and saturating gamut at two light levels. Participants were asked to judge the preference and (un)naturalness of the objects. Each trial lasted 20 seconds, followed by 2 seconds of darkness. The experiment duration was 10 minutes in total. Eye tracking data was collected using Pupil Labs Invisible eyeglasses, which has been used in similar studies [39,40].

Table 1. The lighting conditions used in Study 1

Condition #	Nominal	Illuminance (lx)	Fidelity (R_f)	Gamut (R_g)	T_{cp} (K)	D_{uv}
1	Neutral	400	98.2	100.0	4000	-0.0050
2	Desaturating		79.7	85.0	4000	-0.0050
3	Saturating		72.3	115.0	4000	-0.0051
4	Neutral	50	98.4	99.9	3988	-0.0048
5	Desaturating		82.0	85.0	4000	-0.0050
6	Saturating		72.2	115.0	4009	-0.0049

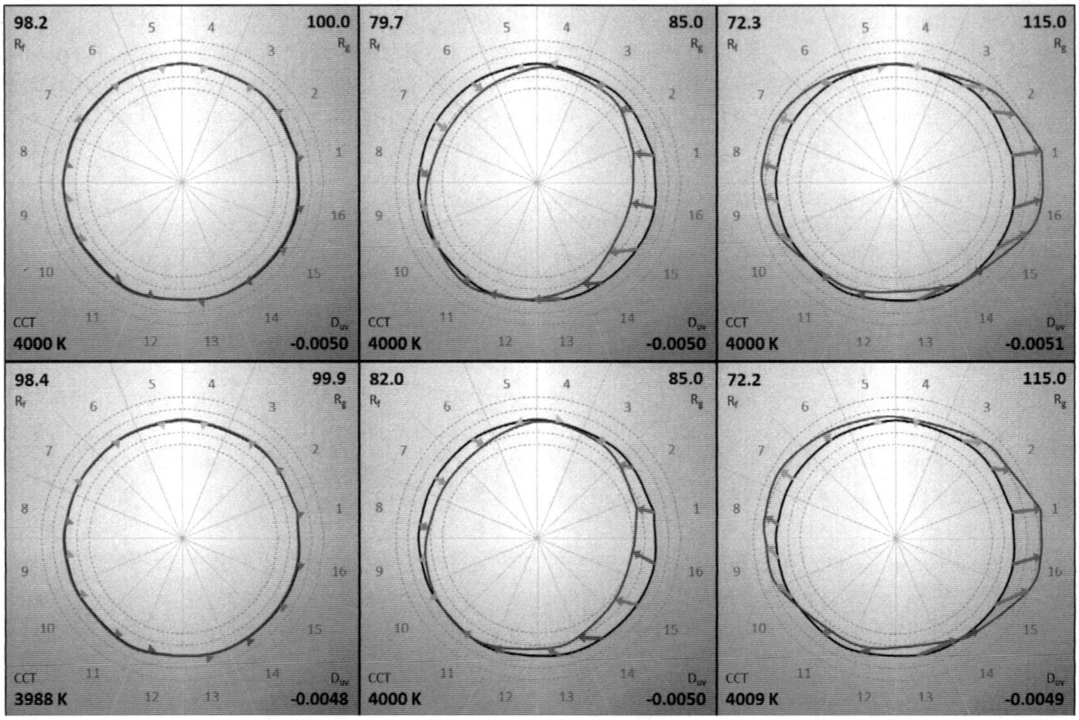

Figure 1. (Top row) Lighting conditions 1-3 at 400 lx and (bottom row) lighting conditions 4-6 at 50 lx used in experiment 1.

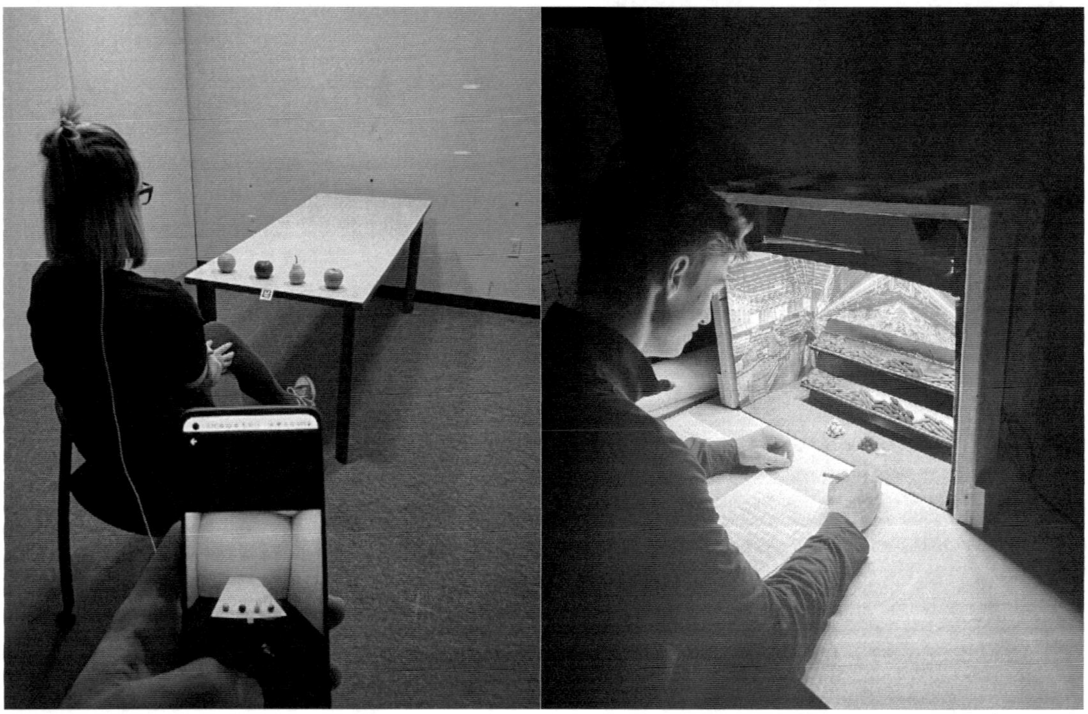

Figure 2. (Left: Study 1) Eye tracking and subjective judgments of fresh produce. (Right: Study 2) Purchase intent and subjective judgments of model fruits shown in a miniature retail setting.

2.2. *Study 2*

A second experiment was conducted to evaluate the impact of illuminance levels, color gamut size and shape in a miniature model of a retail space. Six lighting conditions were generated using a Rockville linear LED fixture that has five LED channels mounted on a small-scale model of a retail store. The lighting conditions varied in terms of color gamut (nominally ANSI/IES TM-30 R_g = 90, 100, 120) and illuminance (nominally 500 lx and 750 lx) to meet the IES retail lighting guidelines [41], as shown in Table 2 and Fig. 3.

Table 2. The lighting conditions used in Study 2

Condition #	Nominal	Illuminance (lx)	Fidelity (R_f)	Gamut (R_g)	T_{cp} (K)	D_{uv}
1	Desaturating	500	70.6	87.0	2920	-0.0015
2	Neutral		78.8	98.1	2940	-0.0010
3	Saturating		67.1	119.5	2940	-0.0021
4	Desaturating	750	62.8	87.0	2920	-0.0015
5	Neutral		78.5	98.0	2910	-0.0014
6	Saturating		66.8	119.5	2940	-0.0019

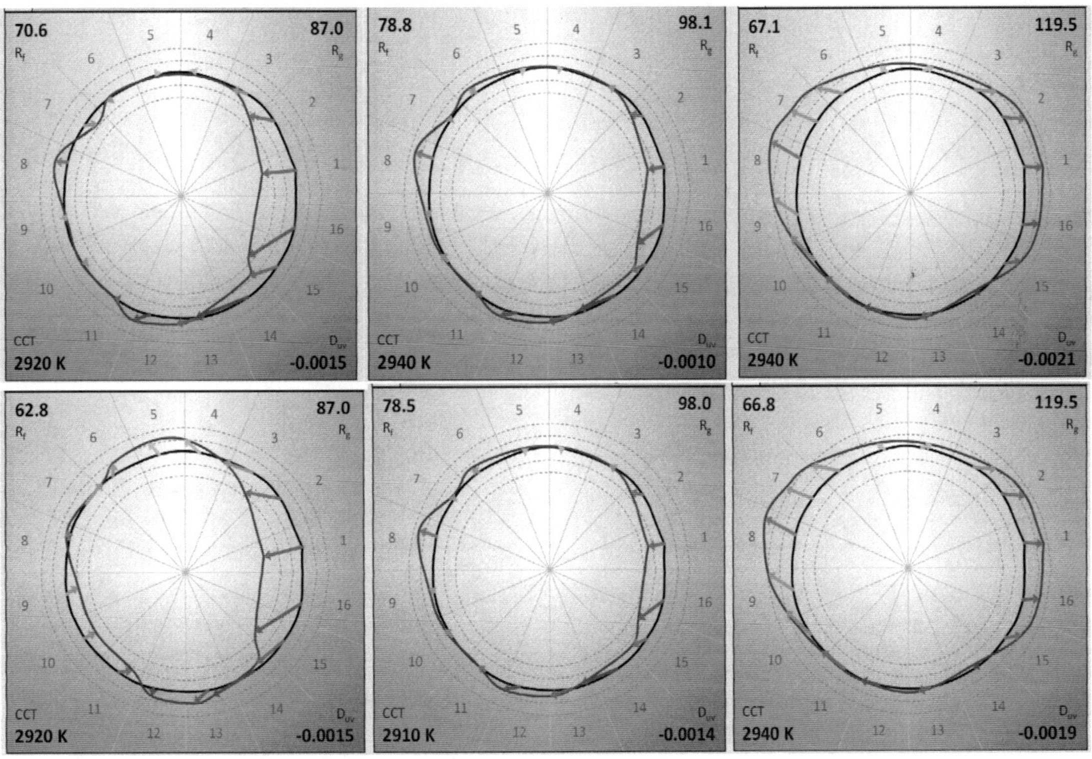

Figure 3. (Top row) Lighting conditions 1-3 at 500 lx and (bottom row) lighting conditions 4-6 at 750 lx used in experiment 2.

The trials were randomized using the William Latin square design [42]. Each lighting condition was repeated three times and lasted for 45 seconds. Between trials there was a two second dark transition period to reduce afterimage effects [43]. The experiment duration was 14 minutes in total. Thirteen American participants (4 males, 9 females) were asked to judge the preference and unnaturalness of the objects, as shown in Fig. 2 (right).

IOP Conf. Series: Earth and Environmental Science 1320 (2024) 012024 doi:10.1088/1755-1315/1320/1/012024

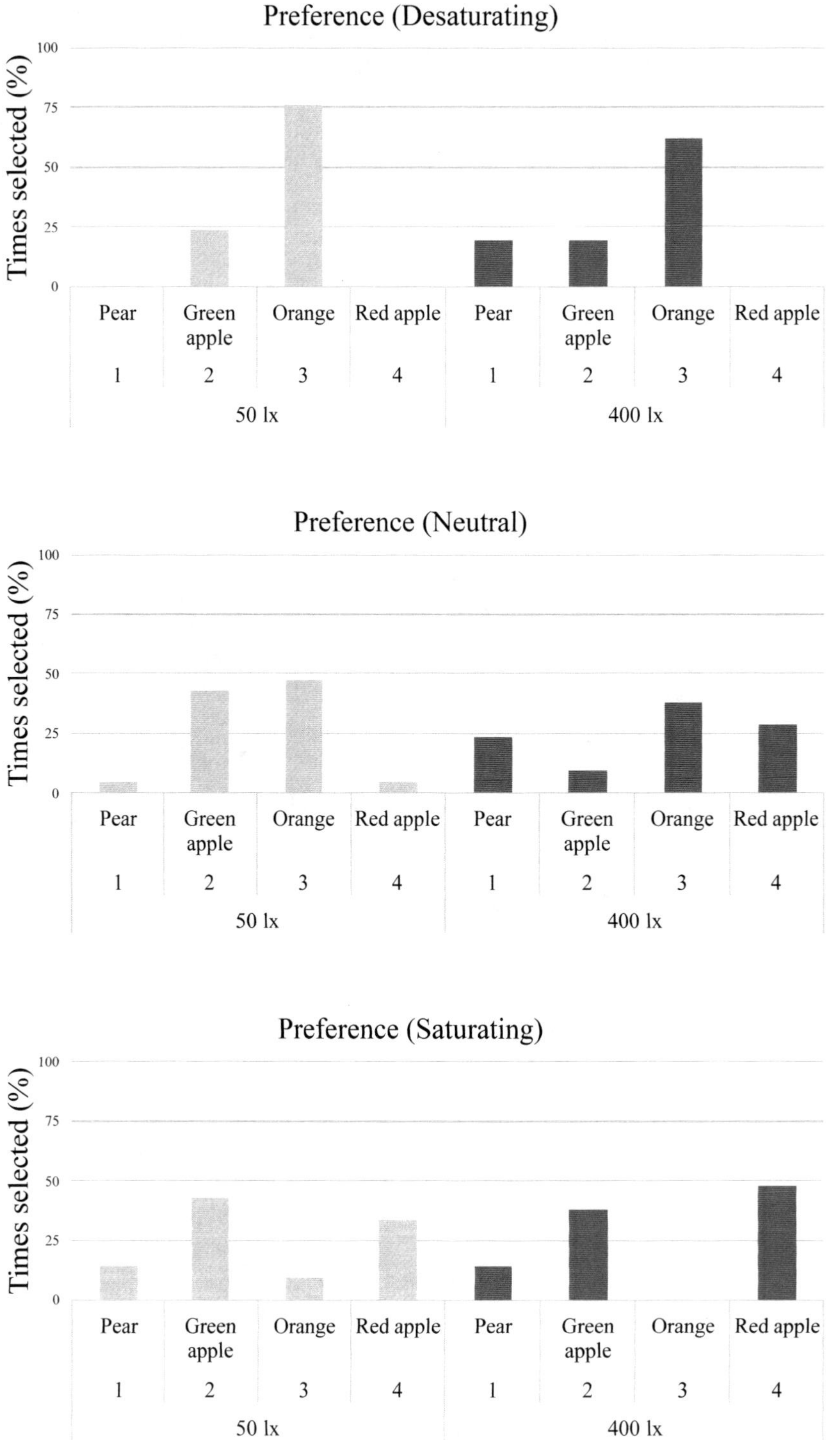

Figure 4. Participants' choices of the fresh produce under desaturating (top), neutral (middle), and saturating (bottom) spectra for 50 lx (light blue columns on the left) and 400 lx (dark blue columns on right).

3. Results

3.1. *Study 1*

Results from study 1 suggest that light source spectrum (color gamut) can impact subjective evaluations of fresh produce more than illuminance, as shown in Fig. 4. Under desaturating light source, the most preferred produce was orange likely due to the already saturated nature of the fruit. The opposite was also true where orange was the least preferred fruit under saturating lighting confitions (i.e., increased saturation made the orange look cartoonish, thus preference reduced in this lighting condition). Green and red apples under saturating lighting conditions of 400 lx were the second most preferred produce. On average, pear was the least preferred produce. While the color appearance can make an impact on preference evaluations, it is important to acknowledge other factors that can influence participants' decisions (e.g., the taste of the fruit).

Unnaturalness responses supported preference results, where the fruit orange under saturated 400 lx lighting condition was perceived to be the most unnatural fresh produce. Red apple in desaturating lighting conditions followed orange in terms of naturalness. Green apple was the least unnatural on average, indicating the robustness of near-neutral (originally less saturated) objects.

Gaze data analysis was limited due to technical difficulties (i.e., sensitivity of surface creation in video recordings). However, the limited information from the gathered data indicated that participants gazed at unnatural or preferred objects most of the time, as shown in Fig. 5.

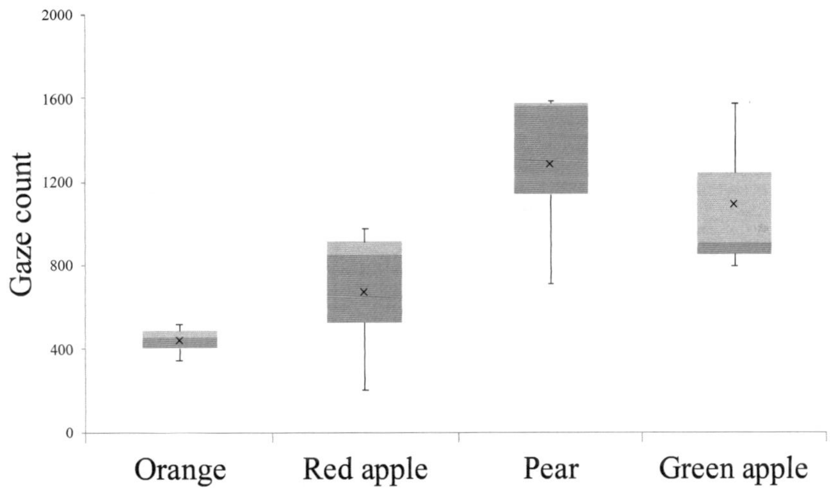

Figure 5. Participant SP's gaze counts were the highest for the most unnatural object (pear) followed by the most preferred object (green apple) under 50 lx $R_f = 100$ and $R_g = 100$ (neutral condition). The error bars show the interquartile ranges.

3.2. *Study 2*

The second study results indicate an increase in willingness to purchase as the lighting level and color gamut increased, as shown in Fig. 6. The higher light level resulted in a higher purchase intention at the same color gamut. The difference between R_g values at 750 for purchase intent was significant (p = 0.003, and $\eta_2 = 0.084$). The lowest purchase intent was for desaturating spectra at 500 lx. There was also a significant difference between light levels. However, the effect of illuminance should be taken with a grain of salt due to the limited number of illuminance ranges that may cause range bias [43,44].

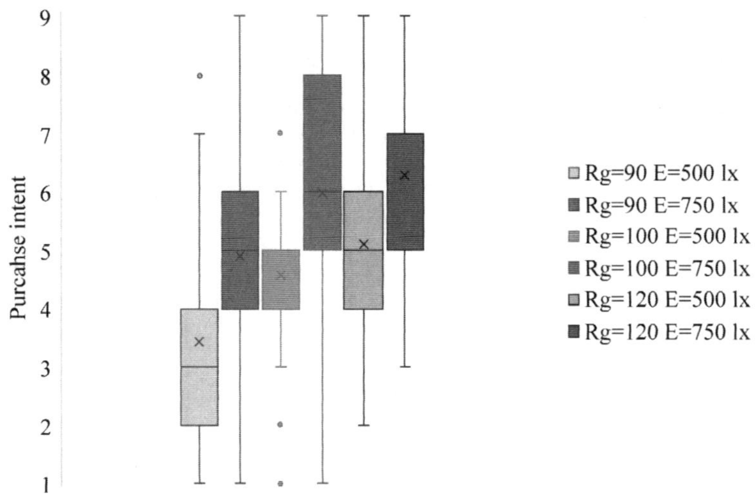

Figure 6. Purchase intent (higher number denotes increased intent) under desaturating (green hue), neutral (red hue), and saturating (blue hue) lighting conditions for 500 lx and 750 lx.

4. Conclusions

In the realm of microeconomics, the physical properties of the environment can significantly impact observers' actions, particularly their intent to purchase goods. Lighting plays a pivotal role in influencing the perceived quality of objects, with illuminance and color being the primary stimuli involved. Therefore, lighting designers hold direct control and responsibility over these key factors, enabling them to shape shoppers' visual perceptions of goods. Although past studies have separately explored the influence of color gamut on preference and purchase intent, a strong connection between these two related phenomena has not been established due to a lack of controlled conditions. This research aims to bridge this gap and establish a robust link between color gamut, preference, and purchase intent, shedding light on the intricate interplay between lighting, perception, and consumer behavior.

Visual experiments were conducted to examine the effect of lighting on purchase intent. The findings indicate that lighting has a significant influence on consumers' perception of products. Participants in these studies demonstrated the ability to discern changes in color gamut and illuminance, highlighting the sensitivity of their visual perception to lighting conditions. However, it is important to note the limitations of the studies, including a small sample size, a restricted range of illuminance levels, and a limited visual field of view in the second experiment. To gain a more comprehensive understanding of the mechanisms underlying purchasing intent, further research is warranted, particularly exploring different price ranges and product types such as fresh produce versus manufactured goods. Further research in this field will enable a deeper exploration of the complex relationship between lighting, consumer perception, and purchase behavior.

Acknowledgments

Authors would like to thank Tony Esposito for inspiration and J. Mundinger for his assistance with the experimental setups.

References

[1] Hussain R and Ali M 2015 Effect of store atmosphere on consumer purchase intention. *International Journal of Marketing Studies*, 7(2).
[2] ANSI/IES 2020 Recommended Practice RP-2-20. Illumination Engineering Society. New York: IES.
[3] Smet K, Ryckaert W R, Pointer M R, Deconinck G and Hanselaer P 2011 Correlation between color quality metric predictions and visual appreciation of light sources. *Optics Express*, 19(9),

pp.8151-8166.
[4] Jost-Boissard S, Avouac P and Fontoynont M 2015 Assessing the colour quality of LED sources: Naturalness, attractiveness, colourfulness and colour difference. *Lighting Research & Technology*, 47(7), pp.769-794.
[5] Dangol R, Islam M S, Hyvärinen M, Bhushal P, Puolakka M and Halonen L 2015 User acceptance studies for LED office lighting: Preference, naturalness and colourfulness. *Lighting Research & Technology*, 47(1), pp.36-53.
[6] Wei M, Houser K W, David A and Krames M R 2017 Colour gamut size and shape influence colour preference. *Lighting Research & Technology*, 49(8), pp.992-1014.
[7] Royer M P, Wilkerson A, Wei M, Houser K and Davis R 2017 Human perceptions of colour rendition vary with average fidelity, average gamut, and gamut shape. *Lighting Research & Technology*, 49(8), pp.966-991.
[8] Durmus D and Davis W 2017 Object color naturalness and attractiveness with spectrally optimized illumination. *Optics Express*, 25(11), pp.12839-12850.
[9] Klabes J, Babilon S, Beck S, Braun S, Johannbroer B, Joppien A M, Trinh Q V and Khanh T Q 2021 Towards a comprehensive lighting-quality model: validation of brightness, visual clarity, and color preference formulae applicability in two realistic mock-up scenarios. OSA Continuum, 4(12), pp.3139-3156.
[10] Wu L, He Z, King C and Mattila A S 2021 In darkness we seek light: The impact of focal and general lighting designs on customers' approach intentions toward restaurants. *International Journal of Hospitality Management*, *92*, p.102735.
[11] Steele K S, Weber M J, Boyle E A E, Hunt M C, Lobaton-Sulabo A S, Cundith C, Hiebert Y H, Abrolat K A, Attey J M, Clark S D and Johnson D E 2016 Shelf life of fresh meat products under LED or fluorescent lighting. *Meat Science*, *117*, pp.75-84.
[12] Hasenbeck A, Cho S, Meullenet J F, Tokar T, Yang F, Huddleston E A and Seo H S 2014 Color and illuminance level of lighting can modulate willingness to eat bell peppers. *Journal of the Science of Food and Agriculture*, *94*(10), pp.2049-2056.
[13] Özkul E, Bilgili B and Koç E 2020 The Influence of the color of light on the customers' perception of service quality and satisfaction in the restaurant. *Color Research & Application*, *45*(6), pp.1217-1240.
[14] Nagyová Ľ, Berčík J, and Horská E 2014 The efficiency, energy intensity and visual impact of the accent lighting in the retail grocery stores. *Potravinárstvo*,1;8(1).
[15] Wu T Y and Wang S G 2015 Effects of LED color temperature and illuminance on customers' emotional states and spatial impressions in a restaurant. *International Journal of Affective Engineering*, *14*(1), pp.19-29.
[16] Chen W, Wu X, Liu Z, Liu Y, Liu Q, Pointer M R, Liang J and Khanh T Q 2022 The impact of illuminance level, correlated colour temperature and viewing background on the purchase intention for bread and cakes. *Food Quality and Preference*, 98, p.104537
[17] Fiore A M and Kim J 2007 An integrative framework capturing experiential and utilitarian shopping experience. *International Journal of Retail & Distribution Management*, *35*(6), pp.421-442.
[18] Ahmadian Tazehmahaleh K, Godazgar H, Smet K A and Hanselaer P 2022 Multi-Channel LED Luminaires: An Object-Oriented Approach for Retail Lighting Based on the SOR Framework. *Sustainability*, *14*(10), p.5994.
[19] Durmus D, Abdalla D, Duis A and Davis W 2018 Spectral optimization to minimize light absorbed by artwork. *Leukos*. 16(1), 45-54.
[20] Durmus D and Davis W 2018 Appearance of achromatic colors under optimized light source spectrum. *IEEE Photonics Journal*, *10*(6), pp.1-11.
[21] Zhang J, Smet K A and Meuret Y 2019 Tuning color and saving energy with spatially variable laser illumination. *Optics Express*, *27*(19), pp.27136-27150.
[22] Lombana M and Tonello G L 2017 Perceptual and emotional effects of light and color in a

simulated retail space. *Color Research & Application*, *42*(5), pp.619-630.

[23] Tantanatewin W and Inkarojrit V 2016 Effects of color and lighting on retail impression and identity. *Journal of Environmental Psychology*, *46*, pp.197-205.

[24] Areni C S and Kim D 1994 The influence of in-store lighting on consumers' examination of merchandise in a wine store. *International Journal of Research in Marketing*, *11*(2), pp.117-125.

[25] Huang Z, Liu Q, Liu Y, Pointer M R, Luo M R, Wang Q and Wu B 2019 Best lighting for jeans, part 1: Optimising colour preference and colour discrimination with multiple correlated colour temperatures. *Lighting Research & Technology*, *51*(8), pp.1208-1223.

[26] Hemalatha K, Chandramathy I and Dugar A M 2022 Effects of lighting conditions on user preferences in retail apparel stores, within the cultural context of India. *Building and Environment*, *221*, p.109270.

[27] Park N K and Farr C A 2007 The effects of lighting on consumers' emotions and behavioral intentions in a retail environment: A cross–cultural comparison. *Journal of Interior Design*, *33*(1), pp.17-32.

[28] Hutchison J, Thomas N A and Elias L 2011 Leftward lighting in advertisements increases advertisement ratings and purchase intention. *Laterality: Asymmetries of Body, Brain and Cognition*, *16*(4), pp.423-432.

[29] Babin BJ, Hardesty D M and Suter T A 2003 Color and shopping intentions: The intervening effect of price fairness and perceived affect. *Journal of Business Research*, 56(7), pp.541-551.

[30] Bilgili B, Ozkul E and Koc E 2020 The influence of colour of lighting on customers' waiting time perceptions. *Total Quality Management & Business Excellence*, *31*(9-10), pp.1098-1111.

[31] Biswas D, Szocs C, Chacko R and Wansink B 2017 Shining light on atmospherics: how ambient light influences food choices. *Journal of Marketing Research*, *54*(1), pp.111-123.

[32] Phumchan N and Tuaycharoen N 2022 Effects of illuminance and correlated colour temperature on emotional state and perception of Thai customers in clothing retail stores. *Asia-Pacific Journal of Science and Technology*, *27*(6).

[33] Tešić D, Bogetić Z and Petković G 2022 The influence of retail facility lighting on shoppers' product perception. *Strategic Management*, *27*(4), pp.41-54.

[34] Liu S, Li L, Tan Y, Xie J L and Lu Q X 2020 A study on the effect of artificial light environment on consumers' purchase intention for fruits. In *Advances in Usability, User Experience, Wearable and Assistive Technology: Proceedings of the AHFE 2020 Virtual Conferences on Usability and User Experience, Human Factors and Assistive Technology, Human Factors and Wearable Technologies, and Virtual Environments and Game Design, July 16-20, 2020, USA* (pp. 308-315). Springer International Publishing.

[35] Durmus D 2022 Correlated color temperature: Use and limitations. *Lighting Research & Technology*, *54*(4), pp.363-375.

[36] Ohno Y 2014 Practical use and calculation of CCT and Duv. *Leukos*, *10*(1), pp.47-55.

[37] Royer M P, Wilkerson A, Wei M, Houser K and Davis R 2017 Human perceptions of colour rendition vary with average fidelity, average gamut, and gamut shape. *Lighting Research & Technology*, *49*(8), pp.966-991.

[38] Esposito T and Houser K 2019 Models of colour quality over a wide range of spectral power distributions. *Lighting Research & Technology*, *51*(3), pp.331-352.

[39] Cohen M A, Botch T L and Robertson C E 2020 The limits of color awareness during active, real-world vision. *Proceedings of the National Academy of Sciences*, 117(24), pp.13821-13827.

[40] Barbierato E, Berti D, Ranfagni S, Hernández-Álvarez L and Bernetti I 2023 Wine label design proposals: an eye-tracking study to analyze consumers' visual attention and preferences. *International Journal of Wine Business Research*, 1751-1062.

[41] ANSI/*IES* RP-2-20 2020 *Recommended* Practice: *Lighting Retail* Spaces. New York *Illuminating Engineering Society*.

[42] Williams E J 1949 Experimental designs balanced for the estimation of residual effects of

treatments. *Australian Journal of Chemistry*, 2(2), pp.149-168.
[43] Royer M, Houser K, Durmus D, Esposito T and Wei M 2022 Recommended methods for conducting human factors experiments on the subjective evaluation of colour rendition. *Lighting Research & Technology*, *54*(3), pp.199-236.
[44] Fotios S A and Houser K 2009 Research methods to avoid bias in categorical ratings of brightness. *Leukos*, 5(3), pp.167-181.

IOP Conf. Series: Earth and Environmental Science 1320 (2024) 012025 doi:10.1088/1755-1315/1320/1/012025

Light across Disciplines and Times: A Walking Seminar at the Stockholm Public Library

Malin Alenius

KTH Royal Institute of Technology, School of Architecture
Osquars backe 9, SE-100 44 Stockholm, Sweden

email:malin.alenius@arch.kth.se

Abstract. Lighting in buildings today is a multi-disciplinary task. Increased complexity and specialization have led to a split process where coordination between disciplines is lacking. Despite extensive technical development, daylight and electric light are still planned in separate processes, often resulting in over-lit environments where energy savings, health and spatial potentials of daylight are not fully utilized. To address these problems, this paper introduces the walking seminar as a method to bridge and increase knowledge across disciplines in the field of lighting. The seminar was held at the Stockholm Public Library, and included invited participants with different professions and expertise: an architect, an electrical consultant, a lighting designer, a librarian, a caretaker, and an architectural historian who shared their perspectives on light. Based on the seminar, findings relevant for today's practice are presented, both for the specific case in itself, a listed building and an active library, and at a more general level concerning lighting standards and the interaction of daylight and electric light. The main future task for the building sector is to preserve and develop our existing building stock, which includes lighting. Sustainable and long-term solutions require that daylight, electric light and spatial lighting design are planned as an integrated whole, based on the characteristics of the specific building. Findings from the walking seminar presented in this paper point at the importance and potential in transdisciplinary dialogue to enable the development of long-term lighting solutions in our existing and future buildings.

1. Introduction

This paper addresses the theme of *knowledge transfer*[1] between different groups of experts—specifically, those working with *lighting* within the building sector. One reason, I hypothesize, for the unsynchronized way in which the planning processes handle lighting at present is the lack of collaboration and coordination occurring between different disciplines. The language available to communicate about lighting, further, is often discipline-specific and as such insufficient in accessing the full potential of the field of lighting design.

In practice, the design of lighting takes place during different stages of the building process. The basic conditions for daylight in new buildings are set early on in the planning stage; these conditions are then developed further by architects and environmental specialists who carry out energy and daylight calculations. In contrast, electrical lighting is planned by the lighting designer and/or by the electrical engineer during the final stage of building design. The maintenance and development of the existing building stock constitutes the main task for the building sector at present and presumably also in the future); because electric lighting is changed more often than other aspects of a building and large transitions like the introduction of LED have occurred, the field has rapidly transformed in recent years.

The (good) intention of creating economic, modern, and energy-efficient lighting, furthermore, often clashes with the original spatial design, which was created under other conditions and to meet different criteria. The prerequisites for developing a long-term, sustainable, and spatially integrated lighting design in buildings lie, I argue here, in an understanding of the different roles that light plays in building design—including spatial, practical, historical, health-related, and energy-saving aspects—and in the ways that the respective professional groups involved in lighting think and talk about light, as well as what is required for these groups to meet and communicate.

Against this background, this paper presents findings from a walking seminar that I organized at the Stockholm Public Library in May 2023. The seminar was designed to explore what might happen when different competences and temporal perspectives were brought together in a conversation about light in a specific, spatial setting of a historical building. In addition to presenting the results of the seminar, I also discuss how methods for transdisciplinary dialogues within lighting can be developed.

2. The Walking Seminar as a Method

In their work on inventive research methods, Cecilia Lury and Nina Wakeford argue that "a method must be made specific and relevant to the problem posed. In short, inventive methods are ways to introduce answerability into a problem."[2] In the case presented here, the walking seminar was chosen as a method in order to make visible different specialist groups involved in lighting in buildings, to understand and identify critical issues in current lighting design, and to highlight and bridge gaps in communication between different perspectives on light.

In existing buildings, electrical lighting is often treated as a secondary technical installation, without considering its spatial and visual effects. If a broader approach is to be developed, multiple competencies and perspectives are required, and (as with other issues in architecture based on teamwork) collective research methods that are designed for group reflection are necessary. Moreover, the visual impact of light must be explored within specific spatial situations. Given these requirements, walking seminars—which involve the design of a series of temporal spatial situations in which participants meet and move together and where each new situation relates to (and is visually impacted by) that which has been experienced previously—are highly relevant.

The Stockholm Public Library, which was chosen as the location for the seminar, is not only an active public library but a well-known cultural historical building designed by Gunnar Asplund, which is often referred to in relation to its lighting design.[3] The library is an interesting example of the encounter between older, daylight-based lighting design and contemporary approaches, whereby the electrical lighting scheme was added over time, giving it an overlapping but not coherent character. Fulfilling contemporary needs in buildings whose lighting stems from another age, when other conditions and ideals prevailed, constitutes a complex task. I argue, in listed buildings, such as the Stockholm Public Library, original luminaires are to be preserved, but the effect of light is not.[4] This raises interesting questions about the role of light and who/what controls and governs its spatial design: What qualities are important to preserve and what may need to change when modernizing lighting in older buildings? How can these values be identified and communicated?

As a doctoral student (DS), I designed and organized the walking seminar as part of my thesis, extending from my previous long experience as a practicing architect and lighting designer. The overall theme of the seminar was "lighting" as it is seen from different perspectives; those perspectives were represented by the invited participants, who were carefully selected based on their diverse experiences and backgrounds. The participants were: a caretaker (C) and a librarian (L), whose long experience working in the library had given them a high level of insight into the daily activities that occur there; a lighting designer (LD) with a background in architectural education and research; an architectural historian (AH) with specialist knowledge of the building; and an architect (A) with expertise in heritage buildings and an electrical consultant (EC) with experience working with heritage buildings, both of whom were involved in ongoing maintenance work at the City Library. The participants were informed and gave their consent via email to be part of the research, and to be recorded and cited in publications. The

IOP Conf. Series: Earth and Environmental Science 1320 (2024) 012025 doi:10.1088/1755-1315/1320/1/012025

recordings were transcribed and edited by me. The text was revised according to the comments from the participants, none of whom requested anonymity. The seminar was designed as a walk through the library with a number of stops linked to various themes that touched on different aspects—technical, practical, and spatial—of light, represented by the different perspectives of the participants. The seminar moved between different time periods, reflecting upon the technical innovations and legislative changes that had affected the building's lighting scheme since its construction in 1928, the same year that the frosted incandescent was introduced to the Swedish market.

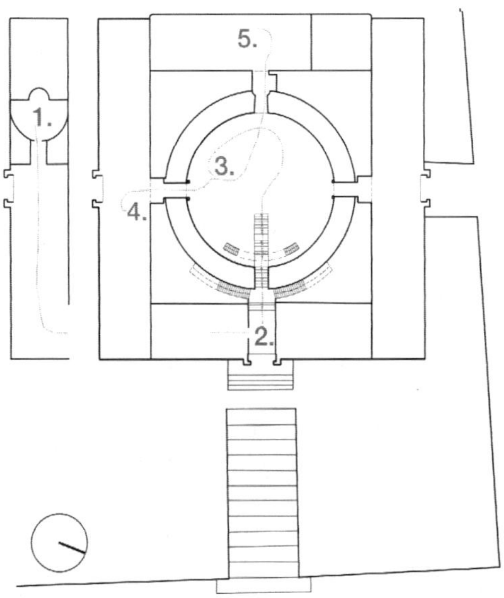

Walking Seminar—Route and Themes
Stockholm Public Library, 23 March 2023.

1. The Storytelling Room—Introduction and Presentation
From your perspective, what is particularly noticeable in the changes in lighting over time?

2. The Entrance Vestibule—Contrast and Movement
Is it possible to preserve contrasts between light and dark in older buildings when lighting is modernised? Is it even desirable?

3. The Rotunda—Spatial Lighting and the Motion of Daylight
What is dominant in the room, daylight or the electric lighting?

4. South Reading Room—Light for Reading
What is important when creating good reading light in libraries today?

5. West Reading Room—Collaboration across Diciplines
What competencies are involved in the work with light in buildings today?

Figure 1: Information sent to the participants showing the route and themes for the walking seminar

3. The Walk

In the following section I present the conversation that occurred on our walk through the library, structured in terms of the five stops that were made, each of which had a different theme, see Figure 1. Themes and questions were sent to the participants in advance for their preparation. Questions were deliberately designed to enable an open dialogue between the participants and allow for each participant's perspective.[5] The conversation is presented as a curated summary based on transcripts, written in quotes in present tense. The ambition is to highlight knowledge that is of interest for the specific case, as well as knowledge that is relevant to lighting in general.

3.1 The Storytelling Room—An Introduction to the Walk and a Presentation of the Participants

(DS): Welcome all participants! The Storytelling Room is the first stop on our tour through the library. During the seminar, we will have the opportunity to share perspectives on light and to move between different time periods, reflecting on historical changes in the lighting of the building. The Storytelling Room has largely remained unchanged since the library was opened in 1928, with the exception of the electric lighting, which originally consisted of only an incandescent floor lamp. From your perspective, what is particularly noticeable in the changes in lighting over time? Is there something special you would like to tell us about the lighting in the Storytelling Room?

[The participants explain their roles in the handling of lighting in the building and show and tell each other about their experiences in the Storytelling Room.]

The caretaker (C): I've been working in the library for almost ten years and lighting in particular has been a journey. When I started here, there were light sources and luminaires from all generations. I had to familiarize myself with which lamps fit where and make sure that everything could be stocked. Now they're being phased out and replaced by LED, and the light sources are integrated in the luminaires, so suddenly it's no longer my responsibility but the Property Office's if the lights don't work. This saves us a lot of time; we used to spend half an hour a day changing lights.

Architect (A): My role in this project is that of a cultural heritage expert and I am in dialogue with the County Administrative Board about all the changes being made as the library is a listed building. The grandiose restoration plans that existed previously have, due to a lack of finances, ended up in a maintenance programme, meaning a limited renovation focusing on technical issues—among other things, the electrical system from the 1950s is worn out. Regarding the electric lighting, the building still has a unique set of original lighting fixtures, which normally disappear rather quickly, and this is of great value to the building.

Electrical consultant (EC): I have been involved in the project since those more grandiose plans; an inventory of the building's luminaires was made at that stage and now the original luminaires have been renovated. It has been important in the project to use existing mountings as much as possible. The energy issue has also been central: we expect to reduce electricity consumption by 65% by changing all lighting to LED. In here, no major changes have been made to the lighting, it basically looks the same as it did when it was installed in the 1970s.

A: What is disturbing in here today is the new emergency light, which is very bright.

Architectural historian (AH): The Storytelling Room is quite different from other rooms in the City Library. It is strongly associated with the Skandia cinema from 1923, which also informed the changes that were made here when the library was renovated 1979–81. It has become a bit of a cinema in here with recessed baffle fixtures similar to those used when Skandia was renovated in 1968.

Librarian (L): I have worked with the youngest children, with songs and rhymes for babies and their parents here in the Storytelling Room. It's always fully booked and there are 12 babies and 12 adults at a time, so it gets quite full. We push the benches out of the way and sit on the floor. The parents sit with their baby in front of them and I sit in the centre. Since the babies get the light right in their eyes, we usually dim in the ceiling and keep a stronger light on me and on the mural. The bright colours of the painting attract the attention of the young children. It is fascinating to see how the children respond to the painting, from the very youngest to the older children; I think it has a lot to do with the lighting—that the painting is illuminated like this.

Lighting designer (LD): If we fantasize freely in this room, the first scene with just a floor lamp by a reading chair is a really prime example of a light zone[6] where all attention is pointing towards one direction. Then we have the second scene that you (L) describe, where the attention shifts to paintings, children, and parents. It would be interesting to experiment with different light scenes in here together with the children and study how their attention in the room changes. A library today contains many different functions with different focuses, there is no single, "right" lighting solution. There are many interesting possibilities with modern lighting; we have technical luminaires that don't glare and we can control the light for different light scenes. These are the good lighting opportunities of our time.

3. 2 The Entrance Vestibule—Contrast and Movement

[The group gathers in the Vestibule.]

DS: In older buildings, light levels are often lower and light distribution less even than in contemporary buildings. In the dark entrance vestibule, contrasts in light and colour are carefully planned to make the rotunda above appear brighter by comparison. With the strong light being situated at the back, the visitor gets used to a lower light level and is drawn up the stairs towards the light of the rotunda.[7] Is it possible to preserve contrasts between light and dark in older buildings when lighting is modernized? Is it even desirable? What requirements are in conflict with a lighting design that includes darkness?

[Two original glass light fixtures on the wall next to the large glass opening of the entrance are lit. Additional ceiling lights are temporarily switched off. We look up towards the space of the black ceiling.]

A: The dark space is suggestive; even more so when it's dark outside.

LD: The large amount of daylight and how it is reflected on the black surfaces in this room is fascinating.

DS: Why do you think the luminaires are positioned next to the window at the entrance?

LD: To put lamps in the windows is a Swedish tradition. The warm light of the wall lamps against the blue daylight is beautiful. It's the contrast in the colour of the light that allows you to see the warm light; if there had been a cold light in the lamps, you wouldn't have seen the light at all.

DS: Look upstairs, the sun is coming into the rotunda. We're lucky!

LD: When the sun comes in like this everything stops for a moment. A little bit of sunlight is so much more interesting than the light from a big glass facade.

DS: Is the additional lighting in the ceiling necessary?

EC: We can't meet requirements in here otherwise, not with the differences in height levels.

L: At night it's completely dark in here without the lights and there's no handrail to hold on to. This poses an accessibility issue for those with poor eyesight. I don't know if it's light-related, but we've had incidents here in the evening—an elderly man fell relatively recently.

DS: But do the lights need to be on at all times, even during the day?

EC: During the day, on the way in and up, it's ok, but when you go out again and are met by the strong backlight, even the contrast markings on the steps won't help, you sort of step out into the air. With the new control system I installed, each luminaire can be controlled individually so it would be possible to control the lighting against daylight sensors outside, but it has not been done.

L: There are many aspects to take into account in this building: the unique architecture and the cultural heritage but also the daily use of the library from the perspective of the staff and primarily the visitors. When I give tours of the building, the group usually gathers in the vestibule, which is a good place to talk about the architectural motif of the long staircase and the democratic idea of access to books for all. But it is difficult to gather here, it is so crowded, so dark and narrow, and people pass by all the time. This entrance can actually be perceived as a barrier for inexperienced library users, the first time they visit the building.

[The conversation in the vestibule exemplifies the contradiction between the original high-contrast lighting with low light levels and today's needs and requirements for accessibility. A number of typical characteristics of older lighting design are identified through the example of the vestibule, such as lowering the light level for adaptation between inside and outside, emphasis on contrast, and a daylight-based lighting design that allows for lighting events.]

3.3 The Rotunda—Spatial Lighting and the Motion of Daylight

[We ascend the stairs up to the Rotunda.]

DS: We will look at the space through two different lighting scenarios: one that mimics the original lighting from 1928, where only wall lamps and the large glass pendant in the centre of the room are lit; and the other representing the current lighting situation with additional uplighting and vertical light on the bookshelves.

To the original scene, we welcome a new participant, Erik Blomberg, who in a review of the library that was published in its opening year, describes how he, after negotiating "the dark and tunnel-like vestibule," experiences a sense of relief in reaching the bright, high Rotunda with its changing light, noting "in the centre of the wall, a ring of narrow windows. It is these windows, with their shifting daylight across the walls, that bring to the interior a warm and vibrant illumination, filling out the large dimensions and somewhat abstract form. [...] The removal of the dome was, from this point of view, a real gain."[8]

AH: The glass dome was an early proposal that would have provided more light from above.

LD: I think it's great that there isn't a dome, and that the light from the windows is instead reflected on the white wall surface.

DS: What do you think, is it perceived as a daylit room and if so, why?

[The group looks up at the drum, where daylight is reflected on the walls, no direct sunlight enters, and it is rather quiet. A short silence, then the conversation takes off. We talk about memories of previous light events in the rotunda, such as when the low winter sun shone straight through the room.]

LD: I call these light events "magic moments". They have a very high quality and are something that I think we should try to reintroduce into architecture again. But that requires a discussion about the role of standards. Standards try to describe minimal nuisance and an acceptable level; they make light seem indifferent and flat. The ever-changing nature of daylight is an amazing quality that we should find ways to recapture; it's sustainable and it connects us to the season and the day. With all the digital tools we have today, we could easily map the "magic moments" that can happen in this building and then try to reintroduce them. The connection to daylight has become a luxury. When this building was constructed, daylight was just a light source and artificial light was expensive. Now it is the other way around and daylight has become a luxury.

L: Those of us who work here are constantly busy meeting all the visitors, so I usually don't have time to think about where I am, but then suddenly something happens with the light—as you say—and I'm reminded that I'm here. I'm also often reminded by visitors what a fantastic environment this is.

AH: It strikes me that in the original scenario the books are in darkness, in the background, and become a bit like wallpaper. The dark bookshelves were a practical problem that was recognised early on. Already in the 1940s, loose luminaires were installed to direct light onto the books; these were later replaced by shelf lighting. I understand the need for light on the bookshelves, I think it is necessary.

DS: But on the other hand, the contrast with the dark bookshelves makes the drum appear brighter.

AH: Yes, and it was even more apparent back then when the books were bound in uniform leather bindings in muted colours.

A: The contrast is also found in the treatment of the materials. Down here and on the way up the stairs everything is so finely chiselled compared to the drum above which has that very rough wall surface.

[C adjusts the lighting to the contemporary scene. Vertical lights on bookshelves and up-lights on the wall of the drum above the edge of the bookshelves are added. The group's intensity level rises, the participants spread out and engage in several simultaneous lively conversations about the difference between the new and old lighting.]

DS: I have to ask the group to gather again. We will miss all the exciting things you're talking about otherwise. How would you describe the difference in perception of light and space between the two scenarios?

AH: In the previous example, the gaze was drawn upwards to the space in the drum. Now the gaze is drawn down to the books. There's a point to that: it's the books you're after here.

A: The first scene was the classic library, a bit like a former mansion, while this scenario seems more fragmented—a bit more like "confetti," as you (AH) said. The uplights are difficult to handle I think, because it so clearly changes the experience of the room.

DS: Yes, and together with the strong vertical light, I think a lot of the daylight feeling is lost. But at the same time, I understand the practical necessity, and perhaps also the symbolic value, of making the books more visible.

LD: The attention in the room shifts. We move from the room and the ceremonial, to the inviting books in all colours, a modern democratic wallpaper.

IOP Conf. Series: Earth and Environmental Science 1320 (2024) 012025 doi:10.1088/1755-1315/1320/1/012025

[C informs the group that the renovation has brought a clear improvement to their work. The large pendulum can now be hoisted all the way down, whereas previously a scaffold was required to reach it. Above all, the switch to LED means that there is basically no need to change light sources.]

C: The worst thing was changing the fluorescent tubes for the up-light above the bookcase. Five years ago, we changed all 144 tubes on the ladder in the narrow aisle, it was a hell to change, so I am happy to avoid that!

LD: I think it's important to discuss light quality now that we have LED. I felt immediately when you switched on the light on the bookshelves that it is a high-quality light. It is a light that tastes and touches. If it had been a standard LED, it would have been a completely different experience.

EC: Yes, the LEDs here have a Ra index of between 92 and 98, a great improvement compared to the previous fluorescent tubes.

[LD measures the illuminance at the shelves with a luxmeter]

LD: There is a lot of light here and strangely enough it still feels good. High levels that one would be afraid of in a calculation can turn out fine when the light quality is good, but also because the lighting experience is logarithmic. This is yet another reason to discuss the standards.

[The effect of the new lighting is the subject of a lively discussion and an improvised test lighting situation arises. C leaves to switch off the uplight and the group studies the difference and the effect it has on the room. We agree that on a day like this, the room is better without the uplight. EC points out that it is different on a dark November evening, then he thinks the uplight is needed. AH adds that the uplight could be switched off during the light season.]

A: What about the light on the bookshelves? Does it need to be so bright on the shelves or can it be dimmed?

EC: Each shelf can be dimmed individually, but we need to have 100 lux on the lowest shelf to comply with the requirements for vertical light from the Cultural Administration.

[LD measures with the luxmeter to 2000 lux at the top and 600 lux on the bottom shelf. C takes the initiative to go and adjust the vertical lighting.]

C: The pre-setting was 90% and now I have reduced it to 50%. Do you think it is better this way? Is it possible to read the backs of the books? I think the room is more pleasant and a little less "confetti," as you called it.

[Everyone in the group agrees.]

EC: But the levels need to be tested on several occasions.

DS: It depends on whether you (C) are still in the building and can control these things.

[Laughter.]

LD: The adjustment of the light, which is one of the most important things, is rarely prioritized in projects. A number of tests are required to cover different situations. It would be interesting to try and do an evening scenario in here. When the sun shines brightly, it is experienced despite modern high light levels, but at dusk a low dimmed general lighting is required. It saves energy and it's also not healthy for people to walk around in too much light all the time.

EC: The large glass pendant can be dimmed and the colour temperature adjusted from 2,700 K to 6,000 K.

DS: But when would you want a cooler light in the glass pendant?

EC: It was a request from the organization. You can have it yellow for Easter and red for Christmas if you want, but whether it will be used I don't know.

A: I would prefer not to use it. For some events it might be ok, but the main setting should be warm.

AH: I think the main point of the new installation is exactly what we are doing now. That the light can be regulated and varied for different occasions and that you can show the library as it was originally lit if you want. Before, it was very static and difficult to make any kind of changes at all.

[In the Rotunda the group visually evaluates and compares two different light scenarios. We discuss how light shifts the attention from the high room in the original scenario, to the level of the bookshelves in the contemporary scene. In the original scenario, the movement of daylight and "magical moments" are recognised as important qualities, a "luxury" that should be reintroduced in building design. In the contemporary scene, we discuss the importance of light quality and the great potential of lighting control to adapt the light to different situations and needs.]

3. 4 South Reading Room—Light for Reading

[The group walks from the Rotunda and gathers in the South Reading Room]

DS: When the library opened in 1928, the reading rooms were heavily criticized for their poor lighting. Blomberg writes in his review that the rooms "do not really come into their own because of the poor lighting. The windows are too high and too small or too few, however one wants to look at it. The reader finds himself in a region below the flow of light, sometimes almost in semi-darkness." According to Blomberg, this is because the architect let the composition of the façade determine the character of the light sources instead of the other way around. "It is strange," Blomberg writes, "that the architect neglected something as fundamental in a reading room as the lighting."[8] What is important when creating good reading light in libraries today? How have the needs and conditions changed?

AH: It was probably deliberate that the reading rooms should be quite dark and that you should concentrate on the book at your reading place. But there is no doubt that the light was inadequate; that has been a recurring criticism. Not only due to the fact that individual artificial light sources were relatively few and weak, but also because the walls and ceiling were covered with matte plaster in a muted tone, which meant that daylight was not reflected very much.

DS: So, there was a stronger contrast in the reading rooms between reading light and a darker room and an overall contrast between the darker reading rooms around the light Rotunda in the middle?

AH: Yes, and a high contrast between windows and walls. The reading rooms were painted white in 1935.

A: That the colour scheme was so short-lived was certainly due to the need for more light, but wasn't it also an architectural fashion that changed and Asplund with it? It became *funkis* [functionalist]. The original colour scheme was typically 1920s classicism, like Thorvaldsen's Museum in Copenhagen, which was a source of inspiration at the time.

AH: It is exciting to see the change that has taken place in relation to the light source. First, there were incandescent lights everywhere in the amps and reading lamps. When the halls were painted white, the reflectors came with a kind of strong discharge lamps and all the reading lamps were cut down and the original glass amps ended up in the basement. Then in the 1940s, fluorescent tubes were added along the edges, and later the reading lamps returned in a new model.

DS [turns to the electrical consultant]: What have you changed with the lighting in here?

EC: It was a precondition that we use the existing ramp fixtures; we have replaced old fluorescent tubes with LED. After a test of the lighting, the screen above the bookshelves was developed by the architect as a way of hiding fixtures. It also serves as an installation zone, a way to get around the space without penetrating with holes. The lighting is not adjusted yet, especially not the uplights.

A: Yes, the uplights look a bit crazy; the light is sharp here and over there it's diffuse and sort of floats…

[DS tries to switch off the uplight, but can't find the right button. She finally succeeds.]

LD: You could feel it immediately: when the uplight was turned off, the room became more pleasant. I think it would be even more inviting if the lights at the reading place were on. They are a central part of the room. Now there is nothing that invites you to read and you should feel tempted to do so.

DS: Today, the digital screens, which are light sources themselves, have partially replaced the books.

L: Yes, but the reading lamps here are used extensively anyway. When I work in the evening, I have to go and switch them off one by one, manually, before the library closes.

[In the South Reading Room, we talk about how the colours and materials used in different architectural styles affect the experience of light. The rapid pace of change of electric lighting is exemplified by all the different types of lights that have been used in the reading rooms. The transition from books to digital screens is briefly discussed. Once again, we talk about how light directs attention in the room, this time exemplified by the uplights and reading lamps.]

3.5 West Reading Room—Shifting Ideals in Lighting and Collaboration across Disciplines

DS: The West Reading Room was built in 1932 and differs in style from the original reading rooms. At this time, Asplund wrote in the journal *ERA* about new lighting technology, which he described as having put "a new and effective constructive element in the hands of the architect." In the article, he emphasized the importance of a close collaboration between the architect and the electrical engineer, stating that:

> The rational use of light, in relation to all of its shifting applications, relies on complex techniques which in turn require a great deal of experience and specialized knowledge; it is therefore important that architects and lighting engineers collaborate closely in solving issues pertaining to lighting technology. The absence of such cooperation can lead to a great deal of laborious fumbling and tinkering, which is also costly and time-consuming, for the intended effect to be achieved.[9]

Which competences are involved in the work with light in buildings today? How do different competences cooperate?

DS: There are many more people involved in the lighting field today, not just the architect and the electrical engineer.

EC: Yes, and then on top of that we have a lot of experts and specifications. Before, when it was just me and A, it was easy, but now we have ten more people to satisfy.

A: Yes, it can get a bit hectic. There is no established hierarchy regarding who is in charge.

LD: Lighting has become more complex: it is no longer only the architect who works with it, engineers do the calculations (one makes a projection regarding sunlight, another daylight, and the architect gets a spreadsheet back). But I'm optimistic, because young people today are so digitally skilled. We in the older generation want simple rules—we design poorly in relation to the possibilities that exist with the new digital tools.

A: I was thinking about Asplund's article. The interest in lighting doesn't exist in the architectural profession in the same way today. This is a problem that has existed for the 40 years I have been working. There are so many luminaires and no overview—you get overwhelmed.

LD: Yes, there's a huge amount. But architects often look at luminaires as an object. If you draw how the light is distributed, however, there are only a certain number of luminaire types.

DS: There is a requirement to preserve original luminaires in a cultural heritage building like this one, but is there also a cultural-heritage value in the effect of the light?

A: Yes, I think so.

DS: Asplund writes about achieving "the intended effect" of light—but what effect do we want to achieve?

IOP Conf. Series: Earth and Environmental Science 1320 (2024) 012025 doi:10.1088/1755-1315/1320/1/012025

A: Yes, and how can we formulate it?

DS: I think the daylight is experienced differently here than in the other rooms, the windows are different and the boundary between inside and outside is less apparent here.

L: Yes, there is a big difference between the colour scheme in here and outside—that's something I also hear from the visitors. The other rooms are perceived as murky and darker. This room opens up when you enter it. The whole wall is covered with windows and you can actually see trees from inside here. But it gets very hot in here in the summer, and very cold in the winter.

AH: So, there's a challenge here in terms of solar radiation and radiators for future renovation work. I like to emphasize this room, which is clearly different from the rest of the building. It's like a pavilion at the Stockholm Exhibition with the rounded shape of the roof and the way the light enters it. Accounts from the employees who worked on the design of this room exist, which include a discussion about how the light should enter and about the curtains, so there was obviously an awareness of the lighting. Something happens after the Stockholm Exhibition and general interest in lighting increases. Before that, I find that Asplund's view of lighting is a bit decorative, he is more interested in how the fixtures look than in the light they provide. [AH directs to DS] You have pointed out the importance of contrast in the lighting of the older parts of the building, but in here there has been a deliberate change of strategy. You get the feeling that it is the general lighting that matters here.

DS: Yeah, for sure. What matters is light and air and the room doesn't have the inward-looking feeling the older parts have.

AH: Yes, and in a way that makes it easier to manage. Even though there are a lot of additions of artificial light fixtures in here, they do not interfere with the overall picture.

DS: That's a really good point. As soon as a light is added to a darker room, the whole lighting scheme changes. It's much more difficult to maintain the character of a darker room than in a room with a higher light level.

[AH and LS agree.]

[In the West Reading Room, the discussion centres on the increasing number of competences and requirements involved in working with light in buildings today. The group agrees that the architect's interest and role in lighting has shifted over time and that the responsibility for lighting design has moved from the architect to other specialists—engineers and lighting designers.]

4. Findings and discussion

Asplund wrote in the aforementioned article in *ERA* 1930, about the importance of collaboration in order to achieve the "intended effect" of light, but what effect do we want to achieve? And as A says: how can we formulate it? One possible answer to this question is demonstrated in the conversation during the seminar and through the improvised test lighting that occurred in the Rotunda.

According to Nel Janssens, a "conversation"—as opposed to knowledge production through discussion—offers a different way of making sense.[10] Conversations use questions to combine different issues and make unforeseen connections. Conversations are "syncretic" because they are about creating a new whole without removing the contradictions between the parts.[10] The description is applicable to the field of lighting where many different perspectives, even if contradictory, need to find ways to interact in order for knowledge to be utilized and developed. Janssens proposes curated conversations using "instructs" as a research tool. As a method, a qualitative conversation requires the design and curation of its material "instructs" and its immaterial instructions in order to provide a structure in which interaction can take place.[10] The walking seminar used the building as a material "instruct" that enabled a variety of experiences through the interaction between the participants, and between the participants and the material structure.

There is no instrument other than the human eye that can be used in order to describe our visual experience of light.[11] In order to develop a common language for the visual experience of light, the specific spatial context is crucial. Observing and discussing light together provides a unique opportunity to develop a common language for what we perceive, and serves as an exercise in articulating and communicating the effects of light. By putting words to what we see, we see more.[12] The fact that the spatial experience of light is situation-dependent—the interaction of light with the space, the objects, and the observer is never quite the same—does not mean that the knowledge gained from such situations cannot be utilized. By attentively and systematically conversing about aspects of light, observations and relationships can be recognised, which builds experience that can be used in new ways in future situations.

During the seminar, a recurring theme was how light directs attention in a room and how attention shifts when lighting changes. Along the walk, participants put into words some of the characteristics that were present in the original lighting design: the lower light level, the emphasis on contrasts in colour of light, and between light and darkness. The ever-changing daylight was considered an important quality and, in contemporary architecture, a luxury, that anchors us in time and season, and allows for light events — "magic moments". The conversation revealed how these characteristics can be connected to the experience of a more ceremonial architecture; the darkness in the vestibule was suggestive and, in contrast to the darkness, the light up in the rotunda appeared even brighter. As the light changed, attention shifted within the room, and consequently, the role of light was also transformed. In the original lighting scene of the Rotunda, attention was directed towards the room, the gaze was drawn upwards to the high bright space. In the contemporary scene, the attention shifted from the room to the inviting books in all colours, and an emphasis was placed on the lower part of the room where people are.

The good lighting possibilities but also the increased complexity in contemporary lighting were exemplified by the building's new control system and its complete transition to LED. The control system enables pre-set lighting scenarios for different situations: for instance, showing the light as it once was, or adapting it to the different needs of a modern library. However, the important act of setting lighting is rarely given priority in practice and the question is rarely asked: Who is responsible for operating the system?

With LED, light quality has become an important issue. The vertical lighting of the bookshelves in the Rotunda has a noticeably good quality, it tastes and touches, and suddenly a high illuminance level that would raise concern in a calculation seems surprisingly good. This observation gives rise to one of several discussions on the role of standards, their legitimacy, and the limiting role that they play in lighting design. With all the digital tools now available, could we find new ways to reintroduce the qualities of daylight, to allow dusk and potential light events to be visible? And how would that interfere with the current standards?

At our last stop, Asplund's article on the collaboration between the architect and the electrical engineer opened up a discussion on the increasing number of competences and requirements involved in lighting. We agreed that working with light has become more complex and that an interest in light does not exist within the architectural profession in the same way that it did during Asplund's time.

One of the participants mentioned the large amount of time saved per day in lighting maintenance after the transition to LED and that the demarcation of who is responsible for lighting has changed.[13] Leaving aside the hindering factors—trends and changing needs in lighting and the industry's desire to sell—what would a slowdown in the rapid pace of change of electric lighting (compared to other changes in a building) mean for lighting design? Could a slower pace of change give electric lighting a more permanent and integrated role in our buildings, where it would be less likely to be perceived as a loose technical installation? And in turn, could a more integrated approach to lighting, based on principles rather than products, revive architects' interest in light and contribute to greater collaboration between disciplines where daylight and electric light were designed together? This hypothetical scenario that would favour a more thoughtful lighting design with the potential to stand the test of time could be seen as a goal worth striving for.

The walking seminar presented in this article is part of a research project aimed at developing a method rather than solving a concrete lighting design task. The method is therefore not directly transferable to practice, but the outcome of the seminar shows the benefits of approaching the lighting task transdisciplinarily also in practice. The seminar exemplifies how new knowledge can be generated by combining different disciplines such as building history, technology, legal frameworks, spatial design, as well as practical and social aspects—new knowledge that is essential for creating long-term integrated lighting solutions. All participants expressed the view that they had gained new knowledge during the seminar, which they would take with them into future work situations. The seminar shows that the building process is in need of a planning step or "category" that could be used in practice in the initial and final stages of projects to ensure integrated lighting solutions and knowledge transfer in the field of lighting, a finding that is useful for future projects.

5. Conclusion

As the conversations between the seminar participants described and summarized in this paper show, the knowledge basis needed for a more informed lighting design already exists, but it is not utilized because it contains different perspectives that rarely meet. Therefore, collaboration and dialogue between different disciplines is a prerequisite for putting knowledge into practice. The outcome of the seminar shows the potential of transdisciplinary methods for dialogues to achieve this. It demonstrates how conversations based on a specific spatial situation make it possible to concretize and share the visual experience of light, and to talk about it. The question of transdisciplinary conversations may be key in all forms of architectural design practices, but in lighting it becomes especially pertinent, as lighting technology develops fast and in renovations is often not treated as an integral part of an existing building. During the transdisciplinary conversation, different perspectives on light, their contradictions, and unforeseen connections became visible, building a meaningful narrative around the many roles of light. This means that, if transdisciplinary dialogue methods are allowed to be developed and implemented in the lighting field, there are good possibilities for a more informed and spatially integrated lighting design.

References

[1] Doucet I and Janssens N 2011 *Transdisciplinary Knowledge Production in Architecture and Urbanism: Towards Hybrid Modes of Inquiry* vol 11 (Dordrecht: Springer Nature)

[2] Lury C and Wakeford N 2012 "Introduction: a perpetual inventory" in *Inventive Methods* (London: Routledge) pp 15–38

[3] Plummer H 2012 *Nordic Light, Modern Scandinavian Architecture* (London: Thames and Hudson) pp 76-77

[4] Vårdprogram Stadsbiblioteket med park 2012 Stockholms Stad: Fastighetskontoret.https://fastighetskontoret.stockholm/hallbarhet/vardprogram/

[5] Kvale S and Brinkmann S 2014 *Den kvalitativa forskningsintervjun* (Lund: Studentlitteratur)

[6] Madsen M 2002 Lysrum *PhD Thesis* Kunstakademiets Arkitektskole

[7] Alenius M 2022 Förändring av rummets och arbetets ljus i tre bibliotek *Bebyggelsehistorisk tidskrift* **83** pp 59-61

[8] Blomberg E 1928 Stadsbiblioteket i Stockholm *Svenska slöjdföreningens tidskrift* pp 7–20

[9] Asplund G 1930 Det elektriska ljuset som konstruktionselement *ERA* **6** p117

[10] Janssens N 2017 Collective sense-making for change: About conversations and instructs in Schalk Kristiansson and Mazé, *Feminist Futures of Spatial Practice: Materialisms, Activisms, Dialogues, Pedagogies, Projections* (Baunach: AADR, 2017) pp 151-156

[11] Liljefors A 1999 Seende och Ljusstrålning: baskompendium. KTH Arkitektur, Stockholm.

[12] Alenius M 2019 "8.2 Visual model for light in space", in Dubois, Gentile, Laike, Bournas and Alenius *Daylight and Lighting under a Nordic Sky* (Lund: Studentlitteratur) pp 196–204

[13] Graham S and Thrift N 2007 Out of Order: Understanding Repair and Maintenance *Theory, culture & society* **24** (3) pp1–25

Acknowledgments

The author wishes to thank all the participants who contributed greatly to make the seminar so rewarding.

IOP Conf. Series: Earth and Environmental Science 1320 (2024) 012026 doi:10.1088/1755-1315/1320/1/012026

Exploring daylight in two different hemispheres: *Sweden and Uruguay: a comparative study of daylight as a formgiver and its impact on visual comfort*

Carolina Borgia

MSc in Architectural Lighting Design from KTH Royal Instiute of Technology (2023)

Email: caroborgia@gmail.com

Abstract. Daylight's critical significance for both the built and natural domains underscores its potential to enhance energy efficiency and environmental sustainability in architecture. This study presents a careful exploration, using a tailored case study, to reveal diverse strategies for leveraging daylight's inherent attributes as a building material. Integrated climate adaptation serves as the bedrock of architectural projects, fostering regenerative development as well as liveable and sustainable spaces. This research delves into daylight's complexities, positioning it as a primary input in early architectural stages. Through comprehensive analysis in Uruguay and Sweden, this study highlights the interplay of daylight's effects. An experimental cabin serves as a focal case study, bolstered by Climate Studio software, yielding a thorough daylight analysis across these locations. The synthesis of academic and practical approaches underscores daylight's pivotal role in shaping sustainable built environments.

1. Introduction

Access to daylight is vital in any dwelling (Mark Rea, 2014), yet effectively integrating it into new projects presents challenges. Prioritizing natural lighting design from the outset is crucial for a regenerative built environment. Designers face the responsibility of crafting spaces aligned with sustainability goals. This thesis explores strategies to master natural lighting design, emphasizing that daylight quantity and quality should not be afterthoughts, as they profoundly impact the built environment's quality. By designing an experimental cabin and using Climate Studio, daylight analysis is conducted in Uruguay and Sweden, showcasing how academic methods enhance design practice. Examining daylight in different latitudes reveals unique qualities and enriches the understanding of its role in sustainable design. As stated by Reinhart (2014) “designers must hence understand how much solar radiation is available at different sites throughout the world”

2. Daylight at different scales

The multifaceted nature of daylight, as emphasized by Peter Tregenza (2014), holds intrinsic qualities of changeability and dynamism. Its significance transcends mere illumination, deeply impacting life and well-being for individuals in terms of both health and emotional satisfaction.

At planetary scale, daylight's interplay with Earth's rotation and orbit results in predictable variations throughout the seasons and the daily 24-hour cycle. However, the unpredictable element of weather adds complexity that designers must address when harnessing daylight's potential.

Zooming in to the urban scale, the atmosphere's role in solar radiation absorption and greenhouse effects becomes paramount, impacting Earth's climate zones. Examining Uruguay and Sweden, distinct geographical locations, illustrates diverse sunlight availability due to hemispheric positioning differences. The interplay of sunlight during solstices and equinoxes showcases seasonal variation magnitude. Climate conditions in these locations underscore the intricate relationship between urban planning, weather, and human interaction.

Transitioning to the building scale, the thesis explores daylight behavior in a one-person cabin within peri-urban settings of Uruguay and Sweden. The cabin serves as a microcosm for understanding interconnections between daylight, architecture, and sustainability.

At the human scale, daylight's profound impact on human psychological, and social aspects comes into focus. Marie Claude Dubois (2019) elucidates its manifold benefits, including visual performance, circadian rhythm stimulation, stress reduction, and heightened productivity.

Visual comfort emerges as a central concern in daylighting design, influencing occupants' satisfaction and well-being. Christoph F. Reinhart's definition of successfully daylit spaces as those balancing visual satisfaction, thermal comfort, and energy efficiency underscores the complex interrelation between these elements.

This comprehensive exploration navigates daylight's intricate journey across planetary, urban, building, and human scales. It underscores daylight's pivotal role in shaping facets of human life and built environments, and the delicate balance designers must strike to enhance visual comfort while considering lighting elements' dynamic interplay.

3. Methodology

The methodology unfolds across four distinct phases (Fig. 1), each contributing to a comprehensive exploration of daylight's impact. First, a dwelling is meticulously designed as a case study. This deliberate choice is driven by the desire to anchor the daylight analysis within a well-structured, square plan footprint. The role of voids within this confined space becomes paramount. Achieving a harmonious equilibrium between window dimensions and potential energy concerns, while addressing visual comfort, constitutes the study's primary objective.

The second phase involves subjecting the case study to scrutiny in four orientations at each location, achieved through a 90-degree rotation of the floor plan. This experimentation introduces varying sun exposure on each facade. Additionally, alterations to interior materials are implemented to gauge their influence on illuminance levels.

Moving on to the third phase, the case study, modeled in Revit and transferred to Rhino, undergoes analysis through the Climate-Studio plugin (v1.9.8389.21977). Employing 30 sensors on the lower floor and 36 on the upper floor, this software affords dynamic daylight metrics such as Daylight Autonomy (DA), Useful Daylight Illuminance (UDI), Annual Sunlight Exposure (ASE), and Spatial Daylight Autonomy (sDA). The analysis encompasses calculations under CIE standard clear sky at 9:00 am, 12:00 pm, and 15:00 pm during equinoxes and solstices for both locations, with Sweden's solstices extended to 6:00 am and 18:00 pm. Further calculations involving CIE standard overcast sky are performed at 12:00 pm.

The final stage involves evaluating visual comfort conditions across the spatial realm. Insights garnered from this assessment inform potential modifications to building orientation and materials, aimed at optimizing performance.

Limitations and Considerations

This study acknowledges four main limitations. Firstly, it addresses climate conditions exclusively, omitting the influence of social, cultural, and physical contexts which significantly shape the built environment in practical scenarios. Secondly, the analysis excludes trees and neighboring structures due to their potential to introduce complex lighting scenarios that could distort the results. The immediate environment's contribution is approximated with grass (R.vis=15.70%) in both locations. Thirdly, while the study focuses on wood-based systems, a thermal analysis with different building systems -other than wood- should be carried out in order to have a more complete picture of what should be the optimal design adapted to each climate. Lastly, quantitative assessment should be complemented with qualitative methods such as the one proposed by Liljefors (1999), for deeper insights into daylight quality. Such analysis would require an in-depth qualitative study of actual living experiences that fall out the scope and extension of this thesis.

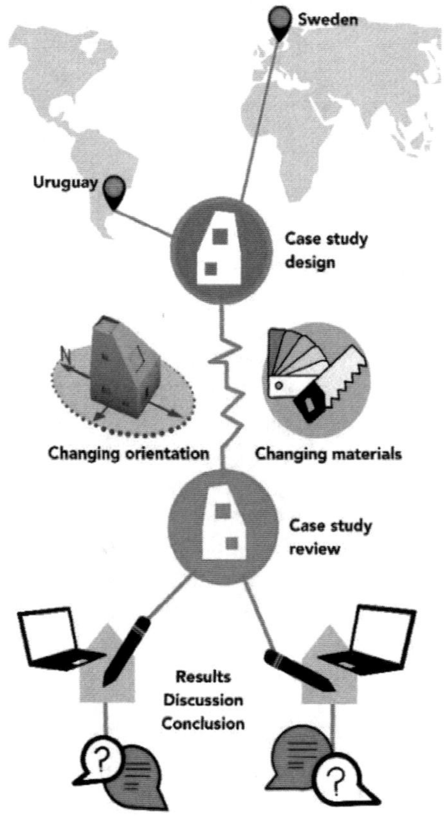

Fig.1 - Methodology scheme. Created by the author.

4. Case study design

The case study serves as an experimental project, offering a flexible platform to assess daylight dynamics. With this objective in mind, project decisions were strategically crafted to facilitate experimentation, featuring a compact 4m by 4m square plan floor (Fig. 2). The rotation of such a building, aimed at exploring its impact across various orientations, ensures clearer outcomes compared to a larger and more intricate layout.

Designed as a one-person cabin, the program is intentionally open-ended, embracing an undefined purpose. This design allows for the examination of annual glare probability, particularly if the cabin were employed as a workspace.

The cabin's sectional composition draws inspiration from contemporary architectural projects such as the Living Spaces pavillion in Copenhagen (Velux, 2023) and Tham & Videgård's Vertical Village II in Gothenburg (Videgård, 2018), which offer intriguing spatial possibilities (Fig. 3). Simultaneously, an exploration of asymmetry is introduced, fostering a play of light that enhances the ambient atmosphere.

Rooted in sustainability, a wooden structural system emerges as an apt choice for this endeavor, built in grey sawn wood façades with natural aesthetic and demanding minimal maintenance. Both Sweden and Uruguay, as timber-producing nations, extensively employ this system. Employing locally-sourced materials stands as a crucial step towards carbon neutrality by

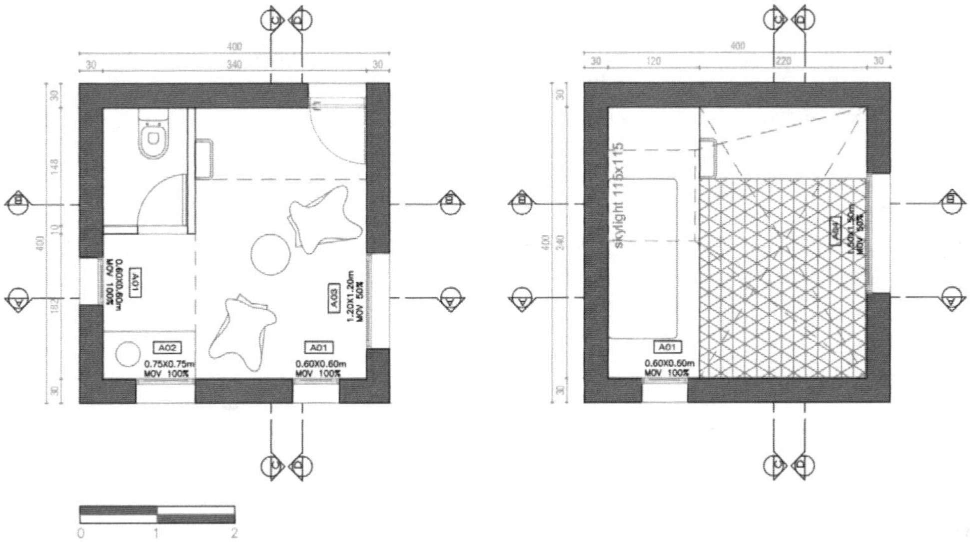

Fig.2 - Plan floors

Fig.3 - Sections

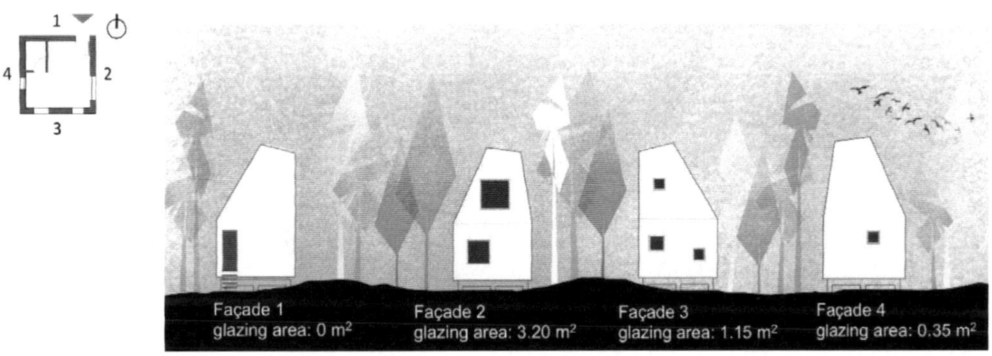

Fig.4 - Facades

minimizing material transportation. Within the cabin's interior, material variations are orchestrated to assess their reflective attributes and their impact on lighting conditions.

The windows design adheres to specific principles: firstly, each façade boasts distinct glazing areas (Fig. 4), cultivating diverse lighting scenarios to enrich the analysis. Secondly, windows are strategically positioned to establish a connection with the external environment, while others artfully frame exterior views, integrating external elements within the cabin's interior (Martin, 2017).

5. Case study review

As mentioned before two main actions will be applied to the case study: rotating the building and changing the interior materials.

All facades have been numbered from 1 to 4, starting from the access facade as 1. In order to change the orientation, the plan has been rotated clockwise, following the same sense that Climate studio uses to rotate the building (Fig.5)

The materials used for the purpose of this research were taken from Climate Studio library and are specified in material packs in Fig.6. The selection criteria consisted of starting with materials with high reflectance properties, as is the case of white plaster walls with a R.vis=89.9%, and then start testing different options according to the resulting analysis. Also, the glazing was modified by adjusting its reflective properties.

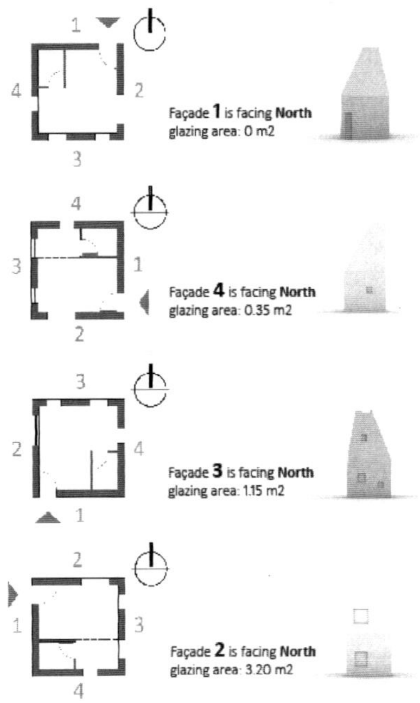

Fig.5 - Floor plan rotation for the analysis

	OBJECT	MATERIAL	R. vis	T. vis
OPTION 1	Grass	Grass 5	15.70%	0.00%
	Walls	Whiteboard Cubicle wall	80.00%	0.00%
	Roof	White plaster wall	89.90%	0.00%
	Door	Wall LM83	50.00%	0.00%
	Floor	Brick Building 5 Old	30.10%	0.00%
	Window (opaque)	Whiteboard Cubicle wall	80.00%	0.00%
	Window (glazing)	Clear-Solarban 60 (3) (Krypton)	11.90%	69.60%
OPTION 2	Grass	Grass 5	15.70%	0.00%
	Walls	Wood Pine Treated	32.00%	0.00%
	Roof	Wood Pine Treated	32.00%	0.00%
	Door	Wooden Door	4.90%	0.00%
	Floor	Wood Parquet Floor	7.70%	0.00%
	Window (opaque)	Wooden Frame	13.40%	0.00%
	Window (glazing)	Clear-Solarban 60 (3) (Krypton)	11.90%	69.60%
OPTION 3	Grass	Grass 5	15.70%	0.00%
	Walls	Wood Panel Wall	54.50%	0.00%
	Roof	Wood Panel Wall	54.50%	0.00%
	Door	Beige wooden door	79.00%	0.00%
	Floor	Concrete Floor	28.90%	0.00%
	Window (opaque)	Wooden bench	9.70%	0.00%
	Window (glazing)	Clear-Solarban 60 (3) (Krypton)	11.90%	69.60%
OPTION 4	Grass	Grass 5	15.70%	0.00%
	Walls	Wood Panel Wall	54.50%	0.00%
	Roof	Wood Panel Wall	54.50%	0.00%
	Door	Beige wooden door	79.00%	0.00%
	Floor	Concrete Floor	28.90%	0.00%
	Window (opaque)	Wooden bench	9.70%	0.00%
	Window (glazing)	Clear-Solarban 90 (3) (Krypton)	18.70%	50.30%

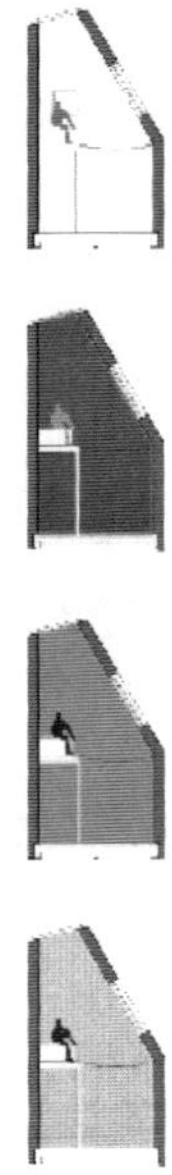

Fig.6 - Groups of materials used in the analysis

6. Results and discussion

The initial phase of analysis involved the utilization of Dynamic Daylight Metrics (DDM) within Climate Studio. For this assessment, the materials chosen corresponded to option 1, characterized by an average reflectance of 51%. Additionally, façade 2, encompassing a glazing area of 3.2m², was oriented towards the East in both studied locations.

Surprisingly, the findings unveiled an unexpected outcome: despite registering a relatively high glare probability (sDG) (53.2% for Uruguay and 60.8% for Sweden), the Useful Daylight

Illuminance (UDIa) metric failed to exceed 52%. This underscored the need for alterations in either the chosen materials, the orientation, or possibly both, to yield an improved performance.

Aligned with the research methodology, a two-fold series of modifications ensued. Firstly, an adjustment in the cabin's orientation was implemented, followed by an alteration of materials deployed within the cabin's interior space.

6.1. Point-in-time analysis

Therefore, an analysis of point-in-time illuminance levels was carried out in Climate Studio for each location. Four orientations were tested by rotating the floorplan clockwise, combined with three different types of materials (option 1-3). These measurements were taken during summer solstice, since it presented the more critical situation according with the preliminary results.

After analyzing the data, we chose two materials, one for each location (Table 1), based on their lower reflectance properties to avoid excessive illuminance beyond 3000 lux. The aim was to strike a balance between the desired illuminance range of 100-3000 lux and maintaining glare below 40%. This selection aligns with Marie-Claude Dubois' (2019) research on optimal lighting activation (100-300 lux), LEED v4's suggested range (300-3000 lux), and Tregenza's task-related recommendations for different lighting levels.

The main reason for the aforementioned criteria is avoiding heat gaining. Furthermore, the use of the cabin is broad, and being a working place is also a possibility, so glare probability should be reduced to lower values than 40%, as mentioned before.

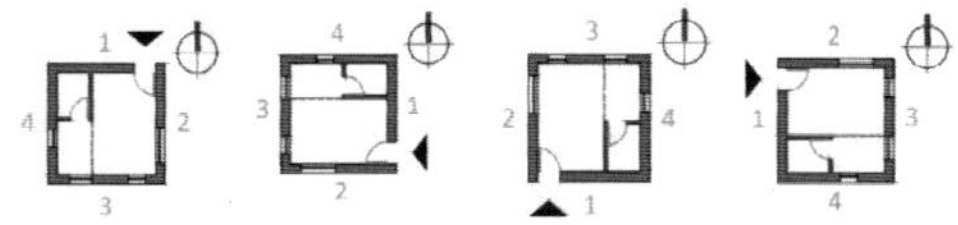

OPTION 1	SUMMER SOLSTICE	URUGUAY	mean	5860	3614	10162	12105
			median	3971	3755	2451	4598
		SWEDEN	mean	5961	8516	2929	2531
			median	3865	3861	2619	2260
OPTION 2	SUMMER SOLSTICE	URUGUAY	mean	2767	**851**	8525	8976
			median	795	**781**	649	1092
		SWEDEN	mean	3092	5529	842	560
			median	788	745	506	518
OPTION 3	SUMMER SOLSTICE	URUGUAY	mean	3561	1627	8992	9779
			median	1585	1576	1109	2003
		SWEDEN	mean	3807	6322	1393	**1049**
			median	1547	1517	1108	**955**

Table 1 - Point in time calculation during summer solstice in Uruguay and Sweden

6.2. Material & orientation

As anticipated, both locations required different materials to achieve the desired range of illuminance levels. A potential solution for Uruguay could involve the use of treated pine wood (option 2). In the case of Sweden, employing wood panel walls (option 3) along with Clear-Solarban 90 glazing (option 4) could prove effective.

While material selection also took orientation into account, our exploration delved deeper into this aspect. Once the chosen materials were established, Climate Studio was employed to compute the Useful Daylight Illuminance (UDIa), a pivotal metric for evaluating the optimal orientation for each location with the designated materials (Table 2).

The UDIa analysis yielded intriguing results. For Sweden, the most effective configuration involved positioning the cabin with façade 4 facing the North. This arrangement meant that the façade with the largest glazed area (3.2m^2) would be oriented towards the South. Similarly, in Uruguay, an analogous conclusion was drawn, suggesting a North-facing façade 4 and positioning the larger glazed area of façade 2 towards the South. These findings are of particular interest as they challenge a simplistic design approach that might have suggested orienting the cabin with its largest glazed area to the north, aligning with the analysed solar path, in the case of Uruguay.

	DA lux	UDIa	ASE	sDA	DGP
URUGUAY	1582	80.10%	1.50%	94.50%	4.50%
SWEDEN	1373	61.40%	33.30%	96.90%	21.60%

Table 2 - Dynamic Daylight metrics with the chosen orientation (facade 4 facing north for both locations). Regarding materials, option 2 was chosen for Uruguay and option 4 for Sweden. Information from Climate Studio.

6.3. *Daylight metrics with chosen materials and orientation*

With the materials and orientation choices firmly established, the subsequent phase entailed conducting fresh measurements within Climate Studio. This encompassed both point-in-time evaluations of illuminance and the computation of daylight metrics to pinpoint critical scenarios.

The outcomes of these analyses unveiled that, despite the adjustments made to achieve the desired levels of visual comfort, certain values still exceeded the intended range. This discrepancy was notably pronounced during Uruguay and Swedish's summer solstice, as well as both equinoxes at 12:00 pm. in Sweden (Table 3-4). These findings underscore the necessity for further refinements in the design.

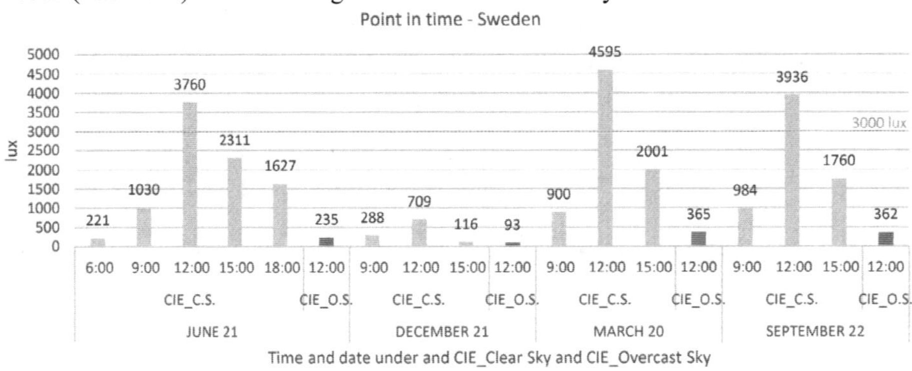

Table 3 - Point in time in Sweden using materials option 4. Facade 4 is facing North (facade 2 with biggest glazing area is facing south). Data from Climate Studio.

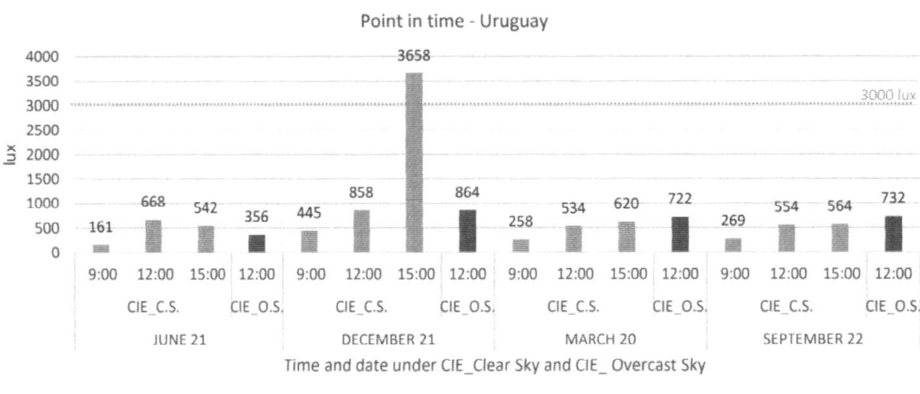

Table 4 - Point in time in Uruguay using materials option 2. Facade 4 is facing North (facade 2 with biggest glazing area is facing south) Data from Climate Studio.

Another crucial facet that merits close attention is the spatial distribution of light. This element serves as a potent design instrument capable of elevating a space from uniform monotony to one enriched with contrasts. The significance of achieving perceptual uniformity in evaluating daylight quality has been stressed by Dubois (2019).

Addressing uniformity emerges as another facet of consideration from these calculations (Table 5). A closer scrutiny of the point-in-time analysis and the correlation between average and minimum illuminance reveals that uniformity occasionally deviates from the recommended range for workspace (0.4-0.8).

		URUGUAY			SWEDEN		
		Emin	E avg	Uo	Emin	E avg	Uo
21 JUNE	06:00				109	221	0.49
	09:00	35	161	0.22	572	1030	0.56
	12:00	116	668	0.17	560	3760	0.15
	15:00	98	542	0.18	519	2311	0.22
	18:00				286	1627	0.18
21 DECEMBER	09:00	192	445	0.43	146	288	0.51
	12:00	204	858	0.24	185	709	0.26
	15:00	315	3658	0.086	33	116	0.28
20 MARCH	09:00	84	258	0.33	417	900	0.46
	12:00	249	534	0.47	575	4595	0.13
	15:00	323	620	0.52	424	2001	0.22
22 SEPTEMBER	09:00	94	269	0.35	489	984	0.5
	12:00	238	545	0.44	661	3936	0.17
	15:00	257	564	0.46	373	1760	0.21

Table 5 - Useful Daylight Illuminance. Facade 2 facing north.

Upon meticulous examination of the results of point-in-time illuminance in Uruguay, it becomes apparent that 58% of calculations fall below 0.4, with 42% falling within the recommended 0.4-0.8 range. In Sweden, 64% of calculations fall below 0.4, while 36% lie within the suggested range. It is essential to note that these guidelines are relevant to workspaces. In the context of residential spaces, the focus lies in maintaining uniformity below 0.8, as exceeding this threshold could result in an overly monotonous ambiance.

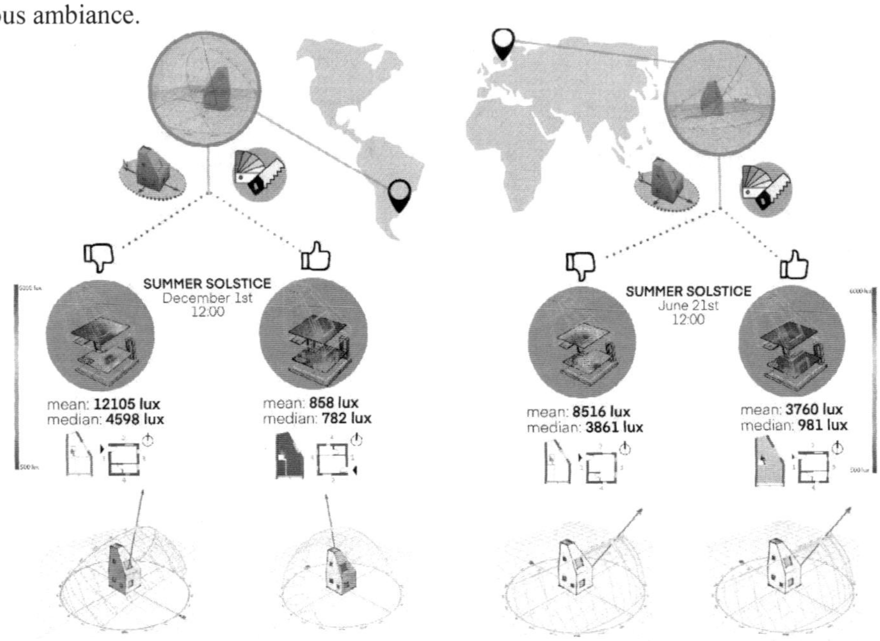

Fig.7- Scheme depicting the optimal scenarios in Uruguay and Sweden. Data sourced from Climate Studio and Dr. Andrew Marsh's blog.

6.4. *Impact of material variation*

To delve into the ramifications of altering the reflective attributes of the interior materials within the cabin, we embarked on calculating the average illuminance for each orientation and material. In the case of Uruguay, the disparities between the highest average (7935 lux - option 1) and the lowest (5279 lux - option 2) amount to 2656 lux. Similarly, in Sweden, a comparable variance of 2479 lux emerged, with the peak average being 4984 lux (option 1) and the nadir at 2505 lux (option 2).

These figures underscore the tangible influence of material selection on the overall illuminance levels. While the anticipation of this impact was rational, what captures our interest is the extent to which illuminance levels fluctuate (fig. 7).

Distinct differences, albeit consistent with the inherent variations between the two countries, become most apparent during the equinoxes. A plausible explanation for this phenomenon may lie in the shared characteristic of both locations wherein the facade with the most expansive glazed area faces south.

Despite endeavors to optimize daylight performance through material and orientation adjustments, the comprehensive results indicate a need for design reassessment. While the outcomes remain within acceptable bounds, the pursuit of preventing illuminance levels surpassing 3000 lux appears to inadvertently lead to significant reductions in levels at specific points or impacts on light dispersion.

These revelations, though insightful, could potentially be confined by the absence of comprehensive thermal analysis employing diverse building materials. To gain a comprehensive panorama of the ideal design for each locale, further investigations are imperative to validate thermal conditions across distinct climates.

7. Conclusions

The aim of the present research was to analyse daylight and utilize it as an input at the early stages of architectural projects. By considering two different latitudes and a wide range of information, the study highlighted the complexity and importance of assessing daylight. Thus, combining academic methods with practical design can offer a more comprehensive approach to architectural projects.

The relevance of analysing daylight with a simulator is clearly supported by the final findings, since it provided interesting data regarding the effect of using certain materials and on the cabin's orientation. The metrics provided by Climate Studio can help to evaluate the overall daylighting within a space at the very outset of the project.

However, this quantitative approach using digital tools must be complemented with qualitative information to consider subjective aspects, as discussed in the limitations. The same applies to thermal analysis with different building materials, which is crucial for evaluating final project decisions.

Nonetheless, the utilization of this digital method has been positive, resulting in a highly recommended methodology for professional practice. Familiarity with these calculations equips architects with tools to meet user needs and promote buildings with responsible energy use.

Lars Courage's perspective, as shared in a conversation with Marilyne Andersen (The Daylight Award, 2023), advocates for a distinctive approach to architectural design, encouraging architects to begin their design process by considering the external environment and strategically incorporating natural light where needed, as opposed to simply adding openings to an enclosed structure. This approach emphasizes a deep comprehension of daylight and its integration into the design process. This viewpoint serves as the driving force behind the motivation for this research. Understanding the differences between both hemispheres and utilizing digital tools to verify and modify designs are assets in this regard.

Daylight is an essential element for both the built and natural environment. A careful design that acknowledges daylight performance can improve energy efficiency, wellbeing and environment protection.

IOP Conf. Series: Earth and Environmental Science 1320 (2024) 012026 doi:10.1088/1755-1315/1320/1/012026

Acknowledgments
The author would like to express her gratitude to Rodrigo E. Muro Avendano and Luke Lowings for their invaluable guidance and support throughout this research project. Their expertise and mentorship were instrumental in the completion of this work.

References
[1] Rea, M. P. (2014). Report on the Development of the Swedish Healthy Home. Stockholm: Lighting Research Center.
[2] Reinhart, C. (2014). Daylighting Handbook I - Fundamentals Designing with the sun. USA: Ria Stein.
[3] Tregenza, P. L. D. (2014). The Design of Light. New York: Routledge.
[4] Dubois, M.-C., & Ng, N. (2019). Daylighting and Lighting under a Nordic Sky. Lund: Studentlitteratur AB.
[5] Liljefors, A. (1999). Lighting Visually and Physically. Lighting department School of Architecture KTH.g. Sweden: White arkitekter AB.
[6] Velux. (2023). Velux. Retrieved from Build for Life: https://buildforlife.velux.com/en
[7] Videgård, T., & Tham, &. (2018). Tham & Videgård. Retrieved from https://www.thamvidegard.se/work/housing/vertical-village-ii/
[8] Martin, B. (2017). Zipped: Arquitectura Residencial Japonesa Contemporanea Producida en Fuertes Marcos de Restricción de Proyecto Derivados de Condiciones Geométricas y Dimensionales. Doctoral Thesis, Facultad de Arquitectura, Diseño y Urbanismo UDELAR.
[9] The Daylight Award. (2023). The Daylight Award. Retrieved from A cross-disciplinary conversation about daylight between Marilyne Andersen and Lars Courage.: https://thedaylightaward.com/community/

IOP Conf. Series: Earth and Environmental Science 1320 (2024) 012027 doi:10.1088/1755-1315/1320/1/012027

Methods for inclusive design processes at the early stages of a research project in School Environments

S Angelaki[1], A Frelin[2], J Grannäs[2], U Besenecker[1] and C B Danielsson[3]

[1] Division of Lighting Design, Department of Architecture, KTH Royal Institute of Technology, 100 44 Stockholm, Sweden
[2] Faculty of Education and Business Studies, University of Gävle, 80176 Gävle, Sweden
[3] Department of Architecture, KTH Royal Institute of Technology, 100 44 Stockholm, Sweden

angelaki@kth.se

Abstract. This paper explores the use of participatory methods prior to designing interventions within a research project at a primary school in central Sweden. The approach presented in the paper is based on the principles of participatory design (PD), to enhance the use of these methods within the areas of educational research (ER), lighting, and architecture. This approach aims to include participants of educational spaces and incorporate their views prior to design interventions. Two workshops were designed to support teachers' participation through hands-on activities. Twenty-eight teachers participated in the workshops. Scale models corresponding to two of the school's classrooms were used to initiate discussions regarding the interconnection between spatial layouts, lighting, and learning activities. The workshops' data collection analysis assisted the research group in understanding the school's spatial and learning characteristics. The information gathered from the workshops provided additional knowledge and informed the research project in a way that allowed for further development and changes within the project related to the additional variables measured along with light. According to the analysis, there is a correlation between the activity and the desired layout of the space, while the type of equipment also varies according to the task.

1. Introduction

As part of a research project studying the opportunities for lighting design to support teaching and learning, the authors' research group conducted on-site workshops at a Swedish school. The school accommodates a testbed for a larger research project exploring innovative school lighting beyond retrofitting solutions. With the following workshop investigation, the research team's goal is to discuss with the teachers and elicit their input for the research project before the data collection and intervention processes. The principle of participatory design (PD) and the benefits of this approach are evaluated and applied in the workshops' development.

Educational research (ER) can be defined as a systematic attempt to gain a better understanding of the educational process while aiming to improve its efficiency [1]. It is referred to as the application of scientific methods to the study of educational problems. Educational research can be divided into several subcategories, although the present study focuses on action research (AR) and participatory design

IOP Conf. Series: Earth and Environmental Science 1320 (2024) 012027 doi:10.1088/1755-1315/1320/1/012027

research (PDR) practices. According to Lewin [2], who used the term *"action research"* for the first time in 1946 it is defined as:

"... a comparative research on the conditions and effects of various forms of social action, and research leading to social action." (Lewin, 1946, p.35). [2]

Social action is achieved by social change, which in this project relates to exchanging information and input so that research is informed by stakeholders and the results produced are valuable for the community. Action research aims to produce social change and management. Thus participants' viewpoints are of great importance for the project. Action research follows the scheme of planning an action (activity or intervention), performing this action, analysing the collected data, and concluding with a result that can be further generalised [3,4]. Participatory design research (PDR) is considered a type of action research methodology that focuses on the participants' feedback [5, 6]. Participatory design focuses on allowing participants to become co-researchers, supporting the principle of listening to others' voices during the design process. It is often referred to as *"co-operative design"* or *"co-design"*, although both terms fall underneath the participatory design methodology.

Among other factors, educational reforms called for redesigning and rethinking educational spaces. Therefore participatory processes were in focus to assist the design processes. The shift in pedagogy from teacher-centred to learner-centred (pupil-centred) approaches [7-9] called for constructing and redesigning school facilities across Europe [10]. To support the educational shift from a teacher-centred approach towards a learner-centred knowledge, spatial changes needed to be reflected in the learning environment (LE) [11], and an approach to that is for learners and teachers to become co-designers of their own LE [12, 13]. In parallel, examples of open-plan schools that came from collaborative design processes between educators, architects and designers revealed a different pattern of use compared to similarly designed schools without users' participation throughout the design process [14]. Research on the various levels of participants' involvement explored at which level each group (teachers, pupils, parents, and school staff) needed to be consulted [15]. Participatory practices were previously used in evaluating and designing office spaces [17], with positive outcomes in employee satisfaction, work engagement and performance [18-20]. Due to the various educational reforms [7,8] and the lack of adequately designed spaces to accommodate these changes [21,22], participatory methods were increasingly incorporated into educational research [13]. Soon enough, it became apparent that design does not necessarily determine how an educational space will be used [15]. Supporting the argument that further research is required to reveal and map the extent of a physical space's influence on changes in educational processes in a school environment [20-24], the presented workshop investigation aims to:

a. Identify potential interrelations between the space, the available equipment, and the current learning activity.
b. Reveal the role of lighting, as an environmental factor, in supporting the various learning tasks based on teachers' proposals and feedback.
c. Explore the integration of participatory processes for case study research within education.

2. Study design

An investigation consisting of a series of workshops was conducted, with the goal being to inform, discuss and gain feedback from the teachers regarding the use of the space and educational facilities. Three workshops were designed, each focusing on a specific characteristic (furniture layout, type of activity, lighting) of the use and design of two primary school classrooms. The workshops focused on understanding:

1) How educational space is used.
2) The type of activities that are performed in the classroom.
3) The teaching and learning processes conducted by the teachers.

The given tasks focused on activities initiated by the teachers and performed by the pupils to provide the researchers with the teachers' perspective. At the same time, researchers had a chance to communicate and discuss the project's goals with the teachers through collaborative and interactive work. The activities targeted to the teachers were designed to provide them with a basic understanding of how: to incorporate/use lighting: a) in their work, and b) to define different spaces and educational activities. The interaction between the researchers and teachers during the workshops aimed to provide both groups (teachers and researchers) with input and knowledge. The workshops had an informative character for the researchers, helping them better understand the needs and challenges of the school environment. Feedback from the teachers suggests that the workshops were also informative to them, in relation to the lighting aspect and the way it can be part of their daily work.

2.1. Workshop process model

A series of workshops were designed as part of a wider research study that aimed to enhance and support collaboration and exchange of input through teachers' participation in the design process. In this paper, two out of the three workshops are presented. Each workshop aimed to answer two questions on spatial or lighting characteristics. In the first two workshops, scale models of existing classrooms were used. Two out of four scale models represented renovated classrooms; the other two corresponded to traditional classrooms. The renovated classrooms resulted from a long-term intervention that aimed to reshape and redesign the school's building facilities. The classrooms are located in different buildings within the school. Different grades use the classrooms, thus certain variations derive from the curriculum and the use of space. The first workshop aimed to answer the following questions:

- Is the layout affected by the different subjects within the weekly curriculum?
- Are there differences between teacher-oriented and pupil-oriented activities in relation to preferred furniture layout?

The outcome of this workshop aimed to understand the classrooms' spatial qualities and the value they held for the teaching and learning processes. For this workshop, four scale models were made in total. The argument for modelling existing spaces was motivated by the teachers' familiarity with these spaces, their challenges, and characteristics. Thus, their feedback would be site-specific and valuable for the workshop and the research project. The first workshop had two working sessions, and both focused on proposed furniture layouts for specific activities that were suggested by the teachers. The suggested activities in the workshop were derived from the school's weekly schedule and a series of observations performed by the research team prior to the workshops.

The second workshop aimed to answer questions about the classroom's lighting, encouraging the teachers to focus on the challenges, issues, and qualities of light. The research team's objective was to answer the following questions after the workshop:

- Should the lighting be designed based on the furniture layout proposals?
- How much flexibility regarding lighting (intensity, placement, color appearance) is required in relation to the teaching activities?

During the second workshop, two working sessions took place. The initial session focused on an additional layout proposal, and the second session focused on the lighting parameters. The same scale models were used for both workshops. The teachers participating in the workshops also taught in the actual classrooms upon which the scale models were based during the academic year.

2.2. Teacher workshops

The workshops were part of a design process where researchers, architects, and pedagogues participated to inform each other and facilitate the project's development following the PD research principles of participants' inclusion and consultation. The first workshop had an informative character and it primarily focused on the research team, while further exploring the everyday use of the space, furniture layouts, and potential spatial issues. The second workshop focused more on lighting, providing mutual input through conversations between the researchers and the teachers. The lighting-related workshop aimed to link spatial characteristics and learning activities to lighting. The goal was to assist the users of the space to understand the reason lighting is important and explore ways to incorporate it into everyday activities. The preparation stage for the workshops included the construction of classroom models on a scale of 1:20. To minimize the effect of texture and color the models were white with light grey floors. The openings and overall layout of the classroom corresponded to the school's physical spaces. The 3D-printed furniture was also white to create contrast between the floor area and tables, chairs, and cabinets. Information regarding the existing furniture in both buildings was collected upfront to have scale models of the same equipment for both workshops. The equipment used in Building C (traditional classroom) was allocated to the model of Building C (Hus C), and the new equipment for Building F (renovated classroom) was allocated to the Building F (Hus F) models.

Four models each including the normal amount of equipment for each room were prepared for the workshop. The number of tables and chairs corresponded to the current teaching technique used in the school: collaborative learning. According to collaborative learning, pupils need to work in groups of four or five. Tables and chairs for twenty-eight pupils were distributed in each model to form groups of four and five pupils per table. Even though not all grades have twenty-eight pupils, all groups worked with the same number of tables and chairs to keep a consistent working pattern throughout the workshop. In both workshops, teachers from all grades participated. Therefore, there were four working groups corresponding to the models. The school administration informed the research team that the teachers already had working groups based on the grades they worked with, therefore, the same groups were used for the workshops.

A summary of the workshop's outline is illustrated in Figure 1. Building F (Hus F) currently hosts kindergarten and grades 1-3. Building C (Hus C) is used for grades 4-6 and grade 9. Corresponding to the number of models, there were four working groups during the workshops. Teachers working in Building C (Hus C) were assigned to the Building C (Hus C) models, and teachers working in Building F (Hus F) worked with one of the Building F (Hus F) models. The teachers whose subjects required the use of a specific space (workshop space, art room, or other spaces) formed a group and worked with Building F (Hus F). Summarizing the group distribution, three groups worked with the spaces daily, and one group was allocated to a different space from the one they use. There were twenty-eight participants in total. Prior to the workshop, consent forms were provided to the teachers so that the results would be further analyzed and used within the larger research project.

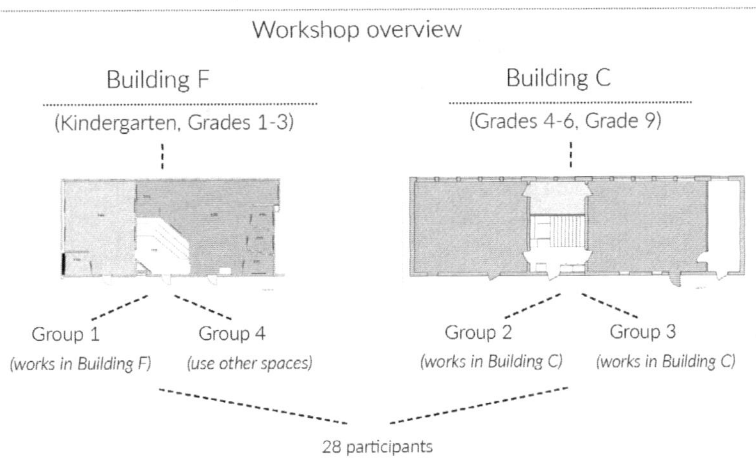

Figure 1. Workshop overview of used spaces and assigned groups for the teachers' workshop.

2.2.1. First workshop. The first workshop took place on the afternoon of the 26th of October, 2022. It was a two and a half hours workshop from 14.00 to 16.30 CET, consisting of two separate working sessions with the teachers. The initial part of the workshop was a ten-minute introduction to the working sessions. The first session focused on teacher-centred activities (e.g. presentation, discussion) and the groups were asked to discuss and suggest a preferred layout for each activity. Worksheets were given to each group, and one of the participants was responsible for filling out their group's worksheet. The groups were asked to work with the model and the 3D-printed furniture to try different furniture configurations before deciding on the configuration that, in their opinion, was the most optimal for the corresponding activity. During the second working session, the groups of teachers focused on pupil-centred activities (e.g. individual and group work between pupils). For every activity, a preferred scenario was proposed by each group. A feedback session including all groups was held at the end of the workshop.

2.2.2. Second workshop. The second workshop was held on the morning of the 1st of November 2022 from 8.00 to 12.00 CET. The initial part of the workshop included a brief introduction to the working sessions and the distribution of the worksheets to each working group. The models were set up following the placement of the first workshop, and the working groups were kept the same. The first session was based on the feedback discussion from the previous workshop; therefore, the target was to propose an ideal furniture layout scenario for two additional activities (test and playtime). During the second session, the furniture layout on the models was set based on the suggestions of workshop 1, and the variable at this point was the type of lighting each group would like to suggest. During workshop 1 each group suggested a specific layout for the same activities. For the second workshop, the furniture layout needed to be the same to work with the lighting. The teachers were asked to vote between the proposals they made during the first workshop. After voting, the furniture were set according to the preferred layout in the models. As a result, each group with the same scale model would work with the same layout. The variable explored during this workshop was the lighting; therefore, the furniture layout needed to be constant. Light fixtures, paper diffusers, and three different light filters were given to each group. At the end of the second session, a questionnaire was given to the teachers summarising all the questions from the worksheets of the two workshops to collect data from all the participants consistently. After the questionnaire, another feedback session took place to discuss the working sessions and the teachers' suggestions and thoughts on the workshop.

3. Results

This section presents the data and analysis of the initial two workshops. The results are presented chronologically following the workshops' order. Data was collected through photographs, smaller group discussions, feedback sessions with all the participants, and a questionnaire. Additionally, facilitators filled out observation logs to monitor the process and workflow during the workshops. This paper focuses on the collected data from three out of four groups. The group of teachers who did not use the chosen classrooms, represented by the scale models, was excluded from the present analysis and will be presented separately.

3.1. Photo registration

As presented in Figure 2, a matrix was created to illustrate the differences between the teachers' groups' suggestions that corresponded to the same activities. The horizontal axis is the activity scenario, and the vertical axis is the group number and the corresponding scale model. The activity scenarios were predefined by the research group and were the same for all working groups. The first and third groups used the same layout for teacher-centred activities (presentation or other activity led by the teacher). The second group proposed two different layouts for the same activities. Regarding pupil-centered activities, all the groups had different suggestions for individual and group work.

The group that worked with Building F (Hus F) included both spaces for each scenario. One of the differences between the four activities in Building F (Hus F) is that the smaller rooms on the right part of the scale model are used in addition to the main space for the pupil-centered activities. For the group scenario, there are more chairs in the smaller rooms compared to the individual activity. Additionally, for the individual activity in Building F (Hus F), the tables in the bigger classroom are not placed side by side but seem scattered over the room, allowing for personal working space for each pupil.

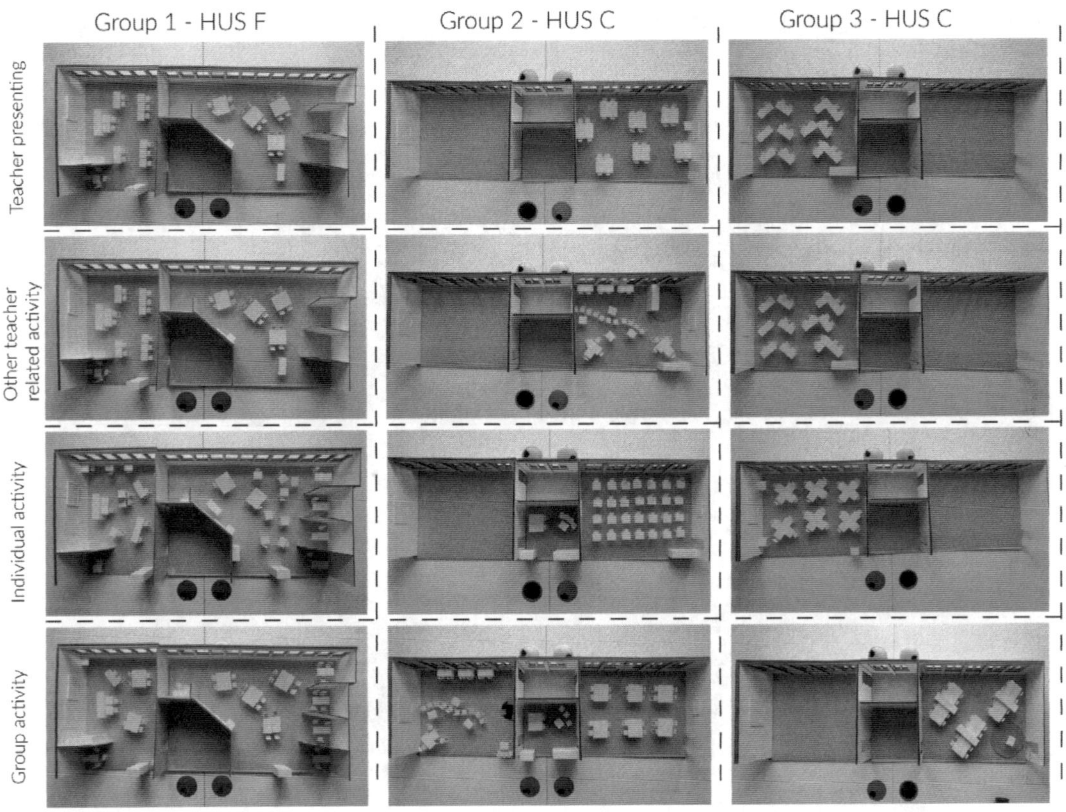

Figure 2. Photo registrations of the preferred furniture layout configurations from the first workshop. Rows 1 and 2 correspond to teacher-centred activities and rows 3 and 4 are related to pupil-centred activities.

The comparison between the two groups that worked with Building C (Hus C) showed that both groups preferred using one of the classroom spaces for teacher-centred activities and the pupils' individual activities. The second group proposed using two classrooms for the pupils' group activities, while group three only used one for the two spaces for the same activity. Both groups' layouts for the "teacher presentation" scenario are not fixed to a single focal point in the space, but the chairs are placed in different orientations. The available equipment in each classroom (monitors, smart boards) facilitates this choice, allowing everyone to see the presentation regardless of their seating orientation. The third group proposed the same layout for the presentation scenario and other teacher-centred activities, while the second group suggested having different layouts for each teacher-centred activity. According to the worksheet notes, the additional teacher-centred activity needs to promote discussions within the group, therefore a more organic layout is proposed by group two. Both groups appear to be consistently using one of the two spaces for three out of four scenarios. This choice corresponds to and is influenced by the current use of the classroom spaces according to the school's schedule. Regarding the pupil-centred activities, the second group suggested having a layout for the individual activity, which was more relatable to traditional classroom layouts having a specific orientation and focal point. However, the tables' orientation is different since they face the wall opposite to the one with windows.

The second workshop focused on pupil-centred activities corresponding to group and individual work. Figure 3 illustrates the proposed lighting scenarios for the group activity in Building C (Hus C). Both groups of Building C (Hus C) suggested using ceiling spotlights. The main difference between the two groups is the spotlights' placement.

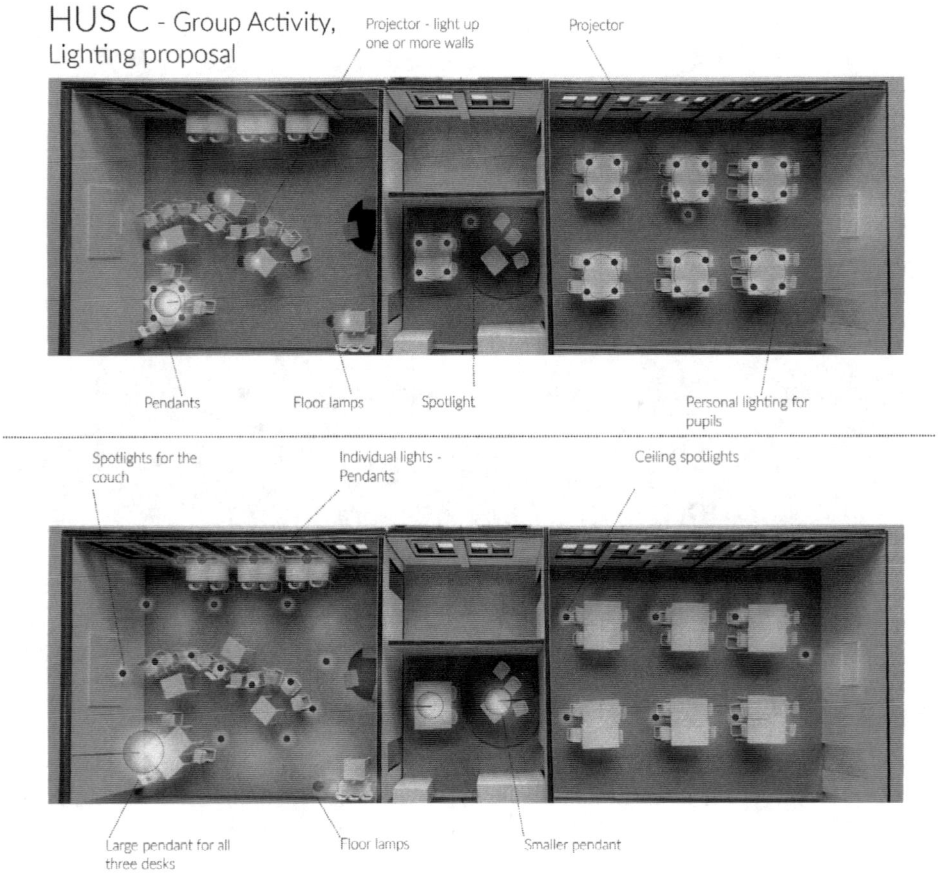

Figure 3. Lighting proposals for the group work activity scenario in Hus C, comparison between the second and third group.

In the bottom illustration, the ceiling spots in the left room follow the furniture configuration. The teachers mentioned that these spots are to be used as personal lighting for the pupils working on the curved seating area below them. They suggested having additional ceiling lights for general illumination. The seating areas closer to the window (bottom illustration) have pendants as individual lights for the pupils working in those areas. Additionally, different sizes of round pendants were proposed for the group tables and the middle space. The classroom space on the right part of the illustration has a series of ceiling lights for general lighting. The teachers suggested using a few floor lamps to move them around the classroom according to their needs. Additionally, they mentioned that floor lamps would be more accessible for the pupils to use, especially considering that they might need to move in the space based on the given tasks during the courses.

3.2. Questionnaire

The questionnaire consisted of eighteen questions in total. Some of the questions were similar to the ones in the worksheets that the teachers needed to fill out during the workshops. Due to variations in the type and number of fully filled-out worksheets after the first workshop, a questionnaire was made for the second workshop to collect participants' responses consistently. There were nine multiple-choice questions, five semantic differential ratings questions, and four open-ended questions. Twenty-eight questionnaires were collected and sorted based on the building each group worked with during the workshop. Seventeen out of twenty-eight responses were used for the analysis. Three responses were eliminated due to lack of information, and the remaining eight responses came from teachers using other spaces within the school facilities. Their responses will be presented separately. The number of participants is defined by the total number of teachers employed at the school. The workshops focused on the understanding and development of activities based on the schools' educational structure, therefore the workshops are site and content-specific.

The first part of the questionnaire focused on the type of equipment used in the three scenarios: a) presentation, b) group activity, and c) individual activity. The multiple-choice questions corresponding to these scenarios are at the top of Figure 4. The answers to the multiple-choice questions are summarized in the Figure below. According to the results, digital equipment (projectors, screens, and smart boards), except laptops, is mostly used for presentations. Analog tools (books and paper) are mostly used for individual activities. Laptops were chosen for group and individual activities even though the number of answers related to the "individual activity" scenario was higher. Screens and whiteboards were also preferred tools for presentation and group activity. Smart boards were chosen for the presentation and group activity scenario rather than for the individual activity. Compared to the rest of the digital tools, smart boards are the least preferred choice. A parameter that should be mentioned at this point is that smart boards are a fairly new addition to the school's equipment and have been used for less than a year, therefore the teachers might not be as familiar with them as the rest of the given choices. Compared to the individual scenario, chalkboards were chosen mostly for the presentation and group activity scenarios.

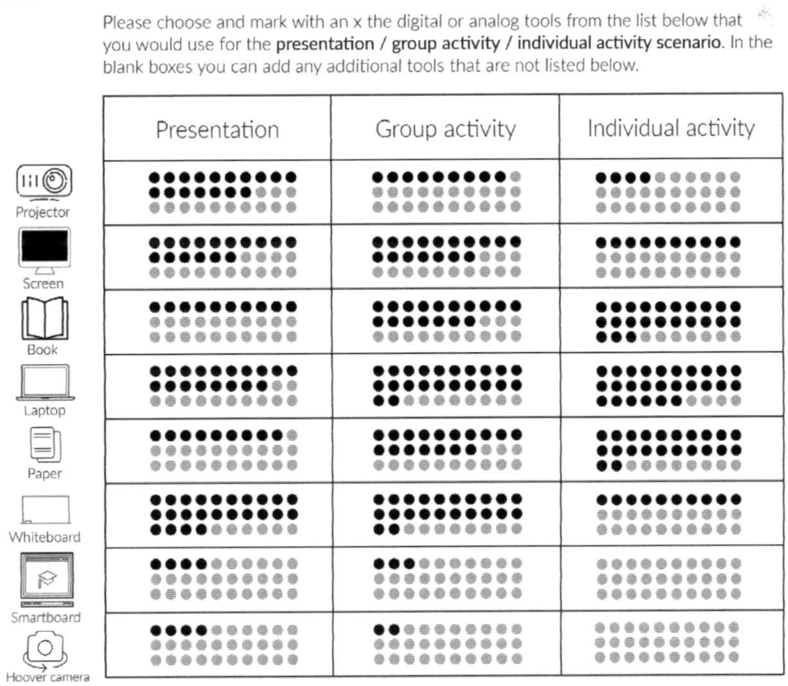

Figure 4. Type of equipment used for each activity based on teachers' feedback. The last choice marked as "Hoover camera" was a suggestion made by one of the groups.

Overall, the responses indicated that tools that require the user to move in order to use them (screens, smart boards, projector - that are controlled via a laptop) are linked to the first two scenarios (presentation and group activity), while tools that can be used in a more static situation were preferred for the individual activities. Both the presentation and group activity scenarios may require the teacher or pupils to move in the classroom.

The second part of the questionnaire focused on the lighting aspect. The semantic differential rating questions were analyzed separately for each building since the lighting equipment and spatial elements differ considerably in Building F, which was recently renovated. One of the primary outcomes from the second part of the questionnaire is that the pupils do not change or choose the lighting in Building C (traditional classroom). The teachers responded that they chose and changed the lighting more often in Building F (Hus F) compared to Building C (Hus C). This outcome might be affected by the newly installed lighting in Building F (Hus F) that allows choosing between different lighting scenes. Teachers from both buildings mentioned that there were issues related to glare in the classrooms in both buildings. Glare in Building C (Hus C) was mentioned in relation to daylight.

The open-ended questions were also linked to lighting and the view from the windows. Most responses mentioned that the view can be distracting, especially if, at the time of the course, another grade is having a break. However, two responses mentioned that if the classroom is on a higher floor the view is not distracting for the pupils due to lack of direct visual contact with the yard. It was mostly when the screens or smart boards needed to be used that the teachers wanted to block the daylight to create the conditions for the screen to be visible. The answers varied quite significantly when asked if they would like to change the lighting and what type of changes they would prefer. Specific attributes were mentioned in some answers, such as *"In the autumn and winter I would change the light to warmer",* or the ability to dim the lights. Almost all responses were positive when asked if they would like to control the lighting. Some even mention the possibility of giving pupils control over the lighting.

3.3. *Feedback sessions*

At the end of each workshop, a feedback session took place, to discuss and get an overview of the working sessions. The feedback sessions were informative and influenced the workshop's flow. In particular, after the first feedback session, an additional working session was added to the second workshop, after the teacher's comments about having an additional scenario for the furniture layout. The second feedback session influenced the outline and preparation of the third workshop, adding presentations and longer working sessions so that the teachers could familiarize themselves with the equipment.

The main points from the first workshop focused on the worksheets' length and the fact that it was not possible for all teams to consistently keep notes. As a result, a questionnaire was made, summarizing the main points and questions of the worksheets from both workshops. Additionally, at the end of the first feedback session, the teachers mentioned that using the scale models could be implemented in their own work since it helps them test different layouts much faster.

During the second feedback session, the discussion focused on the lighting parameter. The teachers mentioned that they would like to have more time for the lighting task. Furthermore, a wider variety of equipment would be preferable during that workshop. All groups focused on the way lights should be controlled, which is consistent with the responses collected from the questionnaire. One of the groups supported the idea that the pupils should also be in control of the lighting, while teachers from other groups focused more on describing fixtures that are not available at this point at the school. Overall, the feedback sessions assisted in summarizing ideas, bringing everyone together to discuss the group proposals, and most importantly their comments regarding the workshops resulted in alterations in the working sessions themselves.

4. Discussion

The scope of this workshop investigation was to explore potential interrelations between spatial elements and learning activities at a primary school in central Sweden. Additionally, the lighting component was presented and discussed in relation to the furniture layout and learning activity. The workshops were initially used as a means of participatory practice within a larger research project that explores lighting interventions in school environments beyond retrofitting solutions.

The aspect of participatory research highlights the importance of inclusivity, collaboration, and interdisciplinarity in research and application, as indicated in post intervention interviews and feedback sessions. The participants' suggestions and comments on further improving the project show their involvement beyond the intervention phase [11,21]. In particular, the extended architectural interventions in Greek schools showed that the pupils' attitude towards the space itself had changed, since they respected and took care of the additions and alterations within their classrooms and school space [12].

The workshops' results, pointed out that the learning environment plays a crucial role in the activity that takes place in the classroom. The spatial characteristics, such as size and height, the architectural characteristics, such as windows, and access to the corridor space, but also the furniture layout, were all mentioned throughout the workshops by the teachers. The teachers' proposals suggested that the furniture layout should not be static since the learning processes have become dynamic, and therefore require more flexibility in terms of spatial design.

In addition to this, the workshops played a central role for the main research project per se, in terms of providing information to the research team and discussing potential issues. Comments and feedback from the workshop were used to re-shape the overall research project itself since the workshops took place before the data collection processes. Additional variables for data collection, such as, for example, noise, were added to the experiment's design for the school based on the feedback session. It was particularly interesting that during the group discussions, some of the teachers referred to issues related to noise, apart from lighting or limited availability of equipment. This investigation showed that initiating discussions at an early phase is important both for researchers and participants since it allows for the creation of a common ground and understanding of everyone's work and role within the project. Furthermore, this investigation underlines the importance of PD within ER and pre-intervention discussions with the participants for an overall understanding of the school's function and needs. Overall, the workshops assisted in shaping the primary research and the data collection variables based on the teachers' comments.

4.1. Limitations

However, there were certain limitations to the approach described here, specifically, regarding the workshops' length and the complexity of activities. There was a difference between the first and second workshop when comparing similar working sessions and a learning curve of the process was obvious. The same activity during the second workshop took less time as the teachers were more familiar with the models and the equipment. Additionally, the level of participation was not always the same within the groups' participants, since some of them were more active than others. Therefore, there were differences regarding the group dynamics among the working groups, which might have potentially affected the results. An attempt to minimize these biases was the use of individual questionnaires, which allowed the collection of feedback at individual, group, and overall discussion level. In relation to the second workshop, which focused more on the lighting component, there were differences in teachers' existing experience and familiarity with lighting parameters, such as: lights' correlated colour temperature (CCT) and dimming. The teachers who worked in Building F (Hus F) were using newly installed lighting with a greater control capability compared to Building C (Hus C) spaces, in which traditional lighting systems are installed. These differences resulted in a variety of comments and suggestions. Furthermore, the workshops were framed around the schools' pedagogical approach and teaching technique and the data are linked to certain spatial typologies, thus a generalisation of these results to other school environments may not be appropriate.

4.2. Future work

The two workshops presented in this paper are part of a bigger research project that focuses on lighting interventions in school environments. The analysis and overall discussions with the teachers resulted in alterations regarding the project itself. There were two main sections of the research projects that were affected by these workshops:

- The outline and preparation of the third workshop.
- The design of the main experiment for the data collection.

The third workshop, focusses on lighting placement in relation to activities for which different tools are used (paper and laptops). Through a 1:1 mock-up, the research team and teachers discussed the interrelation between furniture layout and lighting on a different scale. Furthermore, the main experiment, which focuses on lighting will include additional parameters (noise and temperature) in the data collection phase. These parameters were mentioned in the initial observation phase and further discussed during the workshops. As a result, the research group re-shaped the experiment's design to accommodate the teachers' comments.

5. Conclusion

This paper explores the use of participatory methods within a research project studying the opportunities for lighting design interventions to support teaching and learning at a Swedish school. The presented workshop investigation illustrates the research group's goal to hold discussions with the teachers prior to the data collection and intervention processes. The workshop sessions, discussions and feedback sessions were structured to support the principles of PD within ER and include the teachers at a pre-intervention stage in the project. The analysis showed that the equipment, both the furniture equipment and the digital tools, is used in different ways according to the task that is performed in the classroom. Additionally, the layout can change to support and enhance collaborative teaching. In relation to lighting, some of the teachers were supportive of the idea of controlling the light themselves, but also to give over some control to the pupils to define their own lighting, and therefore their learning environment. There were certain limitations, since the workshops were designed to explore a certain teaching technique and specific classroom spaces. Overall, the initial approach to work along with the teachers at a pre-intervention phase was perceived in a positive way since they had the opportunity of informing the research group regarding current issues and problems.

Acknowledgments

This work is part of a project financially supported by the Bertil & Britt Svensson's Foundation for Lighting Technology and the Swedish Energy Agency.

References

1. Wellington J. Educational Research: Contemporary Issues and Practical Approaches: Bloomsbury Publishing. 2015.
2. Kurt L. Action research and minority problems. Journal of Social Issues. 1946;2(4):34-46.
3. George, T. (2023). What Is Action Research? | Definition & Examples. Scribbr. https://www.scribbr.com/methodology/action-research/
4. Brydon-Miller Mary GD, Maguire Patricia. Why action research? 2003;1(1):9–28.
5. Light A, Akama Y. The Human Touch: Participatory practice and the role of facilitation in designing with communities. 12th Participatory Design Conference: Research Papers 2012; 61–70.
6. Mäkelä T, Leinonen T. Design Framework and Principles for Learning Environment Co-Design: Synthesis from Literature and Three Empirical Studies. Buildings 2021; 11: 581.
7. Istance David, Marco K. OECD Work on Technology and Education: innovative learning environments as an integrating framework. European Journal of Education. 2013;48(1):15.
8. OECD. Education at a Glance 2017: OECD Indicators. Paris; 2017.

9. Pieters JM. Designing artefacts for inquiry and collaboration when the learner takes the lead. European Educational Research Journa. 2004;3(1):77–100.
10. French R, Imms W, Mahat M. Case studies on the transition from traditional classrooms to innovative learning environments: Emerging strategies for success. Improving Schools. 2019;23(2):175-89.
11. Duthilleul W. "Constructing education" : A framework for participation to support the effective planning and use of new school spaceσ. Open Journal of IUL University. 2022;3.
12. Könings, K.D., Seidel, T. & van Merriënboer, J.J.G. Participatory design of learning environments: integrating perspectives of students, teachers, and designers. Instr Sci 2014; 42: 1–9
13. Liapi M, Oungrinis KA. Designed for better learning" – An interdisciplinary program bridging pedagogy and architecture to improve the quality of public education in Greece. 11th annual International Conference of Education, Research and Innovation. 2018, 12th-14th November. Seville, Spain.
14. French R, Imms W, Mahat M. Case studies on the transition from traditional classrooms to innovative learning environments: Emerging strategies for success. Improving Schools. 2019;23(2):175-89.
15. Woolner P. Building Schools for the Future through a participatory design process: exploring the issues and investigating ways forward. BERA Manchester2009
16. Woolner P, Hall E, Wall K, Dennison D. Getting together to improve the school environment: user consultation, participatory design and student voice. Improving Schools. 2007;10(3):233-48.
17. Petersson Troije C, Lisberg Jensen E, Stenfors C, Bodin Danielsson C, Hoff E, Martensson F, et al. Outdoor Office Work - An Interactive Research Project Showing the Way Out. Front Psychol. 2021;12:636091.
18. Ohrn M, Wahlstrom V, Harder MS, Nordin M, Pettersson-Stromback A, Bodin Danielsson C, et al. Productivity, Satisfaction, Work Environment and Health after Relocation to an Activity-Based Flex Office-The Active Office Design Study. Int J Environ Res Public Health. 2021;18(14).
19. Danielsson CB, Bodin L. Office Type in Relation to Health, Well-Being, and Job Satisfaction Among Employees. Environment and Behavior. 2008;40(5):636-68.
20. Danielsson CB. The shaping factors forming contemporary and future office designs Nordic Journal of Architectural Research. 2019;(2):53-79.
21. Frelin A, Grannäs J. Designing and Building Robust Innovative Learning Environments. Buildings. 2021;11(8).
22. Woolner P, Thomas U, Tiplady L. Structural change from physical foundations: The role of the environment in enacting school change. Journal of Educational Change. 2018;19(2):223-42.
23. Gislason N. Architectural design and the learning environment: A framework for school design research. Learning Environments Research. 2010;13(2):127-45.
24. van Merriënboer JJG, McKenney S, Cullinan D, Heuer J. Aligning pedagogy with physical learning spaces. European Journal of Education. 2017;52(3):253-67.

IOP Conf. Series: Earth and Environmental Science 1320 (2024) 012028 doi:10.1088/1755-1315/1320/1/012028

Light Activism: alternative urban nightscapes

Martin Flugelman

Light Designer, Buenos Aires, Argentina.

mflugelman@gmail.com

Abstract. Since their beginnings, artificial lighting technologies have been used as a tool by those in power to express their authority and ensure their control of cities. State and commercial hegemony determine urban lighting, which has a substantial spatial, semiotic, emotional, and physiological effect on us. Lighting conditions imposed from the top down inform our bodily experience of the night.

LED technology is revolutionizing artificial public lighting. Nevertheless, critical inquiry about its potential uses is sorely lacking. This research will explore the capacity of collective light interventions to turn public spaces into more inclusive nightscapes. It will focus on community-driven projects where the use of lighting contributes to the production of alternative urban experiences. The research engages methodological issues by asking how lighting tools can potentially contribute to community empowerment. Can lighting actions in public spaces propose other ways of interacting and other ways of experiencing the night?

This paper revolves around critical lighting practices. It shows how the use of light can foster citizen engagement and ignite new nocturnal imaginaries as the perception of the night is reshaped to restore communities' right to the city. The intention is to bolster critical awareness regarding the use of light and the power underpinning its instrumentalization. This text will also shed light on the effect ephemeral collective light interventions have on public spaces and its perception.

1. Introduction

Darkness is associated with the unknown, with a time and space rife with menace and supernatural forces. Gloom is feared and nighttime is a time to withdraw. Light, meanwhile, represents the godly. It is a manifestation of life and knowledge and a tool that allegedly provides safety and security.

Access to devices and infrastructures that use natural resources to cast light and shadows—that is, lighting technologies—has always been a privilege reserved for the few. Monarchies and independent nations, as well as luxury commerce and entertainment, have been at the forefront of the use of light, establishing a night experience where citizens interact under a sky of branding, marketing and the minimum functional light to stimulate consumerism and work (Slater & Entwistle, 2019).

Lighting technologies are by no means neutral, and as such, they must be understood critically for the sake of more inclusive urban settings. Urban life takes place under lighting conditions imposed from the top down. Two contrasting situations frame urban lighting with, firstly, a panopticon-like approach that seeks to standardize minimum levels of light for the sake of safety and security. Geared to streets dominated by cars, this approach tends to a uniform urban nightscape, or what we might call a city of sameness, having as a result over-illumination and light pollution. Secondly and conversely, lighting is installed as marketing and branding strategy, where powerful commercial forces organize the flow of electrical power to represent themselves and monopolize the image of the night.

If, then, powerful institutions have always determined what is illuminated and that use of lighting has a spatial, semiotic, emotional, and physiological effect on us, how might a bottom-up approach produce more inclusive urban nightscapes? Can ephemeral lighting interventions in public spaces propose other ways of interacting and experiencing the night?

In the first chapter, the paper aims, to make visible the power structures that determine the lighting and darkness conditions in the places we live, promoting critical awareness of the link between light and authority. A map (Figure 1) introduces the invisible spectrum of light as a visual summary of the literature reviewed, a diverse bibliography regarding the history of light, mainly in Western European culture. The main authors referenced come from the global north European and American academic context that contrasts with the global south case studies selected from Buenos Aires, Argentina. In this sense, it is necessary to frame that mainly there is a wide gap of literature available coming from Latin American studies that study topics regarding lighting. These infrastructures developed in Europe and America and were exported, also colonizing the night in foreign territories.

The second and third section introduce and discuss possible alternative ways of building an *other* (Awan, 2011) urban night experience by presenting a case study where communities have been empowered by lighting technologies during a series of ephemeral collective light interventions in public spaces in Buenos Aires, Argentina. And to finalize, the conclusion summarizes the theoretical framework regarding the link between power, light and darkness, its effect on citizens, and stresses the importance for new multidisciplinary tools to engage communities in the discussion regarding lighting within cities.

2. The invisible spectrum of light

The following map is a visual representation of the literature review studied for the paper using the traditional diagram of the spectrum of light. This updated, deviated and hyperbolic map portrays a bigger picture, making visible the hidden layers of light and darkness and their cultural associations. Furthermore, it frames the practices within the lighting discipline and the power structures that are linked to the medium of light. The literature review will constantly be referenced to the map.

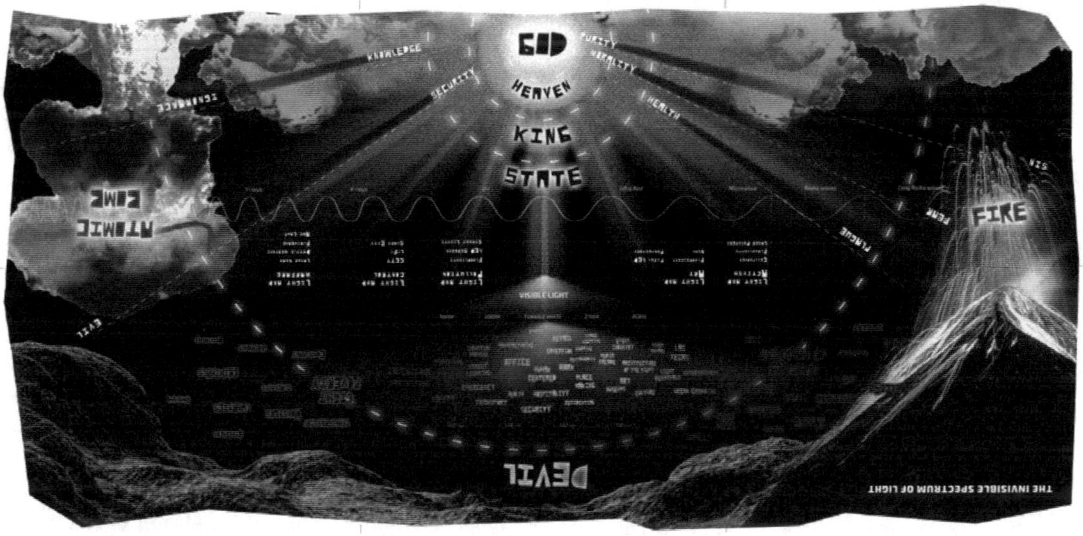

Figure 1- The invisible spectrum of light map. Author-2023.

The map can be read in multiple ways. To begin with, like nighttime, everything is covered by blackness, and light is a fragment that intrudes within it. In a vertical way, from top to bottom, the institutions that historically have control over the visible spectrum of the light are named: God, the

IOP Conf. Series: Earth and Environmental Science 1320 (2024) 012028 doi:10.1088/1755-1315/1320/1/012028

King as its representation and the State. At the bottom, Hell, the Devil, and the Night are the ever-constant menaces that justify the need for illumination.

Inside the visible spectrum of the light - what we can see from light - the lighting design discipline is exaggeratedly depicted. It is possible to polarize conceptually and oppose two types of light. We have, on one extreme, urban lighting infrastructures that use homogenous white floodlights in social housing with the promise of safety and security. And on the other extreme, light as excess, the flickering bling-bling and play of brilliants, the over-illumination of desire and consumerism with Las Vegas as its epitome. In between these extremes, the discipline uses different light strategies, colors and CCT (Correlated Colour Temperature) to define spaces, with more focus on visual perception and well-being.

This map is reversible, it promotes another way of looking at the light spectrum, deepening in the invisible part. When we turn the map around, we see from a different perspective. On one side, we have gamma rays and the atomic bomb: the strongest light source ever invented and the most dangerous one, setting the boundaries of technological exploration. On the other side, fire sets the limit: the primordial and most basic source of energy. In relation to the other wavelengths, the different uses of light are portrayed: light as a weapon in warfare, as an element for control in smart cities, as a pollutant and as an element for art and activism. Each employs a set of particular light technologies related to their respective objectives.

On the margins of the visible light, we can read a set of words that relate to cultural associations with darkness: from cultures who are fascinated by the mysteries and freedom of the night to hygienic enlightened cultures who condemn and shame the darkness, criminalizing certain practices and building images of violence and fear that work as a way of legitimizing the process of illumination in order to expand a 'respectable' nocturnal sociability (Koslofsky, 2011).

The map evidences the political spectrum of the different light technologies and critically exhibits the relationship between power and light. These devices structure and define the nightscapes of our cities and determine the way our bodies are affected. As lighting designers, we must deepen the history of these technologies to trace and understand how this broad medium has been introduced into culture and shaped it. The following parts of this chapter examine the multiple layers of the map, convey the history of light and its relation with power, and explore quests for an *other* history, one of alternatives and resistant lighting practices.

2.1. The conquest of the night

What are the cultural associations with the night? The cultural meanings of lightness and darkness are ambiguous and are subject to constant flux (Meier et al., 2015). The night is a moment related to fear and simultaneously to freedom, an alternative to daylight practices; its routines, obligations, and tasks. It is a time for dissidence, where danger and pleasure are common ground for the definition of this otherness, a time for dark cultures to escape daylight. As such it has always been challenged, legislated, restricted, curfewed, narrowed, illuminated, glared and polluted against.

Night is a vital space-time, crucial to the construction and reconstruction of identities, both individual and collective. The night is tied to notions of transgression and the experiences of minorities and marginalities: the origin of many subcultures lies in nocturnal practices. It is about the emergence of clandestine stories, where human expression tries to escape surveillance (Palmer, 2000). The image of the night is full of contradictions, from the territories of the Devil to a time for praying, from honest nocturnal professions to excesses and transgressions (Figure 1). Nighttime is also free time, when the vast majority of people have a moment for leisure and interact in other ways (Koslofsky, 2011).

After two centuries of saturation of artificial lighting infrastructures, our cultural understandings of darkness are still informed by the fears, legends, myths and imagery of nighttime over the ages (Edensor, 2017). However, nowadays, new conceptual frameworks are emerging from understanding the impact of light as a polluting agent (Cerveira Lima, 2022) repositioning the relationship with darkness. Researchers and practitioners are redefining darkness as a positive condition (Narboni, 2017), an exceptional luxury in an over-illuminated contemporary life.

2.2. Brief history of urban lighting

It is possible to define two differing processes that give origin to artificial lighting in public spaces with two models that were happening simultaneously in the late 17th century in Europe. There is a top-down process that started with Louis XIV of France, the "sun king", who used lighting infrastructure as a way of exhibiting the monarchy's presence and rule over the city at night. He introduced order, control, and safety to the proximity of his palace by hanging '*réverbères*' whose brightness contrasted with the night's darkness. In contrast, a citizen initiative took place in Amsterdam where painter and inventor Jan van der Heyden proposed an efficient lighting pole to the Municipality that emerged from the ground. He designed a network system needed to illuminate the city and redefine the urban nightscape of Amsterdam (Dietrich, 2015).

Koslofsky's seminal work (2011) explains how European courts, between 1600 and 1750, experienced a conspicuous temporal shift in their celebratory practices from diurnal to nocturnal. In a display of symbolic self-identification, monarchs and rulers of this era sought to forge a divine connection, aligning themselves with both God and the celestial body of the Sun, positioning themselves as beacons of illumination (Figure 1). The festivities, spectacles and pleasures were essential tools of government that valued festivals and entertainment to display their majesty. The act of nighttime illuminations emerged as a veritable art form, symbolic of the privileged status of the elites in Europe that portrayed nocturnal emerging sociability. A new way of conveying power, blinding people with light in the deep dark contrast of the night. Crucially, the ostentatious display of majesty not only served as a manifestation of political power but also played an active role in its very construction.

In a context where light and darkness have so much symbolical meaning, the control of these two elements constitutes an image of power. If the night is a land of menaces, having the tools to project light over darkness is a way of showing authority. Not only because of the possibility of using scarce resources, but also because of the capacity to manage a tool that was aligned with the divine: those who manipulate light have power.

2.3. Nocturnalization, regulating for a respectable night

Being in a public space during the night was a restricted activity. An unexplored realm, the urban public night opened for the broader population, with a shift away from curfews to a street lighting system controlled by the authorities. The focus transitioned from the suspicion of night walking and the necessity of carrying lanterns for visibility to the pursuit of beautification and respectability after dark (Koslofsky, 2011). The Nightwatcher's role evolved into proto-police forces and was replaced by more modern forms of night authorities that eventually expanded their activities into daylight. Through the introduction of police-controlled lighting infrastructures in specific urban spaces, the more illumination available the more people ventured outside, increasing mobility and the need for patrolling the growing cities. The utilization of light by authorities was a mechanism for establishing order and worked as an instrument of discipline. Illumination by night is a surveillance tool that enables the regulation of what happens after dark, a disciplinary power, producing subjects that fall under a gaze or glare (Schlör, 1991).

The introduction of artificial lights at night in Western European city landscapes was a diverse process that was mainly developed by a powerful minority allowing the expansion of human activity into night public space mainly controlled by law enforcement departments and police forces. The conquest of the night imposed a nightscape, a way of moving and behaving with values that responded mainly to bourgeois and religious morality.

The invasion of artificial illumination was shaking traditional nightlife and political order too. Street lighting also had increasing utility for businessmen, who advocated for their installation defending that trade follows light (Bouman, 1991). While elite social life was expanding into the

IOP Conf. Series: Earth and Environmental Science 1320 (2024) 012028 doi:10.1088/1755-1315/1320/1/012028

night, its traditional inhabitants - servants, apprentices, students, tavern visitors, sexual workers, and those who sought to escape the social legibility of early modern daily life - saw their space and time threatened by disciplining forces (Figure 1).

The image of the night in cities is characterized by light, a history of increasing illumination, where darkness is expelled into mysticism, prehistory and mythology, and light conquers the streets (Schlör, 2016). The process of nocturnalization (Koslofsky, 2011) is now being looked at with a critical lens: we are entering a decolonization era regarding light, where darkness is paradoxically a symbol of luxury and inequality.

2.4. Opening the lighting Black-box

From the 19th century, cities in Europe began to open, outer walls became obsolete, the intense urbanization and growth expanded, problematizing the city-periphery division, and borders began to dissolve. Likewise, the nocturnal city introduced a new topography in which people started expanding their activities after sunset.

Lighting infrastructures developed at a slow, uneven, patchy and privileged pace until the introduction of gas infrastructure and later the electrical revolution. These innovations spreaded and institutionalized artificial light at night, making it a part of everyday life and producing a black box effect, hiding how the technology works, making invisible the network and distancing users from the light sources. Light turned into an infrastructure of illumination serviced and maintained by professional experts, regulated by administrative structures and shaped by vested commercial interests (Edensor, 2017). The thrill subsumed by the first lights waned to a mundane feeling. A disenchantment process culminated in a technocratically regulated homogeneous over-illuminated nocturnal built environment (Schivelbusch, 1995).

Nowadays, this vital network is defined by normalization organisms that are mainly regulated by the industry, defining the way the world is illuminated, and thus, how our bodies interact and are affected. The implementation of new SMART technologies is opening a new era where infrastructure of outdoor lighting encompasses a combination of elements such as wireless sensor networks, the Internet, cloud computing and even the invisible spectrum of light. Surveillance systems use the infrared spectrum to produce thermal-images(Figure 1). These hyper panoptical techniques are entering the domain of the invisible (Edensor, 2017).

It is imperative to incorporate the extensive historical significance of socio-cultural symbolism associated with urban lighting within the framework of urban lighting design. The adoption of smart lighting has resulted in community unrest due to its consequential utilization for law enforcement and surveillance purposes. Conversely, cities are also actively endeavoring to develop the implementation of these smart technologies by prioritizing urban participation and community values (Oram, 2022).

2.5. Resisting lights: A brief story of lighting alternatives

The night was always a contested territory where lineage, age and gender mark the lines between respectable and prohibited activity after hours. Light was used to free the night from the darkness, from the ghosts and evils that await in the dark space. Nonetheless, the night remained a time to undermine and challenge authority.

During the 18th century, a strong resistance to lighting emerged. Activities like lantern smashing were a traditional dissent action from youth cultures all over Europe, mainly students from male noble privilege, that had a more accessible and inhibited relationship with the night, but also revolutionaires during French revolution. (Koslofsky, 2011). Moreover, during the French Revolution, the lighting infrastructure was a strong symbol of the monarchy, and was therefore constantly smashed and lighting posts were even used to hang people. Later, under the Commune, smashing lanterns and returning the city to darkness was used as a defense mechanism, a wall of darkness that protected themselves from the state authorities. After the French Revolution ended, instead of darkness, there were blazing festive illuminations, where victory was celebrated in the light and with light (Schivelbusch, 1995). It is interesting that people smashed the top-down technologies and then

IOP Conf. Series: Earth and Environmental Science 1320 (2024) 012028 doi:10.1088/1755-1315/1320/1/012028

celebrated with bottom-up lighting, in close relation to the light source, using self-made hand-held torches.

At the same time, in London, crowds gathered at night in urban centers to express with light devices their political opinion. For instance, in several occasions there were illuminations associated to failure of bills, anniversary coronations of King William and Queen Mary, and different political events from mid 18th century. Light, or its absence, had a powerful agency within political discourse. Yet, this sporadic and self-organized way of expressing opinion and dissent slowly shifted into spectacularization and aestheticization, through designed experiences of lighting displays in public space that neutralized the political value of light (Barnaby, 2015).

There has been a constant shift in how light is used: from a way of control, a way of celebration, an effective direct expression of dissent, to the beautification of a city for enchanting. In this sense, it is interesting to look at how art approaches lighting technologies. Art has a way of deviating things from its original objective, changing and modifying how elements are combined and presented to convey and communicate a critical way of seeing. Since the beginning of the 20th century, artists have been using light to address issues of pollution, surveillance, ecology and consumerism. Beitin (2021) states that light is a basic condition of modern social life, and yet, its permanent and excessive use has numerous negative effects. In that sense, it is within our power to limit these effects and to ensure the responsible use of light. We can see that artists have been providing small-scale experimental light setups to question and critique the way we live. From Dan Flavin's fluorescent tubes to Jenny Holzer's LED (Light-Emitting Diode) screens, each technology offers a world of repurposing and expressive opportunities. Artists, technicians and communities use lighting technologies in critically engaged practices that question the status quo. The aesthetic possibilities of light are just being explored (Joelson, 2022).

2.6. From social engaged lighting practices to Light Activism

There is a dispute about the night between public and private lighting, with collective and individual visions of how the urban night landscape should be (Nye, 2015). In the last years, a series of practices that seek to look for other ways of addressing the night and its illumination emerged; understanding the night more as a process, a space and time for cultural practices and not a thing to be floodlighted. Groups like the Light Collective, Social Light Movement, Configuring Light, Guerrilla Lighting, amongst others, develop workshops, manuals and open source information to stimulate a lighting engaged with local social practices, expanding the range of the lighting design discipline away from the traditional luxury lighting projects (Bordonaro, 2018).

In recent years, light has been used in protests and struggles for demanding more access to rights. In those circumstances, it works as a tool for resistance (Caixeta, 2022). Citizens are exploring how light can make problems visible; can express community concerns and political opinions; and can address urban conflicts and inequalities in an aesthetic manner.

Light activism is the appropriation of lighting technologies by citizens and communities (not governments or institutions) that use devices to visualize and problematize conflicts, disputes, abuses of power and political and ideological positions. The medium of light is transformed into a metaphor: to put something under the spotlight, to focus and reveal it. Light activism uses light to develop actions and interventions in the urban space to hack the status quo, ranging from more spontaneous self-organized actions to collective projections and performances that reclaim the political value of light.

In the last decade, from 2010-2020, there has been an incremental tendency in different cities of popular demonstrations using lighting technologies in diverse ways to communicate a message. For example, pro-democracy (Hong Kong, 2019), anti-corruption (Romania, 2017), during election campaigns (Brazil, 2018) and anti-state (Chile, 2019), amongst others (Schielke, 2018; Caixeta, 2021). Currently, the use of low-voltage LED promotes more accessibility and possibility of control. This new tendency is increasing exponentially because of a trending radicalization of inequalities, more collaboration between artists and citizens, and a visual culture that constantly demands new images. It is an alternative way of using light and of communicating non-hegemonic narratives, using light as an attack and defence mechanism. Although related to power struggles, light activism also has a festive agency, gathering people and uniting them.

The installation of light infrastructures in the city has always been followed by a series of objectives for ensuring security, safety and accessibility, while also being intertwined with a complex network of meanings, symbolisms, and exclusions. This alternative use of light at night not only enables the communication of messages and provides alternative night images, it also questions who has the right to use light in the night and affect the senses by doing so (Rancière, 2013).

3. Ephemeral light interventions: citizen empowerment through lighting

This section examines a series of three collective light actions that happened in 2020-2021 in Ciudad Autónoma de Buenos Aires (CABA), Argentina as one case study. They were selected because I participated as co-director of these events, facilitating access to information and stakeholders, while also being examples of alternative bottom-up lighting experiences. This section will give an insight into the ways they have been produced and their impacts. The interventions combine practices from collaborative architecture, participatory design and non-conventional lighting, to build urban actions and interventions through light that promote eco-social integration.

The light interventions were organized by FLUXLIAN, a light research platform that I am co-founder of, and in collaboration with a wider number of co-producers. The actions occured in different public spaces related to the accessibility of green public spaces in a context of dense urbanization and the privatization of public commons driven by the government of the city.

People live their lives in the context of engineering-defined street lighting with minimums that do not contemplate social practices. After sunset, governments try to stimulate the nighttime economy by reducing social interaction into commercially driven activities. The case studies explore other ways of living this standardized night. Through the aid of lighting tools, the interventions created social interaction between diverse groups who had an alternative night experience and built collective images together. The lighting choreographies allowed people to relate materially and symbolically to the places they are denied access to, manipulating the light technologies, affecting, being affected and having an effect on their agency within the city.

This section will continue to outline the background of FLUXLIAN and context of the urban controversies before describing the interventions that took place across the three sites. In this sense, the three actions happened consequently every 2 month, and yet, they are addressed together as one case study due to its methodological similarities. It was a work-in progress where improving methodologies were tested.

3.1. FLUXLIAN

As a collective, we focus on how to use lighting technologies, concentrating on engaged bodily actions that deviate objects from their traditional function and use them in non-conventional ways to research with space, perception, light and darkness. With a multidisciplinary background, we perform a diverse range of actions, from light interventions for live music to stage design, dance shows, light installations, urban performances and collective light choreographies. To build caring relationships and networks, we developed a non-conventional lighting workshop to share a place for testing and drifting around the possibilities of experimental lighting. We address a wide range of urban problems through the prism of light, practically and symbolically, as a tool to affect the immaterial and material urban environment.

3.2. Parque Salguero

In 2020, the Government of CABA organized an architectural contest called "Parque Salguero" to seek ideas for the urbanization of 32 hectares of land on the coast of the Rio de la Plata. However, the project faced criticism as it was perceived to be a plan for another luxury neighborhood for high-income individuals rather than a public green park in public property. The sale of public lands to private stakeholders for real estate developers caused widespread dissent and backlash from various communities. Concerns were raised about the inadequate amount of green space in the city, with CABA having only an average of 6 m2 of green space per inhabitant compared to the World Health Organization's recommended minimum of 10 m2. The contested land included the Arroyo Ugarteche, a water stream, a typical landscape of the region yet that is rarely visible in Buenos Aires because water streams are all piped. This case provided an opportunity to make the riverside accessible to people by night and to engage in building a new night experience in a denied part of the city.

3.3. Playon Colegiales

The Predio Ferroviario de la Estación Colegiales, an abandoned railway workshop and depot, is located in Colegiales, a neighborhood with high population density and minimal green spaces (only 0.7 m2 per inhabitant). Over the years, neighborhood organizations in Colegiales have been mobilizing to protect green and public spaces. Since 2016, residents have advocated for transforming the railway yard into a large, 100% green public park. Despite these efforts, a local government project approved in 2017 designated nine plots for sale to private stakeholders for luxury residential buildings. The vacant railway yard, spanning almost 7 hectares, is seen as a unique opportunity to create a significant urban park and increase green space availability in the neighbourhood and the city. This case offered the possibility of engaging local residents with a territory that is physically inaccessible to them, but by illuminating it, they could appropriate the space they are trying to inhabit.

3.4. Isla de la Paternal

In the neighborhood of Paternal, another vacant public space was subject to privatization. The Albergue Warnes building complex, originally intended to be South America's largest children's hospital, was demolished after a complex political history. The land, previously owned by the city government, was sold at a low price to a retail company, which obtained a controversial approval in 2018 to construct towers on the site. After being vacant for a long time, the space was used by neighbors as a park. A change in building codes facilitated real estate speculation and led to the acceleration of its resisted urbanization. Citizen organizations organized to protect the public space. This case had the potentiality of gathering diverse stakeholders that inhabit the park in a festive, carnivalesque attitude that showcased citizen organization and local empowerment.

3.5. Designing the light interventions

In these contexts, FLUXLIAN and a collective of self-organized researchers, activists and citizens decided to produce a series of light choreographies to defend public commons in the territories under dispute. For each intervention, an open call invited interested people from diverse backgrounds to occupy the public space that is under a privatization process and to build a social bond with their environment using accessible light technologies. Motivated by the need to shed light on urban controversies and engage participants using their bodies, out goals were to make visible the privatization of public lands carried out by the Government of CABA; build a series of images and videos for social media distribution; promote social cohesion after a post-pandemic situation; and explore the role of light to create a bottom-up approach while stimulating a novel nighttime experience.

The light interventions were the final acts of public events organized to defend public commons in the neighborhoods. In each case, during the day, a light stencil workshop was developed where people cut out their own phrases to project later with the light devices. The tools to be used were mainly tactical LED flashlights, green laser pointers and cell phones. The coordination team was divided into a master of ceremonies who communicated the joint actions to be performed via loudspeakers; several lighting assistants who helped people with the devices and instructed them on how to use the light sources; a documentation group who was recording video and taking pictures of the whole event; and a lead photographer who was in charge of making long-exposure photographs.

IOP Conf. Series: Earth and Environmental Science 1320 (2024) 012028 doi:10.1088/1755-1315/1320/1/012028

Activities were designed to occur in the whole spectrum of twilight for people to experience the nighttime. Before sunset, tactical flashlights and laser pointers were given to participants, while others, as instructed, used their own lighting devices. Starting at civil twilight, there was an introductory workshop where participants were instructed on how the lighting devices work and made warm-up movements together to rehearse future movements to be developed in the light choreographies on site. During nautical twilight, the designed actions began with participants walking with their lighting devices and projecting their stencil messages onto the walls and surfaces in the surroundings. At astronomical twilight, when there was enough darkness and people were familiarized and in tune with each other and the site, the light choreographies started in order to produce long-exposure photographs. While the master of ceremonies, with the aid of an audio system, explained and stimulated participants to illuminate different things in the city, the documentation team took pictures and videos. The event finished with a series of portraits made by participants illuminating themselves.

Figure 2 - Open-call flyers and lighting choreography maps. Parque Salguero (right), Playon Colegiales (center), Isla de la Paternal (left).

4. Results and discussions

In Figure 1, light activism is positioned within the electromagnetic wavelength of the microwaves, in this sense the collective light interventions are micro actions that provoke an increased engagement of participants who gather together with flashlights in their hands trying to communicate a message; they illuminate the built and environmental heritage they are trying to defend from privatization. Participants prepared and improved the light stencils, valuing the places they care about. Light projections effectively communicated the messages without harming the infrastructure. (Figure 2)

The projects had a playful and inclusive approach, fostering interactions between different stakeholders. Participants enjoyed a unique nighttime experience engaging with spaces of the city that are characterized by their inaccessibility, darkness and sense of insecurity and safety, sharing images and videos to highlight urban conflicts and defend public commons. The post-pandemic context added significance to these outdoor actions as a way to reclaim nightlife and public spaces after the intense and long insolation from the curfews.

Following local pressure, the government has initiated the construction of a public park in Parque Salguero and two plazas in Playon Colegiales, not meeting the full requirements of the neighbors. The projects still face ongoing challenges, and despite not achieving all the desired outcomes of the light interventions to stop the urban developments and constructions of buildings, the interventions succeeded in building bonds and creating memorable collective experiences.

4.1. Illuminating playfully with our hands

By giving the tools of illumination to the participants, they shift from spectators to active performers of the action. Different light sources, manipulated manually and enlightening collectively the landscape, created multiple points of view, exhibiting the complexity that needs to be assumed when dealing with urban transformations. If light is usually imposed on our bodies from the top, then this is a way of illuminating from beneath, from the ground. Light can be cast to the city by its own citizens: an empowering action that places the power to illuminate within the communities, giving them agency to define the sensory and spatial perception of the city with and without light and its variability in a playful way.

The playful nature of the game is a relevant way to deal with fear, distrust and insecurity within the city. If light should enhance the security feeling in our cities, projecting light in a playfull way clearly increases these feeling of belonging and care, this unexpected methodology allowed improvisation between bodies and light that produced the emergence of spontaneous choreography, building a valuable, memorable event. The collective lighting actions produced an atmosphere that stimulated a shared mood, permeated the bodies of the participants and triggered emotions that strongly marked their memory. The interventions were a collaborative way of producing a new image of the city, where empowered people illuminate the places they want to inhabit, emerging new future imaginaries (Figure 3)

Figure 3 - Isla de la Paternal, children with flashlights, light stencils and choreography images. Author, 2021.

4.2. Light as a mediator

Illuminating the disputed territories allowed their visibility, making tangible the issues, giving them an entity and defining them amid the darkness. Light showed the scale of the conflicts and emphasized the space available for the dreams of local residents. Light worked as a mediating agent, giving the right to the people to appropriate that space for a moment (Figure 4).

The actions co-produced images in territories under conflict. By placing the body collectively in the sites, a new imaginary around the night, light and public spaces was built. In these cases, giving flashlights to participants allowed for an intensified exploration of the darkness and the re-discovery of the built environment (Figure 3).

Figure 4 - Parque Salguero, light choreographies and long-exposure images. Author, 2020.

4.3. Re-enchanting the night, an ephemeral action

These interventions produce zones of autonomous time (Bey, 1991) and illuminations: a series of ephemeral actions that shifted disenchantment to fascination and portrayed the actual power of light.

In contrast to the homogenous, glary and standardized urban lighting, these interventions promote the emergence of new, innovative approaches to lighting, fostering more culturally diverse lightscapes while acknowledging vernacular and site-specificity. In a post-pandemic context, this was a way of re-enchanting the relationship with the night that was being curfewed and privatized.

These actions promote the re-focusing and valuation of the landscape and built environment for a particular moment, for one night. This soft approach allows people to engage and appropriate the space for a specific time frame. During this particular time, light shows an alternative: what would be possible, the imaginary, another option. In this sense, the ephemeral condition enhances the power to observe, analyze and reflect.

To change the city illumination, the first steps are to educate citizens about light technologies and their power to transform and impact human and non-human experiences. Nightlife is fabricated by social practices; it is socially mediated. The history of artificial lighting is not only technical but of changing social practices and imaginations. Therefore, it is not only about changing the lighting infrastructure of the city, but also about changing how people relate to the night, to light and to the city. It is the first step for collectively thinking about our nightspaces. To transform the built environment with inclusive participation it is necessary to develop transdisciplinary mechanisms that integrate, educate and engage with communities, so they can be active designers of their spaces.

5. Conclusion

To summarize, public space lighting usually is either a homogenous uniform functional light that theoretically works exclusively to promote safety, security and mobility, and is deployed using minimum standardized levels of lights and is mainly a state-controlled dispositive. Or it is a private-driven marketing illumination that maximizes the relationship between architecture and light to enhance the communication power of light, resulting in a completely staged urban nightscape.

The introduction of artificial lighting and the shift to nocturnal lifestyles have had profound effects on various aspects of human society and the natural environment. Artificial light sources have shaped human behavior, urban perceptions, and daily routines. Lighting not only facilitates vision but also influences our physiological well being, including sleep patterns, hormone regulation and brain activity. Additionally, lighting plays a crucial role in our perception and emotional responses, creating atmospheres and moods that impact our experiences and interactions with the built environment. It is important to consider the social and cultural dimensions of lighting, together with its biological effects.

Overall, the use of light has evolved from a symbol of power and privilege to a medium of resistance, community expression and aesthetic exploration. It has the potential to shape social practices, challenge authority and create meaningful experiences in the urban environment. However, there is a need to consider the politics of light, its potential for standardization and spectacle and the importance of incorporating diverse perspectives within lighting design and urban planning.

Light can be much more than a governmental control dispositive, and there are more creative and artistic uses to resist that control. To propose an *other* alternative way of thinking about the night and its illumination, this paper discussed how ephemeral light actions can spark discussions about bottom-up inclusive lighting and involve citizens in shaping their nightscapes. Collective lighting interventions have the power to build a temporal new image of the city, to create a zone of autonomous time, where empowered individuals shape their spaces and can project their imagination to new places.

Lighting infrastructures need to be a relevant public issue and citizens should be informed about the impacts of lighting technologies on the environment. Facilitating discussions through light intervention workshops can promote citizen engagement and foster inclusive urban lighting. The bottom-up approach enables diverse perspectives and democratic decision-making. It is possible to use the city as an open-forum to discuss, try and change the shape of a place by night.

Lighting actions in public spaces offer transformative potential for engaging with and experiencing the night. By distributing light sources and enabling social interaction, new forms of communication and diverse nightlife can emerge. Shifting from permanent installations to ephemeral actions creates captivating moments and alternative nightscapes, acting as catalysts for change.

During collective light actions, people communicate with whispers of light, playfully re-enchanting the night and forging connections with their surroundings. Light has the power to change our perception of the world and shape our experiences. The complexity of urban problems in the 21st century requires alternative and interdisciplinary approaches in order to transform public space, its perception and future imaginaries. The need for unconventional, resourceful approaches to urban problems is more urgent than ever. We must educate and communicate about the power of light: power to destroy, to pollute, to convey, to highlight, to disrupt, to affect and to have agency.

References

[1] Awan, Nishat, et al. Spatial Agency: Other Ways of Doing Architecture. Routledge, 2011.

[2] Barnaby, A. (2015). 3 - London. In S. Isenstadt, M. M. Petty, & D. Neumann (Eds.), *Cities of Light: Two Centuries of Urban Illumination*. Routledge.

[3] Beitin, A. F. 2022. Power! Light! Cornerhouse.

[4] Bey, Hakin. 1991. La zona temporalmente autonoma.

[5] Bordonaro, Elettra, and Joanne Entwistle. 2018. *Social Lightscapes Workshops*. Edited by Don Slater, London School of Economics.

[6] Bouman, M. The "Good Lamp is the Best Police" Metaphor and Ideologies of the Nineteenth-CenturyUrban Landscape. American Studies, 32(2), 63-78. http://www.jstor.org/stable/40643595. 1991, Fall

[7] Caixeta, Helena. 2022. *Lights for Democracy*. KTH.

[8] Cerveira Lima, R. 2022. Setting limits to light pollution: a change in strategy is going to come. ROLAN.

[9] Dietrich, N. 9 - Berlin. In S. Isenstadt, M. M. Petty, & D. Neumann (Eds.) 2015. Cities of Light: Two Centuries of Urban Illumination. Routledge.

[10] Edensor, T. 2017. From Light to Dark: Daylight, Illumination, and Gloom. University of Minnesota Press.

[11] Jo Joelson. 2022. "Violence, power, surveillance from the blind lantern to the searchlight for "White Torture."" *Power! Light!*, Kunstmuseum Wolfsburg.

[12] Koslofsky, C. 2011. Evening's Empire: A History of the Night in Early Modern Europe. Cambridge University Press.

[13] Meier, J., Hasenohrl, U., Krause, K., & Pottharst, M. 2015. Urban lighting, light pollution and society. Routledge.

[14] Narboni, Roger. 2017. "Imagining the future of the city at night." *Archlighting.*

[15] Nye, David. 2015. "The transformation of American Urban Space: early electriclighting, 1875-1915." *Urban lighting, light pollution and society*, Routledge.

[16] Oram, T. 2022. Re-engaging with the origins of social public lighting through smart lighting technologies.

[17] Palmer, B. D. 2000. Cultures of Darkness: Night Travels in the Histories of Transgression. Monthly Review Press.

[18] Rancière, J. 2013. *The Politics of Aesthetics* (G. Rockhill, Ed.; G. Rockhill, Trans.). Bloomsbury Academic.

[19] Schivelbusch, W. 1995. Disenchanted Night: The Industrialization of Light in the Nineteenth Century. University of California Press.

[20] Schielke, T. 2018. Urban lighting: The semiotic shift from functionalism to scenography.. *Transitional streets, Narrating stories of convivial streets.*

[21] Schlör, J. 2016. Nights in the Big City: Paris, Berlin, London 1840-1930 - Second Edition. Reaktion Books.

[22] Slater, D., & Entwistle, J. 2019. Making space for 'the social': connecting sociology and professional practices in urban lighting design. *The British Journal of Sociology, 0.*

Socially inclusive Lighting Design: exploring a method to enhance freedom of movement for marginalised groups.

Kaja Glenne Lund[1] and Mette Hvass[1]

[1]Aalborg University, Copenhagen, Denmark. Department of Architecture, Design and Media Technology

Corresponding author: kajagl@gmail.com

Abstract. Based on literature and field research, women and marginalised groups often experience unwanted attention in the public realm worldwide. This attention often produces a gendered fear, which can result in avoidance behaviour of urban public places. Universal design has been implemented to ensure everyone can access urban areas, but this access is still restricted for many vulnerable groups. This investigation aims to find relevant lighting design principles for making public urban spaces more inclusive for all marginalised groups. A literature review, case studies of three urban places, focus group interviews, and an experiment with torches in the field were the chosen methods to gain knowledge. The findings suggest that the participants experience fear in public places when they feel exposed as someone who can become a potential target. This results in a need to see their surroundings and have an awareness of the presence of others. In addition, the experiment showed how they wanted to use warm lighting and beautify the place by enhancing details. The conclusion suggests six different lighting design methods to create inclusive public areas, such as beautifying details and legibility with vertically lit surfaces. These six principles could make urban public places more socially inclusive for marginalised groups, women, and other vulnerable groups. The results should be compared and confirmed by using control groups with other non-marginalised groups.

Keywords: socially inclusive lighting design, marginalised groups, mobility

1. Introduction

The United Nations has recognised unequal access to urban public places as a global problem. Their Sustainable Development Goal 11 has been made to: "*make cities inclusive, safe, resilient, and sustainable" [1]*. Equal rights to the cities have further been included in one of their targets, 11.7. By 2030, the UN aims to: *"provide universal access to safe, inclusive, accessible, green and public places, particularly for women and children, older persons and persons with disability" [2]*. Also, UN Women are concerned with safe mobility and inclusion in the public realm [3].

One possible solution to make cities safer is implementing gender mainstreaming in planning [4]. The city of Vienna is an example of how they included these issues in the planning and created a city for everybody, paying special attention to the needs of women and vulnerable groups. [5].

Groups that break the norm of the majority in our society, generally defined as marginalised groups, typically suffer from unwanted attention, sexual harassment, and misrepresentation in the public realm. Women and minorities, such as racialised, older, and people with disabilities, also commonly defined as vulnerable groups, often share the experience of exclusion from urban public places. This exclusion often leads to a heightened fear of crime, and consequently, their freedom of movement and mobility in the public realm is restricted [6], [10]. This is defined as gendered fear and is a complex phenomenon discussed within the field of human geography since the 1980s; however, without conclusion [11], [12]. A study of young girls in Melbourne in 2022 shows that *"vulnerability in public space after dark are consequences of gender inequality and perpetuate the social exclusion of women*"[10, p. 86]. Their findings show that lighting has the potential to be an element that can have a substantial spatial and social impact. These findings complement previous studies [13]–[15], where gendered fear has been studied as a factor compared to spatial design when creating safer urban places. Our literature review suggests that only a few articles have studied how lighting design can make urban public places more inclusive for marginalised groups.

1.1. The aim

In a master thesis, [16], a study was made to find relevant lighting design principles for making public urban spaces more inclusive for all marginalised groups. This article is based on this thesis. The aim is to investigate the research gap in how lighting design can make more socially inclusive urban places for marginalised groups. By broadening the research to include marginalised groups, more equitable and inclusive insights into how lighting design can contribute to more socially inclusive urban places for vulnerable groups can be gained. Although this article uses members of one marginalised group as a starting point, the findings could potentially be generalised t be valid for other vulnerable groups, such as racialised groups, women and children or older people. The article is based on studies in Oslo. This is in a part of the world where equality and legislation exist, and in Oslo, the public realm is considered safe. Still, as the results from the field study indicate, gaps remain to be filled.

2. Methodology

The chosen methods for this investigation consist of 1) a literature review, 2) a case study of three urban places in Oslo, and 3) a three-part field study. The first two parts create a basic understanding of how lighting design can support social interaction.

The literature review of three subjects laid the foundations for an analytical social inclusivity framework. These three subjects were inclusive urban planning, atmosphere as a phenomenon, and methods used in lighting design to achieve atmospheres. This framework was further used to analyse social inclusivity in the case studies.

In the case studies [17], several phenomenological methods are used when analysing the spatial design, the social aspects, and the lighting design at three different urban squares in Oslo. These sites were chosen because they represent diverse uses and variations in the spatial design. To analyse both spatial and social elements of the squares, both macro [18], [19] and micro levels [20] were chosen. The six lighting qualities as defined by Descotte and Ramos [21] were used to analyse the lighting and luminance maps.

10 members of one marginalised group participated in all of the three parts of the field study. These people identified themselves as members of the LGBT+ community in Oslo. They were selected to learn how marginalised groups in Oslo restrict their movement in urban public places and how lighting can influence their choice of routes. The members gave informed consent to their participation and that the anonymised data would be recorded and published.

The first part was primarily a focus group interview to discuss avoidance behaviour. For the second focus group interview, a method of visual ethnography called walking with video was chosen. This was done to enable the participants to reflect on how the lighting influenced their walks and their embodied experience in the space; this open method of handing over the camera to them was chosen so they could reflect on the felt experiences and used as a method to start a discussion with the participants [22].

The equipment used in the experiment, as the third part of the field study, was handheld torches provided by the lighting department at the University of Southeast Norway (USN) and two work lights. The equipment consisted of 10 tiny LED torches called Ledlenser P7R Core with 1500lm, 6 Ledlenser P18R Signature with 4500lm, and two work lights called Canopus 3000 RE. These two had a colour temperature range of 6500K-27000K and a dimming function. The audio from the whole experiment was recorded using a tape recorder in the author's pocket and documented by a photographer. The photographs and audio recordings from the experiment were analysed and used to develop a proposal for socially inclusive lighting design principles. The interview data were transcribed and analysed through NVivo software. After coding the material, a "Content analyses" method was chosen. This is a text-driven, problem-driven, and method-driven content analysis [23].

3. Literature review

Socially inclusive lighting design is a new direction within the lighting industry [24]. Only a little research has been done on this subject. A review of the first 50 articles through a search on Google Scholar with the search words: *"lighting + marginalised groups + urban public places"* found that only two hits were related to the subject [16]. One article, [10], mentions in their conclusion how marginalised groups should be included in future studies. The other hit was the book where this article was published [25].

Then, the search was broadened to replace "marginalised groups" with "gender": *lighting + gender + urban public lighting."* A new review of the first 50 articles showed that though some of these studies included an overview of gender in their sampling, only six had a gender perspective on perceived safety [16]. A review [26] found that five of the 1315 articles they reviewed focused on how lighting influenced women's perception of safety. The review concluded: *"...important knowledge gap is related to the insufficient representation of cultural or ethnic groups." [26, p. 28].* These reviews of articles on this issue confirm a knowledge gap in how socially inclusive lighting design can enhance freedom of movement for marginalised groups. Yang, Berry, and Kalms conclude that *"other marginalised groups should be included in future studies"* [10, p. 100]. From the research review findings, three subjects were found relevant to further investigation through a literature review: urban planning, atmosphere as a phenomenon, and lighting design.

The literature reviewed on urban planning shows the importance of considering the social aspects of spatial design. Inclusive and more participatory planning processes will help urban planning create urban spaces that are more accessible for everybody. Some ideas for gender-inclusive urban planning have evolved from Jane Jacobs [27] and her thoughts about spatial diversity. Jacobs is, however, most known for her theory of how safety can be achieved by "*eyes on the street"* and that form of social control. William Whyte made many vital observations when studying urban public life and strongly influenced placemaking [28]. His theories show how to turn a space into a place people are attracted to, while Jan Gehl was more concerned about which elements in the spatial design influenced the social life in the cities [20]. His theories have significantly impacted the methods of analysing what makes people stay in urban spaces.

The literature reviewed on the atmosphere as a phenomenon [29]–[31] shows that the atmosphere is complex and cannot create a socially inclusive place alone. Participation, such as light walks with lighting designers and architects, [32]–[35] can help make a foundation for more sociable sustainable places and to find atmospheres that could change the place.

The literature review on lighting design was divided into three subcategories that comply with methods commonly used by lighting designers: 1) reassuring atmosphere, 2) welcoming atmosphere, and 3) inclusive atmosphere [21], [36]–[38].

The findings from the literature review showed how lighting design can contribute to a change by highlighting elements like urban furniture, activities, and dark corners of an urban place. In that way, predict future use and promote diversity and inclusion. In addition, by lighting architectural features, lighting design can enhance public places' perceived safety, beauty, and image so more people are attracted to the place. This, in turn, can attract more people and ensure more eyes on the street, making the place safer and more inclusive.

4. Phenomenological observations as case studies

To make a supportive lighting design, the lighting designer needs to gain a complete understanding of how people would like to use the place to be able to support their social interaction. Observations of three urban public places in Oslo were conducted to gain insights into how to analyse the spatial, social, and potential atmospheres together and to learn how lighting design could help social interaction in urban public places.

The places were chosen because they represented three different uses by the public and had different spatial designs. When using the analytical framework developed through the literature review, Tøyen Torg was the most successful in the social inclusion [16].

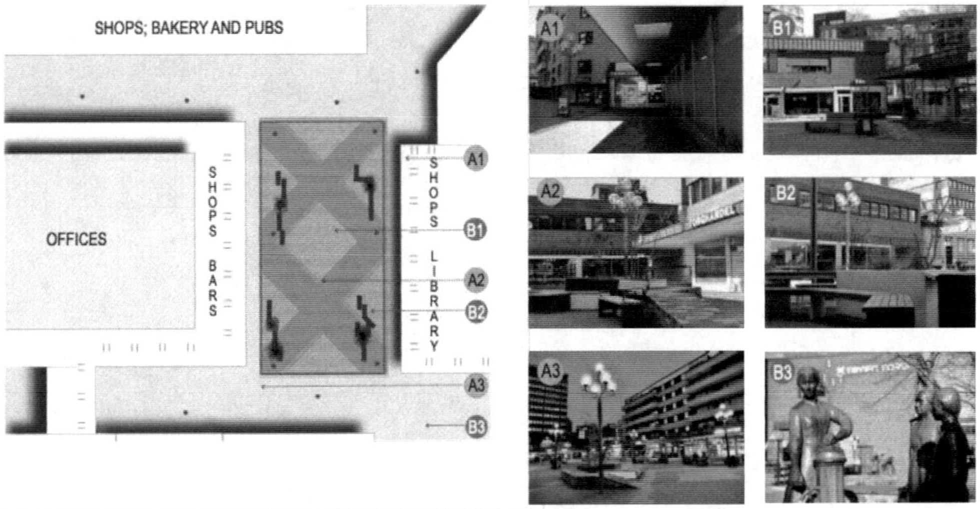

Figure 1: Tøyen torg is located at Tøyen in Oslo and is frequently used by residents and visitors from other parts of Oslo. Zenisk (lighting design firm in Oslo) made the lighting design, a re-design based on the original fixtures. The globes have been made translucent and fitted with gold reflectors. (see A2, A3 and B2)

The phenomenological observations concluded that a welcoming nocturnal atmosphere is present after dark with the pole lighting and the colours in the materials.

Despite these limitations to the spatial layout and the lack of social activities in the evenings, the warm qualities in the square's lighting are very suitable for the square. It provides a warm and welcoming atmosphere and supports social interaction after dark. The square went through a renewal process in 2014, where the residents were invited to participatory workshops. Their point of view was used to create the square as it is today, which might be the key to the success of Tøyen Torg. If urban public places become more accessible and inclusive for marginalised and vulnerable groups, everyone would feel included.

The findings from the case studies show that lighting has a vital role in facilitating this inclusion. This is further used when analysing the findings from the field study.

5. The field study.

To broaden the research within the lighting field to include marginalised groups, members from the LGBT+ community in Oslo were chosen to participate in the field study.

This community consists of people of all different social identities, ages, abilities, backgrounds, and ethnicities. However, they have one thing in common: they often experience unwanted attention, abuse, and personal attacks simply because their presence in public spaces can be provocative for some as they break the norm [39]. Due to homophobia and transphobia, their presence might be perceived as threatening or provoking [40], and being a minority can lead to a double fear [41]. This fear is often based on personal experiences and cannot be solved with lighting design. However, these are the factors which make the participants of the field study a marginalised group and very applicable to participate in the research for this article.

The field study used focus groups of members identifying with the LGBT+ community. The study consisted of three parts. 1) one interview session, 2) one session using a visual ethnographic method, walking with video and 3) one experiment in the field. The following section will present the two first focus group sessions. Then, the experiment will be presented, and the findings from all three parts of the field study will be analysed.

5.1. The focus groups interviews

Ten people were required to form two focus groups. This classified each group as a mini-focus group, as most focus groups typically consist of six to twelve members [23]. The participants were between 32 and 59 years of age with different gendered identities. When quoted, they will be referred to by their identity and age to preserve anonymity. The first session was an interview mainly aimed at understanding their behaviour in urban public places in Oslo. To get further insights into how lighting design could make a difference for this community, the participants were asked to make a video recording of a walk after dark. The recordings were presented, discussed, and analysed. Five participants and one moderator formed each group. The moderators were chosen beforehand. A structured guide for the discussions was prepared, consisting of eight and six questions in each session [16].

5.1.1. The set-up of the two interview sessions. The first interview session lasted for two hours. It started with a preliminary round where everyone introduced themselves, then went more in-depth about which external factors influence their everyday mobility routines and how the lighting influenced their route choice.

The second focus group session lasted for two hours and started similarly as the first session; however, with one change. After the participants had arrived, the researcher had a brief presentation of 15min with some slides that described with a few examples how much difference lighting design could make for the feeling of inclusion in urban public places. This was to create a common language and to inspire the focus group members to think of different solutions. The participants were asked to record a walk in an area they chose. Each participant presented their video, and the moderator asked predefined questions [16]. The questions were asked individually before the group could comment on how they would have wanted the lighting to be different.

5.1.2. Findings from the two focus group interview sessions. The first finding was discovering how much their identities impact their risk avoidance behaviour regardless of whether they were alone, with a partner, or with friends. This was the most substantial outcome and the main finding from the first interview session. The lack of social diversity causes feelings of unsafety and a wish to become less visible. Belonging to a minority often results in risk avoidance and a need to have an overview of the surroundings in the urban environment. As a lighting designer, it is hard to provide for both those needs to provide good visibility and uniformity and to give people the choice of being less illuminated and visible. *"You do not quite know what can happen when facing individuals or homogenous groups, especially in places where it's not that busy. I can think about that a bit about feeling safe" (non-binary, 42).*

The second finding was how the spatial design influenced their mobility choices. This was strongly connected to the sense of safety. One of the participants expressed this: *I didn't feel excluded, but I still did not feel comfortable because I didn't have an overview. It's a shortcut that I want to feel safe on, so that's why I go there. But it doesn't feel safe. And I go there during the day too, but it is scariest at night » (transperson, 32).* Belonging to a minority group because of gender expression, identity, skin colour, gender, or body can lead to a more vital need to control the environment and the presence of other people than the majority experiences.

A third finding is how the lack of spatial diversity causes feelings of unsafety as it influences legibility. It creates legibility by lighting the vertical surfaces, doorways, dark corners, exits, and entry points to a square. And by creating better legibility of space and providing for wayfinding, lighting design can make a more inclusive place. *"I think it's a shame that there are areas in Oslo we don't use. It restricts our freedom of movement. I also walk a long detour to avoid some places because they feel less inviting" (Man, 32).*

5.2. The experiment in the field.

The site for the experiment has suffered much crime. Therefore, a more hidden area underneath a flyover is floodlit with bright spots to ensure facial recognition on the CCTV surveillance cameras. This colour temperature might conflict with making a warm atmosphere. The participants noticed this, and one expressed it very clearly: *"The white light makes it feel creepy, and the warm light we are using feels safer and more welcoming" (non-binary 37).*

IOP Conf. Series: Earth and Environmental Science 1320 (2024) 012029 doi:10.1088/1755-1315/1320/1/012029

It is hard to change the spatial design of a place, and it is also hard to change social diversity, but enhancing the perceived safety and creating atmospheres is a point where lighting design could start to make a change. The chosen site has a bad reputation for crime and crime-related incidents and is rated as the place in Oslo most people feel unsafe. This site was also a topic in the discussions during the focus group interviews.

The experiment aimed to investigate how light could influence the participant's feeling of inclusion at the site and how they would light the areas if they could. Altogether 11 participants showed up to participate in the two-hour experiment. Nine of the participants were from the focus groups, and two additional participants were from an architectural firm. They all aged between 30 and 60 years and identified themselves within the LGBT+ community. None of the participants were lighting designers. Five tasks were given to the participants during the experiment:

1. exploring angles of lighting a person's face
2. exploring the direction and distribution of beams to enhance details and define space
3. exploring colour and indirect lighting methods
4. one group lit a passage with different angles, and the other group walked through
5. light a dark area to make it appear more welcoming.

Facial recognition had been a recurring topic during the sessions, so the first task was to light a person's face using direct and indirect illumination. This part was primarily also a start for the participants to get familiar with the equipment. As they grew more confident handling the equipment, the participants started exploring more details. As an example, one of the participants placed one light source underneath a bench with 2700 Kelvins. This was later analysed as something they experienced as inviting when the participant's vocabularies were developed to express more precisely how the lighting influenced them.

A dark passage became a subject for testing different angles and heights with the torches. The group was divided into two subgroups, with one part of the participants getting the task to find a way to light the passage so it felt more inviting to walk there. The other group was experiencing how it was to walk in different settings (see Figure 2, image 1). This was a relevant experience for many participants, as they better understood the impact different angles and heights could have on the space and their perception of safety and inclusion.

Figure 2: Images 1 (left) and 2 (right) show the participants testing directions and illuminating vertical surfaces. Trying methods using the materials as surfaces that reflected the beam as an indirect light source was explored more in the walkway next to a hotel entrance, as displayed in image 1. Image 2 shows how vertical surfaces were being lit.

The last area that was explored was an area near a flyover next to the water. The municipality has tried many different planning ideas to make the experience safer and more welcoming but has yet to succeed. Instead, they have installed overhead lighting fixtures that floodlight the whole space. Here, the participants were asked to take control of the luminaires and use their experiences to make the place more inclusive and safer.

By this point, the participants had grown more confident with each other and using the equipment, and they used one another to discuss and change to something they felt could work better than the first try. For example, among some things that were tried out was a warm light source lighting a wall, as displayed in Figure 2, image 2.

5.2.1 *Findings from the experiment.* The exercises from the experiment resulted in three main themes: 1) use of warm colours, 2) enhancing details, 3) exploring directions.
In the first finding, using colour, all the participants agreed upon the success of using the luminaires setting on 2700Kelvins, which produces a warm, almost amber colour. They felt the place became more inviting and inclusive by changing the colour. This colour temperature also contrasted the area's general lighting, which is very cold and white.

The second finding was how the participants used the materials and the urban furniture with the beams. The materials could be highlighted by direct illuminance or hidden by using contrast to create a brighter area next to the material. By using light and dark, light and shadow, unexpected corners and darker areas became revealed. This made the place appear more predictable, seeing what was hidden and feeling safer.

For the third finding, exploring directions, they tested how it felt to walk in different settings when the lighting came from another order than they were usually experiencing. Again, when the beams created different visible patterns on the path, it was perceived as very successful and diverse.

6. Analyses of findings from the field study

One obvious limitation of the experiment when using handheld torches is height and variation. The size of the light sources could never be very large, so all the testing was done at lower elevations. The colour temperature on the smaller torches was static, and the distribution remained the same. Only two of the luminaires could have a variable colour temperature, which contributed to a lack of variety in the attempt to create atmospheres using more variations in the colour temperature.

The outcome of the focus group interviews shows how feeling excluded and belonging to a minority result in a strong need to have an overview of the surroundings in the urban environment; by lowering the contrast ratio, balancing lux levels and providing vertical illuminance, this can be achieved. Furthermore, feeling scared, for example, due to bad experiences, leads to a change in behaviour, impacting their daily mobility and the patterns in their walkability routines. Although many participants spoke about how they did not notice the lighting unless it was very uneven, full of contrasts, causing discomfort glare by blinding directly in the eyes, or if it was absent, several factors contributed to their changed mobility routines. But also, the lighting, or the lack of illumination, could change mobility behaviour, as one of the participants said: *"Well, at night- it depends if I am walking with someone or if I'm walking alone. Because I would prefer to take the shortest route home, but if I'm walking alone, I will not go where there is less light. I will not take the shortest street. And I will not walk where I can't run" (woman 42).* The videos showed that the anxiety caused by minority stress is most predominated in the daytime or when other people can be identified. The videos' walking patterns and movements could indicate that they became more aware of their presence than the surroundings. Things change in the dark, and people are often seen as silhouettes; anyone can be a potential threat. The lighting needs to provide an overview, legibility, and wayfinding.

The first finding from the experiment showed that participants missed a warm atmosphere in urban public places. Using a warm colour temperature, something perceived as ugly, and repelling was changed to something more attractive. Moreover, using this warm colour to illuminate materials, walls, greenery, or darker areas, made a more welcoming atmosphere. For many of the participants, the experiment contributed to many surprises:
"It's funny; before this experiment, I always thought the solution was to add more lights. Now I realise it's not that simple. It is by lighting the surroundings that makes me see the environments better" (woman 39).

The second finding showed how lighting materials and landscape elements contributed to legibility and visual variety. By illuminating the trees and the greenery and highlighting the materials, it gave a more substantial definition of space and the vertical planes in the place. Using these techniques to beautify the space also made it feel more inclusive.

7. Discussion

How to create inclusive urban places for everyone without excluding others? In a study from Newcastle about safer city centres by Rachel Pain and Tim Townshend, four focus groups with different social identities and genders were used. The conclusion is: *"The challenge is to achieve this in an inclusive way, which does not further restrict the freedom of those already marginalised in city centre life" [42, p. 117]*. The topics for the discussion are related to freedom of movement, the lack of diversity in urban public places, and how lighting design can make a change.

The consequences of restricted freedom of movement can be greater than just a personal loss and can often result in a natural exclusion of minorities from urban public places. This was also a recurring topic in the focus group discussions as the participants spoke of how they would avoid homogenous urban public places. How gender relations are made in our everyday spatial practices and gendered power relations need to be challenged is something Yasminah Beebeejaun problematises in the article "*Gender, urban space, and the Right to everyday life*" [43]. However, Beebeejaun comes up with some simple spatial solutions, like moving a bus stop away from a pub with a homophobic reputation. Such measures can only be taken through participation, where marginalised groups are invited to participate in the spatial planning.

One of the essential findings from the focus group sessions is the consequences of the lack of spatial and social diversity in urban spaces. The literature review and case study findings implicate that the atmospheres in urban public places consist of all three elements: spatial, social, and lighting. When urban areas become homogenous because of a lack of spatial and social diversity, changing this through lighting design alone is impossible.

However, one of the main findings from the focus group interviews was that most of the focus group members agreed that they felt uncomfortable if they had to walk across very evenly brightly lit spaces with no places to be less visible. The results from the field study suggest that lighting design can do a lot to make urban atmospheres reassuring, welcoming and inclusive. This is where lighting design could start to make a change. This finding corresponds with other recent studies [10]. It contradicts most of the research on perceived safety, concluding that uniformed urban areas with white-coloured lighting appear to be the safest [44], [45]. Although non-uniformity was only one of the findings from the field study, the need for visible surroundings was also an important theme. Like the findings from other studies confirm [46]–[48], having the nearby surroundings lit enhances the perceived safety. Further, as Mette Hvass shows in her doctoral dissertation, [49], the surroundings will become more visible by lowering and balancing the general lighting levels. This method will allow 'room' for other subtle light layers to make a lighting hierarchy.

8. Conclusion

The findings from the focus groups and the experiment are divided into three parts: where the first main finding is rooted in the two focus group sessions, while the second and third finding is rooted in the experiment:

- First and foremost, the participants felt most safe when they could avoid the brightest-lit areas in order to avoid losing the ability to see their surroundings.
- Beautification was seen as making the urban place feel more welcoming.
- Legibility was created by using different techniques to illuminate vertical surfaces.

These findings were essential for creating six principles of inclusive lighting design that conclude this article. When used in a layered lighting design, these six principles can create reassuring, welcoming and inclusive atmospheres. Urban public places could become more inclusive with a lighting design based on these principles. By contributing to more freedom of movement for women, minorities and other vulnerable groups, these principles could create a foundation for making urban public places more socially inclusive for everyone.

8.1. *Principles of inclusive lighting design*

Creating legibility can enhance perceived safety by illuminating vertical surfaces. By illuminating dark areas and corners, it makes it easy to navigate around places of refuge, prospect, or escape.

Height variations can also distinguish between the public and the private and give an extended experience of space. Human-scale luminaires can create feelings of intimacy, and illuminating the greenery and highlighting points contributes to more definition of the place. It is essential to avoid glare. This technique can make materials become a light source when bouncing the luminance of the materials without discomfort glare. Glare can be blinding and contribute to unsafe environments, but with non-uniformly lit paths, people can be more or less visible when walking. By changing the perception of a place, it can make it more inclusive and welcoming. These six principles are based on the work of many other lighting designers, such as Herves Descotte [21]. However, these six principles can create more socially inclusive places when used in this way.

Principle 1: Legibility by illumination of vertical surfaces. Vulnerable and marginalised groups need to feel safe in urban environments, as there is an actual risk of experiencing unpleasant and unwanted attention. Illumination and contrasts between light and dark, light and shadow, can reveal unexpected areas. This made the place appear more predictable, seeing what was hidden and feeling safer.

Principle 2: The Wayfinding by luminaires on a human scale. Using light sources with a human scale in the light topography can help with wayfinding as it does not create any obstacles and provides the opportunity to see the surroundings better. The participants of the focus groups expressed a need to navigate easily in the urban environment and have an overview of the surroundings. Another element to assist wayfinding could be using different heights of the luminaries by creating focal points.

Principle 3: Beautification by highlighting materials and details. Illumination can enhance details, texture, and the colour of the different materials and the materials themselves. Enhancing details and beautifying the space creates exciting and attractive urban places. Minorities, women, and other vulnerable groups are very much aware of their surroundings and the social interaction in urban public places. Homogenous places without diversity are avoided if possible. When beautifying an urban area by highlighting the details and materials, more people can become attracted to them.

Principle 4: Inviting by indirect light sources. Indirect light sources stimulate curiosity and make urban furniture more attractive. It can also become a playful element. For many marginalised groups, the freedom of movement gets restricted due to the lack of spatial diversity and lack of activities in an urban public place. However, using indirect sources on urban furniture invites them to be used and creates a focal point on the action that can enhance social interaction.

Principle 5: Visual hierarchy by balancing contrasts. The members of the focus groups expressed a need to be able to be less visible when they felt unsure of their environments and the other people present. Lighting can define space by the placement of the luminaires, and highlighting the entrance and exits of an urban public place can also help determine the space and create an overview. Using the lighting to divide the place into functional zones of activities can also help with the feeling of privacy and intimacy.

Principle 6: Create identity by warm colour temperature. When used carefully, colours can provide visual cues and enhance a place's identity. In the context of this study, using colours can also attract other people, primarily through warm colours. In the Nordic countries, warm-coloured lighting is perceived as cosy, candlelit indoor atmospheres. This is extended when used in an urban outdoor environment. Marginalised groups can benefit from creating a visual identity that can change the perception of the urban place and make it more inclusive and welcoming. For example, a warm colour can help beautify the place and make it feel safer.

Principle 1. Principle 2. Principle 3. Principle 4. Principle 5. Principle 6.

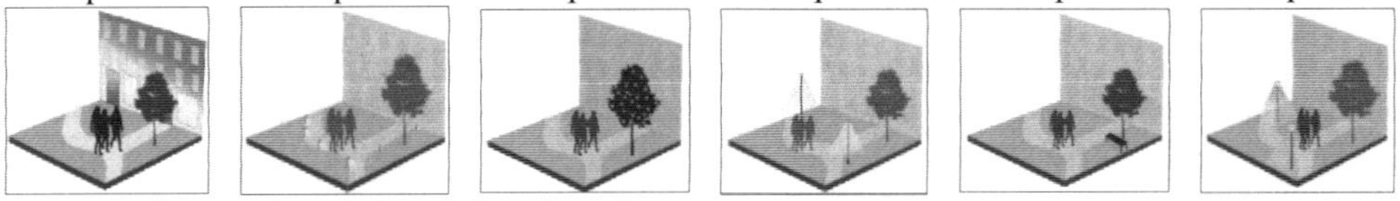

Figures 3: Showing the principles 1-6 as visual diagrams.

9. Future works

The lighting standards restrict much of the lighting design for urban public places. Today, the standards dictate how much illumination they must provide on the horizontal plane and the uniformity rates. However, these standards need to take other factors into account. Like how much light comes from the headlights on a car or how much influence accent light can have on a square. More research is needed to determine how the uniformity rating needs to be to provide enough visibility.

The field study consisted only of 11 participants. The findings could have been different if there had been more participants. Other methods, such as lighting walks or walk-along interviews where participants took pictures, like the methods used by Mette Hvass in her doctoral dissertation. [49] , could have resulted in different outcomes. Some things might be easy to change, but it is impossible to change a place's spatial design with lighting design, nor to change social diversity.

Creating new standards and recommendations for urban lighting is essential to avoid light pollution and be aware of biodiversity. It is also necessary to be mindful of social sustainability and cater for human and social interaction. Future studies should consider the impact of balancing contrasts, warm colour temperatures and uniformity on the perceived safety in urban public places.

Other recommendations for future works are to continue to do research with including other marginalised groups. This can be done through participatory workshops and experiments. Another method could be to go on light walks with the focus groups and look at places they have marked as inclusive or unsafe to look at the differences together. Alternatively, use pictures as a visual ethnographic method instead of video recordings. Using other focus groups with control groups in the study is recommended.

10. Acknowledgements

The author would like to thank the Aalborg University research team for sharing knowledge and dedication to the two-year master's program in Lighting Design. Also, a special thanks to the lighting design department at the University of Sør-Øst Norge for providing the equipment for the experiment.

References:

[1] 'Goal 11: Sustainable cities and communities', *The Global Goals*. (accessed Nov. 06, 2022). https://globalgoals.org/goals/11-sustainable-cities-and-communities.
[2] '11.7 by 2030, provide universal access to safe, inclusive and accessible, green and public spaces, particularly for women and children, older persons and persons with disabilities – Indicators and a Monitoring Framework'. (accessed Jul. 15, 2023) https://indicators.report/targets/11-7.
[3] UN Women, , 2021, 'Safe-City′.
[4] C. Moser and A. Moser, Jul. 2005, 'Gender mainstreaming since Beijing: A review of success and limitations in international institutions', *Gend. Dev.*, vol. 13, no. 2, pp. 11–22, doi: 10.1080/13552070512331332283.
[5]D. Damyanovic, 2013, 'Gender Mainstreaning as a Strategy for Sustainable Urban Planning', in *Fair Shared Cities*, London: Routledge, pp. 177–192.
[6]L. Kern, 2021, *Feminist city: claiming space in a man-made world.* London New York: Verso.
[10] H. Yang, J. Berry, and N. Kalms, 2022, 'Perceptions of safety in cities after dark', in *Lighting Design in Shared Public Spaces*, 1st ed.New York: Routledge, pp. 83–103. doi: 10.4324/9781003182610-5.
[11] R. Pain, 1991, 'Space, sexual violence and social control: integrating geographical and feminist analyses of women's fear of crime', *Prog. Hum. Geogr.*, vol. 15, no. 4, pp. 415–431, doi: 10.1177/030913259101500403.
[12] G. Valentine, 1989, 'The geography of womans fear', *Area*, [Online]. Available: https://www.jstor.org/stable/20000063
[13] A. Fanghanel, 2016, 'The trouble with safety: Fear of crime, pollution and subjectification in public space', *Theor. Criminol.*, vol. 20, no. 1, pp. 57–74, doi: 10.1177/1362480615582097.
[14] C. Keane, 'Evaluating the Influence of Fear of Crime as an Environmental Mobility Restrictor on Women's Routine Activities', *Environ. Behav.*, vol. 30, no. 1, pp. 60–74, Jan. 1998, doi: 10.1177/0013916598301003.

[15] H. Koskela and R. Pain, 2000, 'Revisiting fear and place: women's fear of attack and the built environment', *Geoforum*, vol. 31, no. 2, pp. 269–280, doi: 10.1016/S0016-7185(99)00033-0.
[16] K.Lund, 2023 '*The Rights to The City, Improving inclusivity in urban public places through lighting design for marginalised communities*' MSc Thises, Aalborg University
[17] R. K. Yin, 2018, *Case study research and applications: design and methods*, Sixth edition. Los Angeles: SAGE.
[18] K. Lynch, 2008, *The image of the city*, 33. print. in Publication of the Joint Center for Urban studies. Cambridge, Mass.: M.I.T. Press.
[19] Project for public spaces, (accessed Apr. 14, 2023) 'What Makes a Successful Place?' https://www.pps.org/article/grplacefeat
[20] J. Gehl, *Life between buildings: using public space*, 1987, New York: Van Nostrand Reinhold.
[21] H. Descottes and C. E. Ramos, 2013, *Architectural lighting: designing with light and space*. New York, NY: Princeton Architectural Press
[22] S. Pink, V. Fors, D. Lanzeni, M. Duque, Y. Strengers, and S. Sumartojo, 2022, *Design ethnography: research, responsibility, and futures*. Abingdon, Oxon ; New York, NY: Routledge.
[23] T. Bjørner, Ed., 2015, *Qualitative methods for consumer research: the value of the qualitative approach in theory and practice*. Kopenhagen: Hans Reitzels Forlag.
[24] D. Casciani, 2020, 'From Design Research to Design Practice of Urban Social Lighting', in *The Human and Social Dimension of Urban Lightscapes*, in SpringerBriefs in Applied Sciences and Technology. Cham: Springer International Publishing, pp. 119–140. doi: 10.1007/978-3-030-57165-8_7.
[25] S. Sumartojo, Ed., 2022, *Lighting design in shared public spaces*. New York, NY: Routledge, Taylor & Francis Group.
[26] T. Trop, S. Shoshany Tavory, and B. A. Portnov, 2023, 'Factors Affecting Pedestrians' Perceptions of Safety, Comfort, and Pleasantness Induced by Public Space Lighting: A Systematic Literature Review', *Environ. Behav.*, vol. 55, no. 1–2, pp. 3–46, doi: 10.1177/00139165231163550.
[27] J. Jacobs, 1992, *The death and life of great American cities*, Vintage Books ed. New York: Vintage Books.
[28] W. H. Whyte, 2010, *The social life of small urban spaces*, 7. print. New York, NY: Project for Public Spaces.
[29] G. Böhme, 2013, 'The art of the stage set as a paradigm for an aesthetics of atmospheres', *Ambiances*, vol. 315.
[30] M. Bille, P. Bjerregaard, and T. F. Sørensen, 2015, 'Staging atmospheres: Materiality, culture, and the texture of the in-between', *Emot. Space Soc.*, vol. 15, pp. 31–38, doi: 10.1016/j.emospa.2014.11.002.
[31] J. Pallasmaa, 2012, *The eyes of the skin: architecture and the senses*, 3. ed. Chichester: Wiley.
[32] S. Sumartojo and S. Pink, *Atmospheres and the Experiential World: Theory and Methods*, 1st ed. Routledge, 2018. doi: 10.4324/9781315281254.
[33] S. Sumartojo and S. Pink, 2018, 'Moving Through the Lit World: The Emergent Experience of Urban Paths', *Space Cult.*, vol. 21, no. 4, pp. 358–374, doi: 10.1177/1206331217741079.
[34] L. Schwendinger and P. R. Ross, 2022, 'Envisioning Community Engagement in Smart Lighting Design', *IOP Conf. Ser. Earth Environ. Sci.*, vol. 1099, no. 1, p. 012042, doi: 10.1088/1755-1315/1099/1/012042.
[35] M. Hvass and E. K. Hansen, 2022, 'Balanced Brightness Levels: Exploring how lighting affects humans' experiences of architectural and social urban contexts.', *IOP Conf. Ser. Earth Environ. Sci.*, vol. 1099, no. 1, p. 012017, doi: 10.1088/1755-1315/1099/1/012017.
[36] J. L. Moyer, 2005, *The landscape lighting book*, 2nd ed. Hoboken, NJ: J. Wiley & Sons.
[37] L. Schwendinger, 2021, 'Darkness as canvas', in *Rethinking Darkness*, New York: Routledge, pp. 202–216.
[38] H. M. Brandston, 2008, *Learning to see: a matter of light*. New York, N.Y.: Illuminating Engineering Society of North America.
[39] T. Cresswell, 2015, *Place: an introduction*, Second edition. Chichester, West Sussex, UK ; Malden, MA: J. Wiley & Sons.

[40] K. Beebe and A. Davis, 2015, *Space, place and gendered identities: feminist history and the spatial turn*. London (GB): Routledge.
[41] J. Diplacido, 1998, 'Minority Stress among Lesbians, Gay Men, and Bisexuals: A Consequence of Heterosexism, Homophobia, and Stigmatization', in *Stigma and Sexual Orientation: Understanding Prejudice against Lesbians, Gay Men, and Bisexuals*, 2455 Teller Road, Thousand Oaks California 91320 United States: SAGE Publications, Inc. doi: 10.4135/9781452243818.
[42] R. Pain and T. Townshend, 2002, 'A safer city centre for all? Senses of `community safety' in Newcastle upon Tyne', *Geoforum*, vol. 33, no. 1, pp. 105–119, doi: 10.1016/S0016-7185(01)00025-2.
[43] Y. Beebeejaun, 2009, 'Making safer places: Gender and the right to the city', *Secur. J.*, vol. 22, no. 3, pp. 219–229, doi: 10.1057/sj.2009.4.
[44] I. Rakonjac, A. Zorić, I. Rakonjac, J. Milošević, J. Marić, and D. Furundžić, 2022, 'Increasing the Livability of Open Public Spaces during Nighttime: The Importance of Lighting in Waterfront Areas', *Sustainability*, vol. 14, no. 10, p. 6058, doi: 10.3390/su14106058.
[45] A. Svechkina, T. Trop, and B. A. Portnov, 2020, 'How Much Lighting is Required to Feel Safe When Walking Through the Streets at Night?', *Sustainability*, vol. 12, no. 8, p. 3133, doi: 10.3390/su12083133.
[46] C. Boomsma and L. Steg, 2014, 'Feeling Safe in the Dark: Examining the Effect of Entrapment, Lighting Levels, and Gender on Feelings of Safety and Lighting Policy Acceptability', *Environ. Behav.*, vol. 46, no. 2, pp. 193–212, doi: 10.1177/0013916512453838.
[47] A. Haans and Y. A. W. de Kort, 2012, 'Light distribution in dynamic street lighting: Two experimental studies on its effects on perceived safety, prospect, concealment, and escape', *J. Environ. Psychol.*, vol. 32, no. 4, pp. 342–352, doi: 10.1016/j.jenvp.2012.05.006.
[48] J. Rahm and M. Johansson, 2021, 'Assessment of Outdoor Lighting: Methods for Capturing the Pedestrian Experience in the Field', *Energies*, vol. 14, no. 13, p. 4005, doi: 10.3390/en14134005.
[49] M. Hvass, 2022, 'Balanced Brightness Levels: Exploring human sensory experience of lighting and darkness in uban public contexts.', PhD, AAU. [Online]. Available: www.aau.dk

Situating the light: Methodology for sensory and spatial fieldwork

Turid Borgestrand Øien*[1], Senja Maarit Ruohonen[1], Nanet Mathiasen[1], Anne Kathrine Frandsen[1], and Anette Bredmose[1]

[1]Aalborg University, Department of the Built Environment, A.C. Meyers Vænge 15, 2450 Copenhagen, Denmark

*tuo@build.aau.dk

Abstract. Sensory perceptions are a novel point of departure for lighting research, where human factors have long been approached through isolated variables and controlled environments. However, informed by philosophy and social sciences and supported by ethnographical methods, researchers and practitioners are gaining new ground in understanding human-environment relations by approaching user perspectives in empirical inquiries. Situating the lighting in spatial settings and sensory experiences is crucial when exploring the dynamics of the changing visual perception of aging and vision loss, and how these phenomena can affect everyday life. Although spatial and sensory perceptions have typically been approached from different positions represented by architects or anthropologists, the following question remains: How can we develop a methodological framework for exploring entwined sensory and spatial experiences? This paper presents the method development process for our upcoming sensory and spatial fieldwork in the project "The role of light when vision changes." The process is described in autoethnographic narratives, analysis of the technological frames, approaches, and understandings of light represented in the project, and the knowledge gained from testing the developed tools and schemes hands-on in a home environment. The initial findings indicate that both spatial and sensory experiences are interactional, as experiences situated in a specific body interacting with the specific environment it is situated in. Furthermore, in addition to the changes in visual perception, the spatial and luminous characteristics of the indoor and outdoor environments are also dynamic and changing, making the aspects of transitions and thresholds relevant for our upcoming fieldwork. Our hypothesis is that the shared technological frame developed in the project includes the relevant qualitative and quantitative measures that will allow us to make the knowledge of visually impaired participants explicit in ways that will inform and improve future lighting design.

Keywords: Field work, everyday life, domestic lighting, incipient vision loss, methodological framework

1. Introduction

Light is paramount for humans in terms of both our visual perception and the non-visual support of our health and well-being. However, light is even more crucial for people experiencing visual impairment. On the one hand, light is decisive for their residual vision to function. On the other hand, the threshold between the light working as an enabler or disabler is marginal and dependent on small variations in light quality, position, and direction [1]. The human eye and vision change with age, yet the process of losing sight remains an unexplored phenomenon for most of us, much like in the field of lighting. A 2019 special issue on lighting research methodologies addressed approaches to understanding and working with human responses to lighting. The contributions represented scientific rationales from the laws of physics that have characterized the field of lighting for many years: psychometric principles [2]; photometric and radiometric measures [3]; metrics of theories and techniques for assessing, measuring, and quantifying elements. However, the authors of this special issue also acknowledged the interdisciplinarity of the lighting domain and that light affects humans in multiple ways, both physically and psychologically: "Lighting quality exists in the balance of meeting needs, integrating with architectural elements, and addressing energy and environmental considerations" [2]. The human-experienced light cannot be isolated into single factors and treated individually since it is always "interacting and simultaneous" [4]. "Science, technology, engineering, and mathematics – leave out humanity…[by] not consider[ing] any of the many fields of knowledge that do understand people..." [5]. Consequently, Norman stressed the need to expand or complement the unintelligible and abstract metrics of STEM methods, with meaningful insights relating to humans and their quality of life. Contrary to the averages and accuracy of physical sciences that aim to generalize measures from any specific context, humans are context-sensitive and insights on contextual aspects and meaning can be gained by qualitative approaches [5]. Kelly argued that lighting researchers can learn the practice of qualitative research from the humanities and their methods in exploring complex issues of human behavior in post-occupancy evaluations and case studies [6]. Thus, social sciences, philosophy, phenomenology, and anthropology can provide useful theoretical frameworks for exploring human-environment interactions in architectural research [7]. In line with cross-disciplinary learnings and attempts to cover a more diverse understanding of functional abilities, architectural researchers also borrow concepts or frameworks from other disciplines [8]; however, the methodological considerations of implementing these in a mixed-method practice are seldom described. This paper is an attempt to put these considerations on the table by addressing the following research question: How can we develop a methodological framework for exploring sensory and spatial experiences? This will be tested in the case of understanding the everyday lives of the visually impaired in their luminous home environments. Moreover, we discuss how the findings from this initial inquiry can inform our further work across the quantitative and qualitative approaches to lighting in our empirical inquiries.

2. Background

The research project "The role of light when the vision changes," running from 2023 to 2026, aims to investigate the role of lighting in and around the home in sustaining activity and engaging the everyday lives of visually impaired individuals. For the group of researchers, this involves learning more about enabling and disabling aspects of light in everyday situations while simultaneously challenging our understanding of spatial and sensory perception. Our hypothesis is that translating the findings of this study to the practices of lighting design, architecture, and planning will ultimately support the well-being of citizens experiencing vision loss in the future.

The current project involves three work packages: 1) field studies testing and developing the methodology as well as gathering data on the role of lighting for the visually impaired; 2) design laboratory developing design parameters and scenarios from the empirical material collected in the field work; 3) designing interventions for optimizing light in a housing association. The aim is to translate the findings of the first work package to the practices of lighting design, architecture, and planning in a way that can hopefully support the well-being of citizens experiencing vision loss in the future. The field studies will provide knowledge on the transition between age-appropriate normal sight, incipient vision loss, and impairment and on the role of light as part of everyday life for this target group.

During the winter of 2023/24 and summer of 2024, we will visit 40 participants (55+ years), including a group of 10 normal-sighted participants and 30 participants representing different types of visual impairments (impairment of peripheral vision, central vision, or a combination of both). However, the following question remains: How can we explore this two-fold phenomenon—i.e., the change processes of vision loss and the reconfiguration of everyday life, including the relationships with and use of their luminous home environments—with our participants?

The overall project has been assessed eligible for exemption from ethics review, and informed consent will be obtained from future participants, yet for the initial inquiry presented in this paper, the participants involve the five authors and is therefore not subject to the GDPR regulations. This initial inquiry within the first work package identifies and describes the pool of methodological approaches to light within the project group, and what changes the contexts of housing research and lighting design have initiated.

3. Theoretical framework

In physical science, knowledge has been closely related to the ability to measure. However, contrary to the precise measurements of scientific methods aiming for generalizability, qualitative and more open measures of ethnography aim to understand context-specific aspects. Qualitative insights can be more difficult to measure due to a range of different and dynamic variables, where much of the complexity of human behavior can be lost in an abstract measure. Quantitative and qualitative approaches are fundamentally different, and the act of combining or mixing methods (e.g., numbers and narratives) can be a delicate matter in terms of both how to empirically collect them and how to coordinate their input, with one providing facts and the other providing meaning [5].

Kelly's [6] two first points for good quality qualitative research within the field of lighting concern interviewing techniques and the "critical reflection of interview data and technique documented." He further describes how the iterative process of qualitative research, where a research question might evolve and expand as the research progresses, makes the different parts of this process closely interrelated. As such, the adequacy of methods depends on the research question. Knowing the analytical framework and scope is useful when designing the overall research, while specifying the participants and developing an interview guide are also important. The possibilities of this exploratory approach are massive, but to navigate them, the reflexivity of the researcher is essential in terms of the research, participants, findings, and the researchers themselves.

The concept of technological frames can help us explore the design and development processes of technologies. A technological frame describes a shared framework for how a given technology or artifact is understood in a social group and was first introduced by Bijker in 1987 in an analysis of technological innovation [9]. The technological frames of a given technology structure the use of it; however, they are also at stake across designers, engineers, and other groups in design processes. A technological frame holds a common understanding of goals, key problems, theories, rules of thumb, and testing procedures, thereby structuring the thinking, problem-solving, strategy formation, and design activities of the group sharing the specific frame. Bijker later used this concept to analyze the historical development of fluorescent lighting during the first half of the 20th century [10]: by showing how opposing divisions of the US Government, lamp manufacturers, electricity-producing utilities, and housewives—representing different social groups with various technological frames—played part in negotiating and stabilizing the technology today known as the fluorescent lamp. The technological frames that were established nearly a century ago—with some arguing for the health-improving aspects of the technology—have been refined and developed ever since. In this paper, we use this concept to study the small steps of method development of implementing qualitative measures in a predominantly scientific field.

4. Methods

The methodological development has been the focus of the first work package, which has involved developing a methodological framework that can allow us to explore the sensory and spatial experiences of the visually impaired in their luminous home environments. In addition to studies from the literature, this development process has been facilitated in three steps: 1) assessment of an existing method from low-vision rehabilitation; 2) reflective discussions of individual, autoethnographic narratives of our existing methodological approaches and technological frames; 3) testing our toolbox in a home setting.

One of the project's main objectives has been to refine a method developed in low-vision rehabilitation with further inputs and contributions from the field of lighting design and architecture. The method for lighting assessment and intervention "Better Light, Better Living" (BLBL) was developed and tested by the Center for Special Education, Slagelse from 2017 to 2019 [11], included a narrative interview, vision- or task-related surveys (visual function questionnaire and performance measure), light measurements, and documentation through photos and plan drawings. The initial assessment of BLBL involved the investigation of documents and previous observations [1; 12] evaluating the relevance of the elements from the original method in relation to the fields of housing research and lighting design, and what elements to include in our methodology and toolbox.

In three workshops held during the spring and summer of 2023, the methodology and tools were discussed in relation to our own background practice knowledge, theories, and approaches, including the different approaches to lighting. In preparation for one of the workshops, we drafted individual narratives of our understandings of and approaches to lighting.

To decide what measures and instruments to involve in the fieldwork, as well as how best to operationalize these in the interviews and observations, we conducted a test in the home of one of the group members. With the narrative interview as a point of departure, the adjusted framework was demonstrated in a semi-structured manner by asking and continuously discussing the elements and the process of the interview, such as the introduction, positioning, sequences, and the relationship between the researcher and the participant in this inquiry.

The meetings and the tests were recorded and transcribed. Furthermore, reflections, decisions, and unsolved issues from the meetings were documented in a project log, which, together with PowerPoint files, photos, different versions of schemes, fieldnotes, and drawings have served as the basis for analyzing the methodology development described in this paper.

5. Inquiry on autoethnographic narratives

The project group includes a lighting designer and current Ph.D. scholar, an accessibility consultant with a background as a naval constructor with the lifelong experience of turning blind from retinopathy of premature and later cataracts, and three researchers sharing an architectural background that have different approaches to lighting in their work (i.e., mixed-method measures of daylight and atmosphere; lighting as part of building processes and regulation; lighting and its role in everyday life and among visually impaired). As an initial inquiry into our existing pool of approaches and methodologies, the group members were given the task of describing the individual practice knowledge at stake when working with or experiencing lighting.

Insider knowledge on living with visual impairment was provided by Anette, who stressed that people use their residual vision differently and that even "a visually impaired person sees in two different ways, one with each eye". Her right eye has never been functional, making her unable to see depths. Consequently, adequate lighting and color contrasts have always been essential for her ability to detect level differences. From school age, she used the residual vision in her left eye (6/60 in first grade) and could read normal text and ride a bicycle (on known routes) until 31 years old, when she got cataracts. Reading was assisted by light with high lux levels from a specific direction and reading at a distance of 5 cm or by using a magnifier, reading glasses, or a monocular. Riding a bike was possible on a cycle lane separated from traffic during daylight hours when the lane was visible and contrasted against the pavement. Between the ages of 26 and 31, she could only see the streetlights at night, which she actively used as important direction indicators, supplementing the use of a white cane. After the cataract, her visual acuity of 2/60 decreased due to major complications from surgery. Today, her vision only perceives focused lighting contrasting darker surroundings. Thus, lamps or windows can enable her orientation as a supplement to the use of a white cane.

Visual impairment is further known from a second-person perspective by Turid, who has followed BLBL and the group of low-vision consultants and their clients in developing and testing the lighting assessments and interventions in their practice. They explored the role of lighting in low-vision rehabilitation as it unfolded in the interactions between consultants and participants in the homes and the lighting lab. In participatory observations, ethnographic methods were combined with a spatial/architectural focus by mapping the narratives and scenarios in drawings and photos since the conversations typically addressed spatial, physical, and relational conditions. The observations also included the metrics used for evaluating the physical environment, the interview guide, and the overall assessment design. The low-vision consultants' theoretical frameworks were based on recovery and rehabilitation, and their tools for home visits included surveys, a semi-structured narrative interview, tools for visual performance testing, a spectrometer, a camera, paper, and a pencil. Some of the fieldwork was conducted in the homes of the participants months after the lighting interventions and employed walk-along interviews that involved following the participants as they described the different situations of their homes. Turid's theoretical lens, which was applied in the studies of the low-vision consultants at work, was based on socio-material frameworks such as science and technology studies, practice theory, and architectural anthropology. Spatial assessments were documented in descriptions, drawings, and photos, while sensory perceptions were represented in second- and third-person inquiries.

Another approach to observation is held by Anne Kathrine, which departs in a more architectural assessment of light by including the view of the room, its orientation, rhythm, position, and window distribution, and how the light interacts by and with contrasts, colors, surfaces, and reflections. In these observations, she uses her own perception: How are the contrasts between illuminated objects and the ambient light levels? How do materials and objects look? Are details visible or not? Are there different light characteristics? Soft, sharp, or precise? How do window openings convey the transition between inside and outside? Her theoretical framework has largely involved a phenomenological understanding documented by photography. Interviews have been conducted to understand other people's uses and perceptions of space, architecture, and lighting but with a focus on architectural assessment.

Compared to Anne Kathrines more overall and multiple architectural assessments with user input, Nanet has a more focused approach to lighting, observing light and space. For her, lighting is always dependent on the simultaneous presence of light sources, materials, and the human eye, which is why it relates to physical radiation in the electromagnetic spectrum between 380 and 780 nm. Light is understood as both spatial and material [13], and is explored through light level, light distribution, shadows, reflections, glare, light color, and surface color [14]. However, by drawing on a mixed-methods approach expanding beyond light geometry, Nanet has also drawn on phenomenology and meteorology in her studies. Consequently, this has involved assessing both measurable and non-measurable aspects, such as measuring lux levels, mapping the daylight factor, analyzing the intensity and distribution of light in the room combined with the perceived visual environment. The interaction between the light and the surroundings is assessed by measuring the reflectance of the surfaces and the light transmittance of the windowpanes. Furthermore, her phenomenological observations largely represent a first-person perspective focused on spending time at the location, moving around, and observing the dynamics of the daylight. Photography has been a key mapping tool used to document and maintain the registrations (both single frames and timelapse). By compressing time, the latter has been shown to effectively highlight changes in the light.

Photography has served as a path into the field of lighting for Senja, who was drawn to the different phenomena of light in spaces and their connections to humans' emotional states. Starting as a lighting designer in film and television by creating atmospheres and storytelling, she later started focusing on optimal lighting for different users (e.g., supporting the healthy sleep-wake cycles of the elderly in elderly care facilities). In terms of methodology, user experiences of lighting were assessed by ethnographic interviews and combined with more traditional light measurements such as illuminance on task areas, luminance-based uniformity assessments, and evaluating the quality of the spectral power distribution of light in spaces.

As shown in Table 1, the project group represents different methodological approaches and theoretical framing of lighting, including different positions in first-, second-, and third-person inquiries. The last row represents the low-vision consultants' approach.

Table 1. Paradigms and tools of the individual group members.

	Research paradigms	Tools – spatial perception	Tools – sensory perception
1	Autoethno-graphy	Narratives – Needs and (dis)abilities regarding light levels, direction, magnification, distance, and contrast.	First-person personal observations related to situations and the trajectory of impairment.
2	Science and technology studies practice theory, and architectural anthropology	Observations/descriptions of everyday practices and socio-material interactions. Mapping, drawing, and photos.	Second-person (walk-along) interviews and third-person observations.
3	Phenomenology	Observations and architectural assessments – Orientation, rhythm, context, windows, relationship to space function, contrast, reflections, and photos.	First-person personal observations – How do space and things look? Second-person interviews.
4	Meteorology and phenomenology	Observations and mixed-method lighting assessments. Luminance mapping, daylight factor, lux levels, and photos.	First-person observations of dynamic daylight.
5	Ethnography and lighting engineering	Observation of atmospheres, photography, Illuminance mapping, and luminance-based uniformity assessments.	First-person assessment and design process and second-person interviews.

6. Assembling the empirical toolbox for sensory and spatial fieldwork

6.1. Spatial mapping

In the first phases of the methodology development, the architectural focus was refined in a spatially funded model representing our assessment of domestic lighting (see Figure 1).

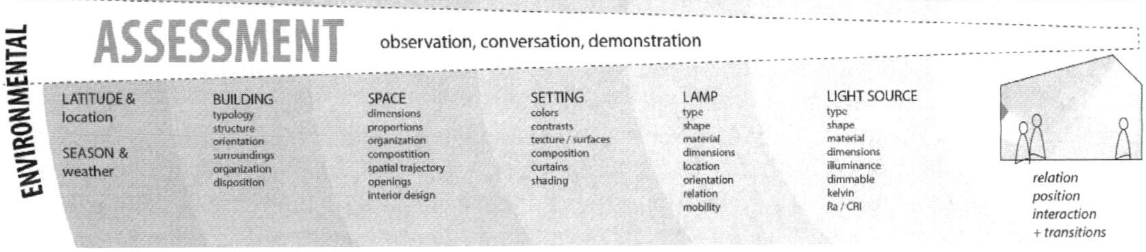

Figure 1. Model for assessing the spatial setting of the building environment and its lighting technologies.

In this understanding, light is not only situated across physical scales from the luminaire to the setting, space, room, building, and close context but also in regard to the personal relationship with and use of the light. The assessment would involve observations, conversations, and demonstrations led by the

IOP Conf. Series: Earth and Environmental Science 1320 (2024) 012030 doi:10.1088/1755-1315/1320/1/012030

participant, as an expert in both this specific environment and of their own condition and everyday life. The personal will be unfolded in the interview but always in relation to the spatial setting. Three categories combining the personal and the environmental—relation position, and interaction—were added as focal points for the assessment.

6.2. Interview guide/mapping the user's everyday practices and sensory experiences

This section raises an important question: How do we get hold of the participant's expert knowledge of perception and everyday life? Our inquiry will focus on understanding the status quo. However, involving the participants as experts in their own everyday lives and with their visual impairment is a key parameter to be taken from BLBL. Although people are often unaware of their behavior, interview questions can enable personal reflections on perceptions and experiences. Knowing that our analytical framework would focus on the role of the home environment and the lighting conditions within the interactions of everyday life, an objective of our design of the interview guide was that it should guide a spatial trajectory from entering the participant's home when seated in a traditional interview-setting or walking across the different spaces in a walk-and-talk interview. The group members' previous experience from conducting walk-and-talk interviews indicated that it was important to be able to "go with the flow" of the conversation and follow up on the way that spatial and material artifacts are put into play by the interviewee. By zooming in and out on details, situations, and activities, as well as the thresholds and transitions between these, we discussed using a diagram to facilitate the narrative interview in order to facilitate a more fluid and less structured scheme, which was inspired by the interview guide of the low-vision consultants.

7. Test: Preparing the toolbox for implementation

The following section concerns our reflections on the different elements of the first test of the methodological framework, on what tools to include in our toolbox, and how it would be appropriate to use them during the visit.

In this test, the toolbox included a practical bag equipped with a clipboard, pens/pencils, a ruler/ distance meter, and squared paper/a sketchbook for taking field notes and making simple sketches. Moreover, we had developed three schemes. Schema A includes a checklist for the visit, listing each step from the greeting and welcome to leaving the place. The checklist also includes a simple guideline for how to use mapping techniques such as signatures, measurements, and photo documentation. The spatial mapping would be documented in schema B and prepared with a plan drawing (specifying scale and direction), and diagrams for mapping weather conditions, time of day, and season. Schema C included the interview guide addressing the participants' narratives, including their descriptions of their sensory experiences. The first draft of the schemes, was discussed during our visit, and a range of issues were processed—the most relevant of which are described in the following sections.

7.1. Scaffolding the narrative interview

The overall objective of the study is to learn more about the role of light in the everyday settings and lives of people experiencing vision loss. Consequently, the main question of the interview guide would be what issues the participant has experienced, first of all regarding their current vision and lighting condition in the specific setting and at the specific time of the interview. Since we expect problems to be related to different situations, such as task-related activities, social settings, and orientation/wayfinding, we need the questions to be rather open. On the contrary, if there is a need to push the conversation forward, sub-questions could expand to past or future conditions and other contexts, and relate to changes made to the physical condition, physical setting, or workarounds in developing new ways or approaches to their activities. When discussing the interview guide in schema C, the group decided to include a more traditional list of questions and a process diagram that—together with the semi-structured interview diagram—could facilitate a more flexible guide to the narrative interview. To allow the flexibility of moving around in the dwelling more freely, a clip-on mini recorder was used for recording the interviews.

Furthermore, field visits have shown that the conversations were largely supported by our presence in the specific environment. In this sense, the interview also draws on schema A, while the spatial mapping and use of the plan drawing could also be incorporated as the participant describes specific issues.

7.2. *Lighting measures*

The methods for conducting the lighting measures should accommodate both the novice and the more experienced practitioner since our experience in assessing light and conducting lighting measurements differs within the group. Simple lux measurements will be conducted in the activities and settings identified in the narrative interview. In a seated activity, the instrument is placed horizontally on a surface (e.g., a table) facing upwards, or at 85 cm height in a spatial setting (the measure is noted in the plan of schema B). If an activity involves social interaction across a table, the light measurement will need to be done vertically, corresponding to the position of the face and torso of the other in order to map the conditions for interpreting body language. If the activity involves a range of positions or viewpoints, it could be relevant to take a series of measurements. The position, relation, and interaction between the participant and the light will be noted in either photos or field notes.

Even for the more advanced lighting measurements, the participant-led decisions on what and where to measure could differ from the regular lighting assessment. Consequently, *how to measure* was discussed in our development. Since the light of interest was contextualized within the participant, the light was not an objective condition. The measure could say something about the physical environment, but not about how the participant perceived this. Therefore, the light measurements will be of secondary importance when compared to the interview. They will be indications of light levels but always accompanied by observations. However, to understand the perception of the participant, we also need to engage our own visual perception of the specific environment in the conversation. Subsequently, in the case analysis, the lighting measurements will support triangulation by providing valuable input to the settings and situations described by the participant while facilitating further assessments across the different cases.

Perhaps one of the most professionally informed understandings of spatial perception among architects and designers is the importance and relevance of transitions and thresholds, which we had recognized during our project development process. In contrast to the scientific approach, the architectural approach is concerned with a holistic whole rather than individual parts, as well as trajectories rather than snapshots, to understand the everyday lives of the participants. We have planned two visits—a winter and a summer visit—to cover the seasonal transition. If these could be arranged so that we covered the day and night as well as natural and artificial lighting without the need to black out the daylight, this would frame different lighting scenarios just from the timing. In the interviews, we would ask about specific activities within the home as well as transitions between activities or the act of moving across the space—or zooming in and out between a focused activity and orientating. Furthermore, exterior settings close to the home would be another transition of relevance. Therefore, taking a series of light measurements representing the movement across different spaces could also be relevant.

Upon discussing the role of the lighting measurements, it was clear that while the lux value was in itself less relevant, it serves an important role as a fragment of the larger situation and setting. Therefore, the position and direction of the measurement would be as important as the measurement itself, while the mapping would involve a plan drawing and the use of photos. During our round of reflections on the measurements, we recognized that taking the measurements could be used as a tool for exploring the settings and issues (e.g., assessing the relational aspects at play in transitions and thresholds). Furthermore, since the luminance photos could potentially be used to support the succeeding assessment of the light distribution in a problematic setting or understand the transitions and thresholds between different positions or activities, in addition to the more detailed luminance mapping, we decided to include a simple version of the luminance photo in the documentation of one or two specific situations.

7.3. Multiple roles of photography
From the beginning, documenting the home environments and specific activities of concern had been an essential part of the method. Since the fieldwork would involve 40 different home environments, photo documentation would allow us to revisit the empirical material by analyzing the contextual complexity of the samples. Photos can support the memory of recalling conversations and the atmosphere of a place. Upon discussing the role of the photos, we recognized that the act of taking the photo—similar to the act of taking light measurements—serves a role in the assessment. As an aspect of the otherwise tacit knowledge of conducting architectural research, observing an environment through the lens of the camera serves an important role. Moreover, from the brief discussion on the role of photography in each of the group members' practices, we seem to have different opinions, objectives, and pay-offs from engaging in this kind of visual inquiry, which will be explored in future fieldwork. A particularly interesting area of exploration is how we can document and represent luminous environments. Prior to our test, we had discussed using a longitudinal section motif to register the transitions across a spatial trajectory, yet the initial experience from testing this out in an actual setting has already provided the first alternatives to this approach.

Beyond documenting the environmental setting, the photography will also be tested and explored to understand the visual perception of the participant across our differences in visual abilities through photo-elicitation. It is thought that the photographic lens does not directly represent the human eye with or without impairment. In a situation chosen by the participant, the interviewer will encourage them to choose a relevant position, direction, and framing of the motif using a tablet. This photo could then be a point of departure for the following conversation: "Can you describe how you perceive this motif?" This is followed by a reflection on how the interviewer interprets the same: "OK, so from what I understand, this part (pointing at something in the picture) with the brighter area gets blurred the most for you…? What about this field?" In this sense, both the photo and the lighting measurements would serve different roles in the field study, thus enabling insights into understanding the other and documenting the specific setting.

8. Discussion
The purpose of the method development process has been to find ways to support the study of lighting when situated in the everyday settings of people experiencing incipient vision loss. Spatial and sensory perception are aspects that are seldom measured and difficult to align or generalize—at least in a traditional scientific manner [5]. Lighting research has suggested photometric principles and measures [2; 3]; however, to understand how people perceive and consequently act and feel in their environment, we must explore the more complex and dynamic relationships between individual, social, and technical aspects, such as contextual aspects and meaning [5]. We argue that positioning the individual as an expert in their situated knowledge and scaffolding a constructive narrative interview—guided by sensitivity as well as our spatial focus—can enable this kind of insight. In the frame of the conversation, we will draw on both the qualitative and quantitative in the assessment of light. By studying light situated in a specific social and physical context, we believe that we will obtain an informed understanding of the phenomenon. In our case of exploring and challenging the role of the lux and luminance mapping within the conversation and the hands-on assessment, mixing and triangulating will contextualize the measures within the experience of activities and spaces situated in the everyday lives of the participants.

The mapping of the different theoretical frameworks and methodological approaches, as well as the first-, second-, and third-person perspectives represented within the group, have been helpful in acknowledging and coordinating our different practice knowledge. Architectural anthropology, ethnography, autoethnography, phenomenology, and practice theory all represent humanistic theories that aims to understand the human perspective. Science and technology studies may be more focused on the technological aspects of human life, while meteorology concerns physics and mathematics. It must be recognized that even if we share many aspects, such as understanding the technology of lighting in a situated manner and focusing on the socio-material relation of lighting and housing, we are also different. As different focuses related to positions in an architectural or lighting design process or in construction of knowledge. Our specificity and generalization vary, as well as the use of context-specific knowledge, rules of thumb, and standards. We represent different technological frames embedded in different fields and practices working and engaging with light. We believe that this will eventually constitute a new shared technological frame as we employ our new spatial and sensory methodology throughout our project. Beyond the overall spatial focus, the focus on transitions and thresholds in both research design and our empirical observations has been refined in our methodological considerations, thus adding a broader and more holistic scope of thresholds and transitions between different scenarios and activities. Furthermore, the approach incorporates different measures of the sensorial in the narratives of the participant, the conversations with the participant and observer, and finally the interaction with the home environment. These sensory aspects are scaffolded by light measures and descriptions in both spoken and visual language. We have already discussed the role of representations and empirical data in different phases of the process—as with the photos and light measures serving a role in the conversation—by informing the personal or shared reflections of the researcher and the participant, as well as representation for subsequent analytical processing. Moreover, it must be recognized that although measurements and photos do not represent what either of us perceive, they preserve a section of time and space that can enable us to describe and explore what we perceive. Combining the quantitative and qualitative requires us to navigate in the specific case and we believe our different first, second-, and third-person approaches will help us triangulate reality.

Qualitative research can be characterized by explorative iterations, where the scope and research question develop in the course of the project, which requires the researchers to reflect on the choices made, as well as their consequences [6]. Some elements of the toolbox will be developed along the way, such as the photo-elicitation and the role of the observational drawings. However, even the interview guide has changed since we conducted the test. We have continued refining the schemes in a series of meetings held after the test. In particular, the interview guide has been adjusted to a more traditional top-down sequential list of questions. It remains to be seen whether we will adhere to this or re-introduce the semi-structured diagrams. In this manner, critical reflections on what tools and approaches support us best in gathering data and exploring the phenomenon are highly related to hands-on testing. On a more abstract level, the reflections of adequacy include considerations of the context-specific fit between the research design (e.g., methods, methodology, and participants), research questions, as well as the transferability of findings [6]. In these initial stages, piloting has been essential, and enabled following reflections: How can we articulate light? Can we explore incipient vision loss using visual methods?

Furthermore, the anomalies and nuances in our background knowledge would not necessarily have been detected if we had not conducted the test together. This is partly because the aspects recognized and discussed are part of the otherwise tacit knowledge of our practices that were made visible by exploring and testing, and partly because the reflections are a result of these actions. The analysis of our technological frames has helped us recognize the differences in our approaches, which can be used in navigating the project while facilitating critical reflections and triangulation as we move on from field observations to the design lab and design processes. Our task will then once again be to find the right measurements to translate the knowledge to design principles or assess whether interventions can improve the quality of our housing guidelines and design parameters. We hypothesize that transitions, thresholds, and trajectories are elements common to both sensory and spatial perception as well as different spaces and diverse bodies and abilities. The spatial and sensory approach and toolbox developed for understanding the phenomenon of visual perception as a dynamic and changing sensorial condition can be seen as a unique case. Yet, we believe this approach can help us recognize important qualitative aspects relevant to all individuals.

9. Conclusion

This paper provides insights into otherwise tacit practice knowledge of methodological considerations regarding the development of spatial and sensory methodologies for exploring the phenomena of aging and changing visual perception, as well as the role of domestic lighting in the everyday lives of the visually impaired. Drawing on the theoretical framework of technological frames, the development process has been analyzed in three iterations. First, in our individual narratives concerning our status quo of how the group members of the project would normally approach lighting in our research. Second, in adjusting the approach for the context specificity of incipient vision loss and the home environment. Third, in testing the new methodological toolkit in a hands-on manner through a shared pilot study situated in a home setting with the insider knowledge of a visually impaired group member.

A method adopted from low-vision rehabilitation has been refined to fit the context and scope of housing and lighting research. The overall focuses on personal motivation and the therapeutic have been replaced by a focus on the situatedness of the individual knowledge that can inform future design in a more generalized matter via the collection across patterns and narratives. Ultimately, this can teach us about the phenomenon of incipient vision loss and aging, however the rich data and methods for collecting them can qualify fieldwork in general, such as providing insights on the relation of environmental factors as light and quality of life within housing and lighting research.

We believe that the findings from the initial phases of this study and the shared space for exploration and reflection will contribute to the further development and implementation of the spatial and sensory approach throughout the project, including the practice of operationalizing the quantitative with the qualitative, and vice versa.

Acknowledgments

The project is funded by the Velux Foundation.

References

[1] Øien T B 2022 A study of environmental factors in low vision rehabilitation *Frontiers in Rehabilitation Sciences* **3** 829903

[2] Veitch J A and Davis R G 2019 Lighting research today: The more things change, the more they stay the same *Leukos* **15** 2-3 pp. 77–83

[3] de Kort Y A 2019 Tutorial: Theoretical considerations when planning research on human factors in lighting *Leukos* **15** 2-3 pp. 85–96

[4] Frandsen S 1985 Lyset i rummet og lyset på tingene *Lampetten* **2** pp. 68–94

[5] Norman D 2023 *Design for a better world: Meaningful, sustainable, humanity centered* (Cambridge: MIT Press)

[6] Kelly K 2017 A different type of lighting research–A qualitative methodology *Light. Res. Technol.* **49** 8 pp. 933–942

[7] Groat L N and Wang D 2013 *Architectural research methods* John Wiley & Sons Incorporated

[8] Øien T B Grangaard S Lygum V L 2023 Exploring the dynamics of architecture with the concept of affordance *Philosophical Psychology* doi: 10.1080/09515089.2023.2273400

[9] Bijker W E 1987 *The social construction of Bakelite: Towards a theory of invention* (The social construction of technological systems) eds W E Bijker, T P Hughes and T Pinch (Cambridge: MIT Press) pp. 159–187

[10] Bijker W 1992 *The social construction of fluorescent lighting, or how an artifact was invented in its diffusion stage* (Shaping technology / Building society. Studies in sociotechnical change) eds W Bijker and J Law (Cambridge: MIT Press) pp. 75–104

[11] Øien T B Jacobsen A M Tødten S T Russotti T Ø Smaakjaer P and Rasmussen R S 2023 Impact of lighting assessment and optimization on participation and quality of life in individuals with vision loss *Occupational Therapy in Health Care* **37** 3 pp. 426–443

[12] Øien T B 2023 Methodological considerations in collaborative processes: a case of ethnographic action research *International Journal of Managing Projects in Business* **16** 1 pp. 165–184

[13] Mathiasen N 2015 *Nordisk Lys og dets relation til dagslysåbninger i Nordisk arkitektur* (Copenhagen: Det Kongelige Danske Kunstakademis Skoler for Arkitektur, Design og Konservering, PhD thesis)

[14] Liljefors A 1999 *Lighting - Visually and physically* (Stockholm: KTH Stockholm)

Empathic Lighting Design for Healthcare Environments

Nanet Mathiasen[1], Turid Borgestrand Øien[1] and Carlo Volf [2]

[1] Department of the Built Environment, Aalborg University, Copenhagen, Denmark
[2] New Interventions in Depression (NID-Group), Psychiatric Center Copenhagen, Copenhagen University hospitals, Frederiksberg, Denmark

E-mail: nam@build.aau.dk

Abstract. Light has the powerful capacity to generate a certain quality and atmosphere within a room. However, lighting design specification is often only characterized via quantitative metrics. In healthcare institutions, such as care homes, hospitals and rehabilitation centres, the use of lighting design may support a positive atmosphere, and potentially also support the health and wellbeing of patients. When people are hospitalised, they are in a difficult and often stressful situation, in which they meet nurses, physicians and relatives within new, unfamiliar surroundings. These situations call for supportive architectural spaces with atmospheres that help people relax and feel safe. Nevertheless, these spaces often lack a lighting design that supports this intent. The objective of this study is to implement two different lighting scenarios and investigate how they affect patients in a psychiatric affective healthcare unit at Copenhagen Psychiatric Centre. Two different lighting scenarios were implemented and tested in a SW-facing group therapy room, during one winter period (November 2022–March 2023). In all, the study included 12 patients. Through questionnaires, the patients were asked how they perceived the luminous environments of two lighting scenarios, one in the existing healthcare lighting vs. a new lighting setup. The patients indicated their preferences and how the lighting design affected and supported heir therapeutic session. The two lighting scenarios are: a) Four high-positioned LED-luminaires (H=2,65 m) in the ceiling; and b) Two low-positioned LED-luminaires suspended above the meeting table (H=1,50 m). The results of the survey showed that patients overall preferred the low-positioned, non-uniform lighting, which scored a total of 60 points in the semantic analysis, compared to 35 points for the high-positioned, uniform lighting. In addition, from a sustainability viewpoint, energy consumption for the low-positioned lighting scenario was only 21 W, compared to 140 W for the high-positioned lighting scenario, providing an 85% reduction in total energy consumption.

1. Introduction

Light has the powerful capacity to generate a certain quality and atmosphere within a room. However, lighting design specifications for institutions, including healthcare spaces, often tend to focus on efficiency, cost-effectiveness and practicality. Traditionally, healthcare spaces have an impersonal atmosphere that tends to reflect functionality rather than empathy and caring [1]. One element designed to support efficiency is a standard lighting design scheme that prescribes a specific amount of lux in the space, often resulting in uniform lighting. However, when people are exposed to healthcare environments, they are often stressed, fragile, and in a vulnerable situation, in need of care and attention. In addition, when relatives visit their loved ones in healthcare institutions, the physical built environment, including the lighting design, is important to the experience. A biography by the Danish author Lise Munk Thygesen describes a visit to a Danish hospital around 2008, where the author meets her ill mother: 'The walls were painted light blue at the bottom, while the rest was white. Black streaks from collisions with rolling hospital beds were drawn across the blue, and then the damned, usual neon light. It wasn't clean and it wasn't private' [2]. She describes quite distinctly how her experience of the environment did not support the difficult situation, nor did the lighting design add positively to visiting a close relative in a vulnerable situation. To highlight the point, she even curses to emphasize the inappropriate nature of the lighting scheme. The stark lighting Thygesen describes is sharp and revealing, in a situation in which the opposite character of light would have felt more supportive [2]. Such accounts of the atmosphere are recognisable in the Danish healthcare context, and unfortunately, are also expected to some extent when visiting a healthcare institution. The functional and efficient uniform institutional hospital lighting does not offer comfort, safety or relaxation [3]. Research shows adjustments to non-uniform lighting support more intimate and home-like atmospheres, can affect the feeling of a place, and can enable acclimation, such as encouraging lower voices and reducing stress in delivery rooms [4]. Similar relationships between the effects of different lighting distributions and behaviours have been observed in other public settings, such as schools [5], workspaces [6] and restaurants [7].

2. Theoretical and methodological framework

Light is a multifaceted element involving many technical and professional skills, and covering a profusion of scientific knowledge branches. Light can be described in precise terms, primarily concerned with its quantitative aspects, traditionally through natural sciences. Perception of light is typically related to the humanistic sciences, spanning multiple topics from healthcare, ethnography, anthropology, and environmental psychology to philosophical theories. Research examining design of electrical lighting can be aligned with research exploring architecture, as both incorporate several branches of knowledge and work across scientific and professional practices, as well as end-users [8]. In order to emphasize the supportive role of lighting in the built environment, including how different characteristics and distribution can support specific functions and scenarios, this project introduces the notion of *empathic lighting design*. Lighting is part of everyday life, and entangled in our practices [9]. In the healthcare setting, it supports the different activities taking place, including interactions and conversations between professionals and their clients in vulnerable situations. In these settings, light is not simply a neutral environmental setting, but a dynamic factor that can impact the trajectory of a given situation [10]. Drawing on the field of empathic design, a context-sensitive design method that combines subjective and objective approaches and design competencies in field studies of everyday life experiences [11], we introduce empathic lighting design as the ability to understand a situation from different user perspectives, accounting for the emotional and cognitive experiences resulting from utilisation of the design. Empathic lighting design focuses on a stimulating visual environment and immersive atmosphere that goes beyond mere functionality.

2.1 Hypothesis

The hypothesis of this study is that current conventional 'institutional' lighting design scheme may be questionable, and in some situations even counter-productive, inappropriate, or insufficiently supportive and empathic for patient care [12]. The objective of this study is to investigate how lighting design can add to a more supportive atmosphere and improve experiences in healthcare environments [13]. This paper presents the preliminary investigations of how vulnerable people experience two different healthcare facility lighting scenarios, installed in a group-therapy room used by patients and their therapist at Copenhagen Psychiatric Centre. The investigations concern what kind of light patients prefer in a group therapy room, to address the following research questions: How can various lighting scenarios change the atmosphere of a space? How do the patients experience different lighting designs? How can this knowledge promote and improve more sustainable lighting design and planning at healthcare units in the future?

IOP Conf. Series: Earth and Environmental Science 1320 (2024) 012031 doi:10.1088/1755-1315/1320/1/012031

2.2 Method

This research is based on a case-study design, using a group therapy room as the case site. The project implements two different lighting scenarios, and subsequently investigates how respondents perceive these two different lighting environments. The group therapy room is located at a psychiatric hospital, where the respondents are patients who attend therapy sessions concerning the challenges in relation to their illness, typically after their discharge from the psychiatric hospital following severe depression. The intervention includes two different lighting scenarios, followed by two semantic questionnaires. The study was conducted during winter months (November 2021–March 2022). The two lighting scenarios respectively consist of high- and low-positioned lighting. The high-positioned light is also the original light setting of the room, and represents the conventional strategy of organising lighting design for a space to support as many different functions as possible. Meanwhile, the low-positioned light represents an alternative lighting scenario that targets the specific function of the space as a group therapy room. The main focal areas in the study are the *distribution of light*, an important aspect when addressing the atmosphere of a room, and the *empathic* potential of the lighting's impacts on the behaviours of the space's occupants [14].

2.2.1 Light measurements

On-site measurements of the two lighting design scenarios are performed through illuminance measurements (Hagner Digital Luxmeter, model EC1-X), photographic documentation, and luminance photographs (SmartBeam). The luminance photographs are generated via a mobile phone camera (Samsung Galaxy S22+) using the SmartBeam mobile application by Fusion Optix to convert the photographs into luminance maps. The SmartBeam illustrations are visually clear, but not reliable in terms of exact measurements [15]; therefore, they only indicate the luminance distribution. The measurements and data documentation for illuminance in both lighting scenarios are measured on-site at a horizontal plane 0.85 m above the floor in a grid of 0.50 m x 0.50 m.

2.2.2 Semantic questionnaire

The respondents are asked to answer a questionnaire related to each lighting design scenario. The questionnaire focuses on respondents' perceptions of the space and preferences regarding the lighting design scheme. The first half of the questionnaire comprises six statements related to the experience of the space in relation to the activity. The respondents rate the statements using a five-point Likert scale ranging from *strongly agree* to *strongly disagree* [16]. The second half of the questionnaire comprises a variety of negatively and positively loaded words, among which each respondent chooses the word(s) describing their experience of the lighting scenario. Data collection via the questionnaire supports respondents remaining anonymous, as no name, age or gender are asked for. At no time was the research team in direct contact with the respondents. Questionnaire distribution is handled by the nurse, who joins the respondents in person and organises the sessions including handout of the questionnaires. This research inquiry is viewed as a quality assessment of the physical environment of the centre, and the project is therefore not required to undergo approvals of The Danish Ethics Committee. As no personal data was included in the study and only the nurses knew the identity of the participants, GDPR regulations do not apply. Before designing the questionnaire, the researchers have held an informal meeting with a relevant nurse, to discuss the patients' typical mental conditions and the organisation of their therapy sessions.

3. Case site

The case site is a group therapy room with dimensions of 4.60 m x 5.90 m, for a total of 27m². The space is on the building's 1st floor, and has an entrance door facing the corridor and a French balcony placed in the middle of the façade-wall that provides daylight, fresh air and a view towards a garden. The ceiling height is 3.0 m. The walls and ceiling are painted white (0.8 reflectance), and the floor is grey linoleum (0.3 reflectance). The two different lighting design scenarios are installed in this room; after installation, the respondents are exposed to one scenario at a time, but sequentially within the same session. The two lighting design scenarios are: A) high-positioned light fixtures, and B) low-positioned light fixtures.

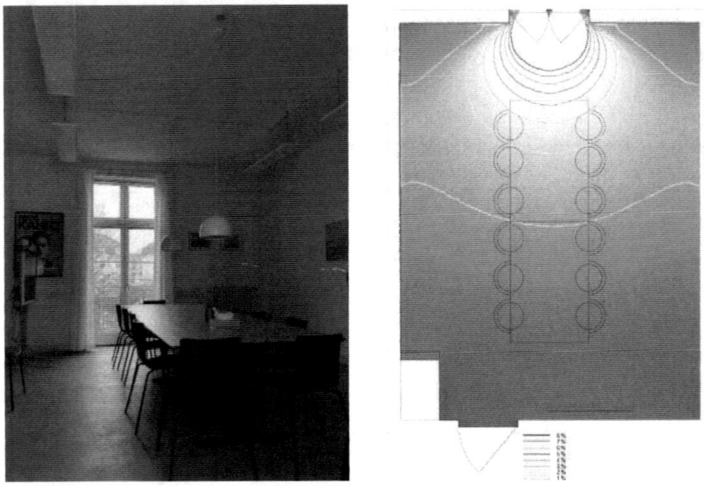

Fig 01 Left to right: Photo of the group therapy room in daylight situation with overcast sky, and plan drawing with DF.

	Specification of luminaire				*Specification of light source*		
scenario	Position of luminaire	Number of luminaires	Size and shape of luminaires	Distribution of light	Light source	CCT	Power consumption
HIGH	2,64m	4	1,20 m x 0,24 m	uniform	LED tube T8, 2 x 17,5 W x 4	3000 K	140 W
LOW	1,50m	2	Ø 0,30 m	non-uniform	LED lamp E27, 2 X 10,5 W	2700 K	21 W

Fig 02 Two scenarios: specification of luminaire and light source for the high- and low-positioned lights

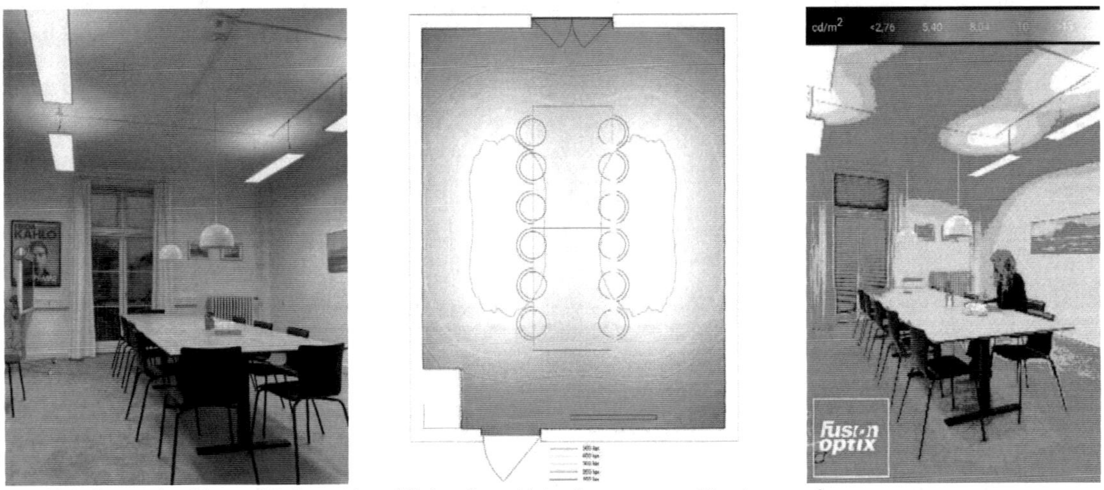

Fig 03 Left to right: Photo of high-positioned light, plan with iso lux curves and luminance photo.

3.1 High-positioned lighting

Four luminaires are positioned in suspension 0.35 m below the ceiling. The luminaires each measure 1.20 m x 0.24 m x 0.025 m. The lamps are positioned symmetrically in the room 1.40 m from the end walls and 1.04 m from the adjacent walls. The luminaires emit light both up towards the ceiling and down towards the table. The luminaires have a matte translucent screen. The light source is LED 140 W, 3000K. The light distribution is uniform, measured on site on a horizontal plane 0.85 m above the floor. The light has its strongest intensity at approximately 500–510 lux, below the two lamps hanging in elongation to one another. From the centre of the space, the intensity of the light diminishes towards the walls, where the light level is approximately 150 lux (Fig 03). The experience of the light distribution in the space is that it is uniform, without any highlights. Even the more intense areas directly underneath the luminaires are not experienced as much brighter.

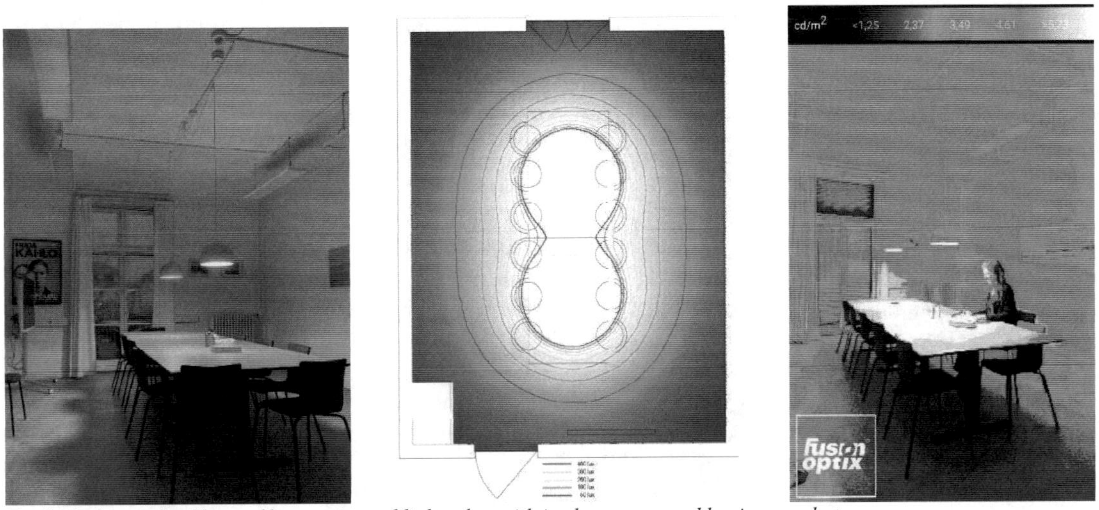

Fig 04 Left to right: Photo of low-positioned light, plan with iso lux curves and luminance photo.

3.2 Low-positioned lighting

Two pendants are suspended above the tables and 1.50 m above the floor. The two meeting tables, with dimensions of 1.20 m x 1.80 m, are placed in the centre of the room. The luminaires are centred 0.75 m above each table. The luminaire is painted white, has the shape of a semicircle, has a matte opaque screen, and measures 0.30 m in diameter. The light source is LED 21 W, 2700 K. The luminaires only emit light downwards. Therefore, their emittance of light is focused on the table surface, leaving the rest of the room darker. This causes a non-uniform light distribution in the room. Measured on site on a horizontal plane 0.85 m above the floor, the hotspots of the luminaires are 550–600 lux beneath the luminaires, fading to 150–200 lux at the edges of the tables. The rest of the room is dimly lit at approximately 20 -30 lux. This non-uniform lighting is clearly experienced when sitting in the room, as shown in the luminance photo (Fig 04).

3.3 The time schedule

The respondents are exposed to one scenario at a time during a single session split into two parts. During the first part of the session, the respondents are exposed to the high-positioned light for two hours. Then, a break is scheduled; after the break, the room lighting is changed to the low-positioned lighting, and respondents are exposed to the second scenario for two hours.

The experiment was conducted during winter, from November 2022–March 2023. To improve the reliability of the results, it was important that the main light source was electrical light, meaning daylight needed to be minimised within the experiment. During the test period (November–December 2023), it was dark and weather conditions were optimal, with overcast skies. The lighting scenario portions of the experiment took place on the 1st and 2nd of February 2023, during the afternoon. These dates were chosen based on the schedules for the participants' group therapy sessions, to allow the research experiments to occur within their usual schedule. A third experiment conducted later was not included in this paper, as the space had too large a contribution of sunlight, giving the daylight too much attention. On both days of the experiment, six respondents attended each group session, and each of them answered the questionnaire related to the two different lighting design scenarios; in all, 12 responses have been collected for each lighting design scheme, for a total of 24 responses to the experiment.

4. Results

The results of the questionnaires show significant differences among patients' perception of the two different lighting designs. In the semantic analysis, the low-positioned lighting is rated as the most pleasant, with respondents rating it 12 out of 18 points, while they rated the high-positioned lighting only six points (Fig 05). Responses indicate the low-positioned lighting works best for activities focused on and around the table, with a rating of 13 points out of 18, compared to the high-positioned lighting with only five (Fig 06). The same applies to how well the lighting supports therapy sessions; the low-positioned lighting receives 14 points out of 21, and the high-positioned lighting only seven (Fig 07). In addition, regarding supporting conversation around the table, respondents rate the low-positioned lighting as better with 12 out of 14 points, and high-positioned lighting with two (Fig 08). Finally, respondents rate the low-positioned lighting better regarding providing desirable levels of light on other participants, with 13 points out of 18, while they only award three to the high-positioned lighting (Fig 09). Only for blackboard education does the high-positioned lighting perform better, with 10 points out of 18, while the low-positioned lighting receives four (Fig 10). In general, the low-positioned lighting performs better (total score of 60 points) in comparison to the high-positioned light (total score of 35 points) (Fig 11).

When selecting words to characterize the different lighting scenarios, the respondents describe the high-positioned light with a diversity of words from the list provided in the questionnaire. Most of the words are highlighted, including both positive and negative evaluations. The statements regarding the high-positioned light can be interpreted as vaguer; overall, the respondents can be divided into two groups, one who dislikes the high-positioned light, and the other who likes it. The lighting design scheme with the high-positioned luminaires is hence a subject of more debate than the other scenario, and yields an unclear impression of the respondents' preferences. Nonetheless, one descriptive word stands out, chosen by most respondents: 'institutional'. This makes a clear statement that respondents interpret the high-positioned, evenly distributed light as creating an institutional atmosphere. Respondents also describe the low-positioned light using a range of different characteristics, but in contrast to the high-positioned light, the characteristics are all related to words typically interpreted as positive feelings, such as 'safe', 'relaxing', 'suitable', 'nice', 'beautiful', 'comfortable', 'homey' and 'cosy'.

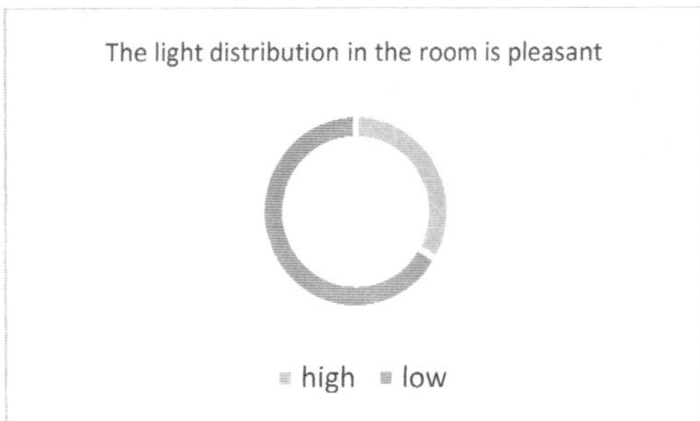

Fig 05 Respondents' preferences.

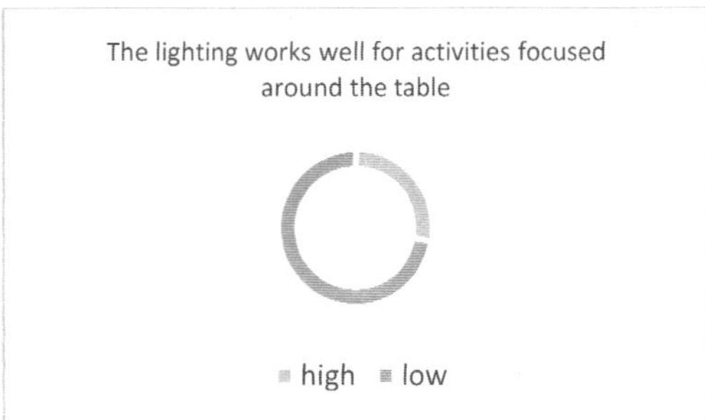

Fig 06 Respondents' preferences.

Fig 07 Respondents' preferences.

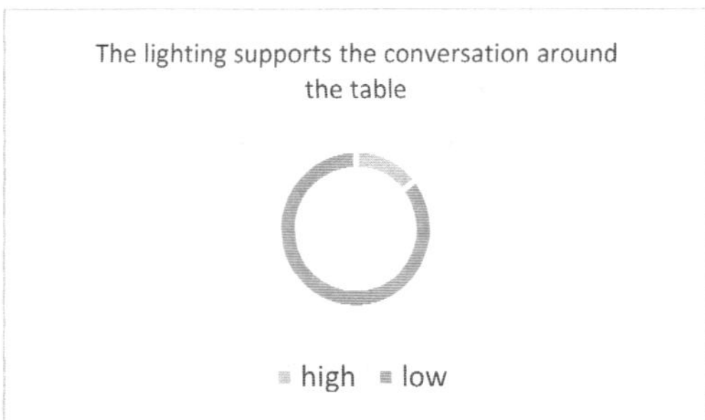

Fig 08 Respondents' preferences.

Fig 09 Respondents' preferences.

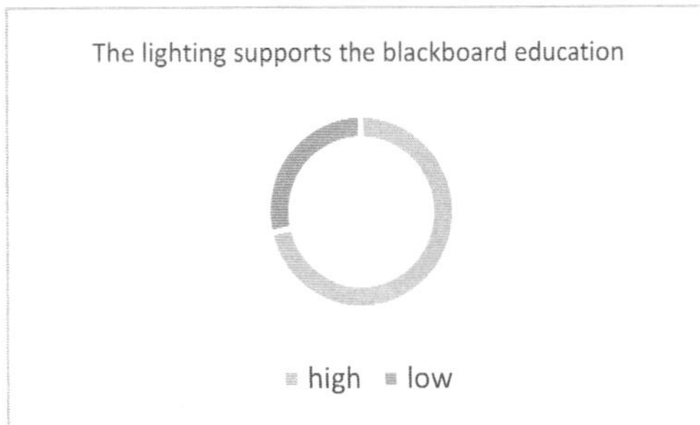

Fig 10 Respondents' preferences.

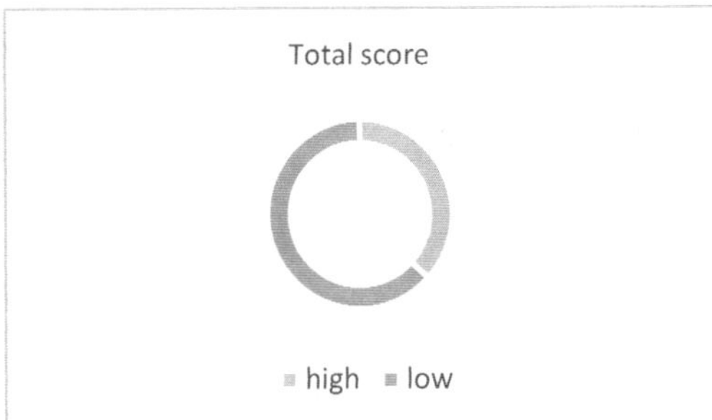

Fig 11 Respondents' preferences between the high- and low-positioned light in total.

The most significant difference between the two lighting scenarios involves the distribution of light, which is shown to have a pronounced impact on the visual environment. In general, light distribution is primarily related to the design of fixtures, and therefore often related to the designer's choice of lamps. Other aspects of lighting are related to the light source, such as light spectrum, colour rendering and the correlated colour temperature. Surveys focusing on aspects of the light source are more common than those examining light distribution. The current research focuses on light distribution, but recognizes the importance of the other aspects related to light source.

5. Discussion

The two different lighting scenarios in this study are fundamentally different in their expression. In addition, the respondents' ratings of the two schemes are markedly different; overall, patients prefer the low-positioned light, which provides a non-uniform light distribution. A potential follow-up question arises: Why has such a lighting design scheme not been widely adopted in practice for healthcare institutions, especially in environments where this type of lighting could support vulnerable users?

Lighting design schemes for institutions, for example healthcare facilities such as hospitals and care homes, have traditionally been planned to be functional and efficient. Due to the need for a space to facilitate different functions, or to be easily reorganizable for additional people or other purposes, lighting design schemes are often created as highly universal: Typically, flexible use of space results in lighting design strategies that promote uniform light distribution. A lighting design scheme without any hierarchy and with approximately 300 lux on a vertical surface 0.85m above the floor [17], provides flexibility and the possibility of positioning desks, meeting tables, and workstations anywhere within the space. The same lighting design scheme is typically replicated within most institutional spaces, as the most effective solution concerning flexible use, cost, and maintaining the light. Focusing on practical solutions tends to emphasise light quantity rather than quality. This approach is widespread within design of electrical lighting schemes for institutions, and has probably given rise to the general understanding of how an 'institutional' environment looks. The current study, however, questions this approach to planning the general lighting for healthcare units, and recommends taking the specific activities in healthcare units more into account – not least due to the needs of a space's occupants, as they typically will be vulnerable individuals, hospitalized or enrolled in a course of treatment during which they are in need of caring attention. The empathic approach can be a means to achieving aspects of social sustainability, including safety and security, sense of place, social interaction, social equity and social participation [18].

In a Scandinavian context, the uniformly lit environment is precisely the opposite of how we traditionally perceive and understand a comfortable, safe, cosy and 'homey' lighting environment [19]. The institutional lighting scheme is uniform, universal and without hierarchy, and the light is not directly linked to the function of the space. In contrast, a 'homey' lighting design scheme is made for specific purposes, with the light fixtures positioned in specific areas at specific heights [20]. Consequently, this generates another distinctive feature of homey lighting: a light hierarchy indicating where the different functions are positioned. These 'isles of light' guide the occupants of the space by pointing out the various functions. Such distinct variations in light can, though, create high levels of contrast, which can lead to visibility problems. Nevertheless, the varied light levels are often intuitively adjusted by the user and are therefore perceived as comfortable, creating a cosy atmosphere in which people typically feel safe and at home. Furthermore, light hierarchies affect the visual environment through the qualities of light: colour rendering, directionality of light and the form-enhancing shadow pattern that makes people and objects stand out more clearly [21, 22]. These findings from prior research are further supported by the results presented above, which show a clear favourite among the respondents, as the majority prefer the low-positioned, non-uniform light for this specific healthcare setting.

While the uniform lighting of many healthcare environments supports functionality and efficiency, non-uniform and intimacy-tailored lighting supports the semi-privacy and intimacy needed for facilitating therapeutic group conversations. Such non-uniform lighting contributes to a focused and shared space where the most important aspects, not only faces and expressions, but also gestural body language, are visible. Not only does the isles of light establish the table as a shared area of common focus, but also the boundary of the non-uniform lit space creates opportunities to participate more or less in the conversation by moving into or out of the isles of light.

6. Context and limitations

This study has only examined Scandinavian settings, and we acknowledge that culture might be a reason for the preferences participants express. The same survey in another cultural setting might lead to different results due to inherent social preferences [23, 24, 25]. Likewise, climate, including the prevalence of low solar altitude, might influence on respondents' preferences, as the relation to daylight can be reflected in electrical lighting design [26]. In addition, electrical lighting includes the light source, and the specifications of the light sources have varied in the two scenarios; consequently, our survey results may include bias, and in future studies the specifications of the light sources must be more uniform.

Furthermore, the low-positioned light follows a non-uniform light distribution. The distribution of light is one of the variables of light, perhaps an overlooked variable that merits further attention [27]. Often the intention is to meet requirements where the light must be as even as possible, whereas this study points towards the preference of a non-uniform light distribution.

Moreover, the survey method could also be discussed as a limitation: we likely would have gained improved understanding of the participants' experience regarding the lighting situation through interviews and participatory observations. However, due to the therapeutic trajectory and the patients' individual recovery processes, the more qualitative approaches were omitted after dialogue with the responsible healthcare professionals. Furthermore, questions related to emotions were also avoided, so as to not spark unwanted reactions among the patient participants. Value-laden spatial characteristics, such as 'safe', 'relaxing', 'suitable', 'nice', 'beautiful', 'comfortable', 'homey' and 'cosy', focus on the environment and social relations, more than on impacts to the individual.

In a broader historical perspective, electrical light is a rather new invention: in particular, the development of the incandescent light bulb and its success in market penetration changed the interior lighting environment. In the Danish context, the architect and designer Poul Henningsen profoundly investigated what was, at the time, brand-new technology, and designed the famous PH luminaire. This was specifically designed in relationship to the emittance of light from an incandescent bulb.

Henningsen wrote intensively about his lighting design concepts, both pedagogically, so that lay-people could understand his points, and more scientifically to confront engineers with knowledge and thoughts on electrical light. In a series of articles published from 1926–1930, he introduces the new lighting technology to the public, and shares his thoughts on the qualities of light and which aspects merit awareness [28], including the position of the luminaire. One of the guidelines he offers is to position light low, both to avoid glare and to support the functionality of the light by directing it onto the subject that needs to be seen [29].

From a more universal perspective, the preference within this study for low-positioned light could also be related to the mythical narrative about how architecture emerges. The low positions of the lamps generate a light distribution type that may be related to the far older central position of the fireplace, which the ancient Roman architect Vitruvius describes in his discussion of the primitive hut and the rise of formal architecture [30]. Gottfried Semper describes a similar progression [31]. Both relate the central fireplace to the feeling of being safe and secure, and to feeling part of a social gathering. Both refer to low-positioned light as offering an atmosphere universally interpreted as positive.

7. Conclusion

This study shows that different light distributions can fundamentally change perceptions of a room. It documents how respondents clearly distinguish between high- and the low-positioned lighting within a specific therapeutic room at a healthcare institution. The respondents characterize and rate the atmosphere of two different lighting design scenarios as being significantly different. The high-positioned uniform light is mostly characterized in negative terms, specifically as 'institutional', by most respondents. In contrast, the low-positioned, non-uniform light is unanimously described in positive terms, including 'cosy', 'homey' and 'comfortable'.

The project is a pilot project investigating how electrical lighting may affect vulnerable people, and how it eventually can support them. In this paper, the results of the user assessments are discussed in relation to the initial hypothesis that the lighting makes a difference. Professional assessment of the lighting settings provides opportunity for these findings to inform future strategies for supportive empathic lighting design in healthcare environments. Lighting a specific space can be conducted in relationship to meeting standards and supporting flexible spatial use. However, supporting the needs of vulnerable people who are hospitalised and in treatment constitutes additional objectives that require other lighting design strategies. Being in a space is more than simply being able to see the area; it also concerns perception of the space and the ability to feel comfortable and secure.

In the space this case study has examined, the respondents participate in sessions in which they are taught by nurses and physicians how to handle challenges related to their diseases. This also means the respondents are examples of vulnerable patients in difficult situations who are encountering a healthcare institution, thereby being exposed to a specific physical environment that includes a specific atmosphere created by the lighting design. Precisely this type of meeting, between vulnerable people and a specific lighting design, has been the core interest of this investigation.

The two lighting design scenarios represent two archetypal lighting schemes, although they also reflect the traditional approaches to lighting design for healthcare institutions and more private environments. A coherent preference for lighting design in a specific culture might reflect relationships to technical developments in the region, climate conditions, or religious traditions and ceremonies. As mentioned above, this experiment has only been executed in a Scandinavian context; cultural preferences might differ if repeated in another context. Furthermore, the survey is a pilot project which has included few respondents; to strengthen statistical reliability of the evidence, future research design should include higher numbers of respondents.

Overall, the distinct interpretations of the respondents, with their apparent preference for the low-positioned non-uniform light, suggests that re-evaluating the visual environment at healthcare institutions could be worthwhile, to focus on creating empathic light design schemes for improved support of vulnerable patients. Such action could be especially worthwhile as the results also document an 85% reduction in energy consumption when using the low-positioned, non-uniform lighting, compared to the high-positioned lighting. Future planning for healthcare lighting design may take these results into account, to incorporate lower light levels and lower luminaire positioning to improve architectural planning of sustainable, supportive, and empathic lighting.

Acknowledgements
The project has been funded by Fonden af 20. December.

References

[1] Volf C Light Air and Natural Surroundings - Different Hospital Typologies *Aarch 17 2017 Copenhagen BUILD* pp 203 –226
[2] Thygesen LM 2023 *Tove Ditlevsen var min mor. En familiehistorie om skam, skæbne og tilgivelse* (Copenhagen: Forlaget 28B ApS) pp 216-218 The quotes are translated by the authors
[3] Nielsen S L 2022 The midwifery feel of life in S Sumartojo Ed *Lighting design in shared public spaces* (New York: Routledge) pp 65–82
[4] Nielsen S L Bille M and Barfoed A B 2020 Illuminating bodily presence in midwifery practice *Emotion, Space and Society* **37** 100720
[5] van Mil I W, Iversen A., Osterhaus W, Garcia Alvarez M, Petersen S and Jeong C-H 2018 Light at eye level is a means to create energy savings and space for learning, focus and concentration *Research Report for the Danish Electricity Research Fund (ElForsk)* Elforsk
[6] Kim D H and Mansfield K 2021 Creating positive atmosphere and emotion in an office-like environment: A methodology for the lit environment *Building and Environment* **194** 107686
[7] Wansink B and Van Ittersum K 2012 Fast food restaurant lighting and music can reduce calorie intake and increase satisfaction *Psychological reports* **111** 1 pp 228–232
[8] Groat LN & Wang D 2002 *Architectural Research Methods* (New York: John Wiley & Sons Inc.) p xi
[9] Pink S and Leder Mackley K 2016 Moving, making and atmosphere: Routines of home as sites for mundane improvisation *Mobilities* **11** 2 pp 171–187
[10] Øien T B 2022 A study of environmental factors in low vision rehabilitation *Frontiers in Rehabilitation Sciences* **3** 829903
[11] Mattelmäki T Vaajakallio K and Koskinen I 2014 What Happened to Empathic Design? *Design Issues* **30** 1 pp 67–77
[12] Stidsen L 2013 *Light Atmospheres in Hospitals Wards* PhD thesis (Aalborg: River Publisher)
[13] Volf C 2011 Light and the Aesthetics of Perception *The Nordic Journal of Aesthetics* vol. **22** No. 40-41 (tidsskrift.dk)
[14] Boyce P 2014 Editorial: Light distribution – a missing variable *Lighting Resource Technology* 46 617
[15] Ruohonen S 2023 Internal Note on Luminance map evaluation BUILD/AAU Denmark
[16] Allen I E & Seaman CA 2007 Likert scales and data analyses *Quality progress* 40(7) pp 64–65
[17] DS/EN 12464 – 1:2021
[18] Larimian T and Sadeghi A 2021 Measuring urban social sustainability: Scale development and validation *Environment and Planning B: Urban Analytics and City Science* 48(4) pp 621–637
[19] Stidsen L, Kirkegaard P H and Thuesen N 2014 Mapping Danish Lighting Trends. Socioculturel aspects of light analysede in Bo Bedre *Nordisk Arkitekturforskning* **1** pp 117–40
[20] Descottes H & Ramos C E 2011 *Architectural lighting: designing with light and space* (New York: Princeton Architectural Press)
[21] Liljefors A 2000 *Seende och Ljusstråling* (Stockholm: Kungliga Tekniska Högskolan)
[22] Frandsen S 1989 The Scale of Light – a New Concept and its Application *2nd European Conference on Architecture* 4–8 December Paris
[23] Bille M 2017 Ecstatic things: The power of light in shaping Bedouin homes. *Home Culture* **14** pp 25–49
[24] Garnet J 2011 On the cultureal history of Nordic light and lighitng *Nordic Light: Interpretations in Architecture* (eds) Sørensen N and Haug P (Odense: Clausens Grafisk) pp 5–13
[25] Munch AV 2018 Nordic Light? Poul Henningensen and the Myths of Scandinavian Twilight *Back*

IOP Conf. Series: Earth and Environmental Science 1320 (2024) 012031 doi:10.1088/1755-1315/1320/1/012031

to the Future The Futurre in the Past ICDHS 10th Barcelona
[26] Mathiasen N 2015 *Nordisk Lys og dets relation til dagslysåbninger i Nordisk arkitektur* PhD thesis (Copenhagen: Det Kongelige Danske Kunstakademis Skoler for Arkitektur Desing og Konservering)
[27] Boyce P 2014 Editorial: Light distribution – A missing variable *Lighting Resource Technology* **46** 617
[28] Jørstian T & Nielsen P E M 1994 *Tænd. Ph lampens historie* (Copenhagen: Gyldendal)
[29] Christensen E, Frandse S, Jørgensen S and Voltelen M 1974 *PH om lys* (Copenhagen: Forlaget Rhodos)
[30] Vitruvius 1960 *The Ten Books on Architecture* (New York: Dover Publication Inc.)
[31] Semper G 1989 *The Four Elements of Architecture and Other Writings* (New York: Cambridge University Press)

Lighting assessment in low-vision rehabilitation: Implementing N-Lited

Turid Borgestrand Øien*[1] and Jeanette Bremer Nielsen[2]

[1] Aalborg University, Department of the Built Environment, A.C. Meyers Vænge 15, 2450 Copenhagen, Denmark
[2] Center for Specialundervisning, CSU-Slagelse, Nørrevangstorvet 6A, 4200 Slagelse, Denmark

*tuo@build.aau.dk

Abstract. Lighting assessments have been part of low-vision rehabilitation for decades, but more structured and evidence-based approaches have emerged in recent years. In Denmark, a mixed-method semi-structured protocol for lighting assessment was developed and tested as a pilot study from 2017–2019, and it showed promising results, in that it improved the visually impaired participants' quality of life. Innovations are often developed in pilot studies, but implementing new approaches across communities of practice poses difficulties due to both individual and organizational norms. N-Lited is an implementation project that entails the dissemination of knowledge and instructions as well as practical trials, and its protocol has been implemented in seven low-vision centers from 2022–2023. Based on participants' observations, document analysis of the framework, and a midway evaluation that assessed the potential for scalability and diffusion, this paper discusses the learning and change processes in the interpretation and role of the lighting measurements and lighting assessments in this project. Beyond mere instrumental learning, the findings reveal that it is crucial to address how individuals learn how to explore and question the difficulties that reveal mismatches or reconfigurations of values, norms, and assumptions to ensure a successful implementation of the method. The changing role of lighting measurements, from instrumental to more reflective, also illustrates the importance of the practical application and learning-by-doing aspects that are essential for learning how to reflect and "think with the metrics." Furthermore, metrics and tools are an important part of the community of practice's shared language. The uncertainty of navigating between the uniformity of the protocol and the flexibility needed for incorporating the method into individual professionals' practices and the various organizational contexts emphasizes the relevance of the process of learning how to manage methodological innovations in practice.

Keywords: Lighting assessment, low-vision rehabilitation, home environment, practical application, implementation & learning process

1. Introduction

The field of lighting research and design is constantly changing, which can be seen in the technological innovations made in terms of light sources, systems, and infrastructures as well as the changes in markets, consumers' needs, and political agendas. The current technological developments and sustainable transitions enabled by emerging and evolving practices constitute comprehensive learning spaces. Moreover, "scaling-out" or "scaling-up" centers around these transitions and learning processes in which individuals, technologies, and practices are engaged. These processes involve alterations to both cognitive (abstract) and physical (contextual) knowledge as well as the reconfiguration of both practices and socio-technical systems [1]. Furthermore, change and transitions can also be driven by individuals gaining inspiration and learning new information from distant fields, as exemplified by the fields of low-vision rehabilitation (LVR) and lighting. Lighting has been an important aspect of LVR for decades [2], and recent developments in rehabilitation practices have been fundamentally influenced by the methodological approaches used in lighting design [3] and the introduction of lighting as a dynamic parameter in rehabilitation processes [4]. From 2017–2019, the Center for Special Education (CSU) in Slagelse developed and tested a new framework for lighting assessment and intervention in the project entitled "Better Light, Better Living" (BLBL). The results were promising, in that the visually impaired participants' visual performance and quality of life improved significantly [5]. Furthermore, in 2020, the low-vision consultants behind BLBL, in addition to a project manager, initiated a new project entitled "N-Lited," which is targeted toward applying the lighting assessment throughout the low-vision rehabilitation community, currently comprising seven of the 23 low-vision centers in Denmark. This project includes providing courses on lighting assessment and intervention, thereby providing extensive knowledge on lighting parameters that are relevant in this approach. However, despite the focused implementation efforts, scaling and sharing the method across the community of practice has been more complex than first expected.

2. Background

2.1. Lighting in LVR

There have been several approaches to lighting assessments among low-vision consultants in Denmark. In the 1980s, the understanding of (residual) vision changed, in that it was previously understood as a more static condition but came to be seen as a more central and dynamic function that needed training and maintenance [6]. The focus on maintaining individuals' visual function resulted in special aids and lighting being made a niche subject within the field in curricula for diploma programs on low-vision services. The knowledge and approaches to lighting have undergone large changes during these 40+ years, starting from a checklist of lux levels that were suitable for the different impairments and the technical lighting assessments that were used to measure the light distribution of rooms in systematized grid measurements. In a large proportion of the early years, lighting for the visually impaired primarily comprised expensive special aids for which these individuals would have to apply and be granted by national health systems, but they had to be classified as eligible. Furthermore, the curricula included theory on lighting physics and technology. However, as the field of rehabilitation has changed from a medical approach with a focus on diagnostics to a field with a broader bio-psycho-social approach that focuses on impairments as a phenomenon situated not in the individual but in their relation to their social and physical contexts, the curricula have changed. This has been a rather slow transition in the niche field of lighting, but courses from the past five years have included more context-specific assessments that are based on the issues experienced by visually impaired individuals in their everyday lives.

2.2. BLBL

The lighting assessment and intervention developed in the BLBL [5] focused on situated individuals and the importance of including the social and physical contexts in the rehabilitation process. BLBL was tested on 60 visually impaired participants from 2017–2019, situated both in the homes of the participants and the lighting laboratory at the clinic. First, the participants each engaged in a narrative interview in their homes, during which they described their everyday lives and the issues they were experiencing concerning the lighting in the different places in their home environments. Using the information attained from these interviews, as well as an occupational performance score, up to three activities were identified as appropriate for assessing lighting. These assessments comprised recording the location and position in which the activity occurred, with the setting being documented in terms of photographs and lux measurements. Moreover, in one of the activities, the participants' visual performance was tested through assessments in the form of sorting colors and tracing lines. These tests were then repeated after the intervention to measure its effect on the visually impaired participants. In addition, the participants completed a questionnaire on visual function before the intervention. Second, a lighting intervention was conducted in CSU's lighting laboratory, in which the participants were shown and tested alternative lighting solutions. The participants' preferences were determined, after which the recommendations were given to the participants for implementation in their homes. Third, a follow-up interview, either in-person or over the phone, was conducted with each participant after two to three months, and the questionnaire on visual function was repeated. The results of this pilot study showed significant improvement in the participants' visual function in several settings as well as an improvement in the performance measures of the chosen activities [5].

2.3. N-Lited

Due to the BLBL's success in significantly improving the participants' quality of life, an implementation project, entitled N-Lited, was launched in 2022 with the objective to implement and test the scalability of the BLBL's approach in seven of the 23 low-vision centers in Denmark. To ensure easy implementation of the method, the BLBL intervention underwent some adjustments: First, the N-Lited project did not include the BLBL's visual performance tests and questionnaires that were used to measure the effect of the intervention on the visually impaired participants; and second, to make the methods more accessible to immobile participants, N-Lited included a "mobile lighting laboratory," comprising a suitcase that was equipped with lamps and luminaires for smaller interventions that were more suitable for one-on-one testing in the participants' homes. The facilitation and education of the method was designed as a course program involving teaching and training the protocol in each of the organizational contexts. Moreover, an external midway evaluation of the course program was conducted to adjust and improve the methods, educational materials, and facilitator-learner dialogues.

This paper investigates the results from the N-Lited project, with a specific focus on the first steps of facilitation, learning, and changing processes within the implementation process as well as how the method was disseminated and received, including the resistance to and challenges in the social and organizational settings identified by individuals. Finally, the aspects relevant to the field of lighting are also discussed more broadly.

3. Theory

Change and transformation processes can be analyzed by investigating a practice as an entity as well as determining the aspects at stake in and between the three elements constituting a practice: material, competences, and meaning [7]. Different practices complement and compete with one another, re-producing or reconfiguring these elements. For example, driving and maintenance are two complementing practices related to cars, but they comprise different materials, competences, and meanings. Likewise, instruments and tools used in professional practices are not neutral aspects as they shape the character of human-world relations through their use and interpretation. These instruments often produce quantitative measures that represent a physical phenomenon, and they are analyzed in terms of statistical estimations of averages, coverages, or accuracy. The development of running practices is currently being impacted by technologies that are developed to monitor performance, impact and be impacted by the meaning of sharing information and motivating others, thereby calling for new competences. These technologies, motivations, and competences that are involved in the use of these new applications can be monitored, the results of which can be shared with other practices, thereby further facilitating change [7]. For the skilled practitioner, the tools used can be the instruments of action, allowing individuals to discern and respond to the ongoing activity by using their perceptual sensitivity [8]. For example, a dentist's probe, serving as an extension of their hands and eyes, is used in examinations to assess the condition and texture of their patients' teeth and mouths [9]. Technical measurements should be easy to conduct, reproducible, and enable these statistical analyses, but to work with measures *qualitatively*, meaning needs to be added, which is dependent on the specific context of interest [10].

To identify or establish this context-specific meaning in professional practices, reflexivity is key, which is defined as reflection as a process of creating meaning and conceptualizing knowledge from past experiences [11;12]. This includes the tacit knowledge embodied in individuals, encultured in organizations, encoded in signs and symbols, and embedded in systemic routines and practices as well as how different types of knowledge are navigated and coordinated [13]. Valuable information can be gained from prototyping or pilot testing. Furthermore, while scientific researchers aim for generalizability, reflective practitioners focus on situation-specific insights in which prototyping can be valuable for the learning process: "The utility of the prototype lies in its ability to generate explanation and experimentation in a new situation" [14]. There are several unknowns in the in-depth investigation of the implementation processes for new technologies in the context of organizational learning and practical knowledge. First, learning processes in professional practices are social. Indeed, as colleagues often inform and educate each other, different generational versions of the same curricula can be used within the same community or even in the same organization. The individuals' different levels of expertise can be accommodated in the framework of participation [15], which entails acknowledging each person's learning style preferences [16] and, beyond the provision of information, providing tools for managing collaboration and learning. Facilitating group dynamics in which different levels of expertise and engagement are represented could necessitate the facilitator having to let go of some of the control to let individuals participate according to their levels as well as take control and responsibility [17]. The success of this co-creation of practical knowledge largely depends on the group's social interactions and negotiations, and it requires the facilitator to be honest with the individuals on the team about the agenda of the course to help the team examine how to approach these collaborative processes [14]. Second, the aforementioned instruments and tools as well as the language and narratives in the practices are relevant to the analysis as these shared technologies engage, focus, and coordinate the social structure's shared memory. Third, paying attention to the implementation process could also provide insights into the transformations that were enacted. The learning process impacts both apprentices and masters due to the individual and shared actions as well as the problem-solving in which they engage. This social community of practice both develops professional identity and reproduces and transforms the very community: "Learning, transformation, and change are always connected, and the status quo requires explanations as well as change" [15].

4. Methodology

To analyze the initial findings of N-Lited, an ethnographic case study was conducted, with the participants being observed and a document analysis of the project's work being conducted. The education materials, particularly the N-Lited manual [18], schemes, and PowerPoints, as well as more project-related documents, such as newsletters, the midway evaluation report [19], and the project description, were investigated. These documents contained text, diagrams, and photographs, being analyzed as inscriptions of "social facts," in that they were produced, shared, and used in socially organized ways [20]. This analysis was conducted to uncover meaning, develop understanding, and determine insights relevant to the research problem [21] by investigating how approaches to lighting were practiced among the participants and in the framework provided. A document analysis entails tracking changes and developments in different versions of the same document [22], such as the different versions of the narrative interview guide developed throughout the project. The changes were identified as the substantive developments that occurred in the project [23] by examining a range of its newsletters.

To triangulate and supplement the project material, reports from a previous mapping of the community of practice [24] and a case study on BLBL [6] were employed as a representation of the status quo in the area, with the project manager's reflections on the low-vision consultants and the course activities they observed being used as supplementary material. Furthermore, both authors have participated in status meetings with the two low-vision consultants who were facilitating the course during this period, during which anomalies and issues were discussed in shared reflections.

5. Findings

5.1. Designing the course

This section discusses the material, demonstration, and testing of the method's first phase, the lighting assessment, and the interview. The first phase of the project entailed making the practice knowledge from BLBL explicit. While BLBL had been described in a more scientific manner in the researcher's articles and reports, this material did not describe the *hows* and *whys* in a way that supported the more tacit and practical knowledge, nor was it in Danish that could be easily interpreted by the practice. The course's curriculum, the presentation materials, and the first version of a manual that was later introduced were all used to disseminate the content in a way that it could be operationalized by the participants. The 40-page manual was drafted during the first months of the project, and it was based on the knowledge gained from the BLBL project and verified in dialogues with the participants. The manual detailed the process of the lighting assessment step by step, which was followed by an introduction to the concept of human lighting [18]. The content was the point of departure for the educational courses, where the methods of lighting measurements were also taught to and tested by the participants.

The first step of the lighting assessment included a narrative interview guide that was used to inform the dialogue with the participants. The interview guide was developed during CSU's practical investigations of the BLBL model and was aimed at facilitating a conversation with the participants that allowed them to describe their experiences and challenges with everyday activities. The guide simultaneously worked as a motivation for making changes. A circular diagram was used to focus on the activities and situations in which the lighting conditions enabled or hindered the participants, and it was also employed to help the participants articulate their experiences and emotional perceptions of these situations in a non-linear manner. The quantitative assessments of the light were based on one of the problems tackled by the participants in the activity in which the lux, color rendering index, and Kelvin were measured using a luxmeter and spectrometer in the position in which the activity occurred. For example, in seated activities, the luxmeter was placed horizontally and pointed upwards on the surface of interest, such as on the dining table where the plate would be, to measure the level of light on this surface, while the light levels in the room would be measured at a height of 85 cm. In situations where the lighting should support a social setting, measurements were taken by aligning the luxmeter with the forehead of the other person in the interaction. The documentation consisted of two photograph types: a) an overview photograph of the scene from chest height at a wide angle if possible, and b) an activity photograph taken from the position of the given activity and at eye level to document the surroundings. The education materials included a template for a presentation application that was to be used on a tablet and on which the photographs could be directly combined with other measurements.

The parameters of human lighting included in the manual were determined according to Christiansen [25], who was a Danish lighting designer who described an approach to lighting that was practiced at the Lighting Laboratory at the Academy of Fine Arts in Copenhagen from the 1950s [6]. In this approach, visual comfort [26] and the knowledge of seeing and assessing light go beyond physics and measurements by involving biology, psychophysics, and perception [27]. The 20 parameters chosen for this project were sorted into three subcategories: a) primary parameters, b) vision discomfort, and c) perception of light impressions [19]. The primary parameters included aspects of vision as an adaptation as well as aspects related to the physics or phenomenon of light (e.g., luminance, color, rendering, spectrum characteristics) and relational aspects of light (e.g., form, shadow, texture, contrasts, direction, and illuminance). The second category of vision comfort comprised glare (whether vision-impairing, uncomfortable, or indirect), flicker, noise, radiant heat, decomposition of material, and pathogenic, while the third subcategory of perception of light impression included the relation or fit of the luminaire and the space or place, spatial trajectory, colors, luminance variation, surfaces and coloration, and perceptual relationships between light, space, and behavior.

The courses included educational materials that described the method and provided instructions for the practical use of the different elements in terms of how to conduct a narrative interview or measure light in the analyzed activities of daily living. The project also involved facilitating the process of determining common incentives by and for the group of low-vision consultants, such as the value of following a protocol for structure and quality assurance as well as the new approach of providing knowledge on and recommendations to the visually impaired according to individual participants' perceptions and experiences with the quality of light in their everyday settings and activities and not just quantitative measures. The narrative interviews were chosen as a methodological tool as they allowed for this investigation.

The 14 low-vision consultants were educated over three days of training, dialogues, and practical lessons. The first two days comprised the participants being introduced to and practicing the narrative interview, thereby allowing for the incorporation of the visually impaired's perspectives on their daily activities and quality of life in relation to light. Moreover, the process for measuring light in specific activities was detailed through examples of where and how to measure it and why the measurements were necessary. Moreover, different lighting solutions to specific cases were presented. The vision consultants concluded the two days with practical training of visually impaired individuals in local nursing homes in the form of a lighting intervention after they learned about the N-Lited method. The third day of training was then conducted after a month of individual practice of the method in their own context, during which the consultants could address any doubts they had concerning the method. Furthermore, alternative procedures were discussed regarding the adaptation to the local setting, considering whether and to what extent the alternatives would still represent the N-Lited method. Throughout the project period, the participants could converse with CSU experts online, which was a service used every 14 days on average.

5.2. The midway evaluation

The objective of the midway evaluation was to support a knowledge foundation that enabled meaningful adaptation of the project throughout its duration and to gain an understanding of the method's dissemination [19]. Six of the centers considered the demand for the service to be indirect when qualifying it, while one of the small centers did not see a need for lighting assessments to be implemented, due to advanced age, lack of action incentives, and energy-saving measures. The participants' personal and professional motivation to adhere to the methodology was evident as they found the approach time-consuming. Furthermore, the project had to be accommodated into the participants' daily schedules, and its nature was unfamiliar to some of them.

At the time of the evaluation that was conducted in 2023, only five of the seven centers had implemented the method in a total of 26 cases, at an average of five cases per center. However, not every center had practical experience. The interviews revealed that the recovery-based approach had gained traction among the centers, and the focus on the visually impaired's ability to act on the recommendations was especially valuable in the approach. Acknowledging that this was related to motivation, help from the social setting, and economic support increased the success of the intervention but also made the lighting assessment and intervention much more complex than previous approaches.

The approach's implementation fidelity, denoting the degree to which an intervention is delivered as intended [19], was determined in relation to the framework and conditions, in which the management and consultants had the correct attitude toward and understanding of the project. Arrangements were put in place that allowed some of the consultants to conduct their initial lighting assessments in pairs and discuss and support each other through their shared experiences. The centers' management structures were allowed to support the consultants in the implementation of the approach in ways that worked for them and their own practices as long as they did not consume inappropriate time or resources from other citizen-oriented core tasks. Furthermore, the centers' management structures acknowledged their overall purpose in contributing to the development of the field of lighting within rehabilitation in terms of a common national practice for lighting assessments. However, the differences in financing between the low-vision centers impacted the organizations' flexibility regarding their core tasks.

5.3. *Learning space*

The learning space within the project was seen as a safe space characterized by an equal and respectful atmosphere in which dialogues between both the consultants and the facilitators as well as among the consultants allowed learning to occur. The practical and experience-based dialogues on the testing were particularly important for both determining the possibilities of luminaires and the demonstrations as well as allowing case experiences to be shared.

The turnover of professional knowledge was ensured when examples of applying the knowledge in practice were provided. This was followed by a proposal that included the facilitators' practical or participatory observations from conducting a real lighting assessment with a citizen. A range of questions arose when the lighting assessment was to be tested in their own practices that had different settings, activities, and impairments, but the demonstration bridged the gap between the facilitators' practical steps and theirs.

Establishing this reflective space was described as important for the implementation project, but the time needed to facilitate these kinds of processes depended on the consultants' existing knowledge as well as the values and norms of the participants, organizations, and communities [14]. The participation framework that had a focus beyond providing information by considering the participants who were testing the protocol in their own settings and sharing their experiences, in addition to the management tools used in the collaboration and for learning, were both well-received. Moreover, the feedback from the midway evaluation indicated that there was a need for a larger focus on practical and experience-based instructions and manuals as well as a forum for continuing the shared reflections on both the contextual and abstract knowledge required [1]. Learning processes, such as the ones used in N-Lited, do not necessarily benefit from questioning *all* its elements. For systems that are working, contextual learning should be prioritized to sustain and continue the instrumental use of the tools, with questions about the meaning [10] of systems and processes that do *not* work within the current context being prioritized.

The tools for managing and navigating these different modes of learning could perhaps have been facilitated more closely as the time needed for learning, understanding, and executing the protocol was determined by each individual. This was also the case in terms of the timing for when these individuals felt ready to move on to the more abstract contributions of reflecting on how well the protocol worked in their specific contexts. However, as this was also a pilot study for the project group, the mere recognition of the difficulties in this progress was valuable knowledge. Furthermore, the first iterations were made using rules of thumb, recognizing the different transition steps between individual and organizational learning, and acknowledging the importance of aligning the participants' and facilitators' expectations at the beginning of the course.

5.4. *Exploring the role of light measurement*

The feedback from the midway evaluation showed that the low-vision consultants were challenged by the amount of information to remember, including the new way of "being present" in the lighting assessment. The close collaborative process of sharing the assessment with the visually impaired could not be learned fully in advance as this was a social process of assessing these individuals' issues in their everyday lives, so the consultants needed to trust the process and know that they could get the support they needed through the tools of inquiry provided. The midway evaluation focused heavily on recovery and rehabilitation and less on light, and we do not know much more about the way the lighting measurements were conducted. This challenge was perhaps also exacerbated by the different elements of the lighting assessment that represented both the quantitative and technical methods as well as the qualitative and social methods. The lux measurements had been a theme discussed in the development of BLBL as the project group was concerned that the measurements would be simplifications of the representation of light. In the transition from the home environment to the lighting laboratory, the lux measurements from the original BLBL were used as a baseline and a point of departure for testing alternative lighting. For the low-vision consultants who were more experienced with lighting assessments and interventions, the role of lux measurements was less important in their core lighting assessments as they employed more qualitative descriptions of the environmental settings. Thus, the luxmeter worked more as an instrument for action and assessment, enabling a reflective dialogue via the consultants' perceptual sensitivity [8]. However, in N-Lited, these instruments and measures comprised an important aspect of the shared language [15] as signs or symbols that represented the physics of lighting and the conversations about the light in the narrative interview were closely intertwined with the specific context. If the participants were able to build the skills they needed to ensure effective perceptual sensitivity, the lux measurement essentially turned into a tool that provided reflections, which would be the basis for relating the quantitative and qualitative data. Therefore, the measurements resulted in facts; the stories provided meaning [10]. Across the organizational contexts, lux measurements were still required by some municipalities as documentation in funding applications, which was why the N-Lited approach included this wording in its methodology, in which the rules of thumb were applied in a way that would not ignore their uniqueness, ensuring robustness for each case that was different [12]. Thus, technologies, language, and narratives were used to engage and focus the participants while also coordinating the shared knowledge of the navigation between the instrumental correction of errors in the technical rationales and the deeper learning and social change processes in rehabilitation.

5.5. Between novices and experts in the community of practice

All communities of practice span from the novice to the expert, but the combination of rehabilitation and lighting adds to the complexity in the spectrum of knowledge. The two low-vision consultants facilitating the course had been exploring lighting within their field for the preceding seven years, from the basic curricula to courses in lighting design as well as their own projects and research collaborations. Moreover, their background as occupational therapists meant that they combined the performance measures, their motivation, and the value of scientific knowledge that was then applied to their practices, even though this did not benefit the visually impaired directly. The consultants' diverse backgrounds, which comprised different levels of expertise in lighting, rehabilitation, or low-vision services, resulted in the need for a safe learning space that allowed open, respectful, and equal dialogue. The midway evaluation indicated that this was achieved in N-Lited and that valuable lessons and new knowledge were created by the consultants, outside of the facilitators' influence.

The missing link between the lighting measurements and the outcomes for the visually impaired participants was an inhibiting barrier for some of the participants: "The lighting assessment has no real value to the visually impaired participants before the recommendations are implemented" [19]. The evaluation proposed that specifying how this project would directly contribute to improving both the consultants' and visually impaired participants' lives as well as providing the empirical documentation needed for the research purpose would counteract this barrier.

In the evaluation, a participant stated that it would be beneficial if the participants had the same level of knowledge due to the effectiveness of the learning situation. Time spent teaching them basic knowledge felt like a waste of time for the participants who already had the requisite knowledge and who were eager to test the method practically. However, as there are different types of knowledge across different specialties, a group of participants that was too homogenous would have perhaps resulted in fewer reflections and less knowledge creation across the team. Furthermore, the consultants acknowledged the need to promote the method among their colleagues who did not participate in N-Lited and to improve the basic knowledge of lighting provided in their basic education.

The implementation fidelity was also assessed in relation to the resources available, and it was found that it was not as much the professionals' engagement and competence but the barriers in the framework that impacted the implementation fidelity in some contexts. The current call for systematized approaches and evidence-based practices in rehabilitation stems from a New Public Management focus on performance measures and management that has been implemented in most public health services. In this political landscape, there are also embedded objectives that call for the opposite, in that they require more systematics and more citizen-oriented approaches [24]. Thus, this is a complex setting, but the open collaborative nature of the project recognized the need for a forum in which joint discussions and sharing can be facilitated, where professionals can place themselves on the spectrum between 1:1 and fragmented implementation, thereby recognizing and navigating their positions in this changing landscape [14]. For most of the consultants, N-Lited constituted both an educational experience and the first steps in organizational change toward a more recovery-based LVR approach by moving from technical rationales or medical models toward more holistic and bio-psychosocial models.

6. Discussion

6.1. The conflicting yet related objectives of N-Lited

BLBL comprises a complex and multi-dimensional understanding of light, and it operationalizes different types of knowledge of light in terms of supporting the rehabilitation process, from a quantitative and measured approach to the tacit knowledge of light that is embedded and embodied in both the visually impaired participants' everyday lives [28] and the consultants' practical knowledge [3]. BLBL represents a transition in lighting approaches within the field of LVR that has moved from understanding light from a technical physics standpoint to explaining it as a sensory and social phenomenon of great importance for perception and participation as well as determining individuals' capabilities within their everyday environments. With the pivot to a recovery-oriented rehabilitation approach, individuals' participation and their knowledge became the center of attention in the rehabilitation processes. Furthermore, examining the materials, competences, and meaning, the elements of practice [7], and the trajectory of change initiated by N-Lited has taught us a lot.

Implementation fidelity was one of the aspects assessed in the midway evaluation, which was measured by specifying the objectives and measures of the intervention. The double objective of both implementing an existing protocol and encouraging critical reflection was incorporated to enable ownership and engagement of the consultants, thereby allowing local adjustments to be made to the approach. However, the findings show that the two objectives also caused confusion, with the protocol, norms, and perhaps the facilitator-learner relationship being questioned. The report from the evaluation discussed the local adaptation in a spectrum between 1:1 execution to a more fragmented and random use of the protocol, which indicates that the evaluation focused on the implementation of the existing protocol. All the consultants followed the protocol quite strictly, at least in the initial phases, and after encouragement from the facilitators, they also reflected on how their approach would change after the project and how these changes would align with their specific contexts [10]. The project group was aware of the organizational contexts' role in the success of the implementation and that the different professional backgrounds could impact how the protocol (and its information from the fields of psychology, occupational therapy, and lighting) aligned with their colleagues' practices. Consequently, after discussing the paradox of the two objectives and how much the protocol could be adjusted and still be considered "the protocol," some rules of thumb were determined for the implementation. Most importantly, the course participants should become intimately familiar with the method before they are allowed to adjust it too much. As a minimum, the entire systematized approach should follow two tenets: a) embrace the recovery-based focus, acknowledge the visually impaired individuals' knowledge as they are the experts in their lives and environments, and support and motivate their change processes; and b) include the alignment of lighting measurements.

These rules of thumb resulted in further discussions on what the minimum implementation measures would be for participants to still claim that they were using this method. For example, was it the narrative interview, the scoring, or the lighting assessment, and should it also include an actual lighting intervention? What if it was more a conversation than an actual assessment of the specific physical home environment? As the participating organizations comprised different organizational levels in the project [13], these adjustments and the subsequent learning processes aligned with different aspects of N-Lited. Therefore, the implementation included: a) a more systematized method (i.e., following a protocol); b) a new recovery-based approach to lighting (i.e., the role of light is seen within the social and physical context of the visually impaired and described in the narrative interview); c) a more collaborative and co-designed approach to recommendations and solutions (i.e., the user was seen as an expert); and d) a learning space and a learning process that had the potential to cause changes in the individual (both the consultant and, hopefully, the visually impaired participant), the organization, and across the community of practice. This change process initiated a reconfiguration of the practice, in which both the required abstract and contextual knowledge were changed [1].

6.2. *Findings relevant to the field of lighting*

Beyond the technical protocol, N-Lited involved a range of organizational shared and individual constructive learning iterations that allowed for the development of new competences and meaning [7]. These elements are all related. Implementing initiatives in a community of practice or sharing and operationalizing knowledge leads to learning processes for individuals and, potentially, their organizations. Individuals represent different levels of knowledge in specific fields. Practical knowledge concerns both explicit and instrumental theory-in-use understanding, procedural knowledge as rules of thumb, and more tacit knowledge embodied in the routines of the practitioner or in the tools and schemas employed in their work. For many years, lighting in LVR has been a mono-discipline representing a technical rationality [12] or a niche for instrumental and technical problem-solving. However, the context of rehabilitation has resulted in new problem-solving techniques needing to be developed as this context ensures that lighting is no longer a solely technical matter but, by and large, a socio-technical issue. By combining new and existing knowledge in reflexive manners [11;12] the context-specific meaning is incorporated [10]. The socio-technical approach formulated within rehabilitation practice in this case is also relevant to other interdisciplinary efforts, such as light and user quality in architecture, applications to user experiences, health and wellbeing, the United Nations Sustainability Development Goals, and the "Leave No One Behind" policy.

The low-vision consultants on each side of N-Lited, the facilitators and participants, all engaged in LVR and the use of lighting in this context. However, this course's objective centered around reconfiguring the competences and meanings of the role of lighting. By the end of the project, some of the participants were well on their way to using the method as an instrument of action [8], while others had found alternative workarounds or implemented certain elements of the method. The quantitative and qualitative approaches to lighting used herein were rooted within the field of lighting design. The facilitators of N-Lited shared a lot of their competence and understanding of meanings in the more qualitative approaches, while most of their colleagues embodied the more quantitative approaches to light, which was also the case for practitioners from other areas within lighting research. We considered bringing the case of N-Lited to the field of lighting research and design, thereby representing a cross-fertilization between otherwise unrelated practices [7] and acting as an example of two practices that share interests in terms of the use of lighting as a professional asset and employing both quantitative and qualitative approaches.

6.3. *Limitations and further research*

As an ethnographic case study of the implementation of new methods for lighting assessments in rehabilitation practices, a limitation of the current study is the lack of observations of the participants conducting and reflecting on their performance of the light assessments, including the role of measurements. The documents in a document analysis represent specific points of time within a dynamic process, but do not provide sufficient detail for exploring the role of these artifacts and instruments [22]. In fact, the mid-way evaluation hardly touched upon lighting at all, and this also illustrates the limitation of the documents, that there is often a thematic focus when knowledge is made explicit. In the evaluation, the rehabilitation and recovery processes of the visually impaired was the main scope, and the use of lighting merely a means for reaching this outcome. However, the participants' call for more one-on-one hands-on trial and error and for sharing case-experiences indicate that they have grasped a much more important lesson on light: understanding the embodied knowledge activated through action, the reflection on what the light 'does' for the other, and what role the measurement plays in this approach. We believe that such observations, across the group of participants could provide valuable insights on how the measurement is individually used within the specific setting and dialogue, its more general role as an inscription representing the current setting, and finally as part of a shared accumulated database of empirical project or research data.

7. Conclusion

This paper investigated the initial findings of the ongoing project, N-Lited, in which a new lighting assessment has been implemented in seven low-vision centers practicing LVR in Denmark. The midway evaluation showed that the participants' approaches to the lighting assessment involved both instrumental and more social inquiries into the physical and social contexts of visually impaired individuals. Beyond the socio-technical protocol and methodology, the project has allowed participants, their organizations, and their communities of practice to gain new knowledge. The facilitators have succeeded in creating a safe learning space across different competences and backgrounds as well as across different levels of background knowledge regarding lighting assessments. The reflectivity established to ensure ownership and facilitate learning among the consultants has enabled the participants to adjust their approaches to their organizational and local contexts. Thus, the reflexivity has also resulted in uncertainty in how to navigate between abstract and contextual learning as well as when to trust instrumental and step-by-step instructions and when to question and reflect on their effects. These findings have provided feedback on the project that has enhanced its determination of how to manage these methodological considerations while still allowing for uniformity and flexibility in future implementation processes. We are hopeful that these discussions on the competences and meanings attributed to lighting as well as the acts of measuring, assessing, and challenging existing practices could enable further cross-fertilization between practices that employ quantitative and qualitative approaches to lighting.

Acknowledgements
The project has been funded by the Velux Foundation

References

[1] Öztekin E E and Gaziulusoy A İ 2019 Designing transitions bottom-up: The agency of design in formation and proliferation of niche practices *Des. J.* **22** pp. 1659–1674

[2] Strandsbjerg N 1998 *Refsnæsskolen: Institut for Blinde og Svagsynede Børn og Unge i Danmark: 100 års jubilæum* (Refsnæs: Refsnæsskolen Trykkeri)

[3] Øien T B 2023 Methodological considerations in collaborative processes: A case of ethnographic action research *Int. J. Man. Proj.* **16** pp. 165–184.

[4] Øien T B 2022 A study of environmental factors in low vision rehabilitation *Front. Rehab. Sci.* **3** 829903

[5] Øien T B Jacobsen A M Tødten S T Russotti T Ø Smaakjaer P and Rasmussen R S 2023 Impact of lighting assessment and optimization on participation and quality of life in individuals with vision loss *Occup. Ther. Health Care* **37** pp. 426–443

[6] Øien T B 2022 *Belysning i synsrehabiliteringen* (Copenhagen: BUILD-Rapport 2022:04)

[7] Shove E Panzar M and Watson M *2012 The dynamics of social practice: Everyday life and how it changes* (London: SAGE Publications)

[8] Ingold T 2000 *The perception of the environment: Essays on livelihood, dwelling and skill* (London: Routledge)

[9] Verbeek P P 2005 *What things do: Philosophical reflections on technology, agency, and design* (University Park: Penn State Press)

[10] Norman D 2023 *Design for a better world: Meaningful, sustainable, humanity centered* (Cambridge: MIT Press)

[11] Schön D 1987 *Educating the reflective practitioner: Towards a new design for teaching and learning in the professions* (Hoboken: Wiley)

[12] Schön D 1983 *The reflective practitioner: How professionals think in action* (London: Temple Smith)

[13] Newell S Robertson M Scarbrough H and Swan J 2002 *Managing knowledge work* (Palgrave)
[14] Savin-Baden M 2004 Achieving reflexivity: Moving researchers from analysis to interpretation in collaborative inquiry *J. Soc. Work Pract.* **18** pp. 365–378
[15] Lave J and Wenger E 2003 *Situeret læring - Og andre tekster* (Copenhagen: Hans Reitzels Forlag)
[16] Felder R M and Brent R 2005 Understanding student differences *J. Eng. Educ.* **94** pp. 57–72
[17] Kolmos A Du X Holgaard J E and Jensen L P 2008 *Facilitation in a PBL environment* (Aalborg: Aalborg University)
[18] Center for Specialundervisning 2023 *N-Lited metoden. Lysudredning hos personer med synsvanskeligheder* (Slagelse: CSU Slagelse)
[19] Astrup C J 2023 *Midtvejsevaluering af projektet N-Lited* (Copenhagen: Ineva)
[20] Atkinson P A and Coffey A 1997 *Analyzing documentary realities* (Qualitative research: Theory, method, and practice) ed D Silverman (Sage) pp. 45–62
[21] Merriam S B 1988 *Case study research in education: A qualitative approach* (San Fransisco: Jossey-Bass Publishers)
[22] Bowen G A 2009 Document analysis as a qualitative research method *Qual. Res. J.* **9** pp. 27–40
[23] Yin R K 1994 *Case study research: Design and methods* (Beverly Hills: Sage)
[24] Øien T B 2022 *Kortlægning af praksis for lysudredning i synsrådgivningen* (Copenhagen: BUILD-rapport 2022:23)
[25] Christiansen A B *Human belysning – Solskinslys på en gråvejrsdag* (Copenhagen: Narayana Press)
[26] Voltelen M 1978 Lux alene gør det ikke *Lampetten* **2** pp. 71–77
[27] Frandsen S 1993 Om belysning og synsvilkår *Arkitekten* **95** pp. 370–373
[28] Øien T B 2021 Housing and low vision rehabilitation–Across theories, practices and everyday settings *Proc. 4th Conf. on Architecture Research Care & Health* (Trondheim, Norway) ed. 5 Academic Press

AUTHOR INDEX